ANNUAL REVIEW OF NEUROSCIENCE

ANNUAL REVIEW OF NEUROSCIENCE

W. MAXWELL COWAN, *Editor*
Washington University School of Medicine

ZACH W. HALL, *Associate Editor*
University of California School of Medicine

ERIC R. KANDEL, *Associate Editor*
College of Physicians and Surgeons of Columbia University

VOLUME 3

1980

ANNUAL REVIEWS INC.   4139 EL CAMINO WAY   PALO ALTO, CALIFORNIA 94306

ANNUAL REVIEWS INC.
Palo Alto, California, USA

REPRINTS The conspicuous number aligned in the margin with the title of each article in this volume is a key for use in ordering reprints. Available reprints are priced at the uniform rate of $1.00 each postpaid. The minimum acceptable reprint order is 5 reprints and/or $5.00 prepaid. A quantity discount is available.

International Standard Serial Number: 0147-006X
International Standard Book Number: 0-8243-2403-X

PRINTED AND BOUND IN THE UNITED STATES OF AMERICA

Annual Review of Neuroscience
Volume 3, 1980

CONTENTS

SOME RELATED ARTICLES IN OTHER *ANNUAL REVIEWS*

From the *Annual Review of Biochemistry,* Volume 48 (1979)

Chemistry and Biology of the Neurophysins, E. Breslow
Epidermal Growth Factor, G. Carpenter and S. Cohen
Surface Components and Cell Recognition, W. Frazier and L. Glaser
Activation of Adenylate Cyclase by Choleragen, J. Moss and M. Vaughan
Peptide Neurotransmitters, S. H. Snyder and R. B. Innis

From the *Annual Review of Biophysics and Bioenginering,* Volume 8 (1979)

Endogenous Chemical Receptors: Some Physical Aspects, F. J. Barrantes
Electrical Properties of Egg Cell Membranes, S. Hagiwara and L. A. Jaffe
Neural Prostheses, F. T. Hambrecht
Three-Dimensional Computer Reconstruction of Neurons and Neuronal Assemblies, E. R. Macagno, C. Levinthal and I. Sobel
Axoplasmic Transport of Proteins, D. L. Wilson and G. C. Stone

From the *Annual Review of Medicine,* Volume 30 (1979)

Efficacy of Biofeedback Therapy, M. T. Orne
Prenatal Sex Hormones and the Developing Brain: Effects on Psychosexual Differentiation and Cognitive Function, A. A. Erhardt and H. F. L. Meyer-Bahlburg
Computed Tomography in Neurologic Diagnosis, A. G. Osborne

From the *Annual Review of Pharmacology and Toxicology,* Volume 19 (1979)

Biology of Opioid Peptides, A. Beaumont and J. Hughes
Physical Mechanisms of Anesthesia, S. H. Roth
β-Adrenoceptor Blocking Drugs in Hypertension, A. Scriabine

From the *Annual Review of Physiology,* Volume 42 (1980)

Binding and Metabolism of Sex Steroids by the Hypothalamic-Pituitary Unit: Physiological Implications, B. S. McEwen
Nervous Control of the Pulmonary Circulation, S. E. Downing and J. C. Lee
The Relation of Muscle Biochemistry to Muscle Physiology, E. Eisenberg
Neuronal Control of Microvessels, S. Rosell

Neuroeffector Mechanisms, T. C. Westfall
Central Nervous Regulation of Vascular Resistance, S. M. Hilton and
   K. M. Spyer
Cardiovascular Afferents Involved in Regulation of Peripheral Vessels,
   H. M. Coleridge and J. C. G. Coleridge
Automatic Regulation of the Peripheral Circulation, D. E. Donald and
   J. T. Shepherd
Neural Mechanisms in Hypertension, M. J. Brody, J. R. Haywood
   and K. B. Touw
Role of Cyclic Nucleotides in Excitable Cells, I. Kupfermann
Biophysical Analysis of the Function of Receptors, C. F. Stevens

From the *Annual Review of Psychology,* Volume 31 (1980)

Chemistry of Mood and Emotion, P. L. McGeer and E. G. McGeer
Spatial Vision, R. L. De Valois and K. K. De Valois
*Neurochemistry of Learning and Memory: An Evaluation of Recent
   Data,* A. J. Dunn

Ann. Rev. Neurosci. 1980. 3:1–22

THE USE OF IN VITRO BRAIN SLICES FOR MULTIDISCIPLINARY STUDIES OF SYNAPTIC FUNCTION

❖11533

Gary Lynch

Department of Psychobiology, University of California, Irvine, California

Peter Schubert

Department of Exp. Neuropathology, Max Planck Institute for Psychiatry, 8 Munich 40, Kraepelinstrasse 2, West Germany

INTRODUCTION

Among the spectrum of experimental strategies used by neurobiologists to promote the understanding of brain function, the in vitro approach has found a wide and constantly growing application. In vitro systems offer a number of opportunities not available with more conventional techniques:

1. a large number of well-defined independent variables can be readily introduced,
2. dependent variables are usually more accessible to measurement and can be monitored with a greater variety of techniques than is the case in vivo,
3. interference from peripheral factors, which often compromises in vivo experiments, is greatly reduced.

The use of cell and tissue culture is well established and, particularly in the hands of biochemists, has contributed considerably to present knowledge of cell function at the single cell level. A complementary technique, which permits the study of the mechanisms of intercellular interactions, appears to be the in vitro brain slice (Yamamoto & McIlwain 1966). These tissue slices are transferred into the artificial environment as intact functioning units in which the complicated "hardware," the interconnections be-

1

0147-006X/80/0301-0001$01.00

tween neurons as well as between nerve and glial cells, are maintained. Slices can be contrasted with "explants" of fetal or neonatal tissue—in the latter the organization of the region in question is allowed to develop in vitro while in the slice it is removed in its "completed" or adult form.

The maintenance of the normal neuronal circuitry and the exposure of its structural orientation to visual control, allows the physiologist to pre- cisely place multiple stimulating and recording electrodes, to analyze signal transmission in identified neuronal systems, and to study the complex in- teractions of different inputs. Further, tissue samples correlated with physi- ological function can be rapidly removed and submitted to biochemical analysis with a minimum number of manipulations. Probably the most important advantage of the slice technique, and one that adds a new dimen- sion to experimental strategy, is that it permits access to the internal milieu of the brain. Thus it becomes possible to measure the effects of ion composi- tion and defined concentrations of drugs on a host of cell functions.

Slice preparations obtained from various areas of the brain already belong to the standard array of techniques used by electrophysiologists and bio- chemists for investigating problems specific to their fields. But the use of the method for multilevel correlations of physiological, biochemical, and anatomical events under controlled conditions has just begun. In view of the anticipated growing importance and need for such studies, this aspect of slice research is particularly emphasized and illustrated in this chapter rather than a comprehensive review of the many valuable data obtained in recent years by using the in vitro slice preparation.

SOME ADVANTAGES OF THE HIPPOCAMPUS FOR SLICE RESEARCH

Among other factors, success with slices will be dependent upon the follow- ing features of the brain region to be studied: (a) the ease and reliability with which the tissue can be prepared; (b) the visibility of structural landmarks; (c) the extent to which relevant fiber systems and synaptic fields can be "captured" in the slice.

In each of these respects the hippocampus has a number of fortunate anatomical properties. The entire structure can be removed from the brain with a minimum of manipulation, while its size is optimal for the slice procedure (taller pieces of brain have not, in our experience, provided satisfactory material). Furthermore, the major landmarks of the hippocam- pus are readily visible even with little magnification and minimal lighting conditions. Specifically, the layers of pyramidal and granule cell bodies can be easily discerned and with a little practice this is also true for several important fiber tracts (perforant path, initial portions of the commissural-

Schaffer projections, mossy fibers). Finally, and most important, hippocampal slices that contain a considerable proportion of the major fiber projections and their attendant synaptic domains can be prepared. Most of the major intrinsic and extrinsic hippocampal fiber systems are organized according to a "lamellar" plan in which they travel at right angles to the longitudinal axis of the structure (Andersen et al 1971b, Blackstad et al 1970). Thus a properly cut cross section of hippocampus will contain a variety of projections whose axons extend for some distance along that cross section. As a result, it is a relatively simple matter to prepare slices that are amenable to the study of several different afferent fiber systems and, in optimal cases, even di- and tri-synaptic circuitries can be analyzed (e.g. Lynch et al 1975).

The hippocampus also possesses a specialized anatomical feature that may be of particular interest to physiologists and biochemists. The fiber systems rigidly laminate themselves and generate enormous synaptic fields in easily identified dendritic target regions. It is possible therefore to stimulate afferents that terminate in very restricted dendritic regions a known distance from the cell soma, as well as to investigate the interactions of inputs acting at separate, defined regions of the cell (Lynch et al 1977, Dunwiddie & Lynch 1978). This should permit analysis of aspects of dendritic integration that have thus far eluded systematic experimental analysis. Beyond this, these projections generate dense fields of *en passage* synapses that extend for millimeters and account for more than 90% of the total synaptic contacts in some regions. If one takes advantage of this structural feature, a powerful synaptic input can be produced, large enough to elicit effects detectable by biochemical methods. By combining focal stimulation with microdissection procedures it should be possible to analyze such effects triggered by activation of an identified homogeneous synaptic system.

PHYSIOLOGY OF THE SLICE PREPARATION

The value of the slice is directly related to the extent to which it preserves the physiological phenomena found in the intact brain. Many of the necessary comparisons have been carried out for the hippocampal slice and the conclusion is gradually emerging that the in vitro preparation is surprisingly normal (Yamamoto & Kurokawa 1970, Alger & Teyler 1976, Schwartzkroin 1975, Dudek et al 1976). In Figure 1 we have used the physiology of the Schaffer-commissural projections to the regio superior to illustrate this point, but it should be appreciated that the general features of this system are found in nearly all of the laminated afferents of hippocampus. The fibers in question arise from the regio inferior pyramidal cells of the ipsilateral and

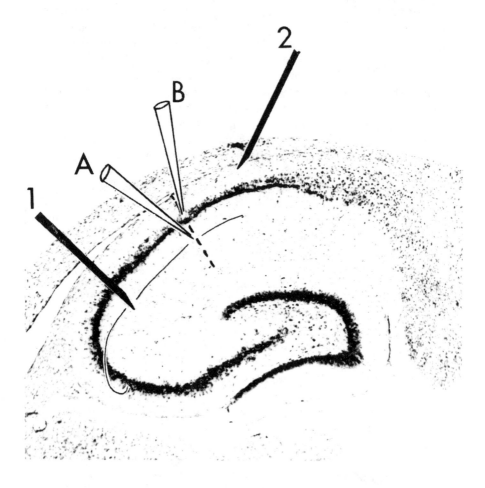

Figure 1 Schematic illustration of the arrangement of electrodes in a typical slice experiment. One of the stimulating electrodes ("1") is placed in the trajectory of the Schaeffer collateral projections (note solid line running under the "electrode") and in a position to activate these fibers and their synapses in cell free dendritic zone of the regio superior. The recording electrode ("A") is located at about 100 μm above the densely packed layer of regio superior pyramidal cell bodies in the dendritic zone, which receives the activated band of Schaffer fibers. The second recording electrode ("B") is positioned in the cell body layer to record the responses of the somata whose dendrites are innervated by the Schaffer projections. The second stimulating electrode ("2") is placed in the most superficial aspect of the hippocampus—the alveus—where it can activate the axons of the regio superior pyramidal cells and hence antidromically stimulate the cells sampled by electrode B. The dotted line indicates the axis of the pyramidal cell dendrite at the recording positions.

contralateral hippocampus and form a band about 250–300 μm high, which transverses the proximal region of the apical dendrites of the regio superior pyramidal cells. These fibers form numerous *en passage* synapses predominantly with the spines of the dendrites. Stimulation of a group of these axons produces the effects shown in Figure 2. An extracellular electrode located in the dendritic target zones records a monophasic negative potential; if the electrode is moved in increments towards the cell body layer this potential grows smaller and finally, as the somata are reached, it turns into an almost mirror image positive response. This is a classic "dipole" response and is thought to be indicative of extracellular inward currents associated with the EPSP in the dendrites and the appearance, at the cell bodies, of an outward current. For the purposes of the present discussion it is enough to note that these effects obtained in slices are virtually identical to the pattern of responses found in vivo following stimulation of the Schaffer-commissural system (Andersen et al 1964, Deadwyler et al 1975b). [When recording in the dendritic zones it is usually possible to detect a small negative response preceding the postsynaptic potentials described above, since at least part of the negative deflection is calcium independent it is thought that this potential represents a fiber volley (cf Andersen et al 1977).] As the stimulation voltage is increased and more axons activated, the slope of the postsynaptic potentials becomes more acute and at some point the positive response recorded at the cell body layer is broken by a very sharp negative deflection. Intra- (Schwartzkroin 1975) and extracellular (Andersen et al 1971a) recording studies have shown that the occurrence of this deflection correlates well with the spiking discharge of the individual pyramidal cells and it has been proposed that the negative potential is a "population spike," in essence the summed currents resulting from the near simultaneous discharges of a group of adjacent somata. This is a reasonable explanation though not definitely proven.

Intracellular recordings have shown, first, that membrane properties of the pyramidal cells in the in vitro slice are in no way exceptional and, second, that the monophasic potentials discussed above correspond in time with excitatory postsynaptic potentials (Schwartzkroin 1975, Dudek et al 1976, Deadwyler et al 1975a). Finally, classical inhibitory potentials (IPSP's) are also found in the in vitro pyramidal cells as they are in vivo and these seem to be relatively effective in blocking cell discharge (cf Deadwyler et al 1975a).

When two closely spaced pulses are delivered in the same afferent of the hippocampus in vivo the response to second pulse is usually markedly facilitated; this effect ("pair-pulse facilitation") is also obtained in the slice preparation and seems to follow about the same rules obtained in vivo. This correspondence between the slice and intact hippocampus also holds true

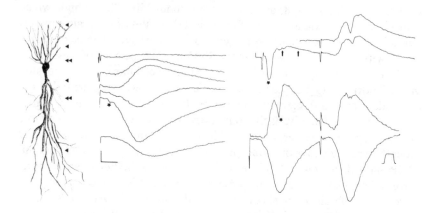

Figure 2 Typical records taken from an in vitro hippocampal slice using the recording arrangements shown in Figure 1. The cell on the extreme left is a camera lucida drawing of a Golgi impregnated hippocampal pyramidal cell. The arrowheads denote positions along the axis of the cell at which the recordings in the middle panel were taken. The double arrowheads signify the sites at which the recordings shown in the extreme right-hand panels were made. The middle panels show the population extracellular responses to stimulation of the Schaffer-commissural projections. The position marked by the lower of the two double arrowheads corresponds to the site marked by electrode A in Figure 1 and the recording made there is the 5th tracing in the middle panel. This zone is in the middle of the projection field of the stimulated fibers and, as can be seen, the postsynaptic potential is acutely negative in slope and maximal in amplitude at this point. Note also the small negative potential (*) preceding the synaptic reponse; this potential is calcium independent and is therefore likely to be the summed action potentials of the stimulated fibers (see Andersen et al 1977). At points in the dendritic field above (6th or bottom tracing) and below (4th tracing) the driven fiber bundles the synaptic potentials are somewhat smaller and less acute in slope. If the recording electrode is placed in the cell layer itself (the upper double arrowhead; see electrode B in Fig. 1), the response appears as a monophasic positive potential (3rd trace); this positivity is maximal in amplitude and slope at the layer of the cell bodies and rapidly diminishes as the recording electrode is moved further into the basal dendritic field (first two traces). Calibration bars: 0.5 mV and 1.5 msec.

The right-hand panels illustrate several aspects of slice physiology mentioned in the text. The lower pair of traces are potentials recorded simultaneously from the cell layer and apical dendrites in response to pair-pulse stimulation of the Schaffer-commissural fibers. Note first that the monophasic dendritic response to the second pulse is larger than that to the first—this is pair-pulse facilitation. The cell body response to the first pulse includes the negative going population spike (marked by the small star) described in the text and, as can be seen, this is inhibited in the response to second pulse. Thus, although the synaptic potentials appear facilitated, the spiking of the target cells is depressed in this paradigm. (Calibration pulse: 0.5 mV and 4.0 msec). The upper pair of traces shows the results of an experiment in which the cell body response to the stimulation of the Schaffer-commissural projections is preceded by antidromic activation of the target pyramidal cells (electrode 2 in Figure 1). The upper of the two traces is the orthodromic response alone—note that the stimulation voltages have been set at a level high enough to elicit a robust population spike. In the bottom trace the antidromic stimulation is given first and elicits a sharp negative deflection (star) followed by a slower, for the behavior of the population spike in this paradigm. At short intervals between the two pulses (5–30 msec) the population spike elicited by the second pulse is reduced, while beyond this time (30–150 msec) it is facilitated (Lomo 1971). The dendritic response to the second pulse is usually larger than its control from 10 msec onwards. The inhibition of the population spike at short interpulse intervals may reflect the activation of recurrent inhibition since this potential is also suppressed by antidromic stimulation of the regio superior efferents. In sum we find that with regard to pair-pulse facilitation the slice behaves in qualitatively the same manner as does the intact hippocampus.

Repetitive stimulation given at moderate frequency ($\sim$5 sec^{-1}) produces a facilitation within the train and often a potentiation that lasts for several seconds afterward. While detailed and quantitative comparisons between slice and in situ hippocampus have not been conducted, frequency dependent effects are in a qualitative sense quite similar in the two preparations. Finally, brief trains of high frequency stimulation in the intact hippocampus produce an increase in the size of the postsynaptic response, which is remarkable in that it persists for extraordinary periods of time (Bliss & Lomo 1973). This effect, long-term potentiation ("LTP"), has been found to endure for days and weeks in experiments using chronically implanted behaving animals (Bliss & Gardner-Medwin 1973, Douglas & Goddard 1975). This unique form of synaptic facilitation also appears to be present in the in vitro slice. Short bursts of high frequency stimulation of the commissural-Schaffer projections induce a robust and non-decremental increase in subsequent responses to single pulse stimulation whether the potentials are recorded extracellularly at the dendrites (Schwartzkroin & Wester 1975, Lynch et al 1976, Alger & Teyler 1976, Teyler et al 1977, and others) or intracellularly from the pyramidal cell bodies (Andersen et al 1977). In a later section of this review we use this effect to illustrate certain advantages of the slice preparation for multidisciplinary investigation but in the present context it provides a further demonstration that slices exhibit many of the physiological properties of the intact brain.

Thus in terms of basic synaptic physiology, as well as of its response to repetitive stimulation, an identified circuit in the slice behaves in very much

----

positive going after-response (marked by the two arrows). It has been found that the first, negative potential is unaffected by replacing the normal bathing fluid with a low calcium-high magnesium medium, while the second positive potential is completely eliminated (data not shown). Hence the first potential represents an antidromic response while the second is a synaptic potential possibly generated by recurrent interneurons that innervate the cell bodies of the pyramidal cells. When the orthodromic stimulation closely follows the antidromic potentials, then the population spike is significantly inhibited.

the same fashion as it does in the animal. There are, to be sure, major differences between the two preparations; in particular, spontaneous discharges of the pyramidal cells are much less frequent in the slice than they are in the animal and the synchronous slow waves characteristic of hippocampus are not to be found. Furthermore, seizures, which often occur in the intact hippocampus due to repetitive stimulation, are relatively rare in the slice although they can be elicited under certain conditions (Okada & Kuroda 1975). But in all it appears that slice physiology is sufficiently like that found in the animal so that the in vitro method can be regarded as a reasonable model system.

THE ANALYSIS OF SYNAPTIC PLASTICITY USING THE HIPPOCAMPAL SLICE PREPARATION

In addition to the better known variants of physiological plasticity, hippocampal synapses also exhibit a very unusual form of synaptic potentiation and the analysis of this is used to illustrate certain advantages of the slice technique. Figure 3 summarizes a typical long-term potentiation experiment using the above described Schaffer collateral-commissural projections to the regio superior. Activation of a discrete population of fibers with single pulse stimulation produces a robust extracellular negative going, monophasic potential. For purposes of control, a second set of commissural-Schaffer inputs to the same region (activated by a second stimulation electrode situated so as to ensure that a separate population of axons is activated; Dunwiddie & Lynch 1978) is also periodically tested. When stable baseline responses are obtained the first pathway is given a 1 sec burst of high frequency stimulation (100 sec^{-1}) and then infrequent single pulse testing of this as well as of the second input is resumed. As shown in the figure, the pathway that received the repetitive train is markedly potentiated and after 2 to 3 min settles at a new baseline considerably above that seen in the control period. Remarkably enough this long-term potentiation (LTP) is extremely stable; it is essentially non-decremental for the remainder of the experiment. Note also that the control pathway is unaffected by the repetitive stimulation of the first set of afferents, which suggests that the potentiation is not a generalized effect but instead is restricted to the activated inputs. [However, the analysis of the possible forms of plasticity in hippocampus has only begun. Thus it should be noted that potentiation induced by lower frequencies also produces generalized effects on the target cells, which can last for several minutes or even longer (Lynch et al 1976, 1977).] Several lines of evidence strongly suggest that LTP is not due to changes in the stimulated axons: (a) the presynaptic volley is unchanged

(Andersen et al 1977); (*b*) antidromic potentials in the repetitively activated pathway are unaffected (Schwartzkroin & Wester 1975); (*c*) high frequency stimulation in zero calcium, which activates the axons but does not cause synaptic transmission, does not alter the postsynaptic potentials obtained when synaptic transmission is restored by infusion of normal media (Dunwiddie et al 1978). Since the effect does not appear to involve any detectable alterations in the axons or target dendrites, it follows that the substrates of the LTP effect are to be found in the terminals, synapses, or dendritic spines.

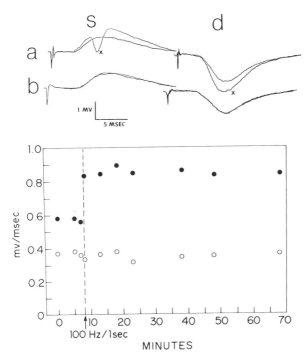

Figure 3   A long-term potentiation experiment in which two distinct collections (*a* and *b*) of Schaffer collateral-commissural fibers to the same dendritic targets are used (see text). The upper panels show the evoked responses obtained in the cell bodies (*s*) to stimulation of the two imputs and in the dendritic regions (*d*) where the fibers terminate. The pairs of traces are responses collected immediately before and 20 minutes after (the small x's) the application of a 100 sec⁻¹ train for 1 second to one pathway (*a*). As is evident, the responses to pathway (*a*) are markedly potentiated—the dendritic potential is enhanced and the cell body response shows a population spike. The second input (*b*) however is unchanged as indicated by the superimposed traces. The graph at the bottom of the figure provides a summary of the slope of the pathways that received the single burst of high frequency stimulation (closed circles) and those that did not (open circles). Note that the potentiation is essentially non-decremental for the duration of this experiment.

What type of substrate could be responsible for so exotic an effect as this extremely persistent, stable form of potentiation? There are at present no clear answers but further experiments using the in vitro slices are beginning to provide some intriguing clues.

We can assume that the enduring changes that are ultimately responsible for the increased synaptic strength must be brought about by biochemical mechanisms, which themselves are responses to transient events evoked by repetitive stimulation. Eludication of the presumably complex mechanisms responsible for LTP will therefore require a multilevel approach that searches not only for enduring biochemical and/or structural changes, but also for transient events that might reasonably serve as intermediates between repetitive stimulation and the final state of potentiation. The series of slice studies described below reflects this point and proceeds from a search for permanent synaptic changes back to an analysis of the events transpiring at the time of stimulation.

Ultrastructural Studies of the Effects of Repetitive Stimulation

The stability and extreme duration of long-term potentiation invite the speculation (Van Harreveld & Fifkova 1975) that the effect is produced by some structural change in the affected afferents or their targets, and recent electron microscopic studies of the regions surrounding the recording pipette tip have provided some evidence in support of this idea. In those experiments two populations of Schaffer-commissural projections to the same dendritic target were given repetitive stimulation at high or low frequency; 15 minutes later the slices were removed, fixed, and prepared for electron microscopy. Measurements were made (in the region of the recording tip) of the numbers of synapses on the spines and shafts of the dendrites as well as of the lengths of the postsynaptic densities (psd's). The size of the population of spine synapses was unchanged but there was a sizeable (40%) increase in the number of shaft contacts; expressed as a percentage of the total synapses, shaft contacts in the high frequency group were 3.6% compared to 2.3% in the control (low frequency) group. The lengths of the psd's of both spines and shafts were not detectably different in the two sets of slices (Lee et al 1979a). (The increase in shaft synapses has now been replicated in experiments using anesthetized rats; unpublished data).

These results provide evidence that anatomical changes do result from stimulation but whether the effects so far obtained are responsible for LTP remains an issue for future research. Studies of the temporal parameters of the observed changes, as well as of the effects of ionic manipulations that prevent the appearance of LTP (see below), should be particularly useful in this regard.

Biochemical Analyses of the Effects of Repetitive Stimulation

The search for biochemical processes that might result in lasting synaptic changes is somewhat simplified by the relatively rapid onset of LTP—any intermediate sequence would necessarily have to be initiated within several seconds of the potentiating train. Protein phosphorylation possesses a number of characteristics that make it an attractive candidate for an intermediate event. Phosphorylation can be initiated within the temporal constraints just mentioned and once accomplished it can last for several minutes or more; thus it persists long enough to participate in even relatively protracted cellular events. Furthermore, protein phosphorylation is one of the primary tools used by cells to regulate a host of intracellular reactions involved in both general and specialized functions. Finally, protein phosphorylation can be produced in synaptic membranes by any number of conditions ranging from electrical or chemical stimulation of slices (Williams & Rodnight 1975) to manipulations of synaptosomes (Johnson et al 1971, Routtenberg & Erlich 1975; see Rodnight et al 1975 and Greengard 1976 for reviews). Unfortunately, it is not a simple matter to measure phosphorylation of individual species of proteins even with a reduced system such as the slice (see Rodnight, Reddington & Gordon 1975 for a discussion). The difficulty lies in adequately labeling the ATP pools from which the phosphate is enzymatically transferred to the target protein. An alternate approach is to perform the experimental manipulation of interest (in the present case high frequency stimulation) on the slice system and afterward assay endogenous phosphorylation in the subcellular fractions where changes in phosphoproteins are expected. This allows the investigator to replace the cell's ATP with a high specific activity pool added to the incubation solution. The assumption behind this approach is that those proteins that were phosphorylated in situ will accept less ^{32}P when subsequently exposed to labeled ATP. That is, if the protein's phosphate binding sites are occupied because of the manipulation (i.e. the protein is phosphorylated) it will be able to accept fewer "hot" phosphate molecules in a later incubation. This assumption holds true in certain model systems (e.g. blood platelets, W. Bennett, unpublished) and has been demonstrated with brain tissue (Williams & Rodnight 1975) but any positive results with post-hoc assays ultimately require checks of the assumption.

Experiments have been conducted using this post-hoc assay to measure the effects of repetitive stimulation of the Schaffer-commissural system on the subsequent endogenous phosphorylation of isolated synaptic membranes (SPM's). In those studies the fibers were stimulated at high frequencies in several slices and then 2 minutes later these slices were removed from the in vitro chambers, pooled, and fractionated so that crude synaptic membranes could be isolated. The SPM's were then incubated in the presence of $AT^{32}P$ for 20 seconds, dissolved, and the resulting material run on a gel electrophoresis system. Autoradiographs were then made of the gels to identify the polypeptides that had incorporated labeled phosphate.

These studies showed that a SPM protein species with a molecular weight of about 40,000 daltons consistently incorporated less phosphate in the slices receiving high frequency stimulation than in the control groups (Browning et al 1979a). Furthermore this effect was not obtained after low frequency stimulation or after high frequency stimulation in low calcium/ high magnesium media, a condition that blocks the induction of LTP (see below). These results provide evidence that high frequency stimulation that produces LTP also phosphorylates a specific synaptic protein or alters the activity of the enzymatic machinery controlling the phosphorylated state of that protein.

While these results provide a first biochemical correlate of the LTP effect they are just that, a correlate. It remains to be established, first and foremost that these effects are causally related to LTP and, second, if so, how they are tied to events surrounding repeated synaptic activation as well as to the quasi-permanent (structural?) effect that is the basis of potentiation. An answer to the second of these questions will probably require extensive pharmacological work to identify agents that may interfere with the phosphorylation of the 40K dalton material. With regard to the links between repetitive stimulation and the observed phosphoprotein effect, it may be necessary to investigate further the biochemistry that underlies specific phosphorylation processes. It is of interest in this context to note that recent work has shown that phosphorylase b kinase added to the SPM incubation selectively phosphorylates the 40K dalton material (Browning et al 1979b). Phosphorylase b kinase is a calcium sensitive enzyme found in brain (Ozawa 1973) and if, as discussed below, repetitive stimulation were to increase intraterminal calcium levels, then a connection between such stimulation and the phosphorylation changes it produces could be readily envisioned (Lynch et al 1979). Relevant to this, Hershkowitz (1978) has reported that a protein of nearly the same molecular weight as that under discussion is phosphorylated in a calcium dependent fashion in incubated synaptosomes.

Possible "Triggers" for Potentiation

This brings us to the question of the initiating events or triggers for long-term potentiation. It hardly need be emphasized that this is an extremely complex problem since several interrelated events are occurring with very brief time courses during repetitive stimulation. This combined with the primitive state of synaptic pharmacology for the major hippocampal pathways means that it is not possible to introduce conditions that will selectively alter a particular step in the sequence of events taking place during repetitive activation. Despite these problems efforts to test for certain possible triggering mechanisms have been started.

The likelihood of inducing LTP as well as the strength of the effect when it is obtained are very much dependent upon the frequency of stimulation used to produce it. Thus a train of 100 pulses delivered at 200 sec^{-1} is considerably more effective than the same number of pulses given at 15 sec^{-1} (Dunwiddie & Lynch 1978). This is interesting because the number of robust postsynaptic potentials elicited during the train is greater in the second case than the first. It seems then that it is not the postsynaptic potentials per se that are responsible for potentiation. Another possibility is that high frequency stimulation somehow alters or disturbs the cellular processes that are needed to maintain ionic balance in either the terminal or spine. Fifkova & Van Harreveld (Van Harreveld & Fifkova 1975, Fifkova & Van Harreveld 1977) have presented an attractive version of such an idea; they suggest that with repeated stimulation the spines accumulate Na^+ ions, which in turn leads to a Cl^- influx to balance the positive charges, this in turn creates a hyperosmotic environment, which is corrected by a transfer of water into the spine. The net result of this sequence, according to their model, is a swelling of the postsynaptic element and this is the cause of potentiation. Another possibility is that Ca^{++} regulation is affected by high frequency synaptic activation. The attractive aspect of this idea is that Ca^{++} is known to modulate a number of important intracellular events—it acts as a second "messenger" in many systems—and thus might well activate biochemical effects of the type discussed above. (Recall that phosphorylase b kinase is a calcium sensitive enzyme.) Experimental tests of these ideas require ionic and pharmacological manipulation of the synaptic environment and this is possible using the slice preparation. Experiments of this type have shown that LTP is significantly retarded in low calcium media; under these conditions pair-pulse and frequency facilitation are readily observed as is post-tetanic potentiation but the incidence and size of LTP is greatly reduced (Dunwiddie & Lynch 1979). As mentioned above, a manipulation such as lowering Ca^{++} concentration can act at many stages in a chain of events and hence it cannot be concluded that Ca^{++} is involved in the induction of LTP; an effect on transmitter release is also a possibility. Nonetheless, these results do encourage the hypothesis that a critical level of some process must be reached during the train since LTP does not develop against the background of reduced transmission seen in a low calcium medium.

Synaptic Modulators and Synaptic Facilitation

Any analysis of the events responsible for triggering changes in synaptic operation must incorporate the possibility that modulators are released by nerve terminals (and/or postsynaptic elements) and hence may play a role in physiological plasticity. One increasingly plausible candidate for a synaptic modulator is adenosine (Schubert, Reddington & Kreutzberg 1979). McIlwain, who was among the first to recognize the utility of brain slices, first raised this idea of adenosine as a modulator; he labeled intracellular nucleotide pools of slices with radioactive adenosine and found that the nucleoside and/or its derivatives were released by electrical stimulation into superfusion medium (McIlwain 1972, Pull & McIlwain 1972). Studies in various brain systems have shown that radiolabeled adenosine (or a derivative) is transported in axons to their terminals and then crosses the synapse into the target cells (Schubert & Kreutzberg 1974, Schubert et al 1977) and that this transneuronal transfer is facilitated by electrical stimulation of relevant afferents (Schubert et al 1976). By using an in vivo-in vitro approach and employing axonal transport in the living animal as a vehicle to load exclusively the entorhinal afferents to the hippocampus, these terminals could be identified as the cellular compartments from which release occurs upon stimulation of the slice (Lee et al 1979b, in preparation). The release of adenosine derivatives seems to be a general phenomenon at the synapse. It is found in several regions of the central nervous system (Hunt & Kunzle 1976, Schubert & Kreutzberg 1974, Wise & Jones 1976) and was shown to occur in addition to the release of the principal transmitter (Rose & Schubert 1977). Other work using in vitro techniques suggests that not adenosine but the nucleotides are the major components released (Barberis & McIlwain 1976, White 1978, Zimmermann, 1978). Since these latter compounds penetrate membranes only poorly, the observed transneuronal transfer of radiolabel implies that extracellular enzymes convert the nucleotides into the very permeable adenosine and such conversions have been demonstrated in brain slices (Pull & McIlwain 1977). The enzyme 5'-nucleotidase, which performs the appropriate reaction, is located throughout the brain, primarily in the glial cells (Kreutzberg, Barron & Schubert 1978), and has a distribution that seems to accord with sites of transneuronal transfer (Schubert, Komp & Kreutzberg 1979). Taken together these data have led to the hypothesis that adenosine levels in the synaptic regions reflect, first, neuronal activity associated with transmission and, second, the distribution of the degradative enzyme 5'-nucleotidase.

Beyond this, there is considerable evidence suggesting that adenosine exerts powerful effects on synaptic transmission. Several studies have shown that adenosine depresses the release of transmitter from peripheral nerves

(Clanachan et al 1977, Vizi & Knoll 1976) as well as of putative transmitters from slices of cerebral cortex (Harms, Wardeh & Mulder 1978). The mechanism by which it produces these effects is unclear, but recent studies have shown that adenosine reduces calcium potentials in the heart (Schrader et al 1975) and also stimulation-induced decrements in extracellular levels of calcium in nervous tissue (Ten Bruggencate et al 1977). Hence adenosine's actions on transmitter release could well be mediated through an interference or alteration of depolarization-induced calcium fluxes.

The possibility exists then that adenosine serves as a "modulator" in brain; further evaluation of this hypothesis requires direct test of the effects of physiological concentrations of the nucleoside on synaptic transmission. Studies of the effects of adenosine on synaptically evoked responses have been conducted using in vitro slices of olfactory cortex (Kuroda, Saito & Kobayashi 1976, Okada & Kuroda 1975) and in the hippocampal formation (Schubert & Mitzdorf 1979) and they have shown that it produces a marked, depressive effect. In the hippocampal slice study (Schubert & Mitzdorf 1979; see Figure 4) it was found that adenosine exerted a significant depression of both the extracellular "EPSP" recorded from the field CA1 dendritic regions (in response to stimulation of the Schaffer-commissural projections) and the population spike sampled at the cell body layer, at a concentration as low as 2.5 μM. Since these effects were obtained when

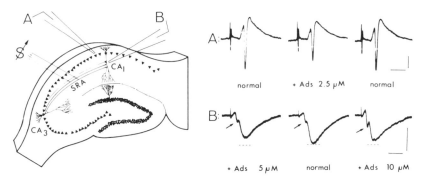

Figure 4   In this experiment hippocampal slices were constantly superfused and after stable recordings were obtained, the effect of several concentrations of adenosine was tested (15 minutes after the addition of the nucleoside to the medium).

A: Recordings from the cell body layer; note the reversible depression of the population spike.

B: Recordings from the dendritic layer; *middle:* field potential obtained in normal medium; *left:* depression of the postsynaptic amplitude by 15–20% in 5 μM adenosine with no change in the fiber potentials (arrows); *right:* afferent stimulation has to be considerably increased in the presence of adenosine (enlarged fiber potential, arrow), to reach approximately the amplitude of the control EPSP. Calibration: 5 msec, 1 mV.

the size of the presynaptic fiber volley was held constant, it is very likely that they reflect an action of adenosine on synaptic transmission. This conclusion is further supported by a current source density analysis. The slice preparation is nearly ideal for the application of this technique, which allows one to demonstrate rather precisely the size and spatial distribution of currents flowing into and out of the neuron during activation. The clearly identifiable synaptic currents were found to be strongly reduced in the presence of adenosine (see Figure 5). Furthermore, the relationship between evoked nerve cell firing (the population spike) and the extracellular EPSP was unchanged by adenosine, a result that strongly suggests that the nucleoside was acting on synaptic transmission and produced little if any general effect on the target cells.

In sum, adenosine, in concentrations that cannot be far removed from the physiological range, exerts a powerful influence over synaptic transmission

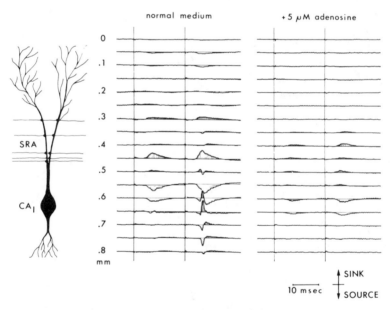

Figure 5 A current source density analysis of the response in the CA1 area to stimulation of the Schaffer collateral projections (SRA). Paired pulse stimulation (see Figure 2) was used here to demonstrate both the synaptic events and those accompanying cell firing. *Left:* In normal medium, large EPSP sinks (inward currents), are evoked in the synaptic region by the first pulse (at position 0.25–0.5 mm). The more effective second pulse initiates an additional high amplitude sink of short duration in the cell body layer (at position 0.55–0.65 mm) reflecting the net inward current that flows during the generation of the population spike. *Right:* In the presence of 5 μM adenosine, the EPSP and cell body layer sinks are drastically reduced (see text for further details).

in hippocampus. This, combined with evidence that it becomes available in the synaptic region during transmission (via the breakdown of adenine nucleotides by 5'-nucleotidase), strongly suggests that adenosine is a true modulator substance (see Figure 6 for a summary of these points). If so what role might it play in the various forms of synaptic plasticity discussed in earlier sections of this chapter?

Pharmacological studies have been conducted in an attempt to provide some insights into this question. Several lines of evidence suggest that theophylline acts as an antagonist of adenosine, possibly by competing for adenosine "receptors." Slice experiments have shown that this drug blocks the adenosine mediated stimulation of c-AMP stimulation (Fredholm 1977, Sattin & Rall 1970) as well as the depressive effects of the nucleoside on release (Harms et al 1978) and synaptic transmission (Kuroda et al 1976,

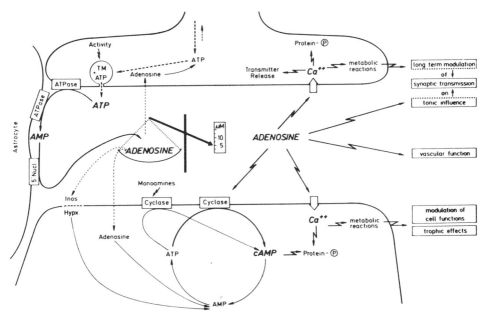

Figure 6   Schematic summary of hypotheses discussed in text regarding the role of adenosine as a modulator of synaptic transmission. The axon terminal is located in the upper part of the figure while its dendritic target is at the bottom—an astroglial process is seen to the left. The transmitter (TM) and ATP complex are seen in the lower left of the "terminal" and are shown to be released by stimulation. This results in the appearance of ATP in the synaptic cleft where membrane bound ATPase converts it (ATP) to AMP. The 5'-nucleotidase situated on glial cells then breaks the nucleotide into adenosine (heavy arrow). The concentration of adenosine (note scale) varies in the low micromolar range according to synaptic activity and the effective 5'-nucleotidase activity. Adenosine then exerts its effects by entering both pre- and postsynaptic elements, as well as possibly acting on the membranes. Several possible roles for the nucleoside are illustrated on the right-hand side of the figure (see Schubert et al 1979 for details).

Phyllis & Kostopoulos 1975, Reddington & Schubert, submitted). It should be noted that theophylline alone augments calcium dependent effects (e.g. the calcium dependent potassium conductance) and the relationship of this action of the drug and its antagonism to adenosine are unclear (Busis & Weight 1976, De Gubareff & Sleator 1965).

Theophylline produces a marked and reversible facilitation of the response of the CA1 zone to activation of the Schaffer-commissural fibers and under these conditions the degree of long-term potentiation induced by repetitive stimulation is reduced (Schubert et al 1979, submitted). This experiment indicates that theophylline's actions on synaptic transmission in brain are comparable to those seen in the periphery; beyond this, and given the data linking the drug to calcium, these findings point to momentary disturbances in calcium fluxes as a possible trigger for the LTP effect. One likely interpretation is that in the presence of theophylline intra-terminal calcium is elevated and therefore repetitive stimulation has smaller than normal effects on calcium fluxes. To the extent that the primary biochemical action of theophylline lies in an antagonism to adenosine, the above pharmacological studies also suggest, first, that the modulator tonically suppresses synaptic transmission in hippocampus and, second, that repetitive stimulation may produce synaptic potentiation via a disturbance or alteration of adenosine modulation. Hopefully further pharmacological work with the slice preparation will provide tests of these admittedly speculative ideas.

CONCLUDING REMARKS

It has been our intention to illustrate some of the unique possibilities offered by the slice technique for interdisciplinary research. Towards this end we selected a synaptic effect—long-term potentiation—and described attempts to uncover its substrates using physiological, anatomical, biochemical, and pharmacological methodologies. This deliberate focusing of the review has precluded any discussion of the specific advantages of slices for physiological work and in particular intracellular recording (the reader is referred to the recent paper by Wong et al 1979 on intracellular recording from dendrites for a fascinating demonstration of these advantages). Furthermore, we did not discuss a number of potential or barely realized applications of the slice. For example, the preparation would seem to be well suited for the application of cell labeling methods to the analysis of local anatomical circuitries and the identification of transmitter substances. With regard to the former, it is possible to replicably inject minute quantities of materials in the vicinity of (e.g. Lynch et al 1975) or inside (Wong et al 1979) cells of interest and then, by using histological methods, to trace dendritic fields and local axonal arborizations. With regard to the study of putative transmitters the slice allows one to perform any number of necessary analyses including: (a) testing the effects of "reasonable" concentrations of potential blocking compounds on identified synaptic responses (e.g. Dunwiddie et al 1978); by using a combination of multiple inputs and antidromic responses tests can be also made of the specificity of the drug of interest; (b) iontophoretic application to various dendritic regions known to contain the synaptic system under investigation (e.g. Lynch et al 1976, Spencer et al 1976); beyond this, intracellular recording will permit comparisons of synaptic and iontophoretic activation of restricted dendritic regions and this could add a new dimension to iontophoretic studies; (c) by using microperfusion of regions containing stimulated afferents it may be possible to collect materials released by the projection of interest. Experiments of this type using preloaded radiolabeled compounds have already been attempted with encouraging results (Wieraszko & Lynch 1979).

Slices should then allow neurobiologists to investigate more thoroughly a number of classical problems and to enter entirely new fields of inquiry especially since they lend themselves so well to both electrophysiological and biochemical studies. Hopefully, the method will provide a meeting ground upon which the various aspects of neurobiology can jointly investigate the common problems that define our science.

Literature Cited

Alger, B. E., Teyler, T. J. 1976. Long-term and short-term plasticity in CA1, CA3, and dentate regions of the rat hippocampal slice. *Brain Res.* 110:463–80

Andersen, P., Eccles, J. C., Loyning, Y. 1964. Pathway of post-synaptic inhibition in the hippocampus. *J. Neurophysiol.* 27:608–19

Andersen, P., Bliss, T. V. P., Skrede, K. K. 1971a. Unit analysis of hippocampal population spikes. *Exp. Brain Res.* 13:208–21

Andersen, P., Bliss, T. V. P., Skrede, K. K. 1971b. Lamellar organization of hippocampal excitatory pathways. *Exp. Brain Res.* 13:222–38

Andersen, P., Sundberg, S. H., Sveen, O., Wigstrom, H. 1977. Specific long-lasting potentiation of synaptic transmission in hippocampal slices. *Nature* 266:736–37

Barberis, C., McIlwain, H. 1976. 5-Adenine mononucleotides in synaptosomal preparations from guinea pig neocortex: their change on incubation, superfusion and stimulation. *J. Neurochem.* 26:1015–21

Blackstad, T. W., Brink, K., Hem, J., Jeune, B. 1970. Distribution of hippocampal mossy fibers in the rat: An experimental study with silver impregnation methods. *J. Comp. Neurol.* 138:433–50

Bliss, T. V. P., Lomo, T. 1973. Long-lasting potentiation of synaptic transmission in the dentate area of the anesthetized rabbit following stimulation of the perforant path. *J. Physiol. London* 232:331–56

Bliss, T. V. P., Gardner-Medwin, A. R. 1973. Long-lasting potentiation of synaptic transmission in the dentate area of the unanesthetized rabbit following stimulation of the perforant path. *J. Physiol. London* 232:367–74

Browning, M., Dunwiddie, T., Bennett, W., Gispen, W., Lynch, G. 1979a. Synaptic phosphoproteins: Specific changes after repetitive stimulation of the hippocampal slice. *Science* 203:60–62

Browning, M., Bennett, W., Lynch, G. 1979b. Phosphorylase kinase phosphorylates a protein affected by repetitive synaptic stimulation. *Nature* 273:273–75

Busis, N. A., Weight, F. F. 1976. Spike after-hyperpolarization of a sympathetic neuron is calcium sensitive and is potentiated by theophylline. *Nature* 263: 434–36

Clanachan, A. S., Johns, A., Paton, D. M. 1977. Presynaptic inhibitory actions of adenine nucleotides and adenosine on neurotransmission in the rat vas deferens. *Neuroscience* 2:597–602

Deadwyler, S. A., Dudek, F. E., Cotman, C. W., Lynch G. 1975a. Intracellular responses of rat dentate granule cells *in vitro:* Post-tetanic potentiation to perforant path stimulation. *Brain Res.* 88:80–85

Deadwyler, S. A., West, J. R., Cotman, C. W., Lynch, G. S. 1975b. A neurophysiological analysis of the commissural projections to dentate gyrus of rat. *J. Neurophysiol.* 38:167–84

De Gubareff, T., Sleator, W. Jr. 1965. Effects of caffeine on mammalian atrial muscle, and its interaction with adenosine and calcium. *J. Pharmacol. Exp. Ther.* 148:202–14

Douglas, R. M., Goddard, G. V. 1975. Long-term potentiation of the perforant path granule cell synapse in the rat hippocampus. *Brain Res.* 86:205–15

Dudek, F. E., Deadwyler, S. A., Cotman, C. W., Lynch, G. 1976. Intracellular responses from granule cell layer in slices of rat hippocampus: Perforant path synapses. *J. Neurophysiol.* 39:384–93

Dunwiddie, T. V., Lynch, G. S. 1978. Long-term potentiation and depression of synaptic responses in the hippocampus: localization and frequency dependency. *J. Physiol. London* 276:353–67

Dunwiddie, T., Lynch, G. 1979. The relationship between extracellular calcium concentrations and the induction of hippocampal long-term potentiation. *Brain Res.* 169:103–10

Dunwiddie, T., Madison, D., Lynch, G. 1978. Synaptic transmission is required for initiation of long-term potentiation. *Brain Res.* 150:413–17

Fifkova, E., Van Harreveld, A. 1977. Long-lasting morphological changes in dendritic spines of dentate granular cells following stimulation of the entorhinal area. *J. Neurocytol.* 6:211–30

Fredholm, B. B. 1977. Activation of adenylate cyclase from rat striatum and tuberculum olfactorium by adenosine. *Med. Biol.* 55:262–67

Greengard, P. 1976. Possible role for cyclic nucleotides and phosphorylation membrane proteins in post-synaptic actions of neurotransmitters. *Nature* 260:101–8

Harms, H. H., Wardeh, G., Mulder, A. H. 1978. Adenosine modulates depolarization-induced release of ^{3}H-noradrenaline from slices of rat brain neocortex, *Eur. J. Pharmacol.* 49:305–8

Hershkowitz, M. 1978. Influence of calcium on phosphorylation of a synaptosomal protein. *Biochim. Biophys. Acta* 542: 274–83

Hunt, S. P., Kunzle, H. 1976. Bidirectional movement of label and transneuronal transport phenomena after injection of ^{3}H-adenosine in the central nervous system. *Brain Res.* 112:127–32

Johnson, E. M., Maeno, H., Greengard, P. 1971. Phosphorylation of endogenous protein of rat brain by cyclic adenosine 3',5' monophosphate-dependent protein kinase. *J. Biol. Chem.* 246: 7731–39

Kreutzberg, G. W., Barron, K. D., Schubert, P. 1978. Cytochemical localization of 5'-nucleotidase in glial plasma membranes. *Brain Res.* 158:247–57

Kuroda, Y., Saito, M., Kobayashi, K. 1976. Concomitant changes in cyclic AMP level and postsynaptic potentials of olfactory cortex slices induced by adenosine derivatives. *Brain Res.* 109:196–201

Lee, K., Schottler, F., Oliver, M., Creager, R., Lynch, G. 1979a. Ultrastructural effects of repetitive synaptic stimulation in the hippocampal slice preparation: A preliminary report. *Exp. Neurol.* In press

Lee, K., Gribkoff, V., Sherman, B., Schubert, P., Lynch G. 1979b. Release of adenosine derivatives from the hippocampus *in vitro* following electrical stimulation. In preparation

Lomo, T. 1971. Patterns of activation in a monosynaptic cortical pathway: the perforant path input to the dentate area of the hippocampal formation. *Exp. Brain Res.* 12:18–45

Lynch, G. S., Smith, R. L., Browning, M. D., Deadwyler, S. A. 1975. Evidence for bidirectional dendritic transport of horseradish peroxidase. In *Advances in Neurology*, ed. G. W. Kreutzberg, Vol. 12. New York: Raven

Lynch, G. S., Gribkoff, V. K., Deadwyler, S. A. 1976. Long-term potentiation is accompanied by a reduction in dendritic responsiveness to glutamic acid. *Nature* 263:151–53

Lynch, G. S., Dunwiddie, T. V., Gribkoff, V. K. 1977. Heterosynaptic depression: a postsynaptic correlate of long-term potentiation. *Nature* 266:737–39

Lynch, G. S., Browning, M., Bennett, W. 1979. Biochemical and physiological studies of synaptic plasticity. *Fed. Proc.* 38:69–72

McIlwain, H. 1972. Regulatory significance of the release and action of adenine derivatives in cerebral systems. *Biochem. Soc. Symp.* 36:69–85

Okada, Y., Kuroda, Y. 1975. Inhibitory action of adenosine and adenine nucleotides on the postsynaptic potential of olfactory cortex slices of guinea pig. *Proc. Jpn. Acad.* 51:491–94

Ozawa, E. 1973. Activation of phosphorylase kinase from brain by small amounts of calcium ion. *J. Neurochem.* 20: 1487–88

Phillis, J. W., Kostopoulos, G. K. 1975. Adenosine as a putative transmitter in the cerebral cortex. Studies with potentiators and antagonists. *Life Sci.* 17:1085–94

Pull, I., McIlwain, H. 1972. Adenine derivatives as neurohumoral agents in the brain: The quantities liberated on excitation of superfused cerebral tissues. *Biochem J.* 130:975–81

Pull, I., McIlwain, H. 1977. Adenine mononucleotides and their metabolites liberated from and applied to isolated tissues of the mammalian brain. *Neurochem. Res.* 2:203–16

Reddington, M., Schubert, P. 1979. Parallel investigations of the effects of adenosine on evoked potentials and cyclic AMP accumulation in hippocampus slices. Submitted for publication

Rodnight, R., Reddington, M., Gordon, M. 1975. Methods for studying protein phosphorylation in cerebral tissues. In *Research Methods in Neurochemistry,* ed. N. Marks, R. Rodnight, pp. 325–67. New York: Plenum

Rose, G., Schubert, P. 1977. Release and transfer of ³H-adenosine derivatives in the cholinergic septal system. *Brain Res.* 121:353–57

Routtenberg, A., Ehrlich, Y. H. 1975. Endogenous phosphorylation of four cerebral cortical membrane proteins: Role of cyclic nucleotides, ATP, and divalent cations. *Brain Res.* 92:415–30

Sattin, A., Rall, T. W. 1970. The effect of adenosine and adenine nucleotides on the cyclic adenosine 3',5'-phosphate content of guinea pig cerebral cortex slices. *Mol. Pharmacol.* 6:13–23

Schrader, J., Rubio, R., Berne, R. M. 1975. Inhibition of slow action potentials of guinea pig atrial muscle by adenosine: A possible effect on Ca^{++} influx. *J. Mol. Cell. Cardiol.* 7:427–33

Schubert, P., Kreutzberg, G. W. 1974. Axonal transport of adenosine and uridine derivatives and transfer to post-synaptic neurons. *Brain Res.* 76:526–30

Schubert, P., Lee, K., West, M., Deadwyler, S., Lynch, G. 1976. Stimulation-dependent release of ³H-adenosine derivatives from central axon terminals to target neurones. *Nature* 260:541–42

Schubert, P., Rose, G., Lee, K., Lynch, G., Kreutzberg, G. W. 1977. Axonal release and transfer of nucleoside derivatives in the entorhinal-hippocampal system: An autoradiographic study. *Brain Res.* 134:347–52

Schubert, P., Komp, W., Kreutzberg, G. W. 1979. Correlation of 5'-nucleotidase activity and selective transneuronal transfer of adenosine in the hippocampus. *Brain Res.* In press

Schubert, P., Lynch, G., Creager, R. 1979. Effect of the presumed adenosine-antagonist theophylline on hippocampal long-term potentiation. Submitted for publication

Schubert, P., Mitzdorf, U. 1979. Analysis and quantitative evluation of the depressive effect of adenosine on evoked potentials in hippocampal slices. *Brain Res.* In press

Schubert, P., Reddington, M., Kreutzberg, G. W. 1979. On the possible role of adenosine as a modulatory messenger in the hippocampus and other regions of the CNS. *Prog. Brain Res.* 50: In press

Schwartzkroin, P. A. 1975. Characteristics of CA1 neurons recorded intracellularly in the *in vitro* hippocampal slice preparation. *Brain Res.* 85:423–36

Schwartzkroin, P. A., Wester, K. 1975. Long-lasting facilitation of synaptic potentials following tetanization in the *in vitro* hippocampal slice. *Brain Res.* 89:107–19

Spencer, H. J., Gribkoff, V. K., Cotman, C. W., Lynch, G. S. 1976. GDEE antagonism of iontophoretic amino acid excitations in the intact hippocampus and in the hippocampal slice preparation. *Brain Res.* 105:471–81

Ten Bruggencate, G., Steinberg, R., Stockle, H., Nicholson, C. 1977. Modulation of extracellular Ca^{++} and K^+ levels in the mammalian cerebellar cortex. In *Iontophoresis and Transmitter Mechanisms in the Mammalian Central Nervous System,* ed. J. S. Kelly, pp. 412–15. Amsterdam: Elsevier

Teyler, T. J., Alger, B. E., Bergman, T., Livingston, K. 1977. Comparison of long-term potentiation in *in vitro* and *in vivo* hippocampal preparations. *Behav. Biol.* 19:24–34

Van Harreveld, A., Fifkova, E. 1975. Swelling of dendritic spines after stimulation of the perforant path as a mechanism of post-tetanic potentiation. *Exp. Neurol.* 49:736–49

Vizi, E. S., Knoll, J. 1976. The inhibitory effect of adenosine and related nucleotides on the release of acetylcholine. *Neuroscience* 1:391–98

White, T. D. 1978. Release of ATP from a synaptosomal preparation by elevated extracellular K^+ and by veratridine. *J. Neurochem.* 30:329–36

Wieraszko, A., Lynch, G. 1979. Stimulation-dependent release of possible transmitter substances from hippocampal slices studied with localized perfusion. *Brain Res.* 160:372–76

Williams, M., Rodnight, R. 1975. Stimulation of protein phosphorylation in brain slices by electrical pulses. Speed of responses and evidence for net phosphorylation. *J. Neurochem.* 24:601–3

Wise, S. P., Jones, E. G. 1976. Transneuronal or retrograde transport of ^3H-adenosine in the rat somatic sensory system. *Brain Res.* 107:127–31

Wong, R. K., Prince, D., Busbaum, A. 1979. Intradendritic recordings from hippocampal neurons. *Proc. Natl. Acad. Sci. USA* 76:986–90

Yamamoto, C., McIlwain, H. 1966. Electrical activities in thin sections from the mammalian brain maintained in chemically defined media *in vitro. J. Neurochem.* 13:1333–43

Yamamoto, C., Kurokawa, M. 1970. Synaptic potentials recorded in brain slices and their modification by changes in tissue ATP. *Exp. Brain Res.* 10:159–70

Zimmermann, H. 1978. Turnover of adenine nucleotides in cholinergic synaptic vesicles of the Torpedo electric organ., *Neuroscience* 3:827–36

Ann. Rev. Neurosci. 1980. 3:23–41

DOPAMINE AND THE PATHOPHYSIOLOGY OF DYSKINESIAS INDUCED BY ANTIPSYCHOTIC DRUGS

♦11534

Ross J. Baldessarini

Department of Psychiatry, Harvard Medical School, and Mailman Laboratories
for Psychiatric Research, McLean Division of Massachusetts General Hospital,
Belmont, MA 02178

Daniel Tarsy

Departments of Neurology, Harvard and Boston University Medical Schools,
and Neurology Division, Deaconess Hospital, Boston, MA 02115

INTRODUCTION

Antipsychotic agents in common use all have unwanted neurological
("neuroleptic") effects as well as beneficial effects in severely ill psychiatric
patients. They include phenothiazines, thioxanthenes, butyrophenones, and
many other compounds proven effective in the management of a broad
range of psychotic symptoms and particularly useful in the treatment of
schizophrenia and mania. The evidence that this class of substances has
selective antipsychotic effects as opposed to merely "tranquilizing" effects
in schizophrenia and other disorders marked by abnormalities of thought
associations, perceptions, and beliefs is now overwhelming (Baldessarini
1977a, 1980). Antipsychotic drugs are highly effective in hastening remis-
sions of acute psychotic illnesses and also seem to prevent later exacerba-
tions of psychotic symptoms in chronic psychotic disorders, leading to their
prolonged use in schizophrenia. The introduction of antipsychotic agents
into European medical practice in 1952 was followed by a virtual revolution
in the theory and practice of modern psychiatry (Swazey 1974). More

23

0147-006X/80/0301-0023$01.00

recently, however, their therapeutic limitations and untoward neurotoxic effects (Duvoisin 1968, Faurbye et al 1964, Marsden et al 1975) have been matters of increasing concern to clinical and basic neurobiologists.

NEUROLOGIC EFFECTS OF ANTIPSYCHOTIC DRUGS

Antipsychotic agents in current clinical use regularly produce a variety of extrapyramidal neurological disorders of the control of posture, muscle tone, and movement. The early, acute, and reversible neurologic effects of antipsychotic drugs include several more-or-less distinct syndromes:

1. *Acute dystonias* involve acute spasms of the muscles of the tongue, face, neck, and back that can be confused with seizures or conversion reactions (Marsden et al 1975, Marsden 1976). Their maximum risk occurs within the first week of treatment, especially with the more potent neuroleptic agents (notably, the piperazine phenothiazines and thioxanthenes, and the butyrophenones), and they may be especially likely to occur at the time of rapid diminution of blood levels of the drug several hours after a dose (Garver et al 1976). Antiparkinsonism drugs, such as benztropine, biperidin, or diphenhydramine (with strong central antimuscarinic actions), given parenterally, are both diagnostic and curative. Nevertheless, the fact that several other forms of treatment (including other central nervous system depressants as well as stimulants) are also helpful on occasion, has led to a confusing pattern of pharmacological responses, and so no coherent theory of their underlying neural mechanisms has been proposed.

2. *Parkinsonism* induced by neuroleptic drugs, similar to postencephalitic or idiopathic forms of Parkinson's disease, includes bradykinesia, shuffling gait, facial inexpressiveness, muscular rigidity, and variable degrees of tremor; these neurological signs are occasionally mistaken for spontaneous behavioral changes that sometimes occur in psychotic patients (Kraepelin 1919, Fish 1967). The period of maximum risk is between the first and fourth weeks of treatment, with a tendency toward gradual abatement (tolerance) over subsequent weeks. Pharmacologic responses are similar to those in other forms of this syndrome (Bruno & Bruno 1966), although the mainstays of treatment are the anticholinergic antiparkinsonism agents since most dopamine agonists (and especially L-dopa) tend to produce too much agitation for use in already psychotic patients (Barbeau & McDowell 1970, Angrist et al 1973). The currently favored hypothesis is that a deficiency of dopamine function in the basal ganglia contributes importantly to drug-induced parkinsonism, as degeneration of the nigrostriatal dopamine pathway is almost certain to contribute to idiopathic forms of the condition (Bernheimer et al 1973). One other syndrome that can

develop months or years later appears to be a localized variant of parkinson-ism-like tremor of the perioral region (the so-called "rabbit syndrome"). This condition is typically spontaneously and rapidly reversible and respon-sive to anticholinergic antiparkinsonism drugs (Jus et al 1974)—all in con-trast to tardive dyskinesia (which is discussed below).

3. *Akathisia* or motor restlessness, marked by severe but not easily de-scribed subjective discomfort and need to move the extremities, is most likely to occur within the first two months of exposure to a neuroleptic drug. Sedatives or antiparkinsonism agents are occasionally partially effective, but often not, and the pathophysiologic basis of this reaction is unknown. The condition is sometimes confused with anxiety or agitation in psychiatric patients, but it abates when the responsible agent is removed (Windelman 1961).

EMPIRICAL DEVELOPMENT OF ANTIPSYCHOTIC AGENTS

A crucial question is whether these and other more subtle, almost routinely encountered neurologic or neuroleptic effects of the antipsychotic drugs are essential to their actions. Ironically, the most potent and selectively an-tidopaminergic of the antipsychotic drugs, while no more efficacious than less potent congeners, are most likely to induce the acute neurologic effects just described. The fact that several effectively antipsychotic drugs (notably, clozapine and sulpiride) have relatively little tendency to induce acute neurologic reactions (dystonias, parkinsonism, and restlessness) now strongly challenges the inevitability of the association of neurologic and antipsychotic effects. The existence of such "atypical" agents also offers some hope that better antipsychotic agents with diminished neurologic side effects can be developed.

An important fact (or artifact) is that the methods of screening new substances for potential antipsychotic utility have essentially involved seek-ing neurologic reactions in laboratory animals, because there are no satisfac-tory animal tests for schizophrenia (Fielding & Lal 1978, Baldessarini & Tarsy 1979). This impasse, coupled with current conservatism of the system for development and testing of new agents, especially in the United States, has contributed to a repeated "rediscovery" of agents with very similar actions and limitations over the past 25 years. Recently, there have ap-peared intriguing suggestions of dissimilarities in the neurobiologic effects of the typical and atypical antipsychotic agents upon various dopamine projection sites in the rodent forebrain (Costall & Naylor 1976), either in the pattern of development of "tolerance" to their presumably antidopa-mine effects (Eichler et al 1977, and personal communication; Bunney &

Aghajanian 1978, Moore & Kelly 1978), or in their interactions with musca-
rinic-acetylcholine and other neurohumoral receptor sites in addition to
dopamine receptors in brain tissue (Creese et al 1978, Snyder et al 1974,
1978). These observations may point the way toward improved methods of
screening for more selective and less neurotoxic antipsychotic drugs.

ANTIDOPAMINE ACTIONS OF ANTIPSYCHOTIC AGENTS

In the past, a number of mechanisms had been proposed to explain the
actions of the antipsychotic drugs. They differ from most other depressants
of the central nervous system in several ways. Thus, in man they have
limited ability to induce generalized sedative effects or coma until enormous
overdoses are taken; in addition, tolerance to their antipsychotic effects is
unknown, and addiction virtually does not occur. Unlike sedatives, they
have been reported to have greater ability to diminish conditional behavioral responses than to depress unconditional responses (Fielding & Lal
1978). Some antipsychotic agents may have a selective ability to dampen the
neurophysiologic effects of peripheral stimuli on the forebrain, while inhibiting to a much lesser extent the effects of stimulating electrodes placed in
the brainstem, although this effect is probably not as selective as formerly
believed (Baldessarini 1980). In addition to these distinctions from sedatives, antipsychotic drugs have striking inhibitory effects on autonomic and
motoric expressions of arousal and presumed strong affect in animals
(Fielding & Lal 1978), believed to be mediated by actions in the limbic
forebrain and hypothalamus. The cellular and biochemical events underlying these behavioral and physiological actions, however, have remained
obscure until recently.

It was proposed by European pharmacologists as long ago as the early
1960s that the neurologic, and possibly also the antipsychotic effects may
reflect the ability of neuroleptic drugs to interfere with synaptic transmission in the brain mediated by dopamine (Carlsson & Lindqvist 1963, Carlsson 1978). This suggestion arose largely from the observation that among
the biochemical consequences of giving a neuroleptic drug to an animal,
there was a consistent increase in levels of the metabolites of dopamine, but
variable effects on the metabolism of other candidate neurotransmitters.
The possible importance of dopamine was given strong support by early
histochemical studies of the normal distribution of amine-containing neurons in the mammalian brain, which indicated a preferential distribution of
dopamine fibers between midbrain and the basal ganglia (notably, the ni-
groneostriatal tract), and within the hypothalamus (Ungerstedt 1971) (Figure 1). More recently, anatomists have come to appreciate the existence of

DOPAMINE PROJECTIONS: RAT BRAIN

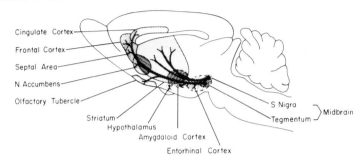

Figure 1   Dopamine-containing neurons in the mammalian brain. The major systems involving dopamine are: the *nigrostriatal* pathway from the zona compacta of the midbrain substantia nigra to the neostriatum (caudate-putamen); *mesolimbic* projections from midbrain tegmentum through the lateral hypothalamus to limbic structures, including the septal nuclei (e.g. nucleus accumbens septi) and olfactory tubercle; and related *mesocortical* projections, also arising in midbrain, and projecting particularly to prefrontal and temporal areas of the cerebral cortex; there is also a *tuberoinfundibular* dopamine-containing (TIDA) system within the hypothalamus. The scheme is based on extensive studies in rat, but the anatomy is believed to be very similar in man. This scheme is based on anatomical studies described in Ungerstedt (1971), Fuxe et al (1974, 1977), and Nauta & Domesick (1979), and prepared with the advice of Dr. V. Domesick.

other dopamine projections from midbrain tegmentum to forebrain regions. This projection to the limbic system, as well as to temporal and prefrontal cerebral cortical areas closely interlinked with the limbic system is probably not primarily related to the extrapyramidal motor system (Fuxe et al 1974, 1977, Nauta & Domesick 1979) (Figure 1). A somewhat simplistic, but attractive concept has been that many extrapyramidal effects of the antipsychotic drugs may be mediated by antidopamine actions in the basal ganglia, and that some of their antipsychotic effects may be mediated by the antagonism of "dopaminergic" neurotransmission in the limbic system, hypothalamus, or cortex. The latter supposition has been given indirect general encouragement by repeated "natural experiments" that have associated psychotic mental phenomena with lesions of the temporal lobe and other portions of the limbic system (Slater et al 1963, Stevens 1973, Baldessarini 1977b).

In recent years, a large body of data has accumulated to support the theory that the antagonism of dopamine-mediated synaptic neurotransmission is an important action of antipsychotic-neuroleptic agents (Baldessarini 1977a, b, 1980, Baldessarini & Tarsy 1979), and only a brief summary of the salient findings can be given here (Table 1). Thus, antipsychotic agents, but not their nonantipsychotic congeners, are reported to increase the rate

Table 1   Effects of neuroleptic drugs on dopamine (DA) neurons in the brain[a]

Secondary or indirect effects:

· DA metabolism increased acutely (increased tyrosine hydroxylation and metabolite production)
· Midbrain cell firing increased acutely.

Primary or direct effects:

· Plasma prolactin increased in rat and man in proportion to behavioral or clinical potency of neuroleptic
· Behavioral or neuroendocrine actions of systemically administered DA agonists blocked (eg, L-dopa, apomorphine, amphetamine)
· Self-stimulation through electrodes in DA-rich forebrain regions blocked in rat
· Arousal in response to local injections of DA agonists in forebrain DA target areas blocked in rat
· Iontophoretic effects of DA (but not of cyclic AMP) blocked in caudate nucleus of rat
· DA-sensitive adenylate cyclase in forebrain homogenates blocked
· Binding of tritiated neuroleptics to membranes in forebrain homogenates antagonized with potency that corresponds closely to behavioral and clinical effects

[a]References not provided in the text are given in Baldessarini (1977b, 1980), Baldessarini & Tarsy (1979), Baldessarini et al (1980).

of production of dopamine metabolites (notably, 3-methoxytyramine, and dihydroxyphenylacetic and homovanillic acids) (Carlsson & Lindqvist 1963, Carlsson 1978), the rate of conversion of the precursor amino acid tyrosine through dihydroxyphenylalanine (dopa) to dopamine and its metabolites (Sedvall 1975, Carlsson 1978), and the firing rate of presumably dopamine-containing neuronal cell bodies in midbrain (Bunney et al 1973, Bunney & Aghajanian 1975, 1978). These effects can be interpreted as secondary or compensatory responses of plastic and adaptive neuronal systems attempting to maintain homeostasis by an increase in firing rate in the face of what is assumed to be a primary interruption of synaptic transmission at the dopamine terminals in the caudate nucleus, septal nuclei, and cerebral cortex. Figure 2 shows the metabolic arrangements at such synapses.

More direct evidence that a crucial primary effect of neuroleptic drugs may be the blockade of postsynaptic dopamine receptor sites includes the ability of small doses of antipsychotic agents to block behavioral or neuroendocrine effects of dopamine agonists (Table 1). Examples are stereotyped gnawing behavior in the rat induced by the putative direct dopamine agonist, apomorphine, possibly acting at the caudate nucleus (Tarsy & Baldessarini 1974); the locomotor excitement induced by the injection of dopamine into the nucleus accumbens septi of the limbic system (Costall

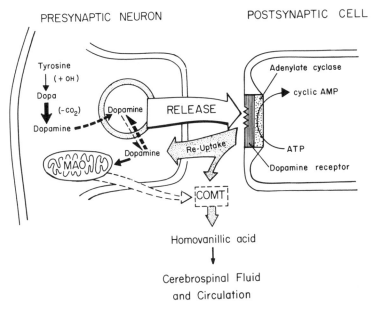

Figure 2 Metabolism at dopamine synapse in the brain. Dopamine is formed from L-tyrosine by hydroxylation (the rate-limiting step) to L-dihydroxyphenylalanine (dopa), which is rapidly decarboxylated. Dopamine is stored in presynaptic vesicles (shaded circle), from which release occurs into the synaptic cleft by neuronal depolarization in the presence of calcium. The released amine has a postsynaptic effect, possibly mediated by a recognition molecule (receptor) associated with adenylate cyclase that converts adenosine triphosphate (ATP) to adenosine-3',5'-cyclic-monophosphate (cyclic AMP), which may, in turn, exert biochemical effects leading to altered neurophysiological sensitivity in the receptive cell. Neurotransmitter is inactivated largely by efficient high-affinity reuptake into the presynaptic terminals; excess dopamine that is not stored can be metabolized by monoamine oxidase (MAO) in the mitochondria and catechol-O-methyltransferase (COMT), largely extraneuronal, to produce homovanillic acid from dihydroxyphenylacetic acid, and 3-methoxytyramine, intermediary metabolites; the metabolites are removed in the cerebrospinal fluid and venous circulation by a probenecid-sensitive uptake, largely at the choroid plexus.

& Naylor 1976); or the prolactin-decreasing response to apomorphine or L-dopa (the immediate precursor of dopamine) believed to be mediated by hypothalamic or pituitary dopamine receptors (Meltzer et al 1978, Sachar 1978). Such "tests" have been proposed as screening methods to detect even more agents of the kinds already available.

More direct evidence of a receptor blockade has been provided by the antagonism of an apparently selective dopamine-sensitive adenylate cyclase in homogenates of caudate or limbic tissue (Clement-Cormier et al 1974, Iversen 1977), and the interference with electrophysiological responses to dopamine iontophoretically applied to receptive cells in the caudate nucleus

—a blockade overcome by presumed circumvention of the receptor sites on the cell surfaces by iontophoresed cyclic-AMP (Siggins et al 1976; see Figure 2).

A more recent development is the application of radioligand binding assays using homogenates of mammalian caudate nucleus and low concentrations (nanomolar) of intensely radioactive [³H]-dopamine, [³H]-labeled neuroleptic drugs (haloperidol or spiroperidol), or [³H]-apomorphine (Creese et al 1978, Snyder et al 1978). Pharmacologic evidence supports the suggestion that the binding of these ligands to brain tissue represents, at least partially, an interaction with a dopamine "receptor" site. Correlations between the in vitro potency of antipsychotic drugs of all types to interfere with the binding of such ligands, and estimates of their potency to block the effects of dopamine agonists in animals or to produce clinical benefits in psychotic patients are impressive (Baldessarini 1977b, Creese et al 1978). Analogues or isomers of the antipsychotic drugs that are clinically inactive lack this potent antagonistic effect against ligand binding. It is particularly interesting that two antipsychotic agents (thioridazine and clozapine) with relatively weak acute neurologic side effects seem to have antidopamine effects in the ligand-binding assays that correlate closely with their clinical potencies. Although their relative lack of extrapyramidal toxicity has been explained by a countervening antimuscarinic (antiparkinsonism?) action of these two drugs, this explanation is not satisfactory for a similar drug, sulpiride.

These findings, together, strongly support the theory that antipsychotic agents interfere with the actions of dopamine as a synaptic neurotransmitter in the brain. Nevertheless, they do not prove that antidopamine effects are either necessary or sufficient for antipsychotic efficacy. They strongly suggest, however, that some of the extrapyramidal neurologic effects of this class of agents may be produced by antagonism of dopamine, largely on the basis of analogy to the demonstrated loss of dopamine in the caudate nucleus and the beneficial responses to its precursor, L-dopa, in idiopathic parkinsonism (Bernheimer et al 1973).

CLINICAL FEATURES OF TARDIVE DYSKINESIA

It is now generally accepted that late-appearing ("tardive") neurological disorders including abnormal oral, facial, and tongue movements, as well as involuntary choreic (quick, tic-like) and athetotic (slower, writhing) movements of the trunk and extremities, first described by Schönecker (1957) and by Sigwald and his colleagues (1959), and called "tardive dyskinesias" by Faurbye and his colleagues (1964), are associated with prolonged clinical exposure to neuroleptic-antipsychotic drugs (Crane 1973,

Tarsy & Baldessarini 1976, Baldessarini & Tarsy 1978, Baldessarini et al 1980). While the evidence for this association is mainly epidemiologic, it is quite compelling (Baldessarini 1974). With increased recognition in recent years, prevalence rates are perhaps 10–20% of patients so treated, with evidently increased risk with advancing age (Smith et al 1978, 1979, Baldessarini et al 1980). This distressing condition challenges the current practice of almost routine prolonged "maintenance" therapy of chronically psychotic patients, mostly schizophrenics (Baldessarini et al 1980). Its cause is not known. The leading hypothesis is that dopamine as a synaptic neurotransmitter in the basal ganglia or limbic forebrain may be overactive, either through directly or indirectly increased availability, or by increased efficacy, possibly by way of postsynaptic receptor supersensitivity (Tarsy & Baldessarini 1974, 1977).

Any hypothesis that attempts to explain the pathophysiology of the tardive dyskinesia syndrome (or possibly, series of related syndromes varying in their timing, duration, and precise clinical manifestations) must take into account a number of salient clinical observations. These include a late onset after many months of treatment with ordinary doses of neuroleptics, usually with some worsening on withdrawal of the drugs; more rapid occurrence on abrupt discontinuation of unusually large doses; close similarity to dyskinesias induced during L-dopa therapy of parkinsonism (Barbeau & McDowell 1970), abuse of amphetamines, or the clinical use of stimulants in children with "minimum brain dysfunction" (Mattson & Calverly 1968). In addition, the clinical "differential pharmacology" of tardive dyskinesia is enlightening (Table 2). These effects include: 1. acute worsening of tardive dyskinesia by L-dopa or stimulants (Gerlach et al 1974), 2. amelioration by small parenteral doses of the partial direct (possibly selectively presynaptic) dopamine agonist, apomorphine (Carroll et al 1977, Corsini et al 1977, Tarsy et al 1979), 3. amelioration following small repeated doses of L-dopa (possibly acting to diminish dopamine receptor sensitivity) (Alpert & Friedhoff 1976), 4. at least partial and temporary suppression by dopamine antagonists (Kazamazuri et al 1973) including receptor blockers (potent neuroleptics), storage blockers (reserpine, tetrabenazine), or a synthesis inhibitor (α-methyl-p-tyrosine), 5. mild and variable effects of cholinergic agents (generally worsening with antimuscarinic-antiparkinsonism agents) and, 6. occasionally, partial improvement with eserine, deanol, or choline or phosphatidylcholine given to enhance the availability of acetylcholine, or putative agonists of GABA (γ-aminobutyric acid) or agonists or antagonists of serotonin receptors (Granacher et al 1975, Growdon et al 1977, Baldessarini et al 1980). In short, the syndrome's differential clinical pharmacology is strikingly opposite to that of parkinsonism, a condition that almost certainly includes a deficiency of dopamine in the basal ganglia as

Table 2 The differential pharmacology of tardive dyskinesia[a]

Agents that may partially suppress tardive dyskinesia

Dopamine antagonists
  Apomorphine (in low dose)
  Butyrophenones
  Clozapine
  Papaverine (mechanism uncertain)
  Phenothiazines
  Pimozide
Amine-depleting agents
  Reserpine
  Tetrabenazine
Blockers of catecholamine synthesis
  Alpha-methyldopa
  Alpha-methyltyrosine
Blockers of catecholamine release
  Lithium salts
Cholinergic agents
  Deanol (mechanism uncertain)
  Choline and lecithin (phosphatidylcholine)
  Physostigmine
Miscellaneous agents
  Baclofen (postulated GABA mechanism unproven)
  Valproate (postulated GABA mechanism unproven)

Agents with variable, negligible, or uncertain effects

Alpha-methyldopa
Amantadine
Antihistamines
Barbiturates
Benzodiazepines
Cyproheptadine
Deanol
Isocarboxazid
Methylphenidate
Penicillamine
Physostigmine
Pyridoxine (B_6)
Tryptophan

Agents that worsen tardive dyskinesia

Anticholinergic agents
  Antiparkinson agents (e.g. benztropine)
Dopamine agonists
  Amphetamines
  L-DOPA (may have opposite effect in small repeated doses)
Other agents
  Phenytoin

[a]Some drugs appear in more than one category, reflecting ambiguity in the literature. Note that while apomorphine is usually classed as a dopamine agonist, it actually has complex mixed actions, may antagonize dopamine at low doses, and has clear antidyskinetic effects. For references, see Tarsy & Baldessarini (1976), Baldessarini & Tarsy (1978), Baldessarini et al (1980).

an important contributing feature (Bernheimer et al 1973); and further-more, it is very similar to that of other hyperkinesias such as Huntington's chorea (Chase 1976), which is believed to represent a state of relative excess of dopamine function (e.g. perhaps by loss of neurons including GABA-secreting cells that modulate dopamine nuerons) (Enna et al 1976). More-over, many of the clinical features of severe forms of tardive dyskinesia are similar to those of Huntington's chorea (Marsden et al 1975).

DOPAMINE OVERACTIVITY: PRESYNAPTIC MECHANISMS

A state of relative excess of dopamine function in tardive dyskinesia, if indeed it is present, could come about through several mechanisms (Tarsy & Baldessarini 1977): (a) presynaptic dyscontrol of dopamine synthesis and release; (b) decreased availability of other modulating systems (including those using GABA or other neuroinhibitory amino acids, acetylcholine, serotonin, substance P, endorphins, or other peptides); (c) increased quan-tity or effectiveness of postsynaptic dopamine receptors or "effectors" (mechanisms that mediate the postsynaptic effects of dopamine, possibly including protein-phosphorylation mediated by cyclic AMP, the synthesis of which can be stimulated by dopamine).

With respect to presynaptic mechanisms, animal data strongly indicate that the initial increase of dopamine turnover and metabolite (dihydroxy-phenylacetic and homovanillic acids) production induced by neuroleptics is short-lasting (Asper et al 1973, Sedvall 1975). The very "tolerance" in-volved in this response to prolonged neuroleptic treatment may be mediated by increased sensitivity to dopamine transmission postsynaptically or at proposed presynaptic "auto-receptors" believed to modulate dopamine syn-thesis and possibly also release (Baldessarini 1975, Bunney & Aghajanian 1975, Corsini et al 1977, Carlsson 1978). Human data on this point do not support the prediction that resting levels of homovanillic acid (or its rate of rise after inhibiting its transport through the choroid plexus) in lumbar cerebrospinal fluid are elevated in schizophrenic patients (Bowers 1974, Post & Goodwin 1975), or in those with signs of tardive dyskinesia (Bowers et al 1978); although increases do occur as an acute response to neuroleptic drugs (Post & Goodwin 1975). This "tolerance" phenomenon is opposite to the result predicted by a presynaptic hypothesis involving excessive production and release of dopamine, but is possibly consistent with a mech-anism involving dopamine receptor supersensitivity. There are also un-confirmed or conflicting reports that the enzyme converting dopamine to norepinephrine (dopamine beta-hydroxylase) may be decreased (Wyatt et al 1975), or that levels of dopamine or its metabolites may be increased in the brains of chronic schizophrenic patients (not necessarily with tardive dyskinesia) post-mortem (Bird et al 1977, see Baldessarini & Tarsy 1979).

It is difficult to comment on other neurotransmitters that may interact with, or act in concert with dopamine neurons since they have been much less intensively investigated than dopamine; furthermore, there are few potent and selective agonists or antagonists for many of these other neurohumors, and those that exist have weak or inconsistent clinical effects on tardive dyskinesia (Baldessarini et al 1980). Data from recent post-mortem studies of brain tissue of chronic schizophrenics exposed to neuroleptics (but not necessarily showing signs of tardive dyskinesia) (Bird et al 1977) include increased or decreased activity of choline acetyltransferase (the acetylcholine-synthesizing enzyme), but fairly consistent, if sometimes small and possibly artifactual, decrease in glutamic acid decarboxylase (the GABA-synthesizing enzyme), as has also been found in brains of patients with Huntington's disease (see Enna et al 1976). These results may possibly suggest neurotoxic effects of neuroleptic drugs on GABA or acetylcholine neurons that may have indirect dopamine-enhancing actions.

DOPAMINE OVERACTIVITY: POSTSYNAPTIC MECHANISMS

Evidence of the existence of dopamine receptor supersensitivity in animals is quite compelling (Table 3) (Tarsy & Baldessarini 1974, Müller & Seeman 1978). It includes direct (same drug given repeatedly and for final test) and crossed (different drug used to test) tolerance to many behavioral (e.g. catalepsy) and biochemical (especially dopamine-turnover-enhancing) effects of neuroleptics (Asper et al 1973). Such effects are not seen with neuroendocrine responses to neuroleptics (e.g. prolactin release), which suggests differences in the regulation of hypothalamic or pituitary vs forebrain dopamine systems (Friend et al 1978, Schyve et al 1978, Reichlin & Boyd 1978). In addition, there is striking and consistently increased behavioral sensitivity to dopamine agonists in many species upon withdrawal from repeated treatment with neuroleptics and other central nervous system depressants (Tarsy & Baldessarini 1974). Nevertheless, these aspects of supersensitivity to dopamine agonists after neuroleptic treatment are generally short-lived (days to weeks) (Müller & Seeman 1978), raising questions about their pertinence to typically long-lasting cases of tardive dyskinesia. Another paradox is that some agonists (including d-amphetamine, apomorphine, ergot alkaloids, and possibly also L-dopa) may also induce supersensitive behavioral responses to themselves or other dopamine agonists ("reverse tolerance") (Martres et al 1977). Moreover, these paradoxical effects (which may reflect presynaptic dopamine turnover reducing actions that outlast their acute stimulant actions [Segal 1975] and so lead to receptor

Table 3 Summary of evidence supporting development of dopamine (DA) receptor supersensitivity following denervation or disuse of dopamine systems[a]

After lesioning DA projections (electrolytically or with 6-hydroxydopamine)

Increased behavioral responses
  · Turning contralateral to lesion after direct DA agonist
  · Turning ipsilateral to lesion with indirect agonist
  · Increased stereotyped behavior to direct, but decreased to indirect DA agonists

Increased biochemical responses
  · Increased response of DA-sensitive adenylate cyclase in brain to direct DA agonists
  · Increased binding of tritiated neuroleptic drugs in forebrain

Increased neurophysiologic responses to direct DA agonists in forebrain

After prolonged neuroleptic treatment (or use of other anti-DA agents)

Increased behavioral responses
  · Tolerance to same or dissimilar neuroleptic in blockade of behavioral responses to DA agonists
  · Increased arousal after amphetamine after withdrawal of neuroleptic
  · Increased stereotyped responses to DA agonists after withdrawal of neuroleptic
  · Tolerance to neuroleptic blockade of self-stimulation in nigrostriatal system and cortex, followed by increased responses after neuroleptic withdrawn

Increased biochemical responses
  · Increased sensitivity of DA-sensitive adenylate cyclase to DA agonists
  · Increased binding of tritiated neuroleptic drugs in forebrain[b]

Increased neurophysiologic responses to direct DA agonists in forebrain

[a] Further details and references not reviewed in the text can be found in Müller & Seeman (1978), Baldessarini & Tarsy (1979), and Baldessarini et al (1980).
[b] Note that while studies of postmortem human brain tissue demonstrating increased binding of ^{3}H-neuroleptics to membrane preparations have usually been interpreted as an aspect of the pathology of schizophrenia, such changes, at least in part, may be due to prior treatment with neuroleptics.

"disuse supersensitivity"), as well as supersensitivity after neuroleptic treatment, have been described following even a single dose of the agent in question (Martres et al 1977).

Other results that derive from frank denervation of dopamine projections to forebrain by selective destruction strongly support the occurrence of dopamine receptor supersensitivity in laboratory animals. These examples of "denervation supersensitivity" include strongly enhanced behavioral responses to dopamine agonists (Price & Fibiger 1974, Thornburg & Moore 1975), inconsistent and small increases in the sensitivity of adenylate cyclase in striatal homogenates to dopamine (Krueger et al 1976), increased sensitivity of striatal neurons to iontophoretically applied dopamine agonists or stimulation of nigrostriatal neurons (Schultz & Ungerstedt 1978), and increases in the binding of ligands ([^{3}H]-neuroleptics) proposed to label "dopamine receptors" in brain tissue (Creese et al 1977). Moreover, most of these observations have more recently been replicated and extended in brain tissues of animals repeatedly treated with neuroleptic agents, with respect to adenylate cyclase (Friedhoff et al 1977), electrophysiologic responses (Yarbrough, 1975), and the binding of labeled ligands (Burt et al 1977, Friedhoff et al 1977, Müller & Seeman 1977). In addition, increased binding of [^{3}H]-spiroperidol is virtually absent in cerebral cortex under such conditions, and is not seen after exposure to the rare antipsychotic agents (such as clozapine) devoid of neurologic effects, in contrast to more typical neuroleptics (M. Herschel, B. Cohen, R. J. Baldessarini, unpublished observations). There are even several recent reports that the post-mortem binding of [^{3}H]-neuroleptics is increased in brain tissue of chronic schizophrenics, possibly at least in part as a consequence of prolonged exposure to neuroleptic agents (Owen et al 1978).

EVALUATION OF THE DOPAMINE SUPERSENSITIVITY HYPOTHESIS

For all these impressive findings, there remain a number of serious problems and shortcomings of the hypothesis that tardive dyskinesia is due to a disuse supersensitivity of the dopamine-receptor. First, the duration of supersensitive responses in most small laboratory animals is relatively brief, usually involving a return to baseline status within a few weeks, at most, after discontinuation of the neuroleptic agent (Tarsy & Baldessarini 1974, Müller & Seeman 1978). While clinical tardive dyskinesia is typically much more enduring, dyskinesia that follows abrupt withdrawal of neuroleptic drugs (Tarsy & Baldessarini 1977, Gardos et al 1978) more closely resembles the experimental situation. On the other hand, newer primate models are more promising in this regard, as long-lasting (months) supersensitivity to dopamine agonists and spontaneous dyskinesias have been produced, particularly in Cebus monkeys (Gunne & Barany 1976).

Second, behavioral supersensitivity to dopamine agonists locally applied to dopamine-sensitive limbic regions of the forebrain of the neuroleptic-pretreated rat has recently been described (Jackson et al 1975, Ungerstedt & Ljungberg 1977). Although these regions are hypothesized to mediate antipsychotic effects of the drugs (Stevens 1973, Meltzer & Stahl 1976, Baldessarini 1977b), increased psychosis following prolonged neuroleptic therapy ("tardive schizophrenia") is unknown clinically. The failure to produce psychosis with prolonged neuroleptic treatment suggests either that some dopamine-sensitive areas (such as the nucleus accumbens septi) that are supposedly limbic in function (Nauta & Domesick 1979) may actually contribute to extrapyramidal function, or that the popular idea that psychoses are mediated by excessive limbic dopamine function may be invalid or too simplistic.

Third, most biochemical and behavioral manifestations of presumed dopamine receptor supersensitivity involve quantitatively small changes (typically 20–30% increases in receptor-ligand binding, and not more than 2- or 3-fold shifts in values of half-maximally effective doses [ED_{50}] for stimulation of iontophoretic responses, adenylate cyclase activity, or of behavior by dopamine agonists). Moreover, these are rapidly reversible, adaptive and plastic, sometimes developing and fading over times as brief as minutes to hours, and seemingly designed to restore dopamine function toward normal; that is, to subserve homeostasis of neurotransmitter function rather than to lead to sustained unbalanced increases of function.

NEUROPATHOLOGY IN TARDIVE DYSKINESIA

For long-lasting or even irreversible changes, one would expect visible neuropathologic changes such as frank cell loss. Nevertheless, available animal neuropathologic studies following prolonged exposure to neuroleptics (Fog et al 1976) do not support that prediction (Baldessarini & Tarsy 1979), and human studies, though rare, have suggested relatively minor changes in the basal ganglia (Jellinger 1977). Possibly, other as yet unknown functionally important but histologically so-far invisible toxic metabolic or structural changes in dopamine-secreting or dopamine-sensitive cells or their membranes can occur, although electron-microscopic studies of brain tissues after prolonged exposure to neuroleptics are almost nonexistent.

CONCLUSIONS

The dopamine-receptor supersensitivity hypothesis may help to explain some features of the pathophysiology of tardive dyskinesia, and is most likely to contribute to its more rapidly reversible forms, such as dyskinesias that follow abrupt withdrawal of neuroleptic drugs (Jacobson et al 1974, Gardos et al 1978). Furthermore, this hypothesis has shed light on mechanisms only recently appreciated that evidently exert important regulatory effects on central synaptic function and that may contribute to neuronal "plasticity" and adaptiveness in the central nervous system. Moreover, the hypothesis suggests new therapeutic strategies, such as, for example, the attempt to prevent or suppress development of supersensitive receptors by treatment with dopamine agonists.

ACKNOWLEDGMENTS

Supported in part by National Institute of Mental Health Research Career Scientist Award MH-47370 and Grant MH-31154. The manuscript was prepared by Mrs. Mila Cason.

Literature Cited

Alpert, M., Friedhoff, A. J. 1976. Receptor sensitivity modification in the treatment of tardive dyskinesia. *Clin. Pharmacol. Ther.* 19:103–4

Angrist, B., Sathananthan, G., Gershon, S. 1973. Behavioral effects of L-dopa in schizophrenic patients. *Psychopharmacologia* 34:1–12

Asper, H., Baggiolini, M., Bürki, H. R., Lauener, H., Ruch, W., Stille, G. 1973. Tolerance phenomena with neuroleptics. *Eur. J. Pharmacol.* 22:287–94

Baldessarini, R. J. 1974. Tardive dyskinesia: An evaluation of the etiologic association with neuroleptic therapy. *Can. Psychiatr. Assoc. J.* 19:551–54

Baldessarini, R. J. 1975. Catecholamine release. In *Handbook of Psychopharmacology,* ed. L. L. Iversen, S. D. Iversen, S. H. Snyder, pp. 37–137. New York: Raven

Baldessarini, R. J. 1977a. *Chemotherapy in Psychiatry,* Cambridge, Mass: Harvard Univ. Press. 201 pp.

Baldessarini, R. J. 1977b. Schizophrenia. *N. Engl. J. Med.* 297:988–95

Baldessarini, R. J. 1980. Drugs and the treatment of psychiatric disorders. In *The Pharmacological Basis of Therapeutics,* ed. A. Gilman Jr., A. Gilman Sr., L. S. Goodman, Ch. 19. New York: Macmillan. In press

Baldessarini, R. J., Cole, J. O., Davis, J. M., Gardos, G., Preskorn, S., Simpson, G., Tarsy, D. 1980. *Tardive dyskinesia,* Task Force Rep. Ser., Washington DC., Am. Psychiatr. Assoc. In press

Baldessarini, R. J., Tarsy, D. 1978. Tardive dyskinesia. In *Psychopharmacology: A Generation of Progress,* ed. M. A. Lipton, A. DiMascio, K. F. Killam, pp. 993–1004. New York: Raven

Baldessarini, R. J., Tarsy, D. 1979. Relationship of the actions of neuroleptic drugs to the pathophysiology of tardive dyskinesia. *Int. Rev. Neurobiol.* 21. In press

Barbeau, A., McDowell, F. H., eds. 1970. *L-dopa and Parkinsonism,* pp. 101–202; 321–47. Philadelphia: Davis

Bernheimer, H., Birkmayer, W., Hornykiewicz, O., Jellinger, K., Seitelberger, F. 1973. Brain dopamine and the syndromes of Parkinson and Huntington: Clinical, morphological and neurochemical correlations. *J. Neurol. Sci.* 20:415–55

Bird, E. D., Barnes, J., Iversen, L. L., Spokes, E. G., Shephard, M., MaKay, A. 1977. Increased brain dopamine and reduced glutamic acid decarboxylase and choline acetyltransferase activity in schizophrenia and related psychoses. *Lancet* 2:1157–59

Bowers, M. B. Jr. 1974. Central dopamine turnover in schizophrenic syndromes. *Arch. Gen. Psychiatry* 31:50–54

Bowers, M. B., Jr., Moore, D., Tarsy, D. 1978. Tardive dyskinesia: A clinical test of the supersensitivity hypothesis. *Psychopharmacology* 61:137–41

Bruno, A., Bruno, S. C. 1966. Effects of L-dopa on pharmacological parkinsonism. *Acta Psychiatr. Scand.* 42:264–71

Bunney, B. S., Aghajanian, G. K. 1975. Evidence for drug actions on both pre- and postsynaptic catecholamine receptors in the CNS. In *Pre- and Postsynaptic Receptors,* ed. E. Usdin, W. E. Bunney Jr, pp. 89–122. New York: Dekker

Bunney, B. S., Aghajanian, G. K. 1978. Mesolimbic and mesocortical dopaminergic systems: Physiology and pharmacology. In *Psychopharmacology: A Generation of Progress,* ed. M. A. Lipton, A. DiMascio, K. F. Killam, pp. 159–69. New York: Raven

Bunney, B. S., Walters, J. R., Roth, R. H., Aghajanian, G. K. 1973. Dopaminergic neurons: Effect of antipsychotic drugs and amphetamine on single cell activity. *J. Pharmacol. Exp. Ther.* 185:560–71

Burt, D. R., Creese, I., Snyder, S. H. 1977. Antischizophrenic drugs: chronic treatment elevates dopamine receptor binding in brain. *Science* 196:326–28

Carlsson, A. 1978. Mechanism of action of neuroleptic drugs. See Bunney & Aghajanian 1978, pp. 1057–70

Carlsson, A., Lindqvist, M. 1963. Effect of chlorpromazine and haloperidol on formation of 3-methoxytyramine and normetanephrine in mouse brain. *Acta Pharmacol. Toxicol.* 20:140–44

Carroll, B. J., Curtis, G. C., Kokmen, E. 1977. Paradoxical responses to dopamine agonists in tardive dyskinesia. *Am. J. Psychiatry* 134:785–89

Chase, T. N. 1976. Rational approaches to the pharmacotherapy of chorea. In *The Basal Ganglia,* ARNMD Publ., ed. M. Yahr, 55:337–50. New York: Raven

Clement-Cormier, Y. C., Kebabian, J. W., Petzold, G. L., Greengard, P. L. 1974. Dopamine-sensitive adenylate cyclase in mammalian brain: a possible site of action of antipsychotic drugs. *Proc. Natl. Acad. Sci. USA* 71:1113–17

Corsini, G. U., DelZompo, M., Manconi, S., Cianchetti, C., Mangoni, A., Gessa, G. L. 1977. Sedative, hypnotic, and antipsychotic effects of low doses of apomorphine in man. In *Advances in Bio-*

chemical Psychopharmacology, ed. E. Costa, G. L. Gessa, pp. 645–48. New York: Raven

Costall, B., Naylor, R. J. 1976. A comparison of the abilities of typical neuroleptic agents and of thioridazine, clozapine, sulpiride, and metoclopramide to antagonize the hyperactivity induced by dopamine applied intracerebrally to areas of the extrapyramidal and mesolimbic systems. Eur. J. Pharmacol. 40:9–19

Crane, G. E. 1973. Persistent dyskinesia. Br. J. Psychiatry 122:395–405

Creese, I., Burt, D. R., Snyder, S. H. 1977. Dopamine receptor binding enhancement accompanies lesion-induced behavioral supersensitivity. Science 197: 596–98

Creese, I., Burt, D. R., Snyder, S. H. 1978. Biochemical actions of neuroleptic drugs: focus on dopamine receptor. In Handbook of Psychopharmacology, ed. L. L. Iversen, S. D. Iversen, S. H. Snyder, 10:37–89. New York: Plenum

Denham, J., Carrick, D. J. 1961. Therapeutic value of thioproperazine and the importance of the associated neurological disturbances. J. Ment. Sci. 107:326–45

Deniker, P. 1960. Experimental neurologic syndromes and the new drug therapies in psychiatry. Compr. Psychiatry 92–102

Duvoisin, R. C. 1968. Neurological reactions to psychotropic drugs. In Psychopharmacology: A Review of Progress 1957–1967, ed. D. Efron, pp. 561–573. Washington: U. S. Public Health Service Publication No. 1836

Eichler, A. J., Antelman, S. M., Kairns, L. 1977. Self-stimulation: evidence for tolerance to neuroleptics in the frontal cortex. Neurosci. Abstr. 3:440

Enna, S. J., Bird, E. D., Bennett, J. P. Jr., Bylund, D. B., Yamamura, H. I., Iversen, L. L., Snyder, S. H. 1976. Huntington's chorea: changes in neurotransmitter receptors in the brain. N. Engl. J. Med. 294:1305–9

Faurbye, A., Rasch, P. J., Peterson, P. B., Brandborg, G., Pakkenberg, H. 1964. Neurological symptoms in pharmacotherapy of psychoses. Acta Psychiatr. Scand. 40:10–27

Fielding, S., Lal, H. 1978. Behavioral actions of neuroleptics. In Handbook of Psychopharmacology, ed. L. L. Iversen, S. D. Iversen, S. H. Snyder, 10:91–128. New York: Plenum

Fish, F. J. 1967. Clinical Psychopathology, Bristol, England: Wright, pp. 84–106

Fog, R., Pakkenberg, H., Juul, P., Bock, E., Jørgensen, O. S., Andersen, J. 1976. High dose treatment of rats wth perphenazine enanthate. Psychopharmacology 50:305–7

Friedhoff, A. J., Bonnet, K., Rosengarten, H. 1977. Reversal of two manifestations of dopamine receptor supersensitivity by administration of L-dopa. Res. Commun. Chem. Pathol. Pharmacol. 16: 411–23

Friend, W. C., Brown, G. M., Jawahir, G., Lee, T., Seeman, P. 1978. Effect of haloperidol and apomorphine treatment on dopamine receptors in pituitary and striatum. Am. J. Psychiatry 135:839–41

Fuxe, K., Hökfelt, T., Johansson, O., Jonsson, G., Lidbrink, P., Ljugdahl, A. 1974. The origin of the dopamine nerve terminals in the limbic and frontal cortex: Evidence for mesocortical dopamine neurons. Brain Res. 82:349–55

Fuxe, K., Hökfelt, T., Olson, L., Ungerstedt, U. 1977. Central monoaminergic pathways with emphasis on their relation to the so-called "extrapyramidal motor system." Pharmacol. Ther. B3:169–210

Gardos, G., Cole, J. O., Tarsy, D. 1978. Withdrawal syndromes associated with antipsychotic drugs. 135:1321–24

Garver, D. L., Davis, J. M., Dekirmenjian, H., Jones, F. D., Casper, R., Haraszti, J. 1976. Pharmacokinetics of red blood cell phenothiazine and clinical effects. Arch. Gen. Psychiatry 33:862–66

Gerlach, J., Reisby, N., Randrup, A. 1974. Dopaminergic hypersensitivity and cholinergic hypofunction in the pathophysiology of tardive dyskinesia. Psychopharmacologia 34:21–35

Granacher, R. P., Baldessarini, R. J. Cole, J. O. 1975. Deanol for tardive dyskinesia. N. Engl. J. Med. 292:926–27

Growden, J. H., Hirsch, M. J., Wurtman, R. J., Wiener, W. 1977. Oral choline administration to patients with tardive dyskinesia. N. Engl. J. Med. 297: 524–27

Gunne, L. M., Barany, S. 1976. Haloperidol-induced tardive dyskinesias in monkeys. Psychopharmacology 50:237–40

Iversen, L. L. 1977. Catecholamine-sensitive adenylate cyclases in nervous tissue. J. Neurochem. 29:5–12

Jackson, D. M., Andén, N.-E., Engel, J., Liljequist, S. 1975. The effect of long-term penfluridol treatment on the sensitivity of the dopamine receptors in the nucleus accumbens and in the corpus striatum. Psychopharmacologia 45:151–55

Jacobson, G., Baldessarini, R. J., Manschreck, T. 1974. Tardive and withdrawal dyskinesia associated with haloperidol. *Am. J. Psychiatry* 131: 910–13

Jellinger, K. 1977. Neuropathologic findings after neuroleptic long-term therapy. In *Neurotoxicology,* ed. L. Roizin, H. Shiraki, N. Grčevic, pp. 25–41. New York: Raven

Jus, K., Jus, A., Gautier, J., Villeneuve, A., Pires, P., Pineau, P., Villeneuve, R. 1974. Studies of the actions of certain pharmacological agents on tardive dyskinesia and the rabbit syndrome. *Int. J. Clin. Pharmacol.* 9:138–45

Kazamazuri, H., Chien, C.-P., Cole, J. O. 1973. Long-term treatment of tardive dyskinesia with haloperidol and tetrabenazine. *Am. J. Psychiatry* 135: 486–88

Kraepelin, E. 1919. *Dementia Praecox and Paraphrenia,* Edinburg: Livingstone. 331 pp.

Krueger, B. K., Forn, J., Walters, J. R., Roth, R. H., Greengard, P. 1976. Stimulation by dopamine of adenosine cyclic 3',5'-monophosphate formation by rat caudate nucleus: Effect of lesions of the nigrostriatal pathway. *Mol. Pharmacol.* 12:639–48

Marsden, C. D. 1976. Dystonia: the spectrum of the disease. In *The Basal Ganglia,* ARNMD Publ. ed. M. Yahr, 66:351–67. New York: Raven

Marsden, C. D., Tarsy, D., Baldessarini, R. J. 1975. Spontaneous and drug-induced movement disorders in psychotic patients. In *Psychiatric Aspects of Neurologic Disease,* ed. D. F. Benson, D. Blumer, pp. 219–65. New York: Grune & Stratton

Martres, M. P., Costentin, J., Baudry, M., Marcais, H., Protais, P., Schwartz, J.-C. 1977. Long-term changes in the sensitivity of pre- and postsynaptic dopamine receptors in mouse striatum evidenced by behavioral and biochemical studies. *Brain Res.* 136:319–37

Mattson, R. H., Calverly, J. R. 1968. Dextroamphetamine-sulfate-induced dyskinesias. *J. Am. Med. Assoc.* 204:400–2

Meltzer, H. Y., Goode, D. J., Fang, V. S. 1978. The effect of psychotropic drugs on endocrine function. In *Psychopharmacology: A Generation of Progress,* ed. M. A. Lipton, A. DiMascio, K. F. Killam, pp. 509–29. New York: Raven

Meltzer, H. Y., Stahl, S. M. 1976. The dopamine hypothesis of schizophrenia: a review. *Schizophr. Bull.* 2:19–76

Moore, K. E., Kelly, P. H. 1978. Biochemical pharmacology of mesolimbic and mesocortical dopaminergic neurons. In *Psychopharmacology: A Generation of Progress,* ed. M. A. Lipton, A. DiMascio, K. F. Killam, pp. 221–34. New York: Raven

Müller, P., Seeman, P. 1977. Brain neurotransmitter receptors after long-term haloperidol. *Life Sci.* 21:1751–58

Müller, P., Seeman, P. 1978. Dopaminergic supersensitivity after neuroleptics: time course and specificity. *Psychopharmacology* 60:1–11

Nauta, W. J. H., Domesick, V. B. 1979. The anatomy of the extrapyramidal system. In *Dopaminergic Ergot Derivatives and Motor Functions,* ed. K. Fuxe, D. Calne. Oxford: Pergamon. In press

Owen, F., Crow, T. J., Poulter, M., Cross, A. J., Longden, A., Riley, G. J. 1978. Increased dopamine-receptor sensitivity in schizophrenia. *Lancet* 2:223–26

Post, R. M., Goodwin, F. K. 1975. Time-dependent effects of phenothiazines on dopamine turnover in psychiatric patients. *Science* 190:448–89

Price, M. T., Fibiger, H. C. 1974. Apomorphine and amphetamine stereotypy after 6-hydroxydopamine lesions of the substantia nigra. *Eur. J. Pharmacol.* 29:249–52

Reichlin, S., Boyd, A. E. III. 1978. Neural control of prolactin secretion in man. *Psychoneuroendocrinology* 3:113–30

Sachar, E. 1978. Neuroendocrine responses to psychotropic drugs. In *Psychopharmacology: A Generation of Progress,* ed. M. A. Lipton, A. DiMascio, K. F. Killam, pp. 499–507. New York: Raven

Schönecker, M. 1957. Ein eigentümliches syndrom in oralen Beireich bei Megaphen-Application. *Nervenarzt* 28:35–36

Schultz, W., Ungerstedt, U. 1978. Striatal cell supersensitivity to apomorphine in dopamine-lesioned rats correlated to behavior. *Neuropharmacology* 17:349–53

Schyve, P. M., Smithline, F., Meltzer, H. Y. 1978. Neuroleptic-induced prolactin level elevation and breast cancer. *Arch. Gen. Psychiatry* 35:1291–1301

Sedvall, G. 1975. Receptor feedback and dopamine turnover in the central nervous system. In *Handbook of Psychopharmacology,* ed. L. L. Iversen, S. D. Iversen, S. H. Snyder, 6:127–77. New York: Plenum

Segal, D. S. 1975. Behavioral and neurochemical correlates of repeated d-amphetamine administration. In *Neurobiological Mechanisms of Adapta-*

tion and Behavior, ed. A. J. Mandell, pp. 247–62. New York: Raven

Siggins, G. R., Hoffer, B. J., Bloom, F. E., Ungerstedt, U. 1976. Cytochemical and electrophysiological studies of dopamine in the caudate nucleus. In *The Basal Ganglia,* ARNMD Publ. ed. M. Yahr, 55:227–48. New York: Raven

Sigwald, J., Bouttier, D., Raymondeaud, C., Piot, C. 1959. Quatre cas de dyskinesie facio-bucco-linguo-masticatrice à evolution prolongée secondaire à un traitement par less neuroleptiques. *Rev. Neurol.* 100:751–55

Slater, E., Beard, A. W., Glithero, E. 1963. The schizophrenia-like psychoses of epilepsy. *Br. J. Psychiatry* 109:95–150

Smith, J. M., Kucharski, L. T., Eblen, C., Knudsen, E., Linn, C. 1979. An assessment of tardive dyskinesia in schizohrenic outpatients. *Psychopharmacology.* In press

Smith, J. M., Oswald, W. T., Kucharski, L. T., Waterman, L. J. 1978. Tardive dyskinesia: Age and sex differences in hospitalized schizophrenics. *Psychopharmacology* 58:207–11

Snyder, S. H., Greenberg, D., Yamamura, H. I. 1974. Antischizophrenic drugs and brain cholinergic receptors. *Arch. Gen. Psychiatry* 31:58–61

Snyder, S. H., U'Pritchard, D., Greenberg, D. A. 1978. Neurotramsmitter receptor binding in the brain. In *Psychopharmacology: A Generation of Progress,* ed. M. A. Lipton, A. DiMascio, K. F. Killam, pp. 361–70. New York: Raven

Stevens, J. 1973. An anatomy of schizophrenia? *Arch. Gen. Psychiatry* 29:177–89

Swazey, J. P. 1974. *Chlorpromazine in Psychiatry: A Study in Therapeutic Innovation.* Cambridge, Mass: MIT Press

Tarsy, D., Gardos, G., Cole, J. O. 1979. The effect of dopamine agonists in tardive dyskinesia. *Neurology* 29:606 (abstract)

Tarsy, D., Baldessarini, R. J. 1974. Behavioral supersensitivity to apomorphine following chronic treatment with drugs which interfere with the synaptic function of catecholamines. *Neuropharmacology* 13:927–40

Tarsy, D., Baldessarini, R. J. 1976. The tardive dyskinesia syndrome. In *Clinical Neuropharmacology,* ed. H. L. Klawans, pp. 29–61. New York: Raven

Tarsy, D., Baldessarini, R. J. 1977. The pathophysiologic basis of tardive dyskinesia. *Biol. Psychiatry* 12:431–50

Thornburg, J. E., Moore, K. E. 1975. Supersensitivity to dopamine agonists following unilateral 6-hydroxydopamine-induced striatal lesions in mice. *J. Pharmacol. Exp. Ther.* 192:42–49

Ungerstedt, U. 1971. Stereotaxic mapping of the monoamine pathways in the rat brain. *Acta Physiol. Scand.* 1971: Suppl. 367, pp. 1–48

Ungerstedt, U., Ljungberg, T. 1977. Behavioral patterns related to dopamine neurotransmission: Effect of acute and chronic antipsychotic drugs. In *Advances in Biochemical Pharmacology,* ed. E. Costa, G. L. Gessa, 16:193–99. New York: Raven

Windelman, N. W. 1961. The interrelationship between the physiological and psychological etiologies of akathisia. In *Extrapyramidal System and Neuroleptics,* ed. J. M. Bordeleau, pp. 563–68. Montreal: Editions Psychiatriques

Wyatt, R. J., Schwartz, M. A., Erdelyi, E., Barchas, J. D. 1975. Dopamine-β-hydroxylase activity in brains of chronic schizophrenic patients. *Science* 187:368–70

Yarbrough, G. G. 1975. Supersensitivity of caudate neurons after repeated administration of haloperidol. *Eur. J. Pharmacol.* 31:367–69

Ann. Rev. Neurosci. 1980. 3:43–75
Copyright © 1980 by Annual Reviews Inc. All rights reserved

BACTERIAL CHEMOTAXIS IN ❖11535
RELATION TO NEUROBIOLOGY

D. E. Koshland, Jr.

Department of Biochemistry, University of California,
Berkeley, California 94720

INTRODUCTION

Bacteria have a simple sensory system and they exhibit behavioral responses. Since they are monocellular organisms it is legitimate to ask whether the study of such a system can be relevant to neurobiology, which, by its nature, involves interaction between many cells. The answer is that relevant information has already been uncovered, which may be helpful in studies on neural networks, and this chapter attempts to describe some of the findings in bacteria and their relationships to more complex systems.

To a first approximation the understanding of a brain involves two extreme problems. On the one hand there is the "wiring problem," and on the other hand there is the problem of the "switch." However, the neuron is not simply a switch. It is a relatively complex cell that adds enormously to the capacity of the brain because it is by itself able to perform a great deal of information processing. If the brain is considered to be analogous to a giant computer, the neuron must be considered closer to a pocket calculator than a simple switch. In such a system, therefore, a fundamental understanding of the role of the individual cell will not only be important per se but will also help in understanding the design of the "wiring diagram."

If one is willing to accept that the chemistry of the individual neuron is relevant to an understanding of the overall system, it remains to decide whether the study of a bacterial cell is relevant to the understanding of the neuron. The answer to this question can be answered in part by historical precedent. At the biochemical level it has been proved repeatedly that all cells have characteristics in common. The same question of relevance of the bacterial system has been asked in relation to the genetic code, carbohydrate metabolism, energy sources, transport properties, etc. In each case the

43

simple bacterial cell has been shown to have a high degree of similarity to its more complex analogues. It would be unusual if behavior were so different that this same analogy would not pertain. In fact already there is substantial evidence that similar relationships exist, and the differences when they occur, can also provide insight into the varied mechanisms involved in behavioral phenomena.

Enough is now known about neurons and bacteria that the simple scheme shown in Figure 1 can illustrate the similarities and differences. A neural cell has receptors from which it can receive information from the outside (usually in the form of chemical signals but sometimes in other forms e.g. light or electrical stimuli) and so does the bacterial cell. In both cases there is a processing system of moderate complexity within the cell and an output response. In the case of the neuron the output can be the release of a neurotransmitter or an electrical voltage; in the case of the bacterium it is a change in the flagellar rotation. Since the neuron is far larger than the bacterium, it must use additional devices such as an action potential to transmit information over much larger distances. And this of course is one of the distinguishing differences between the two cells. On the other hand the bacterial system shows properties of adaptation, memory, receptor function, focusing of signals, etc, which have high degrees of similarity to the equivalent processes of the neuron. Moreover, it is not at all clear whether the individual processing unit is substantially more complex in a neuron than it is in a bacterium. The preliminary evidence for example, that the bacterial processing system utilizes methylation and demethylation is in ways similar to the use of phosphorylation and dephosphorylation in the mammalian system.

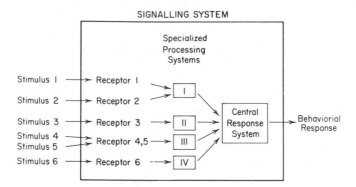

Figure 1 Signal processing in a single cell. External stimuli activate receptors, which convey information to specialized processing parts of the system and then to a central response integrative network. The output is the behavioral response of tumbling in the bacterium and an electrical signal or a neurotransmitter release in the case of a neuron.

If the analogy holds, the convenience of studying the bacteria is considerable. Millions of identical cells can be grown to provide a source of purified protein. Genetic manipulations in bacteria have been developed to a degree so far unparalleled in any higher species. Hence insight into the processing unit of the bacterial system can provide our first foundation for understanding the more complex systems. Because extensive reviews of the bacterial system have been written by this author (Koshland 1977a, 1977b, 1979a,b) and by others (Adler 1975, Berg 1975, Hazelbauer & Parkinson 1977, Macnab, 1978, 1979, Silverman & Simon 1977a), this review attempts to provide an overview of the current status of the field with emphasis on those features most relevant to neurobiology.

BRIEF DESCRIPTION OF THE BACTERIAL SYSTEM

The gram-negative bacteria, *Escherichia coli* and *Salmonella typhimurium,* are the organisms on which the bulk of the chemotactic studies (Adler 1975, Berg 1975, Hazelbauer & Parkinson 1977, Koshland 1977a, 1977b, 1979, Macnab, 1979, Silverman & Simon 1977a) have been made, although chemotaxis has been observed in gram-positive organisms and many other species (Doetsch & Hageage 1968, Weibull 1960, Macnab 1979). *E. coli* and *S. typhimurium* are very similar to each other and are each about 1 μm wide and 2–3 μm long. The receptors, which detect the signals from the external environment, are located either in the inner membrane or in the periplasmic space just outside of the inner membrane. About 30 receptors already have been identified that can detect signals from the external environment and deliver them to the processing system. This processing system requires approximately ten gene products specifically devoted to the transduction and interpretation of the chemotactic response. A wide variety of other components provide a support system either in the form of the energy needed to drive the flagellar apparatus or the level of S-adenosylmethionine involved in adaptation. These support systems have a direct analogy to the energy, oxygen, structural components, etc needed for the operation of the brain, and they can similarly be assumed to be a constant background in the analysis of the specific sensory system. At times of course the interface between the two systems must be taken into account. But in general we can emphasize the properties of the ten proteins that can be considered to be "the brains" of the bacterium and the 30 receptors that represent its "eyes and ears."

The chemotactic response, originally discovered by Pfeffer (1884) and Engelmann (1881) in the 1880s, allows the bacterium to swim towards chemicals and environments optimal for its survival and away from environments that are dangerous or deleterious. It is a highly selected system that has optimized many features for the benefit of the bacterium. In essence it provides for the bacterium a feedback system in relation to the environment equivalent to the pain-pleasure feedback system of higher species.

THE BEHAVIORAL RESPONSE

The behavioral response of bacteria, which allows them to migrate up or down gradients, is achieved by the control of tumbling frequency (Berg & Brown 1972, Macnab & Koshland 1972). When bacteria are progressing in a direction that they "consider" favorable, they suppress tumbling and continue swimming in approximately straight lines. When they are swimming in a direction that their sensory system tells them is unfavorable, they generate tumbling and head in a new direction.

The new direction after a tumble is not oriented specifically to be more favorable. Each tumble causes a random reorientation (Berg 1971). Hence unlike some species, it is not the direct of orientation of the tumble that results in the migration, rather it is the frequency of tumbling. For the bacterium heading in the wrong direction, a tumble can have two possible outcomes. It can by chance continue to swim in the wrong direction, in which case it will not move very far before it tumbles again. If as a result of a tumble it reorients and starts swimming in a favorable direction, tumbling will be suppressed and the bacterium will swim a longer than normal distance. Thus the manipulation of the tumbling frequency is sufficient to bias the random walk of the bacterium to move in a favorable direction.

At first glance such a random walk might appear inefficient. It would certainly be so for a larger organism. If one selects the sensory system so that the bacterium moves very short distances in the wrong direction and quite long distances in the right direction, the net migration can involve little waste.

Not only is the movement efficient in energy terms, but it is efficient in regard to simplicity of behavioral response. Tumbling or not tumbling is the behavioral output of the processing system. Thus, to a first approximation, the output of the cell is an on-off phenomena such as the electrical discharge of a neuron or the secretion of a neurotransmitter.

THE MOTOR APPARATUS

The flagellar structure has been largely elucidated through studies with electron microscopy and with genetic techniques in the laboratories of Iino (1969), DePamphilis & Adler (1971), Silverman & Simon (1973), and others. A series of rings are identified in various positions of the membrane, beginning with the M-ring of the cytoplasmic membrane and ending with the L-ring at the exterior of the lipopolysaccharide layer. Attached to these rings is a "hook" region, which connects to the filament. The filament itself is made entirely of the single protein flagellin.

The energy to drive the motor arises from the energized membrane state, and not from ATP (Larsen et al 1974a, Manson et al 1977). Experiments demonstrated that ATP levels can be altered drastically without loss of cell motility as long as the energy from oxidative phosphorylation is available to the basal structure. In more recent studies using cyanine dyes, it has been possible to alter the membrane potential and demonstrate that a minimum membrane potential is necessary to drive the motor (Miller & Koshland 1977). The source of energy for motility is thus different from skeletal muscle, which utilizes ATP, but is similar to other systems, such as some transport processes, which use the energized membrane state as an energy source (Berger & Heppel 1974).

The flagella drive the bacterium by rotation as a propeller. Silverman & Simon (1974) established the rotation model by tethering a flagellum to the microscope slide with an anti-flagellin antibody and then observing the rotation of the body of the bacterium. Their conclusion that motility is produced by a propeller-like rotation of the flagellum accords well with the biochemical evidence that flagellin is an inert protein, having no ATPase function. The physics of the bacterial motion has been discussed by Berg (1975), Purcell (1976), and many others.

If the driving force for motion is rotation of flagella, the question of tumbling still remains. To understand this process it is necessary to consider the flagellar structure and hydrodynamics. It has been observed that the flagella form a bundle during the process of swimming (Anderson 1975) and that the flagella are attached randomly over the surface of the bacterium (Leifson 1960). Hydrodynamic forces coalesce the individual rotating flagella into a bundle when the bacterium is swimming (Macnab 1977).

The first clue in regard to the mechanism of tumbling came from an observation of Macnab & Koshland (1974) who showed that individual flagella could be seen in dark field microscopy. They noted that the bundle seemed to be formed when the bacterium were swimming smoothly through the water, but flew apart to reveal individual flagella when tumbling occurred. They therefore postulated that the tumbling process involved a flying apart of the flagella, which then caused asymmetric pushing at the various attachment sites of flagella to the bacterial cell body. Return to smooth swimming occurred on reassembly of the flagellar bundle (cf Figure 2).

How could such a separation of the flagella bundle come about? The next step was revealed by Larsen et al (1974b) who utilized the tethering technique

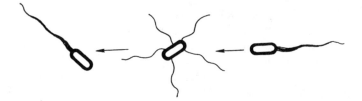

Figure 2 The process of tumbling in bacteria. When the bacteria are swimming smoothly in approximately a straight line, the flagella form a bundle. When the flagella reverse direction the bundle flies apart and the bacterium abruptly changes direction or tumbles for a period. It then resumes smooth swimming and the bundle reforms.

of Silverman & Simon (1974) and noted that mutants that swam smoothly involved counterclockwise rotation of the flagella and mutants that tumbled gave clockwise rotation. The logic of this finding in relation to the flying apart of the flagella seemed obvious. The flagella rotating in a counterclockwise rotation coalesced to form a stable hydrodynamic bundle leading to smooth movement in the medium. Clockwise rotation, however, could not form a stable bundle and hence a reversed rotation should lead to flying apart of the bacterial flagella. However, this simple explanation was not the final word, and the ultimate understanding came from a more detailed knowledge of the bacterial flagella.

The bacterial flagella can exist in several polymorphic forms. There is a normal form, which is a left-handed helix with a pitch of 2.1 Å present in normal bacteria at pH 7 (Asakura et al 1966). However, the work of Asakura & Iino (1972), Kamiya & Asakura (1976), Hotani (1976), and others demonstrated that there was a second form, a "curly" form with approximately half the pitch of the normal form and containing a right-handed helix. Macnab & Ornston (1977) with refined optical methods observed that even in normal bacteria and at normal pH the curly form of helix could be observed when the bacterium was reversing its direction of rotation, and they postulated that a curly helix form might stabilize with consistent clockwise rotation of the flagella.

The conclusion was then combined with studies on mutants showing inverted behavior (Rubik & Koshland 1978) to produce a new model for tumbling (Khan et al 1978). Tumbling is caused by instability of the bundle generated by reversals of the flagella motor. At two extremes in which there is either continuous clockwise motion or continuous counterclockwise motion, smooth swimming results. With continuously counterclockwise rotation of the flagella, the bundle is in a left-handed "normal" helix and the bacteria swim smoothly. With continuous clockwise rotation, they also swim smoothly but in this case the flagella are in the polymorphic form of a right-handed "curly" helix. Reversal of flagella rotation from either extreme causes the bundle to fly apart and generate tumbling. Thus, it is not the direction of rotation but the frequency of reversal of rotation that causes tumbling. If reversal occurs occasionally, random motility is observed. If there are frequent reversals, the flagella never form a stable bundle and constant tumbling occurs (Khan et al 1978).

COMPLEXITY OF THE SENSING SYSTEM

The sensing system per se is composed of two parts, the receptors that receive the signals from the environment, and the processing proteins that interpret those signals. It is now clear that there are approximately 30 receptors and approximately 12 central processing proteins. These have been identified by a variety of biochemical and genetic means, and some of their functions have been identified. The two parts of the system are described briefly below.

Types and Numbers of Receptors

The receptors are designed for a wide range of compounds, mostly carbohydrates and amino acids (Adler 1975, Koshland 1979a). Each receptor is given a name identified with its main activator, although few of the receptors are specific for a single compound. For example in the case of the galactose receptor, galactose and glucose bind strongly (K_d's of approximately 10^{-7}), but a variety of other sugars such as arabinose bind at higher concentrations ($\sim 10^{-3}$) and would be detected by the organisms in the absence of the more strongly binding sugars (Anraku 1968, Boos et al 1972, Zukin et al 1977b). Similarly the receptor for serine has been found to bind cysteine, alanine, and glycine but at lower affinities than serine (Adler 1966). The ribose receptor has been found to bind ribose with a dissociation constant (K_d) of approximately 10^{-7} and the only other sugar that elicits an effect from this receptor is allose, which binds 1000 times less strongly (Aksamit & Koshland 1974). Since allose is not a prevalent sugar found naturally in high concentrations, it is probable that this receptor is physiologically significant for ribose only. Thus the receptors appear to be designed specifically to provide response for a limited group of compounds important to the metabolism and survival of the organism.

Evidence that these receptors are present is obtained by purification studies in the case of the galactose (Anraku 1968), ribose (Aksamit & Koshland 1974), and maltose receptors (Hazelbauer 1975), and by a combination of behavioral studies and genetics for the other receptors (Adler 1969, 1975). Some of these receptors are constitutive in the sense that they are present in every cell of the appropriate species; some are inducible, i.e.

they can be formed if the bacterium is grown under the proper conditions (Adler 1969, Fahnestock & Koshland 1979, Koshland 1979a). This introduces a level of plasticity into the system, which is discussed below.

The Proteins of the Processing System

Bacterial techniques for the selection of mutants are highly developed and have been applied to these systems. A generally non-chemotactic mutant is indicated by the loss of the chemotactic response to all stimuli. This distinguishes mutations in the general processing from mutations in individual receptors, which cause loss of function only for a specific compound (Adler 1969). These genes, together with the phenotypic properties, are shown in Table 1. It can be seen that they fall into two categories. Nine genes are identified with the central processing of all stimuli (Warrick et al 1977). Three of the genes (the *tsr, tar,* and *trg* genes) involve processing of several stimuli but their products can be deleted without damage to other parts of the system (Ordal & Adler 1974, Silverman & Simon 1977b, Strange & Koshland 1976). In a rough analogy therefore the genes that are identified with portions of the responses could be analogous to hearing, seeing, and tasting, and the nine genes of central processing with the central brain (Strange & Koshland 1976). It is intriguing that even a monocellular species develops some compartmentalization of the sensory stimuli.

Although it cannot be said that all the genes have been located, the rather extensive searches that have been made indicate that the additional genes identified with the system will be few in number if indeed they exist (Silverman et al 1977, DeFranco et al 1979). Thus one receives a picture of a

Table 1 Properties of the signal processing genes of *E. coli* and *S. typhimurium*

Gene classification		Molecular weight of peptide product	No-gradient motility pattern
Salmonella typhimurium	*Escherichia coli*		
P	A	76,000	Smooth
Q	Y	8,000	Smooth
R	X	28,000	Smooth
S	—	—	Smooth
T	Z	24,000	Tumbly
U	C	—	Smooth, Random, Tumbly
V	—	—	Smooth
W	W	12,000	Smooth
X	B	38,000	Tumbly
—	*tsr* (D)	60,000	Random
—	*tar*	60,000	Random
trg	*trg*	—	Random

moderately complicated system but one sufficiently simple that all of the components can be identified within a reasonable period of time. In fact most of the gene products in the central processing system have already been identified and their molecular weights are shown in Table 1 (Silverman & Simon 1976, 1977c).

BACTERIAL MEMORY AND THE RESPONSE REGULATOR MODEL

Macnab & Koshland (1972) showed that bacteria respond to gradients by a temporal sensing system in which the "memory" of the bacterium play a significant role. The rationale of the experiments is quite simple. Bacteria subjected to sudden temporal increases in attractant concentration are observed to suppress their tumbling frequency. Bacteria subjected to sudden temporal decreases in attractant concentration increase their tumble frequency. In both cases the bacteria adapt, i.e. gradually return to their normal tumbling frequency. These are exactly the responses required to direct migration favorably for bacteria swimming up spatial gradients of attractants (suppression of tumbling) or down gradients of attractant (increase in tumbling). They swim at uniform velocities and sense spatial gradients by a temporal sensing mechanism. This sensing mechanism integrates the gradient information through a "memory" apparatus within the bacterial cell.

Alteration of the size of the gradients demonstrated that these conclusions applied to the most extreme gradient as well as to the most shallow gradient that bacteria could sense (Berg & Brown 1972, Berg & Tedesco 1975, Macnab & Koshland 1972, 1973). Repellent gradients had the same effect as attractant gradients except for an inverted algebraic sign. i.e. the bacteria tumble more frequently when traveling up gradients of repellent and less frequently when traveling down gradients of repellent (Tsang et al 1973).

The mechanism proposed to explain such a bacterial memory is shown schematically in Figure 3. It has elements of temporal mechanisms proposed for a wide variety of species. It is postulated that some molecule or parameter called the response regulator (X) controls the frequency of tumbling. Bacteria produce a response regulator and decompose it continuously so that X would be present at a steady state level in a no-gradient situation (Figure 3A). If the pool size fluctuated in a Poissonian manner relative to the critical value (signified by X_{crit}) then the random walk behavior observed in the bacteria could be explained. Whenever X (used to indicate the instantaneous value of X) was above this value of X_{crit}, tumbling would be suppressed. Whenever it was below the value tumbling would

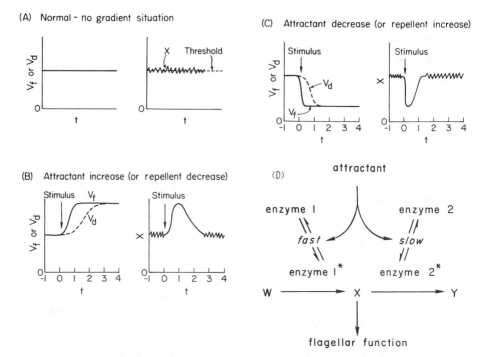

Figure 3 Explanation of response and adaptation at molecular level. (A) Assuming a response regulator is formed at a rate V_f and decomposed at a rate V_d then $V_f = V_d$ on average in absence of a gradient but fluctuations in level of X, the response regulator, relative to threshold produces a random pattern. (B) When attractant is added or repellent removed or diluted, V_f and V_d change but V_f responds more readily than V_d. As a result a transient increase in response regulator occurs followed by adaptation. (C) An attractant decrease or repellent increase again causes a transient change (decrease in X in this case) which adapts back to the no gradient situation in time. (D) One molecular picture of the process in which enzyme 1 catalyzing V_f and enzyme 2 catalyzing V_d are altered by the attractants and repellents.

be generated, and hence the random variation of the tumble regulator could be translated into the random behavior of the bacterium.

This model furthermore explained how increases in attractants or repellents could lead to the observed behavioral responses. Increases in attractants in this model cause increases in the rate of formation of the response regulator and also increases in the rate of decomposition. But the attractant effect on the formation constant (v_f) occurs more rapidly than the effect on the rate of decomposition (v_d) (Figure 3B). Hence the level of response regulator increases transiently. Such a hypothesis can readily be translated into molecular terms, as both rapid and slow conformational changes are known (Figure 3D).

In each of the above cases the model shows a return of the response regulator to the normal state after an appropriate interval of time. It is perhaps worth dwelling on this property for a moment since adaptation is extensive in biological species. There are two types of adaptation. One is the adaptation after a pulsed stimulus in which the stimulus is removed. Such phenomena are relatively easy to explain. The adaptation that is more difficult to explain is the one in which the new stimulus is maintained yet the species adapts. In the bacteria the increase in concentration of ribose leads to suppression of tumbling, but after this initial response the bacteria returns to normal even though the concentration of ribose in the medium remains at the new higher level. Such adaptation to constant stimuli, both pleasurable and noxious, is observed in complex organisms and in individual cells. The mechanism shown in Figure 3 is one that can explain such adaptation in mathematical and chemical terms.

Having thus described the essence of a response regulator model, it is important to state that we are using here a shorthand notation for a variety of models. It is not essential that the tumble regulator itself vary in a Poissonian manner relative to a constant X_{crit}. It is equally acceptable that the level of the tumble regulator has a smooth monotonic value and X_{crit} varies in a Poissonian manner or that both X_{ss} (the value of X at the steady state) and X_{crit} vary in a Poissonian manner. What is important is that the difference $(X_{ss}-X_{crit})$ fluctuates around zero to produce the random walk.

Similarly it is utterly arbitrary to assume that increases in X, i.e. $X-X_{crit}$, greater than zero are essential to suppress tumbling and decreases to generate it. The exact inverse would do equally well if other steps are changed to make the model internally consistent. Such permutations, however, do not in any way affect the basic principles of the model. The precise details must await the elucidation of the biochemical mechanism. The principles, however, can be helpful in explaining the roles of the individual components.

When the non-chemotactic mutants were observed in the microscope it was seen that they usually fell into two classes. Either they were of the smooth-swimming category, i.e. swam in long straight lines with very infrequent tumbles, or they tumbled incessantly. However, it is possible for bacteria to have a normal appearance in the absence of a gradient and yet respond abnormally to a gradient; such mutants were observed also.

The response regulator model could readily explain these observations (Koshland 1977b). An abnormally swimming mutant could be caused by alteration in any of the parameters leading to either formation or degradation of X or the factors involved in X_{crit}. A value of $(X_{ss}-X_{crit})$ not zero but positive, would lead to suppression of tumbling and hence constant smooth swimming. Likewise the always-tumbling mutant could cause $(X_{ss}-X_{crit})$ to be less than zero, thus constantly generating tumbling.

The response regulator model led to two very important predictions of mutant behavior that were tested. These are illustrated in Figure 4. If the

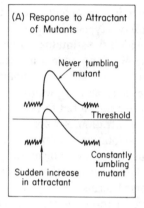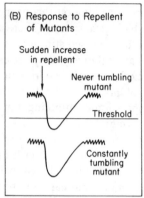

Figure 4 Behavior of non-chemotactic and methionine mutants as predicted by response regulator model.

Predicted behavior which has been observed experimentally:

(A) Never tumbling mutant has $\overline{X}_{ss}-X_{crit}$ value greater than zero in absence of gradient so it is smooth swimming. Attractant increases raises level but there will be no change in response since $(X_{ss}-X_{crit})$ remains in smooth swimming zone. Constantly tumbling mutant has $\overline{X}_{ss}$ level below threshold initially. Attractant raises level into smooth swimming zone temporarily.

(B) Addition of repellent to same mutants as in Part A decreases level of tumble regulator. This can produce tumbling in never tumbling mutant but no behavioral change in constantly tumbling mutant.

steady-state level of the response regulator is below threshold (constantly tumbling) then a sudden *increase* in attractant level (or decrease in repellent) should generate a temporal response that would lead to smooth swimming at least for a short interval (Aswad & Koshland 1974). Similarly a sudden *decrease* in attractant (or increase in repellent) for a smooth-swimming mutant should lead to tumbling for a brief period of time. Both of these predictions were found to be the case. Not only did these findings add important support to the response regulator model, but they eliminated one alternative model for the non-chemotactic mutants, i.e. that they reflected a defect in the mechanics of the flagellar motor, which prevented reversal of rotation. The experiments with temporal gradients established that the defects occurred in the sensory processing.

THE OPTIMIZATION OF MEMORY TIME

The response regulator model readily explains the phenomenon of the bacterial memory (Macnab & Koshland 1972, Koshland 1977b). In essence the bacterial memory is encoded in the enzymes that form and remove the response regulator X. Therefore the bacterium does not have a fixed memory span. Under strong stimuli it will return to normal after a relatively long interval. With a weak stimulus it returns to normal very rapidly.

The bacterial memory should be optimized to select between two opposing commands (Macnab & Koshland 1973, Berg & Purcell 1977). The longer its memory the longer it can integrate information as it swims up a gradient. Since it usually swims through shallow gradients with very small changes in concentration over one body length (in some cases as small as 1 part in 10^4), the longer its memory the better. On the other hand a signal transmitted to its flagella must be communicated rapidly if it is to correlate tumbling with its direction of motion. It is thus faced with the dilemma of the need for a long memory span to improve its analytical accuracy, and a need for a quick response to provide a high correlation with the direction of motion. Obviously it makes a compromise between these two extremes and utilizes two clever devices to do so. In the first place the pool in normal circumstances never rises very much above or below the value of X_{crit}. This means that tumbling is suppressed or generated when the bacterium travels up or down a gradient but returns to normal very rapidly when the gradient changes or disappears. The enzymes of the bacteria have been selected over evolutionary time to optimize their memory span for the gradients to which the bacterium is exposed. Mutations that alter the properties of any of these proteins can perturb the relative rate constants and disturb the system.

The instant the bacterium starts swimming up a favorable gradient the level of X increases relative to X_{crit} and it begins to suppress tumbling. If it should continue to head in the right direction it will maintain a continual positive value of $(X-X_{crit})$, which would continually suppress tumbling. Only if its trajectory is slightly curved or the gradient vanishes will it start to tumble. For example if it were swimming towards a decaying fish it would keep swimming towards the attractants until it arrived at a place where the gradient no longer changed or its receptors were saturated and then would swim in a random manner in that locality. Conversely, the minute the bacterium swims in the wrong direction, it will increase its probability of tumbling and do so in a very short interval. Its random walk is therefore heavily biased towards movement in a favorable direction and its probability of responding in an appropriate manner is biased the instant it starts up or down a gradient.

FUNCTIONS OF THE COMPONENTS

Some, but not all of the functions of the various components of the sensory system have been identified.

Three of the receptors have been purified to homogeneity, the galactose receptor, the ribose receptor, and the maltose receptor. They are proteins of 30,000 molecular weight and are present in the periplasmic space between the outer and inner membranes of the bacterium. They appear to be water soluble and to attach to a membrane component during the transmission of the signal. By attaching fluorescent markers to the protein it was possible to show that the binding of galactose induces a conformational change in the galactose receptor (Zukin et al 1977a) and similar studies on the ribose receptor show that it also undergoes a conformational change. This alteration in the shape of the protein allows it to bind to the next component of the signalling system (Strange & Koshland 1976; cf Figure 5).

The mechanism shown in Figure 5 establishes a weighting of values just by the innate properties of the system. First, the specificity of the proteins provide a hierarchy of stimuli. Thus, the galactose receptor binds galactose and glucose strongly and other sugars more weakly. Likewise the ribose receptor has a much stronger predilection for ribose than allose and essentially receives signals from no other physiological sugar. Second, the size of the signal conveyed from these sugars to the processing system depends on the number of receptor molecules (cf section on Plasticity of the Bacterial Cell) and their affinity for the next component of the system. Some chemoeffectors compete with each other and some chemoeffector-chemoreceptor complexes compete with each other.

It has been established that there is no need to metabolize the sugars in order to observe the chemotactic response (Adler 1969); therefore this response is similar to higher systems of taste and odor which do not require metabolism of their effectors. On the other hand some of the proteins involved as receptors apparently have a dual function since the galactose and ribose receptors can serve for the transport as well as for the sensing system. However, mutants in the transport system can eliminate the transport function without affecting chemotaxis, and mutants in the chemotaxis system can eliminate chemotaxis without affecting transport, which indicates the functional independence of transport and sensing.

Two of the genes in the central processing system have been identified as enzymes. One of these genes, the *che*R gene of *Salmonelia typhimurium* (and the equivalent *che*X gene of *Escherichia coli*), codes for a transferase that methylates the membrane-bound proteins (Springer & Koshland 1977). Similarly, the *che*X gene of *Salmonella typhimurium* (and the *che*B gene of *Escherichia coli*) codes for a protein that has esterase activity, and removes the methyl group from these proteins in a hydrolytic reaction (Stock & Koshland 1978). Other proteins in the sensing system appear to have some interaction with these genes. The *che*T gene of *Salmonella* (corresponding to the *che*Z gene of *Escherichia coli*) appears to interact with the esterase, and deficiencies in this protein can cause loss of esterase function in some cases (Stock & Koshland 1978). Moreover, complementa-

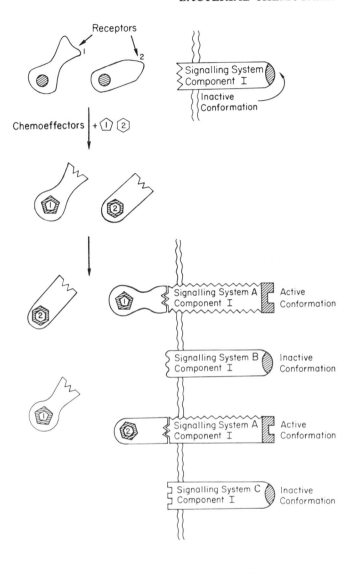

Figure 5 Induced association model. Receptors are initially in conformations that are not attracted to component I, but are induced into new conformations by the chemoeffectors. As a result, individual chemoeffector receptor complexes are induced to encounter and associate with the first component of the signaling system. If one binds there, it induces a conformation change, which activates the signaling system and begins a signal that can be amplified in a cascade process. If two receptor-chemoeffector complexes compete for the same site, one stimulus can diminish or completely block another.

tion studies show an interaction between the *che*B and the *che*Z genes (Parkinson 1975). Other genes also appear to have some supporting role but it is not apparent whether they are enzymes or cofactors. A list of the mutants in each gene and the resulting phenotype of the bacterial system is shown in Table 1.

A function has also been assigned to the *che*U gene product of *Salmonella typhimurium* (Khan et al 1978, Rubik & Koshland 1978). It appears to be a flagella protein that senses the signal released by the signaling system. Evidence that it may be directly involved in the final stage of signal transduction is strong. Certain mutations in this gene prevent proper flagellar assembly and other mutations cause simple changes in the sensing system without destroying the ability of the flagella to rotate (Silverman & Simon 1973). Thus it would seem most likely that this is a protein in which severe changes cause a loss of flagella function and more minor changes allow flagella function but damage the transfer of information between the sensory apparatus and the motor response (Khan et al 1978).

The three proteins that exist in the membrane (the *tar, tsr,* and *trg* gene products) and are methylated operate to sense more than one signal and to focus the initial receptor stimuli. Thus mutants in the *trg* gene eliminate responses to ribose and galactose (Ordal & Adler 1974, Kondoh et al 1979). Mutants in the *tar* gene eliminate responses to aspartate and various repellents, and mutants in the *tsr* gene eliminate responses to serine and other repellents (Hazelbauer et al 1969, Parkinson 1974). Following the lines of the model shown in Figure 5, it would appear then that these are focusing points for interactions of some receptors with the ultimate signal transduction scheme (Silverman & Simon 1977b). However, these three proteins are parts of specialized pathways not essential for all stimuli unlike the nine central genes. On the other hand, there has been a report (Springer et al 1977) that loss of two of these functions by a *tar-tsr* double mutant has produced smooth swimming bacteria that are incapable of responding to all other stimuli. This does suggest that too much damage to any one part of the system may result in effects on other parts of the system that are normally not related.

ADDITIVITY, DESENSITIZATION, AND POTENTIATION

A number of studies have been performed on the integrative capacity of the processing system. Tsang et al (1973) showed that the bacteria could integrate the response from two stimuli with proper use of algebraic signs. Responses to increases in attractants showed a rough additivity with other favorable increases, and offset unfavorable stimuli, i.e. increases in repellents or decreases in attractants. This was also demonstrated by capillary assays (Adler & Tso 1974). When more quantitative procedures became available (Spudich & Koshland 1975) these relationships were examined more closely and a spectrum of exact additivity, desensitization, and potentiation was found (Rubik & Koshland 1978).

With like stimuli, exact additivity was found in each case. A given stimulus, from 0 to 0.5 mM L-serine, was broken up into two smaller stimuli, from 0 to 0.02 mM and from 0.02 to 0.5 mM serine. The recovery times of the two small stimuli added up to the recovery time of the large one (Spudich & Koshland 1975). Thus $\tau_{13} = \tau_{12} + \tau_{23}$ where τ_{13} is the average time of recovery to the overall stimulus and τ_{12} and τ_{23} are the average times of recovery to the fractional stimuli. This same experiment was performed with other attractants such as aspartate and ribose and also with different fractional increments. In all cases the results were the same: exact additivity was observed as long as the stimulant was a single chemoeffector (Spudich & Koshland 1975).

When similar experimental techniques were applied to stimuli of two different chemicals, the results were not strictly additive (Rubik & Koshland 1978). In some cases there was exact additivity as in the case of the like stimuli, i.e. the combination of two stimuli such as aspartate and ribose, which give 1.74 sec and 1.02 sec each, showed a recovery time of 2.84 which is within experimental error of the two components presented separately. On the other hand, in many other cases exact additivity was not found. In the wild type there is almost invariably a desensitization in the combined stimuli, i.e. a sum of the individual responses is greater than the combined response. In certain types of mutants, there is a potentiation, i.e. the combined stimuli are far greater (sometimes 20–30 times greater) than the sum of the two individual stimuli.

For unlike stimuli, exact additivity is only found in mutants in which one of the enzymes of the processing system is removed. When all of the enzymes are operating properly, the system shows desensitization. Hence, it would seem that desensitization is a key part of the regulatory machinery, and was selected over evolutionary time to be of advantage to the organism. The elimination of one of the key enzymes allows the system to fall back onto a simpler interrelationship that gives exact additivity.

Elimination of a different enzyme can, for some of these responses, provide potentiation. It would seem that such potentiation does not usually have great survival value for the bacteria, since so far it has not been observed in wild-type cells. The fact, however, that a simple modification of the chemotactic machinery can produce the kind of potentiation that is observed in neural systems is intriguing. It shows that potentiation can arise from the biochemistry in a single cell. It does, of course, also occur in neuronal systems involving complex networks, but these results indicate that such complexity is not required.

A change in integrative processing in these cases was observed in mutants of the bacterium that are probably lacking a specific protein. The loss of an enzyme function could be achieved by phosphorylation as well as by genetic mutation. Hence the responses of the cell could be altered by environmental conditions such that some cells would amplify responses, other cells desensitize responses, and the third type of cells give pure additivity. This type of plasticity is exactly what one would look for in a neural network involving learning.

METHYLATION AND DEMETHYLATION IN THE ADAPTIVE RESPONSE

A key component of the bacterial regulatory system is a reversible methylation reaction. This reaction appears to be analogous to the phosphorylation-dephosphorylation in neuronal tissue. The levels of methylation change as a result of stimulations by attractants or repellents and the level of methylation bears some relationship to the size of the stimulus (Goy et al 1977, Paoni & Koshland 1979). It has been established that this methylation phenomenon is related to adaptation (Aswad & Koshland 1974, Goy et al 1977) and that it is caused by methylation of the *tar, tsr,* and *trg* gene products by S-adenosylmethionine (Kort et al 1975, Silverman & Simon 1977b). It is catalyzed by the transferase enzyme (Springer & Koshland 1977) and it is removed by the esterase enzyme (Stock & Koshland 1978). A removal of the attractant or repellent returns the methylation level to its initial value (Goy et al 1977). Methylation occurs on a glutamic acid residue of the membrane protein (Kleene et al 1977, Van der Werf & Koshland 1977).

The absence of the enzymes catalyzing this reversible methylation has severe effects on the response characteristics of the system. The absence of the methyltransferase produces a smooth swimming mutant that cannot respond to attractants and will give only a modified response to repellents (Springer & Koshland 1977). The absence of the esterase demethylating enzyme gives a permanently tumbling mutant that can respond abnormally to attractants and appears not to respond to repellents (Stock & Koshland 1978).

It has been observed in neuronal cell lines by Nirenberg and co-workers (Sharma et al 1975) that stimulation of cells by effectors such as morphine increases the level of adenyl cyclase enzyme molecules to a new value that does not decrease unless the drug is removed. Thus the adaptive response, which allows the cell to return to a normal relationship of response regulator in the presence of a steady stimulus, does so by a permanent change in biochemistry. In the bacterial cell this change is in the level of methylation.

MUTANTS THAT GIVE DRAMATICALLY ALTERED BEHAVIOR

An unusual class of mutants was found that gave completely or partially inverted responses (Rubik & Koshland 1978). Normal wild-type *Salmonella* swim toward the attractant serine and away from the repellent phenol. The mutant *Salmonella* strain labeled ST120 was found to swim toward phenol and away from serine. This inversion of behavioral response can be shown quantitatively by study of the tumbling frequency when bacteria are exposed to chemoeffectors. A sudden increase in the attractant serine produces suppression of tumbling in the wild type, which eventually returns to random tumbling in 1.6 minutes. Addition of the repellent phenol causes immediate tumbling, which again returns to random motion after 2.6 minutes. These same additions to the (ST120) mutant produce tumbling in the case of serine and smooth swimming in the case of phenol addition. The recovery time for serine is 1.6 minutes and for phenol is approximately 10 minutes.

This complete inversion of bacterial behavior posed a serious theoretical question. The response regulator model described above showed nine genes cooperating in a central processing system. Since the ST120 was identified as a point mutation in the single *che*U gene, it is extremely difficult to explain a complete inversion of function. One simple explanation would be an inversion of a mechanical function, analogous to a gear shift in which the signs "reverse" and "forward" were interchanged. If so the time constant of the sensory system should remain unchanged. However, the recovery time of phenol was different in the mutant from the wild-type. It was therefore concluded that the *che*U gene product must be involved in both switching and sensory processing.

A model for *che*U function was developed that explains these mutants and may have implications for neuron and mammalian cell function (Khan et al 1978). In that model, the response regulator binds to the *che*U gene product, inducing a change in conformation, which alters flagellar rotation from clockwise to counterclockwise. The behavior is determined by the level of the tumble regulator, designated as X, and some critical value, X_{crit}. It is postulated as discussed above that there is a certain Poissonian variation in $\overline{X}_{ss}$, the steady state level of X, and/or X_{crit}, the critical threshold value. It is further postulated as a feature of the response regulator model that values of $X - X_{crit}$ greater than zero lead to counterclockwise rotation, whereas values less than zero lead to clockwise rotation. (X refers to this instantaneous value and $\overline{X}$ to the average value over time.) Values of $(\overline{X} - X_{crit})$ that are very large, either positive or negative, will lead to smooth swimming because there will be either constant clockwise or constant counterclockwise rotation. On the other hand, values of $\bar{X}-X_{\text{crit}}$ that are close to zero will, because of the Poissonian variation of X or X_{crit} around this value, lead to frequent reversals of rotation and hence constant tumbling. Values of $X_{\text{ss}}-X_{\text{crit}}$ that are moderately large will lead to random swimming since the bundles will form for an appreciable period, but will be interrupted by occasional reversals of rotation leading to random tumbling. Random tumbling can occur both by reversal of the counterclockwise mode to clockwise (normal random) and by reversal of a predominant clockwise mode to counterclockwise (inverse random).

When this model for behavior was applied to the *che*U mutants, their strange and bizarre behavior could be explained in a very straightforward manner (Khan et al 1978). The modification in the *che*U gene apparently alters its affinity for the tumble regulator, and thus the steady-state value $X_{\text{ss}}-X_{\text{crit}}$ becomes appreciably less than zero instead of appreciably greater than zero (Figure 6). The behavioral observation that both the mutant ST120 and the wild-type underwent a random tumbling pattern, which appeared remarkably similar under the microscope, actually obscured a profound difference. The wild-type was swimming randomly because the smooth swimming mode was counterclockwise and was interrupted by occasional reversals to the clockwise mode. The mutant, on the other hand,

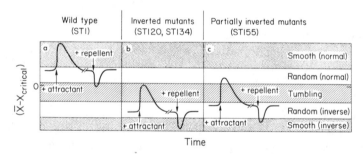

Time

Figure 6   Model for chemotactic response of wild-type and mutant cells. The observations on inverted-behavior mutants are explained by identical changes in the average response regulator concentration, $\bar{X}$, by assuming that the relative value ($\bar{X}_{\text{ss}}-X_{\text{crit}}$) is different in the mutants. The instantaneous value, X, is presumed to fluctuate around the average value, $\bar{X}$, in a Poissonian manner. In each case $\bar{X}$ increases on addition of attractants and decreases on addition of repellents before adapting back to its original level. (a) In wild type, attractant increases $\bar{X}$ from normal random to smooth zone, whereas repellents decrease it to tumbling zone. (b) Inverted mutants have $\bar{X}_{\text{ss}}$ initially in an inverse random zone in which an increase in $\bar{X}$ causes tumbling, whereas a decrease causes smooth swimming. (c) Partially inverted mutant has $\bar{X}_{\text{ss}}$ in upper part of inverse random zone, so that a sudden increase moves it to normal random zone. This would be interpreted as no change in behavior, as was observed. Adaptation leads to drop in $\bar{X}$ through tumbling zone to inverse random, explaining the observed delayed response. (Details of patterns are described in Khan et al 1978.)

was swimming randomly because its predominant rotation was in the clock-wise mode, with occasional interruptions to the counterclockwise mode. Once this point was understood (and confirmed by tethering experiments) it was easy to explain the entire response behavior of the mutant, and the experimental and theoretical analyses as shown in Figure 6.

This basic explanation made it possible to explain other bizarre mutants that show partial inversion effects and also mutants that appear to be insensitive to all chemoeffectors. All of these behavioral patterns could be explained by assuming that the processing machinery to a stimulus remained the same in all of these mutants except for the cheU gene product, whose affinity for the tumble regulator was altered in different ways in each mutant. In the ST120 mutant, the change in affinity made the starting point of $(X_{ss}-X_{crit})$ be in the lower part of the inverse normal zone (Figure 6). In the ST155 partial inversion mutant, the $(X_{ss}-X_{crit})$ value placed it in the upper part of the inverse normal zone. In the completely insensitive mutants that were smooth swimming (not shown in figure), it was postulated that the affinity was so altered that $(X_{ss}-X_{crit})$ was a very large positive value and no changes in the sensory signal were strong enough to generate a negative value.

These mutants have been discussed at length because of their importance in understanding regulatory phenomena. In this case the damage was to a single protein and was probably a single amino acid change in that protein. The complex behavioral response arises not from the fact that the protein changes structure drastically; in fact it changes properties in ways to be expected from other mutants in protein structure, i.e. by a simple change in affinity for its ligand. This single result at the physico-chemical level becomes very complex at the behavioral level. The explanation adds strong support to the response regulator model because it is so readily accomodated by the protein chemistry and the hydrodynamics of flagella. Moreover, if we assume that neurotransmitter release is as complex as flagellar rotation (a quite reasonable estimate of complexity) similar bizarre behaviors could result from alterations in a single gene within the neuron.

ROLE OF THE MEMBRANE POTENTIAL

The relation of the membrane potential to the chemotactic response is an intriguing one since the membrane depolarization is identified with signaling in excitable cells. Responses of membrane gradients cause changes in the chemotaxis of protozoa (Eckert 1972). Alteration of the membrane potential can affect the swimming pattern of bacteria as indicated by the finding that a number of membrane-active drugs as well as inhibitors and uncouplers of oxidative phosphorylation can influence motility (Ordal &

Goldman 1975, 1976). Ordal & Goldman (1975) showed in *Bacillus subtilis* that uncouplers of oxidative phosphorylation cause behavioral changes analogous to those caused by repellents in the gram-negative bacteria. DeJong et al (1976) also concluded that transient changes in potential cause swimming behavior alterations. Szmelcman & Adler (1976) studied permeant cation distribution using triphenylmethylphosphonium ion and found transient changes in membrane potential. A blue light effect caused changes in the oxidation machinery (Macnab & Koshland 1974, Taylor & Koshland 1975) that might be expected to cause changes in membrane potential.

Alteration in membrane potential in *Bacillus subtilis* was studied for correlation with changes in tumbling frequency (Miller & Koshland 1977). The net result of these studies was to demonstrate that the changes in the membrane potential produced changes in the behavioral response similar to those produced by attractants and repellents. Increases in membrane potential cause suppression of tumbling, decreases in membrane potential cause generation of tumbling. However, additions of attractants and repellents did not cause detectable changes in membrane potential as measured by these cyanine dyes. Thus, membrane potential is not the response regulator but it can feed signals to the central processing system. It was concluded that the alteration in membrane potential was detected by the chemotactic system in the same way that alteration in levels of nutrients or toxic substances was detected. A specific potential, associated with a particular ion or a particular combination of ions is not yet excluded.

Work on proton gradients has indicated a similar conclusion. Alterations in the proton gradient can be sensed by the bacterium in the same way that alterations in attractants and repellents can be observed, but the pH gradient is not the response regulator (Miller & Koshland 1979). The external pH can be established at a variety of different values, and the bacterium will stabilize to normal tumbling. However, a sudden perturbation of the pH gradient away from optimal conditions creates tumbling and a change towards optimal conditions leads to smooth swimming.

A blue light effect (Macnab & Koshland 1973) and the effect of electron acceptors such as O_2, nitrate, and fumarate may also work through membrane potential (Taylor et al 1979).

OVERRIDE MECHANISMS

It has been observed that bacteria deprived of methionine swim smoothly in straight lines without tumbling. At first glance this would seem to be a malfunction of the sensory apparatus, which the bacterium would avoid at all costs. In retrospect, such an override mechanism can be likened to responses to hormones, such as epinephrine, which override feedback from metabolic signals. If a bacterium is in an environment hostile to its survival, the best way to leave that environment rapidly is to shut off all responses of turning and tumbling and go in a straight line in any random direction (Koshland 1977a, 1979a). Since S-adenosylmethionine is a key compound in a wide variety of metabolic reactions, a decrease in its level would suggest that the nutrient environment or energy supply was severely depleted. Hence, the override mechanism is an appropriate response.

A decreased membrane potential would also be indicative of depleted oxygen sources or depleted nutrient supplies, which lower the energy sources of the cell. Hence, the evidence that a minimum membrane potential (S. Khan & R. M. Macnab, in preparation; B. L. Taylor, in preparation) could cause smooth swimming would give to the bacterium the same type of override mechanism as depletion of S-adenosylmethionine. It is intriguing, therefore, that the chemotactic system of bacteria has a hierarchy of values analogous to higher systems in which there are feedback mechanisms and also override mechanisms that can nullify the feedback responses.

PLASTICITY OF THE BACTERIAL CELL

The total complement of receptors is not preordained by heredity alone in the bacterial cell lines. Certain receptors are induced. For example, the galactose receptor is essentially at levels too low to elicit much galactose response in the wild type *Salmonella typhimurium*. If the *Salmonella* is grown in the presence of fucose, however, substantial amounts of the galactose receptor are produced and the "adult" bacterium responds to galactose as readily as ribose. Some receptors such as the glucose and ribose receptors of *Salmonella* can be repressed in certain growth conditions (Anderson & Koshland 1979). The serine receptor is constitutive, however, and changes in growth conditions have no effect on its synthesis. Table 2 shows a list of inducible and constitutive receptors developed for the species *Escherichia coli* and *Salmonella typhimurium*. Thus the responsiveness of the grown bacterial cell is plastic, a product of its growth conditions as well as its genetic makeup.

INDIVIDUALITY AND CHANCE

The response of a cell to an environmental stimulus is affected by hereditary, environmental, and probabilistic factors. Separating the contribution of each of these is difficult in mammalian species, since differentiated cells inevitably appear in different environments and have different histories.

Table 2 Induction of receptors for chemotaxis in *Escherichia coli* and *Salmonella typhimurium*

Chemotaxis receptor	Location	*E. coli* induction	*S. typhimurium* induction
ribose	periplasm	ribose	ribose
galactose	periplasm	galactose, fucose	fucose
glucose	membrane	constitutive	glucose, mannose
aspartate	membrane & periplasm	constitutive	constitutive
serine	membrane	constitutive	constitutive
maltose	periplasm	maltose, trehalose	no maltose chemotaxis
trehalose	?	trehalose	no trehalose chemotaxis
N-acetylglucosamine	membrane	N-acetylglucosamine	N-acetylglucosamine
mannitol	membrane	mannitol	mannitol
sorbitol	membrane	sorbitol	sorbitol
fructose	membrane	fructose	constitutive

Bacteria have advantages for such a study since a cloned selection of genetically identical individuals can be grown in a swirled culture in which environmental factors and individual histories can be made identical.

In the chemotactic response, stimuli are processed by a cell machinery that has certain constitutive factors and certain modifiable factors, such as induced receptors. But an interesting feature of bacteria is that even when the heredity and the environment are held constant, the cells appear to show individual variations. The assays developed for evaluating chemotaxis provide an unusual tool for investigating the latter phenomenon and distinguishing it from the former (Spudich & Koshland 1976).

The bacterial properties were measured quantitatively (Spudich & Koshland 1976) in two ways: (*a*) the sensitivity to stimuli as measured by recovery from temporal gradients, and (*b*) the tumbling frequency as an indicator of the steady-state level of tumble regulator. It was found that the tumbling frequency and the response to stimuli were characteristics that differed from cell to cell even though the cells were identical in heredity and environment. Moreover, the individual characteristics remained for the lifetime of the cell.

How could such individuality be produced? A bacterium is a small cell. In the process of division and replication of the cellular components there may well be certain molecules present in such low amounts that they are subject to Poissonian variation. In general, the standard deviation of a Poisson distribution is the square root of n, where n is the mean number of events. If there are 10^4 molecules of a particular type, the standard deviation would be 100 molecules, giving a 1% deviation. If, on the other hand, only 100 molecules are produced, 10% s.d. would arise. In many cellular processes less than 100 molecules participate (for example, there are only 10–20 molecules of *lac* repressor per cell and enzyme cofactor concentrations are extremely low). In other processes there may be more total molecules but their ultimate number may have been determined by a small number of generating molecules (for example, there are only 6–14 *trp* mRNA molecules per bacterial cell, and each mRNA molecule is translated an average of 20 times).

In the case of chemotaxis, it has been postulated that the tumble regulator is produced and destroyed, implying that some enzymes or structural molecules control the velocities of the steps represented by v_f and v_d. If a small number of mRNA molecules code for one of the enzymes in the sensing system, quite different total numbers of molecules of this enzyme would be present in different individuals. This would affect not only the steady-state tumbling pattern, but also the rate of recovery from a stimulus (Aswad & Koshland 1974). Conversely, if a particular cellular component were produced in a much larger number of copies, Poissonian variation would be relatively small and one would not expect much variation in this property from cell to cell.

The existence of such variability leads one to ask whether it has some advantage to the organism or whether it is a disadvantage that must be surmounted in the development of the species. Although there may be no significant selective pressure against nongenetic variability, one can certainly make an argument that nongenetic variability aids in the survival of a population subjected to widely varying conditions during its lifetime. If there were individual bacterial variation in a genetically homogeneous population, a few of the bacteria might either be supersensitive or insensitive. These bacteria would normally not survive as effectively as the bulk of the population, in most circumstances an unimportant loss if the main body of the colony survived. If, on the other hand, unexpected toxic and lethal situations arose, a small fraction, on the wings of the probability distribution, might constitute the only bacteria to survive and would then reproduce the species for future generations. Thus nongenetic variability would be a preferred mechanism for accommodation to random fluctuations in the environment and genetic variability the preferred mechanism for accommodation to long-lasting environmental changes. Induction and repression of enzyme synthesis reflect intermediate accommodations to fluctuations of a lengthy but impermanent type.

It would be intriguing if certain properties were found to be insulated against chance by the large number of copies in the cell at all crucial stages and other properties subject to nongenetic diversity due to small numbers of starting molecules. This would have particular and important applicability to the nervous system. It would allow some predetermined "wiring" as a result of expected learning patterns, and some pure "chance" perturbations.

CONCLUSIONS AND EXTRAPOLATIONS

Although many factors must be worked out in the bacterial system and some surprises may well be in store, many of its features are now clear. The system is composed of about 30 receptors, most of which are produced uniformly through hereditary genes. Some of the receptors, however, are induced by environmental growth conditions. The signals from the initial receptors are processed through some focusing elements and are then integrated into a central processing system that is coded for by nine distinct genes. Each of these genes codes for a protein product. The output response can be explained by postulating a response regulator whose level controls the output response. It seems quite possible that the bacterial processing is similar in principle to the processing within a neuron and that the output system, the flagellar motor, is similar in complexity to release of a neurotransmitter.

With those analogies in mind, some comments on the bacterial system that may have special relevance to the neuron and neural networks seem appropriate.

Integration in the Processing System

The bacterial processing system not only can give additive responses to combinations of like stimuli, but it can integrate the effects of several different stimuli in an algebraic manner. Clearly such a property is similar to that of a neuron, which receives excitatory and inhibitory signals and must have the ability to integrate this information. In the case of the neuron the processing may occur electrically, chemically, or by a combination of the two. In the bacteria the integration is largely chemical but the electrical signals from the membrane potential can also be integrated with other stimuli. In neurons also the properties of the membrane may be modified by the chemical environment. Thus the processing system within the individual cell can do appreciable decision-making by itself. This does not remove the need for a complex neural circuit but it does mean that the neuron is far more complex than a simple conducting element.

Adaptation

Adaptation is a universal mechanism of biological systems and one that is not easy to understand at either the mathematical or chemical level. What is observed in some instances is that a change to a new level of stimulus—chemical in the case of the bacteria, light in the case of the human eye—is detected and then forgotten. In a sense this is discarding information, but it is an extremely clever trade-off for the ultimate purposes of the organism. In essence it foregoes the ability to detect absolute levels in exchange for increased accuracy in detecting increments in levels. Other cells with diminished adaptive properties can be used for absolute levels.

From the bacterium we begin to see how this works and what the trade-off is chemically. The bacterial memory system enables it to detect changes of 1 part in 10^4, i.e. the difference between 9999 and 10,000. It would be almost impossible to devise a detection system that could measure the absolute value of 9999 to 0.1 of a decimal point and 10,000 to 0.1 of a decimal point, another system to subtract the two, and then a device to erase the entire information. It is simpler chemically to devise an enzymatic memory system that produces an amplified transient signal that fades over time. If absolute values must be detected, a separate system with lower accuracy for increments can be devised, as in the eye.

The manner in which this is done chemically in bacteria is by an altered level of methylation. The effect of this changed methylation pattern is to return the response regulator to its initial level relative to its detector. Such a mechanism can be modeled on the computer to give the appropriate characteristics. The price of adaptation of the response regulator is a change in the biochemistry of the cell. In the case of heroin addiction, the adaptive response of the neuron to the increased stimuli produces the property of drug tolerance by apparently increasing the levels of adenyl cyclase (Sharma et al 1975). The inhibition of adenyl cyclase by the morphine stimulus is compensated for by the increased adenyl cyclase levels. When the stimulus is removed, the well-known withdrawal effects are observed until the cell adapts to the new levels of stimulus by reducing its adenyl cyclase level. On a more pleasant and familiar level the response of our eyes to changing levels of light intensity or our ears to altitude pressures are undoubtedly organismic behavioral responses that reflect similar changes in adaptive properties.

Relative Contributions of the Genes

Although there are nine genes, each of which is essential for normal functioning of the chemotactic system, it is not true that all of them contribute equally. Two of the genes, which have been identified with the methyltransferase function and the demethylase function, respectively, are quite clearly essential. Loss of these functions prevents any significant chemotactic response. On the other hand, some of the auxiliary genes seem to play supporting roles and diminution of one gene product can be compensated for by increased amounts of the other gene product. Hence they can be considered not as essential for the operation of the entire chemotactic signaling system. This may be important in the analysis of the inheritance of mental disorders in genetics. Diseases such as schizophrenia and mania have strong indications of being inherited. Yet those who oppose this interpretation clearly point out that the inheritance patterns are confused. The concordance among identical twins are not precisely what one would expect from a Mendelian inheritance, and comparison of frequencies in adoptive children studies also are not as clean as Drosophila mating experiments. If, however, a number of different genes can contribute to the same syndrome, and do not contribute in equal proportions, the inheritance patterns do not appear so mysterious. If one adds the further possibility that even in essential genes, point mutations can occur that change function partially, the prediction of inherited properties becomes even more difficult. In the chemotactic system different single point mutations in the same *che* U gene product lead to very different types of behavior. Since in every case the bacterial "disease" is loss of the chemotactic response, it seems likely that a process such as dopamine release from a neuron will be polygenic and that all genes will not contribute equally. Moreover, different mutations in the same gene may give quite diverse symptoms even at the cellular level. From this point of view schizophrenia in humans may eventually be traced to a single response system despite the wide range of diagnoses. Since it would be logical that mania or schizophrenia is at least as complicated as bacterial chemotaxis, the complex patterns of the genetic studies make considerable sense.

Heredity, Environment, and Chance

In the chemotactic system, heredity, environment, and chance each play a significant role. Certain parts of the system are inherited and are the same in every bacterium unless some mutational deficiency appears. Certain other components of the system, particularly certain receptors, are induced by the growth conditions of the cell. Finally, there is an element of chance in the development of the system, which leads to different behavioral responses. In any neuron it thus seems likely that there would be some elements of heredity, some influence of growth conditions and learning, and some incidence of pure chance. Diseases such as schizophrenia could then well be in the category that in certain cases heredity alone would bring on the disease. In certain other cases, heredity plus conditions during growth that modify the cell would lead to the disease. And in other cases a particularly undesirable chance of events, e.g. the development of a stress situation at a time of severe hormonal changes, could bring on the disease. Similarly, chance factors could help restore the cell to its normal responses. The variability in concordance of identical twin studies and the peculiar patterns of drug and placebo effects in mania and schizophrenia could perhaps be explained by cellular effects of this type.

Long-term and Short-term Memory

The bacterium has essentially only short-term memory. It has a memory that is very sophisticated in that it is optimized for the survival of the bacterium. A memory appreciably longer or appreciably shorter than that observed would be less effective. It would seem that each individual neuron would in the same way select over evolutionary time for a response regulator signal that would be appropriate for its proper functioning.

If one carries the optimized time concept one step further and remembers that potentiation and desensitization are also possible within a cell as a result of alterations in individual protein products within the cell, a mechanism for short-term and long-term memory can be envisioned. Loss of enzymatic function or acquisition of that function can be achieved in a cell by such devices as phosphorylation or by proteolytic cleavage. A timing device in which an initial stimulus creates an influence on a second neuron for a given interval of time could be coupled with the induction of a new protein as in the case of the galactose receptor described above. This modification might itself rise and fall within an interval of time. The cellular system would then respond differently if a second signal was received during the interval of the protein modification as compared to one received after the memory span of the neuron had been exceeded. Thus reinforcement by repetition within a prescribed limit of time could lead to the induction of a permanent modification. Once this permanent change occurs the appropriate feedback between neurons or within the same neuron could sustain this difference and lead to the permanent differentiation of the cell. The computer simulations for the bacterial memory are the same whether the responses are in terms of milliseconds, minutes, or hours. The length of memory will thus depend on the kinetic properties of the relevant enzymes and can be adjusted to any time span.

The sensory system of bacterium is a relatively simple input-output system with a processing capability that is moderately simple. It is in no way as complex as the human brain and it could be argued that it is appreciably simpler than an individual neuron. Certainly it does not have the action potential for distributing a message over a very large geographical environment. On the other hand the neuron does not need to adapt to the variety of stimuli of a bacteria. The central processing of the bacterium may be as complex as the processing of an adenyl cyclase system within a neuron. It has the capacity for adding effects from excitatory stimuli, subtracting effects from inhibiting stimuli, and integrating the two algebraically. Thus any single neuron may not require an individual processing system (e.g. the adenyl cyclase system) any more sophisticated than that of the bacterial cell. The brain acquires its enormous capacity from the enormous complexity of the wiring diagram, and the fact that each individual "switch" is far more than an on-off switch.

A particularly interesting feature of the bacterium is that it encompasses many of the principles of higher behavioral systems within a single cell. It has specialized response systems that ultimately lead into a central system. It can integrate signals algebraically and show potentiation, desensitization, and exact additivity. It can modify behavior through experience. Perhaps the classical epigram "ontogeny recapitulates phylogeny" can be extended into neurosciences as "enzymology recapitulates neurobiology."

Acknowledgment

The author would like to acknowledge the research support by grants from the National Institutes of Health, #AM09765, and from the National Science Foundation, #PCM75-16410.

Literature Cited

Adler, J. 1966. Chemotaxis in bacteria. *Science* 153:708–16

Adler, J. 1969. Chemoreceptors in bacteria. *Science* 166:1588–97

Adler, J. 1975. Chemotaxis in bacteria. *Ann. Rev. Biochem.* 44:341–56

Adler, J., Tso, W. W. 1974. "Decision"-making in bacteria: chemotactic response of *Escherichia coli* to conflicting stimuli. *Science* 184:1292–94

Aksamit, R., Koshland, D. E. Jr. 1974. Identification of the ribose binding protein as the receptor for ribose chemotaxis in *Salmonella typhimurium*. *Biochemistry* 13:4473–78

Anderson, M. J., Koshland, D. E. Jr. 1979. Effect of catabolite repression on chemotaxis in *Salmonella typhimurium*. *J. Bacteriol.* Submitted

Anderson, R. A. 1975. Formation of the bacterial flagellar bundle. In *Swimming and Flying in Nature*, ed. T. Y. -T. Wu, C. J. Brokaw, C. J. Brennen, 1:45. New York: Plenum

Anraku, Y. 1968. Transport of sugars and amino acids in bacteria. I. Purification and specificity of the galactose- and leucine-binding proteins. *J. Biol. Chem.* 243:3116–22

Asakura, S., Eguchi, G., Iino, T. 1966. *Salmonella* flagella: *in vitro* reconstruction and over-all shapes of flagellar filaments. *J. Mol. Biol.* 16:302

Asakura, S., Iino, T. 1972. Polymorphism of *Salmonella* flagella as investigated by means of *in vitro* copolymerization of flagellins derived from various strains. *J. Mol. Biol.* 64:251–68

Aswad, D., Koshland, D. E. Jr. 1974. Role of methionine in bacterial chemotaxis. *J. Bacteriol.* 118:640–45

Berg, H. C. 1971. How to track bacteria. *Rev. Sci. Instrum.* 42:868–71

Berg, H. C. 1975. Chemotaxis in bacteria. *Ann. Rev. Biophys. Bioeng.* 4:119–36

Berg, H. C., Brown, D. A. 1972. Chemotaxis in *Escherichia coli* analyzed by three-dimensional tracking. *Nature* 239:500–4

Berg, H. C., Purcell, E. M. 1977. Physics of chemoreception. *Biophys. J.* 20:193–219

Berg, H. C., Tedesco, P. M. 1975. Transient response to chemotactic stimuli in *Escherichia coli. Proc. Natl. Acad. Sci. USA* 72:3235–39

Berger, E. A., Heppel, L. A. 1974. Different mechanisms of energy coupling for the shock sensitive and shock resistant amino acid permeases of *E. coli. J. Biol. Chem.* 249:7747–55

Boos, W., Gordon, A. S., Hall, R. E., Price, H. D. 1972. Transport properties of the galactose-binding protein of *Escherichia coli.* Substrate-induced conformational change. *J. Biol. Chem.* 247:917–24

DeFranco, A. L., Parkinson, J. S., Koshland, D. E. Jr. 1979. Functional homology of chemotaxis genes in *Escherichia coli* and *Salmonella typhimurium. J. Bacteriol.* In press

De Jong, M. H., Van der Drift, C., Vogels, G. D. 1976. Proton-motive force and the motile behavior of *Bacillus subtilis. Arch. Microbiol.* 111:7–11

DePamphilis, M. L., Adler, J. 1971. Attachment of flagellar basal bodies to the cell envelope: Specific attachment to the outer, lipopolysaccharide membrane and the cytoplasmic membrane. *J. Bacteriol.* 105:396–407

Doetsch, R. N., Hageage, G. J. 1968. Motility in procaryotic organisms: problems, points of view, and perspectives. *Biol. Rev.* 43:317

Eckert, R. 1972. Bioelectric control of ciliary activity. *Science* 176:473–81

Engelmann, T. W. 1881. Neue Methode zur Untersuchung der Sauerstoffausscheidung pflanzlicher und thierischer Organismen. *Pfluegers Arch. Gesamte. Physiol. Menschen Tiere* 25:285–92

Fahnestock, M., Koshland, D. E. Jr. 1979. Control of the receptor for galactose taxis in *Salmonella typhimurium. J. Bacteriol.* 137:758–63

Goy, M. F., Springer, M. S., Adler, J. 1977. Sensory transduction in *Escherichia coli:* Role of a protein methylation reaction in sensory adaptation. *Proc. Natl. Acad. Sci. USA* 74:4964–68

Hazelbauer, G. L. 1975. The maltose chemoreceptor of *Escherichia coli. J. Bacteriol.* 122:206–14

Hazelbauer, G. L., Mesibov, R. E., Adler, J. 1969. *Escherichia coli* mutants defective in chemotaxis toward specific chemicals. *Proc. Natl. Acad. Sci. USA* 64:1300–7

Hazelbauer, G. L., Parkinson, J. S. 1977. Bacterial chemotaxis. In *Receptors and Recognition: Microbial Interaction,* ed. J. Reissig. London: Chapman & Hall

Hotani, H. 1976. Light microscope study of mixed helices in reconstructed *Salmonella* flagella. *J. Mol. Biol.* 106:151–66

Iino, T. 1969. Genetics and chemistry of bacterial flagella. *Bacteriol. Rev.* 33:454

Kamiya, R., Asakura, S. 1976. Helical transformations of *Salmonella* flagella *in vitro. J. Mol. Biol.* 106:167–86

Khan, S., Macnab, R. M., DeFranco, A. L., Koshland, D. E. Jr. 1978. Inversion of a behavioral response in bacterial chemotaxis: Explanation of the molecular level. *Proc. Natl. Acad. Sci. USA* 75:4150–54

Kleene, S. J., Toews, M. L., Adler, J. 1977. Isolation of glutamic acid methyl ester from an *Escherichia coli* membrane protein involved in chemotaxis. *J. Biol. Chem.* 252:3214–18

Kondoh, H., Ball, C. B., Adler, J. 1979. Identification of a methyl-accepting chemotaxis protein for the ribose and galactose chemoreceptors of *Escherichia coli. Proc. Natl. Acad. Sci. USA* 76:260–64

Kort, E. N., Goy, M. F., Larsen, S. H., Adler, J. 1975. Methylation of a protein involved in bacterial chemotaxis. *Proc. Natl. Acad. Sci. USA* 72:3939–43

Koshland, D. E. Jr. 1977a. Sensory response in bacteria. In *Advances in Neurochemistry,* ed. B. W. Agranoff, M. H. Aprison, 2:277–341. New York: Plenum

Koshland, D. E. Jr. 1977b. A response regulator model in a simple sensory system. *Science* 196:1055–63

Koshland, D. E. Jr. 1979a. Bacterial chemotaxis. In *The Bacteria,* ed. J. R. Sokatch, L. N. Ornston, 7:111–66. New York: Academic

Koshland, D. E. Jr. 1979b. A model regulatory system—bacterial chemotaxis. *Physiol. Rev.* In press

Larsen, S. H., Adler, J., Gargus, J. J., Hogg, R. W. 1974a. Chemomechanical coupling without ATP: the source energy for motility and chemotaxis in bacteria. *Proc. Natl. Acad. Sci. USA* 71:1239–43

Larsen, S. H., Reader, R. W., Kort, E. N., Tso, W. -W., Adler, J. 1974b. Change in direction of flagellar rotation is the basis of the chemotactic response in *Escherichia coli. Nature* 249:74–77

Leifson, E. 1960. *Atlas of Bacterial Flagellation.* New York: Academic

Macnab, R. M. 1977. Bacterial flagella rotating in bundles: a study in helical geometry. *Proc. Natl. Acad. Sci. USA* 74:221–25

Macnab, R. M., 1979. Chemoresponsiveness in bacteria and unicellular eukaryotes. I. Bacterial chemotaxis. In *Encyclopedia of Plant Physiology, New Series, Physiology of Movements,* ed. W. Haupt, M. E. Feinleib. Heidelberg: Springer-Verlag. 7:310–34

Macnab, R. M. 1978. Bacterial mobility and chemotaxis. *Crit. Rev. Biochem.* 5:291–341

Macnab, R. M., Koshland, D. E. Jr. 1972. The gradient-sensing mechanism in bacterial chemotaxis. *Proc. Natl. Acad. Sci. USA* 69:2509–12

Macnab, R. M., Koshland, D. E. Jr. 1973. Persistence as a concept in the motility of chemotactic bacteria. *J. Mechano-chem. Cell Motil.* 2:141–48

Macnab, R. M., Koshland, D. E. Jr. 1974. Bacterial motility and chemotaxis: light-induced tumbling response and visualization of individual flagella. *J. Mol. Biol.* 84:399–406

Macnab, R. M., Ornston, M. K. 1977. Normal-to-curly flagellar transitions and their role in bacterial tumbling. Stabilization of an alternative quaternary structure by mechanical force. *J. Mol. Biol.* 112:1–30

Manson, M. D., Tedesco, P., Berg, H. C., Harold, F. M., van der Drift, C. 1977. A protonmotive force drives bacterial flagella. *Proc. Natl. Acad. Sci. USA* 74:3060–64

Miller, J. B., Koshland, D. E. Jr. 1977. Sensory electrophysiology of bacteria: relationship of the membrane potential to motility and chemotaxis in *Bacillus subtilis. Proc. Natl. Acad. Sci. USA* 74:4752–56

Miller, J. B., Koshland, D. E. Jr. 1979. The protonmotive force and bacterial sensing. *J. Bacteriol.* In press

Ordal, G. W., Adler, J. 1974. Properties of mutants in galactose taxis and transport. *J. Bacteriol.* 117:517–26

Ordal, G. W., Goldman, D. J. 1975. Chemotaxis away from uncouplers of oxidative phosphorylation in *Bacillus subtilis. Science* 189:802–5

Ordal, G. W., Goldman, D. J. 1976. Chemotactic repellents of *Bacillus subtilis. J. Mol. Biol.* 100:103–8

Paoni, N., Koshland, D. E. Jr. 1979. Use of permeabilized cells for studies on the biochemistry of bacterial chemotaxis. *Proc. Natl. Acad. Sci. USA* In press

Parkinson, J. S. 1974. Data processing by the chemotaxis machinery of *Escherichia coli. Nature* 252:317–19

Parkinson, J. S. 1975. Genetics of chemotactic behavior in bacteria. *Cell* 4:183

Pfeffer, W. 1884. Locomotorische Richtungsbewegunge durch chemische Reize, *Untersuch Bot. Inst. Tübingen* 1:363–482

Purcell, E. M. 1976. Life at low Reynolds number. *Am. J. Phys.* 45:3–11

Rubik, B. A., Koshland, D. E. Jr. 1978. Potentiation, desensitization, and inversion of response in bacterial sensing of chemical stimuli. *Proc. Natl. Acad. Sci. USA* 75:2820–24

Sharma, S. K., Klee, W. A., Nirenberg, M. 1975. Dual regulation of adenylate cyclase accounts for narcotic dependence and tolerance. *Proc. Natl. Acad. Sci. USA* 72:3092–96

Silverman, M., Matsumura, P., Hilmen, M., Simon, M. 1977. Characterization of lambda *E. coli* hybrids carrying chemotaxis genes. *J. Bacteriol.* 130:877–87

Silverman, M., Simon, M. 1973. Genetic analysis of bacteriophage Mu-induced flagellar mutants in *Escherichia coli. J. Bacteriol.* 116:114–22

Silverman, M., Simon, M. 1974. Flagellar rotation and the mechanism of bacterial motility. *Nature* 249:73–74

Silverman, M., Simon, M. 1976. Operon controlling motility and chemotaxis in *E. coli. Nature* 264:577–79

Silverman, M., Simon, M. I. 1977a. Bacterial flagella. *Ann. Rev. Microbiol.* 31:397–419

Silverman, M., Simon, M. 1977b. Chemotaxis in *Escherichia coli:* Methylation of *che* gene products. *Proc. Natl. Acad. Sci. USA* 74:3317–21

Silverman, M., Simon, M. 1977c. Identification of polypeptides necessary for chemotaxis in *Escherichia coli. J. Bacteriol.* 130:1317–25

Springer, M. S., Goy, M. F., Adler, J. 1977. Sensory transduction in *Escherichia coli:* two complementary pathways of information processing that involve methylated proteins. *Proc. Natl. Acad. Sci. USA* 74:3312–16

Springer, W. R., Koshland, D. E. Jr. 1977. Identification of a protein methyltransferase as the *cheR* gene product in the bacterial sensing system. *Proc. Natl. Acad. Sci. USA* 74:533–37

Spudich, J. L., Koshland, D. E. Jr. 1975. Quantitation of the sensory response in bacterial chemotaxis. *Proc. Natl. Acad. Sci. USA* 72:710–13

Spudich, J. L., Koshland, D. E. Jr. 1976. Non-genetic individuality: chance in the single cell. *Nature* 262:467–71

Stock, J. B., Koshland, D. E. Jr. 1978. A protein methylesterase involved in bacterial sensing. *Proc. Natl. Acad. Sci. USA* 75:3659–63

Strange, P. G., Koshland, D. E. Jr. 1976. Receptor interactions in a signaling system: Competition between ribose receptor and galactose receptor in the chemotaxis response. *Proc. Natl. Acad. Sci. USA* 73:762–66

Szmelcman, S., Adler, J. 1976. Change in membrane potential during bacterial chemotaxis. *Proc. Natl. Acad. Sci. USA* 73:4387–91

Taylor, B. L., Koshland, D. E. Jr. 1975. Intrinsic and extrinsic light responses of *Salmonella typhimurium* and *Escherichia coli. J. Bacteriol.* 123:557–69

Taylor, B. L., Miller, J. B., Warrick, H. M., Koshland, D. E. Jr. 1979. Electron transfer pathways are essential for tactic responses to light and electron acceptors by *Salmonella typhimurium* and *Escherichia coli.* In preparation

Tsang, N., Macnab, R., Koshland, D. E. Jr. 1973. Common mechanism for repellents and attractants in bacterial chemotaxis. *Science* 181:60–63

Van der Werf, P., Koshland, D. E. Jr. 1977. Identification of a γ-glutamyl methyl ester in a bacterial membrane protein involved in chemotaxis. *J. Biol. Chem.* 252:2793–95

Warrick, H. M., Taylor, B. L., Koshland, D. E. Jr. 1977. The chemotactic mechanism of *Salmonella typhimurium:* Preliminary mapping and characterization of mutants. *J. Bacteriol.* 130:223–31

Weibull, C. 1960. Movement. In *The Bacteria,* ed. I. C. Gunsalus, R. Y. Stanier, 1:153. New York: Academic

Zukin, R. S., Hartig, P. R., Koshland, D. E. Jr. 1977a. Use of a distant reporter group as evidence for a conformational change in a sensory receptor. *Proc. Natl. Acad. Sci. USA* 74:1932–36

Zukin, R. S., Strange, P. G., Heavey, L. R., Koshland, D. E. Jr. 1977b. Properties of the galactose binding protein of *Salmonella typhimurium* and *Escherichia coli. Biochemistry* 16:381–86

Ann. Rev. Neurosci. 1980 3:77–95

DEMENTIA OF THE ALZHEIMER TYPE

❖11536

Robert D. Terry and Peter Davies

Department of Pathology, Albert Einstein College of Medicine, 1300 Morris Park Avenue, The Bronx, New York 10461

INTRODUCTION

Dementia can be defined as a progressive loss of cognitive and other "higher" intellectual functions as a result of organic brain disease. There are many causes of dementia: some of them associated with extraneural systemic disease, others primarily in the cerebrum. Storage disease, nutritional deficiency, intoxications, infections, physiologic abnormalities of pulmonary, circulatory, renal or hepatic systems, neoplasms, and cerebral degenerative disorders can all induce progressive mental deterioration.

By far the most common cause of dementia is a primary cerebral disorder called Alzheimer's disease. It was originally described as a presenile disorder, that is occurring before age 65, but in recent years most investigators have agreed that it is essentially identical to the most common form of senile dementia, which is thus called senile dementia of the Alzheimer type, or SDAT.

PREVALENCE

Signs of dementia are very common among the elderly. About 15% of those over 64 years are said to be involved (Wang 1977) and the prevalence is much higher in the eighth and ninth decades of life. About 5% of the population older than 64 are severely demented, and this amounts to more than a million people in the United States.

Autopsy studies by Tomlinson (B. E. Tomlinson, personal communication 1977) show that about half of the patients with senile dementia have Alzheimer's disease. SDAT is thus a major public health problem in regard

77

0147-006X/80/0301-0077$01.00

to the numbers of patients, and as to the economic and emotional cost of their care.

CLINICAL ASPECTS

The onset of Alzheimer's disease is usually subtle, with lessened initiative and some forgetfulness. A decline in judgment, progressive difficulty with word finding, and loss of memory (especially recent) soon become evident. Parietal lobe signs such as tactile agnosia and apraxia may be found. Habits involving personal hygiene, interest in the environment, and insight are lost. Inanition is common. Ultimately the patient requires total nursing care. Death is usually the result of an infection; but rates for accidents, vascular disease, and neoplasia are also elevated in these patients. The course of Alzheimer's disease averages ten years in the younger patients, but only three years in the senile group (Nielsen, Homma & Biorn-Henriksen 1977), so that the disorder can well be regarded as malignant (Katzman 1976). Some of these secondary diseases may be directly related to the cerebral changes of Alzheimer's disease through brain mechanisms involving lowered resistance to stress.

GENETICS

That there is a genetic component to Alzheimer's disease cannot be disputed. Numerous pedigrees display a dominant mode of transmission, but this is more common among the presenile group than in older patients. The great majority of cases are sporadic, but even here there appears to be an increased risk among first order relatives of patients (Slater & Cowie 1971). Concordance rates are high among both monozygotic and dizygotic twins (Kallman & Sander 1949). Women are more often affected by SDAT than are men, and there is an increased incidence of hypodiploidy in senile women, but not in men (Nielsen 1968). Haptoglobin type Hp^1 is more frequent than expected in Alzheimer's disease (Op den Velde & Stam 1973).

PHYSIOLOGIC ASPECTS

A high degree of correlation has been achieved among physiologic, structural, chemical, and clinical aspects of Alzheimer's disease.

Electroencephalography

Symmetrical slowing of the dominant rhythm is characteristic of SDAT, as theta and delta components become quantitatively more important (Obrist 1976). These changes might well be associated with the loss of cortical neurons found in normal aging (Brody 1955), in addition to the decrement in dendritic population that is exaggerated in SDAT (Scheibel 1978).

Myoclonic activity is not rare in Alzheimer's disease, especially later in its course, despite earlier writings to the contrary. Cortical intraneuronal neurofibrillary tangles are probably responsible.

REM sleep is significantly reduced (Feinberg, Koresko & Heller 1967), and this is probably related to the loss of pigmented neurons in the locus ceruleus (Brody 1978, Forno 1978). Neurofibrillary tangles are present in the pontine tegmentum in the disease.

Cerebral Blood Flow

Ingvar's studies (Ingvar et al 1978) demonstrate a significant drop in cerebral blood flow especially in posterior temporal and inferior parietal regions in patients with Alzheimer's disease. This correlates well with the analyses of his colleagues Brun & Gustafson (1976), who found particularly intense tissue changes in these areas, and correlates also with the clinical findings of sensory apraxias. The rCBF in such patients does not respond to sensory stimuli as it does in the normals in Ingvar's study, which involves intracarotid injection of the gamma-emitting tracer.

Focally or asymmetrically diminished cerebral blood flow is more characteristic of multi-infarct dementia as demonstrated by Hachinski et al (1975). Most workers in this field agree that the altered blood flow in either situation is not the cause of the dementia, but is rather the result of diminished metabolic call from the diseased parenchyma.

STRUCTURAL CHANGES OF THE BRAIN

Gross Alterations of the Brain

It has been thought for many decades that the brain is atrophied in senile dementia and in Alzheimer's disease. It is certainly true that in very advanced cases, especially in the presenile form, the brain is much smaller than normal, the sulci are wide, the cortical ribbon is narrow, and the ventricles are enlarged. These alterations are not so apparent, however, in SDAT. We have found in a relatively small series of autopsies that the mean brain weight in SDAT is not significantly different from that of age-matched normals. Furthermore, our measurements of cortical thickness in frontal and superior temporal regions are also similar in the two groups. It begins to appear that this form of dementia is not the result of a major, diffuse loss of brain tissue.

The senile form of the disease, coupled with or even accelerating the other common frailties of old age, has a relatively rapidly fatal course. Nielsen, Homma & Biorn-Henriksen (1977) have shown that 50% of such patients are dead within three years, and 95% have died by five years. Presenile Alzheimer's disease often lasts considerably longer. This prolonged course may explain why the cerebral degeneration may be considerably more severe in the younger group.

Cell Counts

Neuronal counts of cerebral cortex in normal aging have been shown by Brody (1955) to decline markedly. His work indicates a particular loss of small neurons, and the total count of neurons (perikarya greater than seven microns in diameter) in the most affected neocortical areas of the normal elderly brains is down to about 50% of the number found in the young adult specimens.

An even greater loss was predicted in SDAT, but this has not been borne out by our counts utilizing image analysis (Terry et al 1979). This apparatus (the Cambridge/Imanco Quantimet) has been designed to permit editing with a light pencil on the video screen, so that vascular cells can be eliminated, and contiguous cell bodies can be divided. It has been programmed to measure the cross sectional area of each cell body within a rectangle 600 microns along the pial surface and through the full thickness of the cortex. The perikarya are classified into ten groups according to their size. The two smallest classes, 5 to 30 and 30 to 40 μm^2, have been found to correspond almost entirely to glia, while those larger than 40 μm^2 are almost all neurons, as best as can be determined by microscopic inspection. Sections of mid-frontal cortex and superior temporal cortex have been taken from a group of 18 autopsied patients aged 70 to 90, who were neurologically and histologically normal, for comparison with such specimens from 20 patients who were clinically demented and displayed microscopic changes qualitatively and quantitatively characteristic of SDAT, and were in the same age range as the normals. The differences between the cell counts are minimal in each size class above 40 μm^2. The number is slightly smaller in the demented group than in the normal group in each size class, but the difference rarely reaches the .05 level of significance. The total number of cells above 40 μm^2 is not significantly different in the two groups. Furthermore, the number of cells per average mm^2 is unchanged and the neuropil/cell body ratio is also unchanged.

We are again led to the conclusion that dementia is not due to a massive, diffuse loss of neocortical parenchyma. The small decrease in neuron numbers, if it is significant, might indicate a loss of a critical class of cells, small in number, but including neurons of all sizes above 40 μm^2. Conceivably this might correspond to the loss of choline acetyltransferase discussed below.

Dendrites and Dendritic Spines

The Scheibels (1975) have demonstrated with the Golgi technique that in normal aging there is an extensive loss of dendrites, especially those coming from the base of cortical pyramidal cells. This has also been shown in aged subhuman primates (Mervis 1978), and to some extent in aged rats (Feldman 1974). The dendritic atrophy, according to Scheibel (1978), is greater in senile dementia than in age-matched controls. Other investigators (Mehraein, Yamada & Tarnowska-Dziduszko 1975) have counted dendritic spines in such cases, and have found a decreased concentration of these postsynaptic elements in demented patients, as compared with normal elderly. Cragg (1975) has presented some contrary data, having counted synapses in electron micrographs, where he found that the number per unit area is not changed.

It is difficult to correlate the loss of dendrites with our finding that the neuropil ratio is maintained in SDAT. It would seem possible, however, that a proliferation of astrocytic processes might replace the lost neuronal processes in terms of volume, and thus maintain the neuropil ratio.

Lipofuscin

Although this yellow, insoluble material does increase inside neuronal perikarya as a function of age, there is still no evidence at this time that it is harmful to cells or even that there is an increased amount of it in dementia (Mann & Sinclair 1978).

Neurofibrillary Tangles

Alzheimer's neurofibrillary tangles and neuritic (senile) plaques are the hallmarks of Alzheimer's disease. Both may be found in small numbers in the hippocampus of most elderly patients, but their frequency increases markedly in both neocortex and the H_1 region of the hippocampus in Alzheimer's disease. There are frequent examples of SDAT in which both kinds of lesions are very common in the hippocampus, while the neocortex has many plaques but only few tangles. The concentration of these lesions is strongly correlated with the degree of dementia (Tomlinson & Henderson 1976), and also with the deficiency of choline acetyltransferase, according to Perry et al (1978). These lesions are therefore not simply epiphenomena, but are an integral part of the disease as well as being of diagnostic significance.

Neurofibers in mature neurons are of two principal classes, readily apparent to the electron microscopist. These are (a) neurotubules, which are identical in every way to microtubules in other tissues; and (b) neurofilaments, which are members of the class of intermediary filaments. The latter measure 100 Å in diameter, the former 240 Å in diameter, and both have numerous short side-arms, which probably correspond to Tau protein. The neurofibrillary tangle is an argentophilic mass of neurofibers, coursing through the perikaryon, and often extending out into one or more neurites. The neurofibers that make up the tangle are different from the normal types in that they are paired helical filaments (PHF), each member of the pair measuring 100 Å in diameter, and crossing each other about every 800 Å (Wisniewski, Narang & Terry 1976). The maximum width of the pair is 200 to 250 Å, and the minimum width is 100 Å, at the cross-over points (Terry, Gonatas & Weiss 1964). In some tangles there is an admixture of normal appearing filaments, and occasionally normal appearing tubules with the PHF. Frequently there are normal concentrations of tubules and filaments in the neuronal cytoplasm adjacent to the bundle of PHF making up the tangle.

Neurofibrillary tangles made up of these characteristic PHF are also found in the residual neurons of the substantia nigra in postencephalitic Parkinson's disease (Hirano et al 1968). They are also common in the neocortex in dementia pugulistica (Corsellis, Bruton & Freeman-Browne 1973), Guam Parkinson dementia complex (Hirano et al 1968), and in occasional cases of subacute sclerosing panencephalitis (Mandybur et al 1977).

De Boni & Crapper (1978) have reported that treating tissue cultures of human fetal cortex with an aqueous extract of Alzheimer's disease brain induces the formation of PHF in the cultured neurons. This has not yet been confirmed in other laboratories, but the electron micorgraphs are very pursuasive. This would seem to be an indication of a transmissible agent being involved in Alzheimer's disease.

Typical neurofibrillary tangles made of PHF are not found in other animals, even senile dogs and aged primates. In the elderly monkeys (Wisniewski, Ghetti & Terry 1973) we have occasionally found small clusters of PHF in neuronal boutons. These have a crossover about every 400 Å and are thus, at least in this respect, different from the human structure. At any rate, in that these monkey lesions are so rare, they are an unsatisfactory model.

Investigations of the molecular nature of the neurofibrillary tangle have to date been concentrated on comparisons between the structure of the protein(s) of the PHF and those of the normal neurofilament and microtubule (Iqbal et al 1978). Although the data are far from conclusive, the bulk of the evidence suggests a closer similarity between a paired helical filament protein (PHFP) and proteins of the microtubule. This work, which has involved both chemical and immunochemical techniques, depends on the isolation and identification of the PHFP. Although early results showing chemical and immunological similarities to tubulin seemed promising, recently Shelanski (1978) has questioned the identity of the putative PHFP, and appears to believe that the protein isolated was derived from glia. More recently, a microtubule-enriched fraction prepared from a normal human brain has been used to raise an antiserum in a rabbit, and this has been used in immunohistochemical studies on material from SDAT patients. The serum reacts weakly with sections of autopsied normal human brain, but gives clear staining of neurofibrillary tangles in similar sections from SDAT cases (Grundke-Iqbal et al 1979). The interpretation of these data is that an antigen found in microtubule-enriched fractions from normal human brain appears to be concentrated in neurofibrillary tangles. Work directed at identifying this antigen is now in progress.

Neuritic Plaques

While the neurofibrillary tangle is primarily within the perikaryon, the plaque occupies the neuropil, where it is often adjacent to a capillary. Light microscopy indicates that it has a central core of amyloid, surrounded by argentophilic rods and granules. An astrocytic component in the form of a more or less concentric web has recently been demonstrated (Schechter, Yen & Terry 1979) by immunohistochemical methods utilizing an antibody against astrocytic protein.

It has been demonstrated histochemically that the lesion of the plaque is hyperactive in terms of oxidative enzymes (Friede 1965), and also as to acid hydrolases (Josephy 1949).

These findings have been confirmed and amplified using electron microscopic techniques. The core of the plaque is indeed made up of amyloid, which is largely extracellular and present in the form of 90 to 100 Å filaments, quite different in texture and density from normal neurofilaments and from the individual members of the PHF. Surrounding this core are numerous enlarged unmyelinated neurites, which contain large numbers of lamellar dense bodies and many mitochondria in varying stages of alteration (Terry, Gonatas & Weiss 1964). The lamellar dense bodies have been shown to be of a lysosomal nature (Suzuki & Terry 1967). The mitochondria often have a contracted appearance, and this might indicate an altered ATP/ADP ratio. At any rate, the enrichment in both oxidative and hydrolytic enzymes are accounted for by these structures. Astrocytic processes and occasional "microglia" are present in and around the plaque. Many of the neurites within the plaque are in fact axonal boutons (Gonatas, Anderson & Evangelista 1967). The synaptic gap is normal. Usually the postsynaptic bouton is also unremarkable, although occasionally they have been seen to contain PHF. Synaptic involvement is thus frequent, but usually limited to the presynaptic side.

Plaque amyloid reacts with anti-human IGG, and it may, therefore, be that sort of amyloid that Glenner (Glenner, Terry & Isersky 1973) described as being made up of fragments of light chains. Histochemical studies of Powers & Spicer (1978), however, provide some indication that this amyloid is more closely related to the APUD type. Isolation and characterization will be necessary to provide certainty in this regard. A potential difficulty arises in the fact that not infrequently some small blood vessels within the parenchyma are also infiltrated with amyloid. While this might be identical in composition to that of the plaque, it need not necessarily be so. Nikaido et al (1972) have shown that formalin-fixed plaques contain abnormal quantities of silicon, but that whole brain tissue in Alzheimer's disease does not have elevated levels. The precise location of this silicon is not yet known. Lipofuscin-like bodies in the microglia and astrocytes, as well as in neurons, are probably the focus of iron concentrations, which are known to be somewhat higher than normal in senile dementia.

Neuritic plaques are quite commonly found in the neocortex of aged dogs (Wisniewski et al 1970) and old monkeys (Wisniewski, Ghetti & Terry 1973). These plaques are very similar to those in the human, but they lack the PHF in the involved neurites. This sort of neuritic plaque, without PHF, has been induced in certain strains of recipient mice inoculated intracerebrally with certain varieties of the Scrapie agent (Bruce, Dickinson & Fraser 1976). Wisniewski (Wisniewski, Bruce & Fraser 1975) found that he can cause particular concentrations of plaques near cerebral stab wounds in these infected animals. He interprets this finding to indicate again that there may be an infectious agent in Alzheimer's disease, and that the lesions come from an original deposit of intracerebral amyloid that has passed through the wall of the vessel.

Granulovacuolar Degeneration

This change was first described by Simchowicz (1911) and is found almost exclusively in hippocampal pyramids. When many such granulovacuolar bodies are present, there is a very strong correlation with dementia (Woodard 1962). Because of their location, these lesions have not been studied in fresh brain tissue, and therefore electron micrographs are of poor quality. The ultrastructural studies do, however, reveal that the clear vacuole is bounded by a unit membrane and contains a cluster of dense, finely granular material (Terry & Wisniewski 1971). The nature of these lesions is otherwise unknown, but they do to some extent resemble an abnormal endocytotic vesicle, and to some extent they might also be said to resemble a residual body.

THE POSSIBILITY OF INFECTION

Two sorts of evidence have been indicated above in regard to a transmissible agent as a cause of Alzheimer's disease: The induction of PHF in neurons cultured from human fetus, and the induction of neuritic plaques in mice inoculated with Scrapie. A third remains to be cited. Traub, Gajdusek & Gibbs (1977) have reported that twice in six attempts they have caused spongy encephalopathy in primates injected with tissue from human brains affected by familial Alzheimer's disease. More than thirty such trials failed with tissue from sporadic cases. There are several possible explanations for these results. It could have been a cross-infection of the recipients by material already in the laboratory from cases of Creutzfeldt-Jakob disease or from Kuru, or the donor specimens might have been misdiagnosed. Also possible, however, is that whatever causes plaques and tangles in humans affects other primates differently, inducing the spongy change in the latter.

BIOCHEMICAL STUDIES ON SENILE DEMENTIA OF THE ALZHEIMER TYPE

Neurotransmitter-Related Parameters

A summary of the neurotransmitter-related parameters that have been investigated in brains from cases of SDAT is presented in Tables 1 through 7. The majority of the cited studies used samples of brains removed during routine autopsies, although a few utilized biopsies of neocortex. This discussion deals with some of the problems with these investigations, and with the significant points that have emerged. The reader is urged to consult the literature for detailed descriptions of individual studies.

Only one of the five neurotransmitter systems studied to date has consistently shown evidence of gross abnormality. There are now four independent research groups agreeing that the activity of choline acetyltransferase (ChAT) in the cerebral cortex of cases of SDAT is reduced to 10 to 30% of the level found in age-matched normal individuals (Davies & Maloney 1976, Perry et al 1977, White et al 1977, Reisine et al 1978, Davies 1979). It is widely accepted that within the mammalian brain ChAT is found only in cholinergic neurons (Kuhar 1976), and the sizable deficits in its activity have been interpreted as indicating the loss of such cells from the cerebral cortex in SDAT (Davies & Maloney 1976, Bowen et al 1979). Acetylcholinesterase (AChE) activity is also reduced in cortical samples from SDAT cases (Pope, Hess & Lwein 1965, Bowen et al 1976a, Davies 1979), although this enzyme is probably not unique to cholinergic neurons (Kuhar 1976). The present lack of a reliable histochemical or immunohistochemical technique to visualize ChAT hinders attempts to define more precisely the abnormality in cortical cholinergic neurons.

Table 1 Neurotransmitter-related enzymes

Enzyme	Brain region	Percentage of activity in age-matched normals	References
Choline acetyltransferase	Frontal cortex	10 – 30	Davies & Maloney 1976,
	Hippocampus	10 – 30	Perry et al 1977, White et
	Caudate	40 – 80	al 1977, Davies 1978a, Bowen et al 1979, Davies 1979
Acetyl cholinesterase	Frontal cortex	10 – 50	Pope, Hess & Lewin 1965,
	Hippocampus	10 – 50	Davies & Maloney 1976,
	Caudate	30 – 60	Davies 1978a, Perry et al 1978
Pseudocholinesterase	Frontal cortex	130 – 180	Perry et al 1978
Glutamic acid decarboxylase	Frontal cortex	60 – 130	Bowen et al 1976a, Perry
	Hippocampus	60 – 120	et al 1978, Davies 1979
	Caudate	60 – 110	
Tyrosine hydroxylase	Frontal cortex	80 – 100	Davies & Maloney 1976
	Hippocampus	80 – 100	
Aromatic amino acid decarboxylase	Frontal cortex	100 ± 20	Bowen & Maloney 1976
	Whole temporal lobe	100 ± 20	
Dopamine-β-hydroxylase	Frontal cortex	90 – 130	Davies & Maloney 1976
Monoamine oxidase	Frontal cortex	80 – 150	Adolfsson et al 1978

Table 2 Neurotransmitters and metabolites

Transmitter or metabolite	Brain region	Percentage of concentration in age-matched normals	References
Dopamine	Frontal cortex	70 – 100	Adolfsson et al 1978,
	Caudate	70 – 100	Davies 1978b
Homovanillic acid	Frontal cortex	90 – 100	Adolfsson et al 1978,
	Caudate	60 – 100	Davies 1978b
Dihydroxyphenylacetic acid	Frontal cortex	100 ± 20	Davies 1978b
	Caudate	100 ± 20	
Noradrenalin	Frontal cortex	50 – 100	Adolfsson et al 1978,
	Hypothalamus	60 – 100	Davies 1978b
	Hippocampus	100 ± 20	
5-Hydroxytryptamine	Frontal cortex	60 – 100	Adolfsson et al 1978
	Hippocampus	60 – 100	
	Caudate	60 – 100	
5-Hydroxyindoleacetic acid	Frontal cortex	60 – 100	Adolfsson et al 1978
	Hippocampus	60 – 100	
	Caudate	60 – 100	

Table 3   Neurotransmitter receptors

Receptor	Ligand	Brain region	Percentage of concentration in age-matched normals	References
Muscarinic cholinergic	Atropine and QNB	Frontal cortex Hippocampus Caudate	100 ± 20 50 – 100 100 ± 20	Davies & Verth 1977, Perry et al 1977, White et al 1977, Reisine et al 1978
GABA	GABA	Frontal cortex Hippocampus Caudate	50 100 ± 20 50	Reisine et al 1978
Dopamine	Spiro-peridol	Frontal cortex Hippocampus Caudate	100 ± 20 100 ± 20 50	Reisine et al 1978
Opiate	Naloxone	Whole temporal lobe	100 ± 20	Bowen et al 1979
Noraldrenalin (β)	Dihydroal-prenolol	Whole temporal lobe	100 ± 20	Bowen et al 1979
5-Hydroxy-tryptamine	LSD	Whole temporal lobe	60 – 70	Bowen et al 1979

Table 4   Enzymes related to energy metabolism

Parameter in whole temporal lobe	Percentage of activity in age-matched normals	Refs.[a]
Hexokinase	"May be altered"	4
Phosphohexose isomerase	55 – 56	4
Phosphofructokinase	"May be altered"	4
Aldolase	55 – 56	3, 4
Phosphotriose isomerase	73 – 74	4
Glyceraldehyde-3-phosphate dehydrogenase	77 – 81	4
Phosphoglycerokinase	100 ± 20	4
Phosphoglyceromutase	60 – 70	4
Enolase	100 ± 20	4
Pyruvage kinase	77 – 81	4
Lactate dehydrogenase	77 – 81	4
Alcohol dehydrogenase	73 – 74	4
Succinate dehydrogenase	60 – 80	1, 2
6-Phosphogluconate dehydrogenase	100 ± 20	4
Total ATPase	77 – 81	5
Mg^{2+}–ATPase	73 – 74	5
Na^+, K^+–ATPase	100 ± 20	4
Total protein kinase	100 ± 20	4
cAMP-independent protein kinase	"May be altered"	4

[a] References: 1, Bowen et al 1974; 2, Bowen et al 1976b; 3, Bowen et al 1977; 4, Bowen et al 1979; 5, Pope & Embree 1976.

Table 5 Metabolic functions

Metabolic functions in biopsied frontal cortex	Percentage of activity in age-matched normals
Oxygen consumption	100 ± 20
Glycolysis	100 ± 20
Lactate production	100 ± 20
Acetate incorporation (into lipids)	100 ± 20
Amino acid incorporation (into proteins)	100 ± 20

[a]Information taken from Korey et al 1961, and Suzuki, Katzman & Korey 1965.

Table 6 Miscellaneous enzymes, proteins, lipids, and nucleic acids

Parameter	Brain region	Percentage of concentration in age-matched normals	Refs.[a]
Total protein	Neocortex	100 ± 20	2, 5, 6, 8
DNA	Neocortex	100 ± 20	4, 5, 6
RNA	Neocortex	100 ± 20	4
DNA/RNA ratio	Neocortex	100 ± 20	4
Total phospholipids	Neocortex	100 ± 20	5
Cholesterol	Neocortex	100 ± 20	5
Cerebroside	Neocortex	100 ± 20	5
Ganglioside N-acetyl neuraminic acid	Neocortex	60 – 70	1, 4, 6, 8
Acid polysaccharides	Neocortex	"High in neuritic plaques"	3
Neuronin S-5	Whole temporal lobe	100 ± 20	4, 7, 8
Neuronin S-6	Whole temporal lobe	100 ± 20	4, 7, 8
Acid phosphatase	Neocortex	"High in neuritic plaques"	3, 6
β-Galactosidase	Whole temporal lobe	100 ± 20	2, 6, 8
β-Glucuronidase	Whole temporal lobe	100 ± 20	2, 8
Dipeptidase (L-alanylglycine)	Neocortex	100 ± 20	5
γ-Glutamyl transpeptidase	Whole temporal lobe	100 ± 20	4, 8
Carbonic anhydrase	Whole temporal lobe	60 – 70	4, 7, 8
2′,3′-Cyclic nucleotide 3′-phosphohydrolase	Whole temporal lobe	60 – 70	4, 6

[a]References: 1, Korey et al 1961; 2, Bowen et al 1974; 3, Suzuki, Katzman & Korey 1965; 4, Bowen et al 1976a; 5, Pope & Embree 1976; 6, Bowen et al 1977; 7, Bowen & Davison 1978; 8, Bowen et al 1979.

Table 7 Elemental analyses

Elemental analyses in neocortex	Comments	References
Aluminum	Elevated in a group of pre-senile cases	Crapper, Krishnan & Dalton 1973
	Not confirmed in a group of older patients with seemingly identical pathology	McDermott et al 1977
Silicon	Focal elevations in neuritic plaques, and smaller increase in neuro-fibrillary tangles; normal elsewhere	Nikaido et al 1972, Austin et al 1973
Iron	Small increase in tangle-bearing neurons	Nikaido et al 1972, Austin et al 1973

Although the most obvious explanation for the deficit in ChAT activity is the loss of cholinergic neurons, other possibilities should be considered. It may be that cholinergic neuron cell bodies are intact, and that their terminals, which contain a large proportion of the ChAT (Kuhar 1976) are abnormal. It is also possible that the synthesis of ChAT is disordered, such that fewer or abnormal enzyme molecules are produced. Future studies will have to distinquish between these alternate explanations.

Recently the studies of Perry et al (1978) have revealed that there is a relationship between the extent of the cortical ChAT deficit in SDAT cases and the numbers of neuritic plaques. Further, a significant correlation was found between the degree of intellectual impairment and the extent of the ChAT deficit in the cortex. Thus the loss of ChAT activity appears to be related both to the presence of one of the pathological hallmarks of SDAT, and to the loss of functional capacity.

That at least one type of receptor for acetylcholine is not affected in SDAT has been shown by three independent groups. The muscarinic receptor concentrations are normal in all cortical regions of brains from cases of SDAT examined (Perry et al 1977, Davies & Verth 1977, White et al 1977). A loss of about 50% of muscarinic receptors from the hippocampus was noted in the study by Reisine et al (1978), but this does not appear to be a consistent finding (Davies & Verth 1977). There is some evidence for losses of GABA (Reisine et al 1978), 5-hydroxytryptamine, and dopamine receptors, but not of opiate or α-noradrenergic receptors (Bowen et al 1979). Whether or not these receptor losses relate to the cholinergic system abnormalities is not yet clear.

Consistent evidence for major abnormalities in SDAT of any of the parameters related to neurotransmission by dopamine, noradrenalin, 5-

hydroxytryptamine and GABA has not been found (Davis 1978a, Bowen et al 1979). There are data from the group led by Gottfries suggesting that minor deficits in the dopaminergic system may occur in some SDAT patients (Adolfsson et al 1978). These have been difficult to confirm (Parkes et al 1974, Davies 1978b), and are of such minor proportions as to raise serious doubts concerning their functional significance. These small apparent deficits are also found only in areas such as putamen and thalamus, regions that rarely show more than a few neuritic plaques or neurofibrillary tangles. At present it seems likely that what minor dopaminergic system problems there are in SDAT may be due to abnormalities of the cholinergic system or are the result of the presence of coincidental, sub-clinical Parkinson's disease.

Of the five neurotransmitter systems discussed so far, there is evidence indicating that only the cholinergic system shows major deficits in SDAT. The hypothesis that in SDAT there is selective loss of cholinergic neurons therefore seems a reasonable first approximation. This is not to deny that there may be changes in neurons using either amino acids or peptides as transmitters. These have not yet been investigated in any detail in the human brain, and not at all in tissues from SDAT cases. The data available at this time do seem to justify therapeutic attempts to restore the apparent deficiency of acetylcholine. Preliminary trials with choline, lecithin (Boyd et al 1977, Etienne et al 1978, Signoret, Whiteley & Lhermitte 1978), and dimethylaminoethanol (Deanol) (Ferris et al 1977) have been reported, without major improvement. These attempts to elevate levels of acetylcholine by administering presumed precursors may be doomed to failure if there are major losses of either cholinergic neurons or their synaptic terminals. It is quite possible that what little ChAT activity remains is already fully saturated with substrate. Another hindrance to successful precursor therapy is suggested by the studies of Bowen & Davison (1978), who have shown that the activity of the high affinity choline uptake system is markedly reduced in biopsy samples of neocortex from SDAT cases. Alternate therapeutic possibilities, such as the use of centrally active cholinesterase inhibitors or cholinergic agonists may be more productive.

Parameters Not Specific to Neurotransmission

Tables 3 through 7 are concerned with listing the biochemical parameters studied in brains from SDAT cases but which are not directly related to neurotransmission. One of the broad aims of these studies has been to detect enzymatic or other changes in the cerebral cortex that could be correlated with the reportedly widespread neuronal loss (but see above). The assumption behind many of these studies is that certain parameters are found predominantly in specific cell types (e.g. neurons, astrocytes, or microglia), or in cellular specializations (e.g. myelinated axons or synaptic terminals). The most productive group in this field is that led by Bowen & Davison, and much of the data in the Tables derive from their investigations.

The results obtained by Bowen & Davison seem to indicate that in the whole temporal lobe of patients with SDAT, there is a 25–35% reduction in the number of neurons (Bowen et al 1976a, b, 1979). Biochemical parameters thought to be specific to glia are not significantly abnormal. Whether or not these results can be confirmed by morphometric studies is not yet clear (see above). It must be stressed that the interpretation of biochemical results relies heavily on the assumption that the parameter under investigation is localized uniformly to specific cell types. For example, ganglioside N-acetyl neuraminic acid (ganglioside NANA) is regarded as a marker for neuronal membranes (Bowen et al 1979), and is reported to be reduced 30 to 40% in temporal lobes from SDAT cases. If significant amounts of ganglioside NANA are found on glial membranes, the neuronal loss calculated on the basis of this parameter will be overestimated. Similarly, if synaptic terminals contain higher concentrations of ganglioside NANA per unit weight than do axons, cell bodies, or dendrites, the interpretation becomes even more complex. The situation is the same for enzymes related to energy metabolism: we simply do not know how these parameters are proportioned between different cell types, or between different regions of each cell type, and therefore we find it very difficult to interpret changes in their activity in cases of SDAT.

CONCLUSIONS

Senile dementia of the Alzheimer type (SDAT) is a major medical problem that presents a number of fascinating biologic aspects. Psychologic, genetic, physiologic, structural, and chemical features are beginning to come together to form an increasingly complete picture. At this time, several important leads are apparent involving fibrillar proteins, a possible transmissible agent, and a deficient enzyme relevant to a major transmitter.

In regard to the first, we may ask, which normal protein is the major constituent of the PHF, and how is it modified to form the abnormal structure?

As to the "infectious agent," can it be verified, and if so, how is it related to other unconventional viruses?

Several biochemical parameters show statistically significant differences from normal, age-matched controls. In our opinion, only one of those reported to date is likely to be of functional significance to the SDAT patient. Both in consistency and extent, the deficit of ChAT activity observed in the cerebral cortex of SDAT patients is by far the most striking finding. Future work must concentrate on answering the following questions:

1. Does the ChAT deficit indicate selective loss of cholinergic neurons, or some more subtle abnormality?
2. How many of the other less extensive biochemical changes reported are a consequence of the abnormality in cholinergic neurons?
3. How is the ChAT deficit related to the presence of the neuropathological hallmarks of SDAT, the neuritic plaque and the neurofibrillary tangle?
4. Are there measures that can be taken to restore the apparent deficiency of acetylcholine, and how effectively can they alleviate the symptoms of SDAT?

ACKNOWLEDGMENTS

This work has been supported in part by NIH grants NS-02255, NS-03356, and an equipment grant from the Kresge Foundation.

Literature Cited

Adolfsson, R., Gottfries, C. G., Oreland, L., Roos, B. E., Winblad, B. 1978. Reduced levels of catecholamines in the brain and increased activity of monamine oxidase in platelets in Alzheimer's disease: Therapeutic implications. In *Alzheimer's Disease: Senile Dementia and Related Disorders,* ed. R. Katzman, R. D. Terry, K. L. Bick, pp. 441–51. New York: Raven. 595 pp. *Aging,* Vol. 7

Austin, J. H., Rinehart, R., Williamson, T., Burcar, P., Russ, K., Nikaido, T., Lafrance, M. 1973. Studies in aging of the brain. III. Silicon levels in post mortem tissues and body fluids. *Prog. Brain Res.* 40:485–95

Bowen, D. M., Davison, A. N. 1978. Changes in brain lysosomal activity, neurotransmitter-related enzymes, and other proteins in senile dementia. See Adolfsson et al 1978, pp. 421–24

Bowen, D. M., Flack, R. H. A., White, P., Smith, C. B., Davison, A. N. 1974. Brain-decarboxylase activities as indices of pathological change in senile dementia. *Lancet* 1:1247–48

Bowen, D. M., Smith, C. B., White, P., Davison, A. N. 1976a. Neurotransmitter-related enzymes and indices of hypoxia in senile dementia and other abiotrophies. *Brain* 99:459–95

Bowen, D. M., Smith, C. B., White, P., Davison, A. N. 1976b. Senile dementia and related abiotrophies: Biochemical studies on histologically evaluated human postmortem specimens. In *Neurobiology of Aging,* ed. R. D. Terry, S. Gerson, pp. 361–78. New York: Raven. 440 pp. *Aging,* Vol. 3

Bowen, D. M., Smith, C. B., White, P., Goodhardt, M. J., Spillane, J. A., Flack, R. H. A., Davison, A. N. 1977. Chemcial pathology of organic dementias. I. Validity of biochemical measurements on human post-mortem brain specimens. *Brain* 100:397–426

Bowen, D. M., Spillane, J. A., Curzon, G., Meier-Ruge, W., White, P., Goodhardt, M. J., Iwangoff, P., Davison, A. N. 1979. Accelerated ageing or selective neuronal loss as an important cause of dementia? *Lancet* 1:11–14

Boyd, W. D., Graham-White, J., Blackwood, G., Glen, I., McQueen, J. 1977. Clinical effects of choline in Alzheimer senile dementia. *Lancet* 2:711

Brody, H. 1955. Organization of the cerebral cortex. III. A study of aging in the human cerebral cortex. *J. Comp. Neurol.* 102:511–56

Brody, H. 1978. Cell counts in cerebral cortex and brainstem. See Adolfsson et al 1978, pp. 345–51

Bruce, M. E., Dickinson, A. G., Fraser, H. 1976. Cerebral amyloidosis in scrapie in the mouse: Effect of agent strain and mouse genotype. *Neuropathol. Appl. Neurobiol.* 2:471–78

Brun, A., Gustafson, L. 1976. Distribution of cerebral degeneration in Alzheimer's disease. *Arch. Psychiatr. Nervenkr.* 223: 15–33

Corsellis, J. A. N., Bruton, C. J., Freeman-Browne, D. 1973. The aftermath of boxing. *Psychol. Med.* 3:270–303

Cragg, B. G. 1975. The density of synapses and neurons in normal, mentally defective and ageing human brains. *Brain* 98:81–90

Crapper, D. R., Krishnan, S. S., Dalton, A. J. 1973. Brain aluminum distribution in Alzheimer's disease and experimental neurofibrillary degeneration. *Science* 180:511–13

Davies, P. 1978a. Studies on the neurochemistry of central cholinergic systems in Alzheimer's disease. See Adolfsson et al 1978, pp. 453–59

Davies, P. 1978b. Biochemical changes in Alzheimer's disease - senile dementia: Neurotransmitters in senile dementia of the Alzheimer's type. In *Congenital and Acquired Cognitive Disorders,* ed. R. Katzman, pp. 154–60. New York: Raven. 314 pp. *Association for Research in Nervous and Mental Disease,* Vol. 57

Davies, P. 1979. Neurotransmitter-related enzymes in senile dementia of the Alzheimer type. *Brain Res.* 171:319–27

Davies, P., Maloney, A. J. R. 1976. Selective loss of central cholinergic neurons in Alzheimer's disease. *Lancet* 2:1403

Davies, P., Verth, A. H. 1977. Regional distribution of muscarinic acetylcholine receptor in normal and Alzheimer's type dementia brains. *Brain Res.* 138:385–92

De Boni, U., Crapper, D. R. 1978. Paired helical filaments of the Alzheimer type in cultured neurones. *Nature* 271: 566–68

Etienne, P., Gauthier, S., Dastoor, D., Collier, B., Ratner, J. 1978. Lecithin in Alzheimer's disease. *Lancet* 2:1206

Feinberg, I., Koresko, R. L., Heller, N. 1967. EEG sleep patterns as a function of normal and pathological aging in man. *J. Psychiatr. Res.* 5:107–44

Feldman, M. L. 1974. Degenerative changes in aging dendrites. *Gerontologist* 14: Suppl. p. 36

Ferris, S. H., Sathananthan, G., Gershon, S., Clark, C. 1977. Senile dementia: Treatment with Deanol. *J. Am. Geriatr. Soc.* 25:241–44

Forno, L. S. 1978. The locus ceruleus in Alzheimer's disease. *J. Neuropathol. Exp. Neurol.* 37:614. (Abstr. 100)

Friede, R. L. 1965. Enzyme histochemical studies of senile plaques. *J. Neuropathol. Exp. Neurol.* 24:477–91

Glenner, G. G., Terry, W. D., Isersky, C. 1973. Amyloidosis: Its nature and pathogenesis. *Semin. Hematol.* 10:65–86

Gonatas, N. K., Anderson, W., Evangelista, I. 1967. The contribution of altered synapses in the senile plaque: An electron microscopic study in Alzheimer's dementia. *J. Neuropathol. Exp. Neurol.* 26:25–39

Grundke-Iqbal, I., Johnson, A. B., Wisniewski, H. M., Terry, R. D., Iqbal, K. 1979. Evidence that Alzheimer neurofibrillary tangles originate from neurotubules. *Lancet* 1:578–80

Hachinski, V. C., Iliff, L. D., Zilhka, E., DuBoulay, G. H., McAllister, V. L., Marshall, J., Russell, R. W. R., Symon, L. 1975. Cerebral blood flow in dementia. *Arch. Neurol.* 32:632–37

Hirano, A., Dembitzer, H. M., Kurland, L. T., Zimmerman, H. M. 1968. The fine structure of some intraganglionic alterations. *J. Neuropathol. Exp. Neurol.* 27:167–82

Ingvar, D. H., Brun, A., Hagberg, B., Gustafson, L. 1978. Regional cerebral blood flow in the dominant hemisphere in confirmed cases of Alzheimer's disease, Pick's disease, and Multi-Infarct Dementia: Relationship to clinical symptomatology and neuropathological findings. See Adolfsson et al 1978, pp. 203–11

Iqbal, K., Grundke-Iqbal, I., Wisniewski, H. M., Terry, R. D. 1978. Neurofibers in Alzheimer dementia and other conditions. See Adolfsson et al 1978, pp. 409–20

Josephy, H. 1949. Acid phosphatase in the senile brain. *Arch. Neurol. Psychiatry Chicago* 61:164–69

Kallmann, F. J., Sander, G. 1949. Twin studies on senescence. *Am. J. Psychiatry* 106:29–36

Katzman, R. 1976. The prevalence and malignancy of Alzheimer's disease: a major killer. *Arch. Neurol.* 33:217–18

Korey, S. R., Scheinberg, L., Terry, A., Stein, A. 1961. Studies in presenile dementia. *Trans. Am. Neurol. Assoc.* 86:99

Kuhar, M. J. 1976. The anatomy of cholinergic neurons. In *Biology of Cholinergic Function,* ed. A. M. Goldberg, I. Hanin, pp. 3–27. New York: Raven. 716 pp.

Mandybur, T. I., Nagpaul, A. S., Pappas, Z., Niklowitz, W. J. 1977. Alzheimer neurofibrillary change in subacute sclerosing panencephalitis. *Ann. Neurol.* 1: 103–7

Mann, D. M. A., Sinclair, K. G. A. 1978. The quantitative assessment of lipofuscin pigment, cytoplasmic RNA and nucleolar volume in senile dementia. *Neuropathol. Appl. Neurobiol.* 4:129–35

McDermott, J. R., Smith, A. I., Iqbal, K., Wisniewski, H. M. 1977. Aluminum and Alzheimer's disease. *Lancet* 2: 710–11

Mehraein, P., Yamada, M., Tarnowska-Dziduszko, E. 1975. Quantitative study on dendrites and dendritic spines in Alzheimer's disease and senile dementia. In *Physiology and Pathology of Dendrites*, ed. G. W. Kreutzberg, pp. 453–58. New York: Raven. 511 pp. *Advances in Neurobiology*, Vol. 12

Mervis, R. 1978. Structural alterations in neurons of aged canine neocortex: A Golgi study. *Exp. Neurol.* 62:417–32

Nielsen, J. 1968. Chromosomes in senile dementia. *Br. J. Psychiatry* 114:303–9

Nielsen, J., Homma, A., Biorn-Henriksen, T. 1977. Follow up 15 years after a geronto-psychiatric prevalence study. *J. Gerontol.* 32:554–61

Nikaido, T., Austin, J., Trueb, L., Rinehart, R. 1972. Studies in ageing of the brain. II. Microchemical analyses of the nervous system in Alzheimer patients. *Arch. Neurol.* 27:549–54

Obrist, W. D. 1976. Problems of aging. In *Handbook of Electroencephalography and Clinical Neurophysiology*, ed. A. Remond, 6: Pt. A, 275–92. Amsterdam: Elsevier. 342 pp.

Op den Velde, W., Stam, F. C. 1973. Haptoglobin types in Alzheimer's disease and senile dementia. *Br. J. Psychiatry* 122:331–36

Parkes, J. D., Marsden, C. D., Rees, J. E., Curzon, G., Kantamaneni, B. D., Knill-Jones, R., Akbar, A., Das, S., Kataria, M. 1974. Parkinson's disease, cerebral arteriosclerosis and senile dementia. *Q. J. Med.* 43:49–61

Perry, E. K., Perry, R. H., Blessed, G., Tomlinson, B. E. 1977. Necropsy evidence of central cholinergic deficits in senile dementia. *Lancet* 1:189

Perry, E. K., Tomlinson, B. E., Blessed, G., Bergmann, K., Gibson, P. H., Perry, R. H. 1978. Correlation of cholinergic abnormalities with senile plaques and mental test scores in senile dementia. *Br. Med. J.* 2:1457–59

Pope, A., Embree, L. J. 1976. Neurochemistry of dementia. In *Handbook of Clinical Neurology*, ed. P. J. Vinken, G. W. Bruyn, Vol. 27, *Metabolic and Deficiency Diseases of the Nervous System*, Pt. I, ed. Harold L. Klawans, Chap. 21, pp. 477–501. Amsterdam: North Holland. 554 pp.

Pope, A., Hess, H. H., Lewin, E. 1965. Microchemical pathology of the cerebral cortex in pre-senile dementias. *Trans. Am. Neurol. Assoc.* 89:15–16

Powers, J. M., Spicer, S. S. 1978. Histochemical similarity of senile plaque amyloid to apudamyloid. *Virchows Arch. A* 376:107–15

Reisine, T. D., Yamamura, H. I., Bird, E. D., Spokes, E., Enna, S. J. 1978. Pre- and postsynaptic neurochemical alterations in Alzheimer's disease. *Brain Res.* 159:477–81

Schechter, R., Yen, S.-H., Terry, R. D. 1979. Immunohistochemical localization of anti-glial protein P51 in neuritic plaques. In preparation

Scheibel, A. B., 1978. Structural aspects of the aging brain: Spine systems and the dendritic arbor. See Adolfsson et al 1978, pp. 353–73.

Scheibel, M. E., Scheibel, A. B. 1975. Structural changes in the aging brain. In *Clinical, Morphologic, and Neurochemical Aspects in the Aging Central Nervous System*, ed. H. Brody, D. Harman, J. M. Ordy, pp. 11–37. New York: Raven. 221 pp. *Aging*, Vol. 1

Shelanski, M. L. 1978. Discussion. See Adolfsson et al 1978, pp. 425–26

Signoret, J. L., Whiteley, A., Lhermitte, F. 1978. Influence of choline on amnesia in early Alzheimer's disease. *Lancet* 2: 837.

Simchowicz, T. 1911. Histopathologische studien über die senile demenz. In *Histologie und Histopathologische Arbeiten Über Die Grosshirnrinde*, ed. F. Nissl, A. Alzheimer, 4:267–444. Jena, Germany: Fisher

Slater, E., Cowie, V. 1971. In *The Genetics of Mental Disorders*, pp. 122–59. Oxford: Oxford Univ. Press. 413 pp.

Suzuki, K., Katzman, R., Korey, S. R. 1965. Chemical studies on Alzheimer's disease. *J. Neuropathol. Exp. Neurol.* 24:211–24

Suzuki, K., Terry, R. D. 1967. Fine structural localization of acid phosphatase in senile plaques in Alzheimer's presenile dementia. *Acta Neuropathol.* 8:276–84

Terry, R. D., Gonatas, N. K., Weiss, M. 1964. Ultrastructural studies in Alzheimer's presenile dementia. *Am. J. Pathol.* 44:269–97

Terry, R. D., Peck, A., Millner, J., Fitzgerald, C., Horupian, D. S. 1979. Neocortical cell counts in senile dementia of the Alzheimer type and in the normal elderly. In preparation

Terry, R. D., Wisniewski, H. M. 1971. Ultrastructure of senile dementia and of experimental analogs. In *Aging and the Brain,* ed. C. M. Gaitz, pp. 89–116. New York: Plenum. 231 pp. *Advances in Behavioral Biology,* Vol. 3

Tomlison, B. E., Henderson, G. 1976. Some quantitative cerebral findings in normal and demented old people. See Bowen et al 1976b, pp. 183–204

Traub, R., Gajdusek, D. C., Gibbs, C. J. Jr. 1977. Transmissible virus dementia: the relation of transmissible spongiform encephalopathy to Creutzfeldt-Jakob disease. In *Aging and Dementia,* ed. W. L. Smith, M. Kinsbourne, pp. 91–172. New York: Spectrum. 244 pp.

Wang, H. S. 1977. Dementia of old age. See Traub et al 1977, pp. 1–24

White, P., Goodhardt, M. J., Keet, J. P., Hiley, C. R., Carrasco, L. H., Williams, I. E., Bowen, D. M. 1977. Neocortical cholinergic neurons in elderly people. *Lancet* 1:668–71

Wisniewski, H. M., Bruce, M. E., Fraser, H. 1975. Infectious etiology of neuritic (senile) plaques in mice. *Science* 190: 1108–10

Wisniewski, H. M., Ghetti, B., Terry, R. D. 1973. Neuritic (senile) plaques and filamentous changes in aged rhesus monkeys. *J. Neuropathol. Exp. Neurol.* 32:566–84

Wisniewski, H. M., Johnson, A. B., Raine, C. S., Kay, W. J., Terry, R. D. 1970. Senile plaques and cerebral amyloidosis in aged dogs. A histochemical and ultrastructural study. *Lab. Invest.* 23:287–96

Wisniewski, H. M., Narang, H. K., Terry, R. D. 1976. Neurofibrillary tangles of paired helical filaments. *J. Neurol. Sci.* 27:173–81

Woodard, J. S. 1962. Clinicopathologic significance of granulovacuolar degeneration in Alzheimer's disease. *J. Neuropathol. Exp. Neurol.* 21:85–91

Ann. Rev. Neurosci. 1980. 3:97-139
Copyright © 1980 by Annual Reviews Inc. All rights reserved

DEVELOPMENTAL NEUROBIOLOGY OF INVERTEBRATES

❖11537

Hilary Anderson, John S. Edwards, and John Palka

Department of Zoology, University of Washington, Seattle, Washington 98195

INTRODUCTION

The past decade has seen a revival of interest in the developmental neurobiology of invertebrates. After the major features of their neural development were described around the turn of the century the subject generally languished, while other areas such as the role of hormones in development and the cellular physiology of adult neurons prospered. In many ways the advances in these other fields have laid the groundwork for current developmental studies. New techniques in electrophysiology, dye injection, serial section reconstruction at the electron microscopic (EM) level, and the very substantial catalogue of identified neurons, have made it possible to do what had formerly been only "Gedankenexperiments." What is more, the whole corpus of invertebrate neurobiology has shown time and again that the principles of neural function are common across the Animal Kingdom, and that much is to be gained from the detailed analysis of favorable special cases no matter what phyletic group they happen to occur in. Thus, the present emphasis in invertebrate developmental neurobiology is less on the animals themselves than on the general principles of neural development, the origin and basic properties of developing neurons and glia, their many interactions, and the progressive assembly of functioning neural circuits.

A number of pertinent reviews of both classical and contemporary studies have appeared very recently (see below) and we do not attempt to cover the same ground in any detail. Rather, our goal is to present an impressionistic picture of current approaches, questions, and results—a series of snapshots in which not all the characters are present and some of those who are have

97

their faces turned or are squinting into the sun, but the overall nature of their play is apparent and one can guess the direction in which they are moving. By avoiding an encyclopedic coverage we have undoubtedly omitted reference to papers that should be represented: for our omissions, as well as for possible misrepresentations, we offer our apologies.

We have ordered our snapshots in roughly chronological fashion, starting with the embryonic origin of neurons and ending with their regeneration.

Synopsis of Reviews

An indication of the explosive growth of the field of invertebrate developmental neurobiology is the number of recent reviews. Among them are the following, in which many details only briefly alluded to here may be found: Young (1973)—The first symposium volume entirely devoted to the developmental neurobiology of arthropods. It includes Meinertzhagen's classic synthesis of the literature on the development of the insect visual system. Guthrie (1975)—A discussion of much of the literature on neuron growth and specificity of regeneration in the invertebrates. Many comparisons are made with vertebrates that point to fundamental similarities. Bentley (1977) —A review focussing on the development of functional properties of neural networks. Ward (1977)—Emphasizes neurogenetics, especially in nematodes and insects, but reviews much information useful for the developmentalist. Bate (1978)—A detailed account of the development of sensory systems in arthropods, emphasizing insects and especially strong on the visual system. Pipa (1978)—Treats selected topics in the postembryonic development of insects. Quinn & Gould (1979)—Neurogenetics of animals ranging from *Paramecium* to *Felis*. Palka (1979a)—An essay on insect neural development including a summary of general theories of morphogenesis (gradients, compartments, etc) and a review of the development of sensory systems, asking whether the theories help understand the empirical observations. Kankel et al (1979)—A description of the nervous system of *Drosophila* and a review of developmental analyses. Hall & Greenspan (1980)—Developmental neurobiology of *Drosophila* as analyzed with genetic tools. Includes thoughtful evaluations of genetic techniques and raises many testable questions. Govind (1980)—Developmental aspects of nerve-muscle relationships of crustaceans. This list is by no means exhaustive, and does not include reviews that cover primarily studies from the author's own laboratories.

ORIGIN OF NEURONS

Preneural Events

The literature of comparative embryology contains many detailed descriptions of early cleavage stages, the processes of blastulation, gastrulation, neurulation, and organogenesis in a wide variety of invertebrate as well as vertebrate animals (Kume & Dan 1968, Anderson 1972, Jacobson 1978). In the past few years modern methods have been applied to species of special interest to neurobiologists, with the goal of determining the cellular ancestry of fully differentiated neurons, especially the identified neurons that have become a hallmark of invertebrate neurobiology.

In general, more recent studies have confirmed the classical work, but at a new level of resolution and confidence. For example, Fernández (1979) has studied the early embryonic development of leeches, starting with the egg and proceeding to an embryo containing ganglia, and including a sequence of carefully described embryonic stages. Weisblat, Sawyer & Stent (1978) have injected HRP and other markers into single cells at various stages of development. This enzymatic marker persists after injection, does not spread from cell to cell, and is readily detectable even after a long series of cell divisions. Thus, it reveals the progeny of single cells or, conversely, the cell lineage or developmental history of cells from more advanced stages. One of their many elegant findings is that each side of the central nervous system (CNS) arises largely from a single progenitor cell. Injecting a particular single cell in an early embryo leads to the staining of the full chain of hemiganglia on the injected side, while the contralateral chain is totally free of stain. This result directly confirms an inference drawn from simple observational evidence by Whitman in 1887. Futhermore, it shows that cells do not migrate across the midline of the developing CNS, and other evidence is given that antero-posterior migration is also not a common event.

The nematode worm *Caenorhabditis elegans* has become one of the principal model systems for the study of neurogenesis. This species offers many advantages for genetic manipulation (Brenner 1974). Its tiny size makes electrophysiology and dye injection impossible, but makes reconstruction from serial electron micrographs feasible (reviewed in Ward 1977). The laboratory of von Ehrenstein has applied two of the most powerful modern microscopic techniques to the study of the embryogenesis of *C. elegans.* Deppe et al (1978) observed living embryos under Nomarski optics and obtained the complete cell lineage tree from the egg to the 182 cell stage (at hatching this worm consists of only about 550 cells, and the adult of about 800, excluding the gonads). Krieg et al (1978) have carried out an EM reconstruction of a 24 cell embryo in which the past history and future fate of every cell were known. It appears very likely that in a few years the complete cellular anatomy and history of the first stage larva will be known. Much of the postembryonic cellular history has already been described (Sulston 1976, Sulston & Horvitz 1977). The value of such descriptions in the causal analysis of neural development is pointed out in several sections below.

Neuroblast to Neuron

At present the cellular history of mature neurons starting with the recognizable neuroblasts from which they stem is known in two systems, which offer very different insights into the process.

Upon hatching from the egg, a larval nematode already has 33 cells in the ventral nerve cord. Thirteen blast cells give rise, by specific series of divisions, to 65 additional nerve cells, of which 9 die, giving an adult total of 89. The ganglia of the head and tail follow a similar pattern of development, but the number of cells is somewhat greater and cell death does not occur (Sulston 1976, Sulston & Horvitz 1977). Not only are the cell lineages of all the neurons that develop postembryonically from blast cells known, but in fact the lineages of all of the several hundred nongonadal cells are known in similar detail. Only the gap between 182 cells in the embryo and the 550 cells present in the hatching larva remains to be filled.

The wiring diagram of the CNS of *C. elegans,* in the sense of a map of the morphologically recognizable synaptic connections among all neurons, has been determined from serial section EM reconstructions for most of the ventral nerve cord (White et al 1976), the anterior nerve ring (Ware et al 1975), and many sensory cells (Ward et al 1975). This is an heroic task and not many individual worms have been fully reconstructed, but in the best known region of the CNS, the ventral nerve cord, it is apparent that the sets of neurons derived from individual blast cells form a repeating series. The forms of these neurons are very simple; they are bipolar cells with virtually no branching. The techniques available for reconstruction are now reasonably automated so that selected but substantial portions of the ventral nerve cord can be reconstructed in a matter of a few months when an experiment requires it, a fact that makes the analysis of the effects of mutation and of single-cell lesions feasible (see below).

These elegant studies of neural development have provided us with complete cell lineages and complete morphological characterizations of the fully assembled CNS. However, Nomarski optics do not reveal the neurites of individual cells, and EM reconstructions cannot at present provide a dynamic picture of the differentiation over time of single cells. For this exciting inquiry we turn to another system, the embryo of the locust *Schistocerca.*

The separation of the neuroblasts of the locust ventral nerve cord from the general ectoderm and their arrangement into plates corresponding to future segmental hemiganglia was described as early as 1891 and 1893 by Wheeler from histological sections of embryos. A more detailed description, based on careful observation of living embryos was given by Bate (1976c), who recognized that each ganglion is formed by a characteristic and fixed number of neuroblasts arranged with sufficient regularity that at least some of them can be identified in animal after animal.

In the past two years, Goodman & Spitzer (1979) and Goodman et al (1979) have succeeded in penetrating particular neuroblasts with microelectrodes filled with the fluorescent dye Lucifer yellow (Stewart 1978) and studying both the development of the form of particular neurons and their acquisition of physiological properties. The best studied cell thus far is DUMETi (the Dorsal Unpaired Median neuron supplying the Extensior Tibiae muscle). The lineage by which it arises from the DUM neuroblast is known, and its anatomy and physiology in adult animals have been studied in some detail. Some highlights of its development are as follows: (a) Initially all neuroblasts are electrically coupled together, and in fact are coupled to all other cell types of the embryo. (b) There is a progressive electrical uncoupling of neural from non-neural cells, and later of neural cells from each other. (c) Neuroblasts are dye-coupled to each other but not to non-neural cells, as tested with Lucifer yellow. (d) There is progressive dye-uncoupling among neural cells, the coupling persisting the longest between a given neuroblast and its progeny. (e) A neuronal soma begins to develop neurites after it becomes dye-uncoupled from other somata derived from the same neuroblast but while it still remains electrically coupled. (f) In every embryo a particular daughter cell of a given neuroblast acquires the same form and presumably becomes the same identified neuron in the adult. (g) During its morphological development, a cell produces many supernumerary branches, which later disappear. These include neurites in the neuropil and axons extending into "inappropriate" segmental nerves. (h) The onset of electrical responses to bath-applied transmitters coincides with initial neurite outgrowth. (i) The onset of regenerative electrical activity coincides with the arrival of the axon at its target muscle and with electrical uncoupling of the neuron from its siblings. (j) The appearance of the cell's own transmitter (octopamine) coincides with the rich spread of axonal branches over the surface of the target muscle. (k) Spikes are first Na^+ dependent, then convert to the combined Na^+ and Ca^{2+} dependency of the adult. (l) The supernumerary central neurites start to disappear early, at about the time the axon(s) first make their exit from the ganglion. The supernumerary axons disappear much later after the "correct" muscle is extensively innervated.

These events and the temporal relationships among them invite experimental manipulation. With such a detailed description of normal development in hand, it is also feasible to ask whether this sequence of events is peculiar to these particular cells, or whether it is common to many classes of developing neurons. In fact, there are many parallels as well as some differences between the development of locust DUM neurons and other cells both in invertebrates and vertebrates (Spitzer 1979).

Embryonic Neuropil Formation

Glimpses into the generation of neuropil are afforded by light and electron microscopic studies of neurogenesis in the insect embryo. The neuropil develops as a network of processes overlying the ventral neuroblast-ganglion cell mass. These first processes are predominantly orthogonally oriented, foreshadowing the future longitudinal connectives and transverse commissures (Bate and Goodman, personal communication). At that time the ultrastructural picture reveals a predominantly anteror-posterior alignment of fasciculated axons from which axopods extend perpendicularly to give an orthogonally oriented meshwork (Edwards and Chen, manuscript in preparation). Lacunae persist until the last phase of embryogenesis, when the arrival of afferents from sensilla on the body surface fills the remaining space to produce a solid neuropil (Edwards & Chen 1979).

General Chronologies of Embryonic Development

The increased attention now being paid to the earliest events in neurogenesis in a range of invertebrates has emphasized the need for more refined staging as a context for neuroembryology. Examples of such chronologies are available for the nematode (Deppe et al 1978), leech (Fernández 1979), *Aplysia* (Kriegstein 1977a,b), house cricket (Edwards & Chen 1979), and locust (Bentley et al, manuscript in preparation).

Pioneer Fibers

Many invertebrates, including most of those described in this review, undergo substantial postembryonic development during which neurogenesis continues. With each successive stage (instar) a population of new sensory fibers must find its way from cell bodies in the integument to synaptic terminations in the central nervous system. At least for arthropods it is well established that the sensory neurons find their way from the epidermis to the center by fasciculation with existing sensory nerves (Edwards & Palka 1976, Bate 1978, Palka 1979a). Recent work has been directed to the question of how the initial sensory contacts are established between periphery and center. Neuron-like processes have been observed within developing insect appendages prior to the appearance of the first functional sensory cells (Bate 1976a, Sanes & Hildebrand 1976, Edwards 1977a, Edwards & Chen 1979). These processes have cell bodies at the apices of the appendages and extend proximally along the inner surface of the epithelium forming the earliest observable link between center and periphery. These "pioneer" fibers are associated with a row of cells that become glial sheaths; these may precede the outgrowth of the pioneer fibers, and have been referred to as "stepping stones" (Bate 1976a). Do the pioneer cells arise from epidermal cells at the tip of the appendage or do they arise from the central neural ectoderm and migrate outward to the tip? Their cell lineage is not yet known; the cell bodies lie in contact with the inner surface of the epidermis rather than within it as is the case with cell bodies of subsequently formed functional sensilla. It is possible that at the point in embryogenesis when the ventral nerve cord neuroblasts are formed by invagination of midventral epithelial cells, certain related cells are segregated and migrate to the periphery to become part of sensory primordia; a similar suggestion has been made for leeches by Weisblat, Sawyer & Stent (1978). Then the initial central terminations of the first afferent neurons may be determined to a degree by clonal recognition.

During embryogenesis, ganglia are initially part of a continuous chain of neuroblasts and neuroblast progeny. They segregate and move apart only after initial contacts are established, so that the connectives are drawn out between the ganglia (Edwards and Chen, manuscript in preparation). A general feature of the developmental patterns described above is that initial contacts are made over short distances, and usually in a simple direct fashion (Bate 1978). It is only after initial contacts are set up as guides, that subsequent differentiation sets greater distances and sometimes elaborate convolutions between the ends of neurons.

In vitro studies have shown the capacity for outgrowth over longer distances than are initially necessary in vivo (Provine, Seshan & Aloe 1976). The outgrowths were, however, independent of the original axial orientation of the ganglia and while the resulting connections appear to be similar in gross appearance to connectives, their neuronal composition is unknown. The fibers arise from ganglia that have already established connections in vivo, and grow out over distances of about 500 μm, traversing originally homogeneous glass surfaces between the explanted ganglia. The development of sensory nerves within isolated wing and leg imaginal disks of *Drosophila* that metamorphosed while being cultured in vitro show that a connection with the CNS is not necessary for the organization of the nerve bundle (Edwards, Milner & Chen 1978).

What determines the orientation and direction of outgrowth of the first embryonic neurons? The terms used to describe the process are hypotheses in themselves: axonal guidance, pathfinding, navigation. Does the growth cone search? Is it pulled? Does it follow trails? As with animal navigation, multiple processes and diverse cues are surely involved.

The pioneer fibers of arthropods seem to follow specific pathways. In the abdominal cercus of the cricket *Acheta domesticus* they grow along the dorsal and ventral midlines of the cylindrical cercus (which are also the sites of narrow bands of transversely oriented sensilla) and they may play a role in establishing the polarity of the appendage. Specification of the presumptive path of future pioneer axons may occur during the determination of the dorso-ventral axis of the cercal anlagen as they proliferate from the posterior angles of the germ band. The behavior of pioneer fibers resembles the behavior of axons in other distantly related groups. In the tunicate larva, for example, cells of the prospective notochord form aligned embayments along which motor fibers supplying the tail musculature will grow (Cloney 1978). The parallel even extends to the vertebrates, for the first axons to grow into regenerating newt tails seem to follow preformed channels in the ependyma (Nordlander & Singer 1978), and comparable morphogenetic relationships are implicit in the reported role of glial cells in organizing neuronal migration during vertebrate brain development (e.g. Rakic & Sidman 1973).

What determines the direction of growth of pioneer fibers in the embryonic appendage? Insect epidermal cells that have been deprived of their tracheal oxygen supply can form neurite-like processes over 200 μm long, which serve as "grappling hooks" that capture tracheae and pull them toward the deprived area (Wigglesworth 1977). They perhaps respond to an oxygen gradient. This raises the possibility that the growth cones of earliest axons follow linear gradients of substances arising from central target cells and diffusing across distances that fall well within the plausible limits for diffusion phenomena as guidance mechanisms (Crick 1971).

The elements of such a system, a contact pathway and a chemical gradient, essentially combine the concepts of contact guidance and chemospecificity (Hamburger 1962). It is a plausible step to propose that the mid-dorsal and mid-ventral inner margins of the cercal epidermis and the accompanying glial cells provide a pathway of higher adhesivity that enhances the elongation and alignment of the growth cone of the pioneer fibers by a mechanism similar to that shown by Letourneau (1975) in vitro.

Current evidence indicates that all additions to, and the sometimes elaborate changes in, sensory projections that occur during postembryonic development are based on the continuity between center and periphery.

METAMORPHOSIS AND THE NERVOUS SYSTEM

The pattern of growth of the nervous system of hemimetabolous insects (Johannson 1957, Gymer & Edwards 1967, Sbrenna 1971), of a spider (Babu 1975), and presumably of all arthropods with direct development can be simply stated as follows: motoneuron and interneuron populations are established during embryonic development. With the exception of specific centers where cell proliferation persists, postembryonic development is essentially characterized by an increase in cell volume and an increase in the ramification of arborizations. In metamorphosing (holometabolous) insects on the other hand, in which two distinct body forms with different functions (the larva and the adult) are bridged by a metamorphic phase (the pupa), the adult structures are carried through the larval phase as groups of embryonic cells, the imaginal disks, from which the adult sensory systems develop. Neuroblasts retained within the larval central nervous system provide in large part for the adult central nervous system (Edwards 1977b).

The advent of selective filling techniques has made possible a detailed description of neurons that persist during metamorphosis. Taylor & Truman (1974) for the first time established which motoneurons in a larval moth abdominal ganglion die at metamorphosis, which survive, and which adult motoneurons are added. Subsequently, Truman & Reiss (1976) studied the anatomy of particular neurons that first innervate larval muscles and later connect with a new set of adult muscles. They found that the most thoroughly studied cell had a single extensive ganglionic arborization in the larva; this was drastically resorbed during pupation; and it reached new glories in the adult, which, however, it shared with a second, wholly new and even richer arborization on the other side of the ganglionic midline. Pipa (1978) has reviewed many of the cellular phenomena of neural reorganization in insects at metamorposis.

It is not necessary for a whole animal to undergo a profound metamorphosis in order to find profound changes in neuronal structure and function, for White et al (1978) have described remarkable changes that occur in cells belonging to one of the classes of motoneurons (class DD) in the nematode *C. elegans*. These cells are present at the time of hatching, at which time they receive synaptic connections in the dorsal nerve cord from two other classes of motoneurons and form neuromuscular junctions in the ventral nerve cord upon ventral muscle cells. These same cells, by the end of the first larval stage, reverse their polarity, receiving synaptic input from other motoneurons ventrally and forming neuromuscular junctions dorsally.

During this period no obvious metamorphosis of the animal occurs, but many new motoneurons are added from the postembryonic cell divisions described above. In the adult, it is the postembryonically formed cells that provide synaptic input to the DD motoneurons. Mutants are known in which post embryonic cell divisions do not occur and consequently the new synaptic inputs cannot form. In such mutants, the change of polarity of the DD cells occurs anyway: ventral neuromuscular junctions are lost and dorsal ones are formed. This suggests that the change in polarity is not triggered by, nor is it otherwise dependent on, the presence of the postembryonically developed cells. However, some input synapses persist on the ventral side where they would not be found in wild-type animals; possibly the restructuring of the network cannot go to perfection when many network elements are missing.

Postembryonic development, especially when it involves a series of molts accompanied by varying degrees of morphogenetic change, entails specific and highly stereotyped behaviors involving discrete motor patterns that are played out from a central nervous system that is itself undergoing developmental changes. Recent studies of the adult molt of a cricket (Carlson 1977a,b) and of the emergence (eclosion) of a sphinx moth (Truman 1976) show that the motor patterns are composed of discrete and stereotyped bouts of activity separated by periods of quiescence. Sensory feedback (Carlson 1977b) and specific behavior releasing hormones (Truman 1976) are involved in the recruitment and termination of these motor patterns.

Parallels from a distant branch of the invertebrates on the chordate line are afforded by the metamorphosis of tadpole-like tunicate larvae in which a series of coordinated morphogenetic movements rearranges organs, tissues, and cells during conversion of the newly settled larva to a sessile adult (Cloney 1978). During these events larval neural structures, the visceral ganglion and sensory organs of the cerebral vesicle, degenerate and are entirely devoured by phagocytes. The cerebral ganglion and subneural gland of the adult are new structures derived by cellular proliferation from the walls of the so-called larval neurohypophysis, which may be likened to an imaginal disk. The replacement of the nervous system of ascidians thus resembles that of higher holometabolan insects.

Metamorphosis in a more limited context is illustrated by claw regeneration in snapping shrimp. Determination of form in the asymmetric chelipeds of the snapping shrimp *Alpheus* is influenced by cutting the cheliped nerve (Mellon & Stephen 1978). As in other asymmetric Crustacea the loss of the larger of a pair of appendages may result in the transformation of the opposite limb. In *Alpheus* the larger cheliped forms the snapper, the smaller a pincer. Loss of a pincer leads to regeneration of a pincer, loss of both maintains the *status quo,* but loss of the snapper induces transformation of the pincer in the course of two molts to a new snapper, while the old snapper limb regenerates a pincer. When the nerve to the snapper is cut, leaving the appendage intact, the contralateral pincer is transformed during the next two molts to give a symmetrical pair of snappers. There must be ample opportunity to effect at least partial regeneration of sensory and motor contacts in the operated snapper; but such contacts, if they do arise, do not alter the commitment to alter the form and function of the snapper-to-be. The sensory and motor nerves associated with the limb are said not to cross the midline; the morphogenetic response may perhaps be initiated by the altered input of proprioceptors as has been proposed for alteration of the timing of the molting cycle in limb regeneration (Kunkel 1976).

Summary

This section and the previous ones provide glimpses of neuronal development from the egg to the adult, a development characterized by fixed lineage sequences, changes in form (supernumerary neurites being formed and lost, and overall architecture subject to change in some cases), and changes in function (Na^+ to Na^+ and Ca^{2+} dependent spikes, input to output synapses). Different systems are especially suitable for the study of different aspects of neuronal development, but they appear to have many important features in common. Do these features extend to vertebrates? Many do. The review of Spitzer (1979) gives a number of examples from the study of membrane properties. The coming and going of branches in vertebrate neurogenesis has been known at least since Cajal and programmed cell death has also been much studied (e.g. Prestige 1970). The basic events of nerve cell development appear to be just as universal across phyla as are the basic events of membrane function. Variants are found, and differences in emphasis, but the basic principles are the same.

CONSTANCY AND VARIABILITY

One of the most publicized characteristics of many invertebrate nervous systems is the presence of identified neurons, cells that can be recognized from individual to individual in a species on morphological and functional criteria. Many of these cells have been studied intensively because they provide some degree of assurance that the question under study, whatever it may be, is always being examined in the same material and that no sampling problem is confusing the issue.

It is important to know, however, how constant the properties of single cells are (for a brief review of variability in a different context, see Palka 1977a). Morphologically, for example, one might wonder whether cell A and cell B always form the same number of synapses with each other; functionally one might wonder whether the size of the postsynaptic potentials elicited by B in A is constant. There must be some variation even among identified neurons, and we want to know how much, what causes it, and why is there just so much and not more or less.

A classic step towards specifying the amount of variability when genetic factors are controlled was taken by Macagno, LoPresti & Levinthal (1973) in a study on the visual system of *Daphnia*. These planktonic crustaceans reproduce parthenogenetically by a mechanism believed to result in truly isogenic clones. Any variation among corresponding neurons belonging to individuals of such a clone should not be attributable to genetic factors. The conclusion of the study, based on computer-aided reconstructions of serial

EM sections, was minor variation confined to tertiary branches. The relative numbers of synapses between particular pairs of neurons were always the same, though the absolute numbers varied somewhat.

This picture of a high degree of reproducibility of the branching patterns of identified neurons was also reported for locust motoneurons (Burrows 1973), crayfish motor and interneurons (e.g. Selverston & Remler 1972), cockroach motoneurons (Pitman, Tweedle & Cohen 1973), cricket abdominal giant fibers (Mendenhall & Murphey 1974), fly retina-to-lamina projections (Horridge & Meinertzhagen 1970, Meinertzhagen 1973), and other systems. Benjamin (1976) found that soma position in a mollusc, the pond snail *Lymnaea*, could vary widely with no detectable effect on the branching pattern. In nematodes, cells could be assigned to particular classes on the basis of branching patterns (White et al 1976). Even in the interbreeding and genetically heterozygous populations on which most of these and other studies were based, the variability seemed to be very low.

In a series of papers Goodman (1974, 1976, 1977, 1978) has published the most detailed account of substantial neuronal variability, in a study of the morphology of locust ocellar interneurons. This system has the advantage that as many as 17 large and 61 small cells can be individually identified and traced in each animal, and the cobalt filling technique works with high reliability so that many individuals can be sampled. Goodman found that the branching patterns of certain interneurons were markedly different in some animals from those in the majority of the population, and that some neurons could be absent or extra ones present. Locusts can be induced to reproduce parthenogenetically, and it proved to be possible to demonstrate that the differences are inherited. Isogenic clones were produced in which phenotypes that were rare exceptions in the interbreeding population became instead the dominant ones and only particular cells among the 19 best studied large interneurons were affected in each clone. Clearly this constitutes powerful evidence for genetic influences on cell morphology.

Ocellar interneurons of locusts are not unique examples of cell duplication or substantial variability; others include duplications in a leech (Kuffler & Muller 1974) and in *Aplysia* (Treistman & Schwartz 1976); primary afferent neurons taking two alternative routes to the same destination (Altman & Tyrer 1977); substantial anatomical variation in a ganglion whose cells show very consistent functional relationships (Selverston & Mulloney 1974); and many others. In none of these cases, however, have genetic and nongenetic factors been evaluated.

This section, as well as the history of the subject, began with examples of constancy at least to the level of secondary branching. Consider, in contrast, this astonishing example of variability. Pearson & Goodman (1979) have analyzed the branching pattern and the physiological connections formed in the thoracic ganglia of locusts by a much-studied visual interneuron, the DCMD (Descending Contralateral Movement Detector). They found that this cell forms a number of branches in characteristic locations, associated with the locations of its postsynaptic partners. However, the full set of branches is found to be present in only a few individuals (6 out of 26 animals of two related species in their sample). In the majority of individuals one or several, up to one half, of the branches are missing. To quote their summary description: "In both species the striking characteristic of the structure was the variability. In fact, the variability was so great that it is not possible to describe a 'typical,' or 'normal,' pattern of branching."

Thus, even within a single species there exists a range of variability from the minute, limited to the details of tertiary branching, to the enormous, including the presence or absence of major branches, the occurrence of supernumerary cells and the absence of identified cells. What lessons for developmental neurobiology are contained in this data? Thoughtful considerations of this question are to be found in all of the studies that we have cited, and we present here only a few of the ideas that have been proposed.

Goodman has suggested plausible evolutionary explanations both for increase in cell number and for variation. Duplicated cells might well serve the same function as duplicated genes—one of the two cells could be modified for a new role while the other continued to serve the old. Genetic variability at the neuronal level may well have the same significance as at the molecular level—natural selection can only bring new phenotypes to prominence or produce increasingly perfect adaptation to environmental requirements if a population contains heritable genetic variants.

The powerful evidence in both nematodes and insects that each interneuron and motoneuron arises as the product of a fixed lineage implies that supernumerary cells most probably arise from one extra division in the lineage that produces the normal set of cells. If a single cell is missing, the last division in a lineage might have failed to occur. However, since undivided cells in nematodes tend to show the characteristics of both of the daughter cells they should have formed (see below), and thus to fit neither morphological class, it seems more likely that cells in the adult are missing because they die at some early stage.

Goodman (1978) found that in certain isogenic lines of locusts particular ocellar interneurons had a strong tendency to terminate in abnormal areas, but that these areas were consistent from one individual to the next and often included the contralateral homologue of the normal area. To account for this phenomenon, he proposed the existence of "multiple epicenters," areas in the brain supposed to produce a substance which, in low concentration, promotes the growth of particular axons towards it and, in high concentration, arrests growth but promotes branching. The system would have the attribute of influencing the growth of axons at some distance, as well as in the specific area where it is produced. Abnormal branching would result from alterations in the normal balance of production of the hypothetical substance by the several centers that produce it.

At the present time we have several hypotheses to account for abnormalities in neural development but explicit hypotheses for constancy are more difficult to formulate and the virtual perfection of the retina-to-lamina projection of the thousands of receptor cells in a fly's eye is as much of a mystery as ever.

DEVELOPMENTAL INTERACTIONS BETWEEN NEURONS

The harmonious development of the nervous system requires spatial and temporal coordination of the proliferation and differentiation of sensory neurons, interneurons, motoneurons and their targets. Many early studies (reviewed for insects by Edwards 1969, Meinertzhagen 1973, Bate 1978, Palka 1979a) indicate that experimental manipulation of one set of neurons may have considerable effects on another set. In an attempt to unravel as precisely as possible the complex developmental interactions between these neurons or their precursors, we here confine our attention to experiments that have specifically examined particular aspects of neuronal interaction, namely effects upon neuron proliferation, degeneration, differentiation, growth, and physiological maturation.

Autonomy of the Periphery

The insect visual system has been the favored experimental material for examining whether sensory neurons can develop independently of the CNS. Experiments involving transplantation of the retina or its rudiment to sites away from the optic lobe, extirpation of the optic lobe, and preventing or severing connections between the retina and optic lobe, have been performed on a wide variety of species at various developmental stages. In most cases retinal development proceeded normally, the exceptional cases almost certainly resulting from degeneration due to surgical damage. Two recent studies have conclusively shown that the retina can both proliferate and differentiate new ommatidia in the absence of connections with the optic lobe (Mouze 1974, Anderson 1978a). Similarly, sensory receptors on the antenna of the moth *Manduca* continue to be proliferated and to differentiate in the complete absence of the brain (Sanes, Hildebrand & Prescott 1976).

Evidence that the form of the central arborizations of sensory neurons is governed by the nature of the epidermis from which the neurons arise, and not by signals influencing the axons after they leave the epidermis, has been provided in experiments on the locust (Anderson & Bacon 1979). Neurons associated with wind-sensitive hairs on the head of *Schistocerca gregaria* normally form one of two distinct projections in the suboesophageal ganglion according to their location on the head. Neurons developing from epidermis transplanted between locations and entering the CNS via a nerve normally used by neurons forming the alternative projection, nevertheless always exhibit a central arborization appropriate to their epidermis of origin.

Neuron Number: Proliferation and Death

Arthropod sensory receptors are produced throughout embryonic and post-embryonic life by the proliferation and differentiation of epidermal cells. Almost all central neurons by contrast are produced during embryogenesis by the proliferation of sets of stem cells, the neuroblasts, in a stereotyped sequence of asymmetrical and symmetrical mitoses. The neuroblasts degenerate at about the time of hatching. Exceptions are neuroblasts in the association centers of the brain, e.g. the optic lobes and corpora pedunculata, which persist postembryonically and continue to produce neurons (Panov 1960, 1961, Nordlander & Edwards 1969, 1970) and those neuroblasts in the ventral ganglia of holometabolous insects which lie dormant in the larval stages, only producing new neurons during metamorphosis (Heywood 1965, White & Kankel 1978).

The control of neuron proliferation and death has not been studied in embryos or received much attention in the ventral ganglia of holometabolous insects. A single study suggests that peripheral input may regulate central cell numbers. Removal of the mesothoracic leg disc from third instar blowfly larvae effectively deprives the CNS of sensory input from receptors on the imaginal leg and also deprives leg neurons of their muscle targets (Chiarodo 1963). The effect on the thoracic ganglion is a 29% reduction in the number of cells in the cortex. Chiarodo interpreted this loss as due to degeneration of motoneurons whose connections had already been established and were therefore severed at the time of operation. However, there is no evidence that motor axons are present in the stalk of the leg disc at this time; their innervation appears to be composed solely of afferent fibers (Van Ruiten & Sprey 1974). Neuroblasts are known to be present in the ganglion up to the second day of pupariation in *Drosophila* (White & Kankel 1978). It seems much more likely that incoming sensory neurons have a regulatory effect on cell proliferation or death of interneurons within the ganglion, especially since Chiarodo himself reported a loss of small rather than large cell bodies.

The visual system has been relatively thoroughly studied in this context. Indeed, it has been the material for so many investigations in neural development that it seems useful to provide a brief description of the wiring diagram of its most peripheral layers at this point. The many studies on the development of the retina itself are not summarized here; they are discussed in many of the previously listed reviews.

The first-to-second order connections of the visual systems of insects and at least some crustaceans occur in one of two alternative patterns (e.g. Meinertzhagen 1976, Stowe 1977). In species in which the pigment-bearing surfaces of all the receptor cells in one ommatidium, the rhabdomeres, are fused into one central light guiding and trapping structure, the rhabdom, the axons of the retinula cells project to a single cluster of second order cells. The resulting synaptic complex is the optical cartridge. Since light crosses freely from one rhabdomere to another in such a fused rhabdom, all the retinula cells look in one direction in space, and the output cells of the corresponding cartridge look in the same direction. In the alternative case, studied especially well in flies, the rhabdomeres of an ommatidium are separated from each other by a fluid-filled space and each retinula cell looks out through the common ommatidial lens at a slightly different angle. Correspondingly, the axons of the retinula cells project not to a single cartridge but to a small array. Each cartridge receives retinula cell axons from different ommatidia, but these are oriented so that they look in the same direction and again the output of the cartridge represents a single direction in visual space.

A quantitative comparison of neuroblast and ganglion mother cell mitoses in the proliferation zone of the locust optic lobe during normal postembryonic development and during experimental interruption of retinal growth, has shown that new cells continue to be supplied at normal rates to the lamina in the absence of new innervation from the periphery. These neurons fail to differentiate and ultimately degenerate (Anderson 1978a). Even during normal development a small number of degenerating cells may be found among the newly proliferated cells (Nordlander & Edwards 1968, Anderson 1978a, Mouze 1978) and this number may be reduced if the optic lobe is innervated by an experimentally increased number of receptor neurons (Mouze 1978). It thus seems that cells are produced autonomously in the optic lobe and in slight excess of requirements, their exact number being regulated by the number of incoming fibers, those not utilized later degenerating.

Growth of the retina and the lamina could be simply coordinated if they were proliferated from common precursors, as has been suggested for crustaceans by Meinertzhagen (1973) on the basis of the claim by Elofsson & Dahl (1970) that a single proliferative zone gives rise to both arrays. However, the evidence (Elofsson 1969, Dahl 1959) presented in support of this claim is not convincing and it is likely that separate though adjacent proliferation zones exist for the retina and optic lobe as described by earlier authors (Moroff 1912, Peabody 1939, Debaisieux 1944) in a manner exactly analogous to the insect visual system.

Preliminary experiments, in which single cells were ablated by a laser microbeam in the nematode CNS, suggest that regulation of cell number does not occur in this system; no extra divisions occurred to replace the missing cells (Sulston & Horvitz 1977).

Initiation of Neuron Differentiation

Evidence that the differentiation of neurons may be triggered by contact with ingrowing axons has been provided in an elegant description of the developing visual system of the crustacean *Daphnia* (LoPresti, Macagno & Levinthal 1973, 1974). Comparison of three-dimensional reconstructions from serial sections at various stages of development has allowed the determination of the sequence of events occurring in the lamina, after the arrival of the bundles of ommatidial axons from the retina. The first axon of each bundle to reach the lamina is termed the lead fiber and forms an extensive growth cone as it contacts a series of five undifferentiated lamina cells lying at progressively deeper levels. The first two lamina cells to be contacted have been seen to wrap temporarily around the lead fiber and gap junctions are formed with it (LoPresti, Macagno & Levinthal 1974). These could be the site of information transfer for triggering differentiation, for after contact with the lead fiber the lamina cells begin to differentiate, the five sending out their axons in the sequence in which they were contacted. Thus the arrival of each axon bundle results in the formation of a cartridge in the lamina. This view is confirmed by the finding that after destroying some ommatidia before axon outgrowth, the resulting adult animals always show the same reduction in the number of cartridges as in the number of axon bundles (Macagno 1977, 1979).

Many descriptive studies of the insect visual system have shown that axon bundles from the retina arrive at the area of lamina cell proliferation, and there form expanded growth cones (Meinertzhagen 1975). However, the cells in this area densely intermingle with one another and it has proved impossible to reconstruct the area of initial contact between retina bundles and lamina cells (Hanson, unpublished work, reported in Kankel et al 1979) as was done for *Daphnia*.

In flies, retinula cell axons from a single ommatidium initially project to a single cartridge, as is the case for insects with fused rhabdomeres. Only later do they undergo a second phase of axonal growth in which the axons diverge laterally and form their final terminations in separate surrounding cartridges (Trujillo-Cenóz & Melamed 1973, Meinertzhagen 1973). At the ultimate edge of the lamina there are diverging axons without adjacent cartridges to terminate in. These fibers become associated with extra cartridges that develop later than the others (Hanson, unpublished work, reported in Kankel et al 1979). Thus in flies there are more cartridges than there are ommatidia, leading one to consider the existence of a secondary triggering effect upon undifferentiated lamina cells by axons at the diverging axon stage of development.

An EM study of an identified, undifferentiated cell in the abdominal ganglion of *Aplysia* has shown that its first morphological signs of differentiation and axon formation occur after the appearance, at a particular developmental stage, of a synaptic contact on a specific region of the cell body by the axon of a neuron from another ganglion (Schacher, Kandel & Woolley 1979). Process initiation always occurs adjacent to this synapse, which later disappears.

Growth and Maturation of Neurons

After initiation of their differentiation, neurons develop their characteristic patterns of dendritic and axonal arborizations, form appropriate synapses, and acquire their mature physiological properties. These events may be completed fairly rapidly as in the case of the arrays of cartridges in the lamina of the visual system where each cartridge may achieve its final pattern of connections within a few hours (Meinertzhagen 1975). The unique neurons of the CNS however may continue to develop throughout the juvenile stages (e.g. Altman & Tyrer 1974 for motor and sensory neurons; Murphey 1978 for interneurons) or even undergo quite spectacular rearrangements of their processes at a late stage in their development (see above; Truman & Reiss 1976).

The exquisite precision of the retinotopic projection from the retina to the lamina of the visual system has stimulated the investigation of the formation of specific neuronal connections. How do retinula cells form connections with appropriate cartridges? Experiments on fused-rhabdom eyes involving partial removal of retinula input to the lamina (Anderson 1978b, Macagno 1978), inversion of all or part of the retina with respect to the lamina, and transposition of older and younger retina with respect to the lamina (Anderson 1978b), all point to a lack of selectivity by retinula axons for particular cartridges. Rather it seems that the most important factors underlying the establishment of the orderly projection are the well-defined temporal sequence of development in the retina and the guidance of growing axons from the retina to the lamina by tracking along pre-existing axon pathways; neurons growing out from the retina in an orderly sequence are thus delivered to the lamina in an orderly sequence where triggering interactions with immature lamina cells result in an orderly sequence of cartridge production.

Evidence that the organization of the lamina into a well-defined pattern of cartridges does indeed depend upon information provided by the retina has been obtained in *Drosophila* (Meyerowitz & Kankel 1978). Flies were generated that were mosaics of wild-type tissue and tissue of a mutant genotype known to cause abnormal ommatidial arrays in the retina and abnormal patterns of neurons in the lamina. Where patches of mutant retina projected to wild-type lamina, the lamina was profoundly disorganized. Conversely, where patches of wild-type retina projected to a mutant lamina, that area of the lamina was normal in appearance, while the remainder was disorganized. The nature of the information provided by the retina is at present unknown; the intriguing possibility remains that in the mutant lamina disorganization results from a failure of the mechanisms that normally deliver axons to the lamina in an orderly spatio-temporal sequence.

The initial projection of open-rhabdom eyes is like that of fused-rhabdom eyes and probably results from the same mechanisms. The subsequent differential lateral growth of fly retinula axons to produce the final asymmetrical pattern of connections with several cartridges is unlikely to be a simple consequence of fiber following, since occasional errors in the final termination site of retinula axons are not propagated through the remainder of the projection (Meinertzhagen 1972). Various mechanisms have been proposed, ranging from simple mechanical models to complex models invoking intrinsic position-specific programs for the directed outgrowth of individual axons along gradients (see discussions in Meinertzhagen 1973, 1975), but a definitive answer remains to be found.

Interweaving patterns of retinula axons beneath the retinae of several decapod crustacea have suggested the existence of projection patterns similar to those of flies although these eyes are of the fused-rhabdom type (Hámori & Horridge 1966a,b, Meyer-Rochow 1975, Nässel 1976). Careful examination of serial sections in the crab, *Leptograpsus variegatus,* has shown that the axons from one ommatidium pass through one of four holes in the basement membrane at the corners of the ommatidium and that they share these holes with neighboring ommatidia (Stowe 1977). Axons from several ommatidia that exit through the same hole may subsequently travel together for some distance below the basement membrane, giving the appearance of an open-rhabdom type of projection. However, more proximally, the fibers regroup so that all eight axons from one ommatidium project to a single cartridge (Stowe 1977). It is possible that this also occurs in the other decapod Crustacea studied. The observation that axons may not only have different neighbors at different levels of the projection, but also accurately regroup with axons from their own ommatidium, suggests that factors other than fiber following are involved in the production of a retinotopically ordered projection in this species. This certainly merits further investigation.

Cartridge patterns produced in the lamina after perturbation of the retina have also been studied in *Daphnia* (Macagno 1977, 1979). After the differentiating retina has been subjected to ultraviolet radiation, some ommatidia degenerate completely, resulting in loss of a complete bundle, while others degenerate partially to produce incomplete bundles, and in some cases axon bundles join together to form abnormally large ones. Examination of cartridges associated with abnormal bundles revealed some interesting anomalies. Partial bundles may be associated with cartridges containing the normal complement of five lamina cells, showing that a complete bundle is not essential for normal cartridge formation. However some partial bundles are associated with fewer than five lamina cells. One possible explanation for this phenomenon is that UV treatment delays receptor maturation. If undifferentiated lamina cells have limited life spans as has been suggested for the insect optic lobe (Anderson 1978a), then fewer lamina cells might be available for triggering by the time the delayed receptor axons arrived. Alternatively radiation could affect triggering ability.

Bundles containing more than the normal eight axons are associated with cartridges containing only the usual five lamina cells, but these cartridges are invaded by processes from the lamina cells of adjacent cartridges almost as if to compensate for the increase in the ratio of axons to lamina cells (Macagno 1977, 1979). Similarly in one case where a bundle of five axons was not associated with any lamina cells of its own, the bundle was invaded by processes extending from a neighboring cartridge on the other side of the midline and synapses were formed between them. Clearly, cartridge formation is a complex interactive process and further clues about the interactions involved will undoubtedly be found when these bizarre situations are examined while they are developing rather than at the adult stage.

The adult antennal lobes of holometabolous insects are largely or entirely composed of cells derived from a set of postembryonic neuroblasts (Panov 1961, Nordlander & Edwards 1970) and develop a characteristic arrangement of the neuropil into discrete glomeruli where synapses occur between chemosensory afferents and interneurons. After removal of the antennae from pupal stages in *Tenebrio,* distinct glomeruli failed to develop (Panov 1961). This is also the case in *Manduca,* even though bungarotoxin binding acetyl choline receptors develop normally both in quantity and in location (Hildebrand, Hall & Osmond 1979). This has also been shown to be the case for deafferented optic lobes at the level of the medulla and lobula (Maxwell & Hildebrand 1978).

Removal of excitatory inputs to identified motoneurons in juvenile crayfish failed to produce any detectable changes in their morphology (Wine 1973). Removal of both target swimmeret muscles and of afferent input from swimmerets failed to have any effect upon the acquisition of a normally patterned motor output by swimmeret motoneurons in the developing lobster (Davis & Davis 1973), suggesting that at the physiological level their central connections had developed quite normally. In the locust, removal of input from the wing sense organs to the flight motoneurons in young adults when the motor neurons have not yet acquired their mature firing pattern, also failed to have any effect on their development (Kutsch 1974). While these studies have emphasized the insensitivity to perturbation of the physiological and morphological maturation of central neurons at late stages in their development, early stages are much more susceptible to change.

The identified neurons that are known best from a developmental viewpoint are the two largest abdominal giant interneurons of crickets, the lateral giant interneuron (LGI) and the medial giant interneuron (MGI). Their electrophysiological responses may be recorded both extracellularly and intracellularly, and it has been shown that their major source of input is from sensory hairs on the two abdominal appendages, the cerci (Palka, Levine & Schubiger 1977, Murphey, Palka & Hustert 1977, Palka & Olberg 1977, Matsumoto & Murphey 1977).

The arborizations of the LGI and MGI within the abdominal ganglion grow throughout postembryonic development by the increase in length and diameter of dendrites and addition of dendritic spines (Murphey 1978). Deafferentation during development has no effect on the general morphology of the neurons or upon the density of dendritic spines (Murphey & Levine 1979) but does reduce the overall growth of dendrites and dendritic spines (Murphey et al 1975, Murphey & Levine 1979). These deafferentation effects are dendrite-specific, unilateral deafferentation affecting only dendrites on the deprived side of the ganglion (Murphey et al 1975). They may be reversed by allowing the cerci to regenerate, more complete recovery being obtained with longer times of regeneration (Murphey, Matsumoto & Mendenhall 1976). Sensory deprivation of the neurons, by blocking the movement of cercal hairs and thereby preventing their neural activity, has no effect upon their morphology (Murphey 1977). In contrast, sensory deprivation produced in mutants whose cercal hairs lack shafts, and are therefore electrically silent, results in reduced diameter and length of dendrites (Bentley 1975) but in this study possible pleiotropic effects of the mutation on the structure of the CNS have not been eliminated.

Both deafferentation and sensory deprivation have dramatic effects upon the physiology of the giant interneurons. Ipsilateral deafferentation of the MGI results in enhancement of the normally weak contralateral excitation (Palka & Edwards 1974), and increased sensitivity to substrate vibrations mediated by noncercal receptors (Palka & Edwards 1974). Detailed studies of the mechanisms of these effects are now being conducted (Murphey, Matsumoto & Mendenhall 1976, Murphey & Levine 1979). Unlike the morphological responses, these altered physiological response properties of the interneurons never completely recover regardless of how short the period of deafferentation or how long the period of recovery (Murphey, Matsumoto & Mendenhall 1976).

Ipsilateral sensory deprivation of the MGI by immobilization of the sensory hairs also leads to enhancement of contralateral inhibition and reduces the effectiveness of ipsilateral excitation (Matsumoto & Murphey 1977). Sensory deprivation is only effective when applied early in postembryonic development (Matsumoto & Murphey 1977) and some recovery may be obtained when normal sensory input is restored (Matsumoto & Murphey 1978).

Further evidence for experience-dependent properties of the giant interneurons has been obtained by rearing crickets in the presence of continuously repeating tone pulses, i.e. continuous cercal stimulation. The interneurons of treated animals were reported to be much more resistant to habituation than those of control animals (Murphey & Matsumoto 1976).

The cercus-to-giant-interneuron system of crickets thus demonstrates morphological and physiological plasticity in postembryonic development, the two being at least partially independent.

An even more dramatic effect of deafferentation upon interneuron morphology has been reported recently for the auditory system of crickets (Hoy, Casaday & Rollins 1978). Interneuron-1, an identified auditory interneuron in the prothoracic ganglion (Casaday & Hoy 1977), was deprived of its normal auditory input by amputating the ipsilateral foreleg immediately after hatching and before the ear and its associated neurons have developed on the tibia (Ball & Young 1974). The interneuron in the adult showed a reduction in the diameter and/or length of its lateral dendrite, which would normally lie among the auditory afferents. This result is comparable to that obtained in similar cercal-to-giant-interneuron experiments. However, its medial dendrites, which would normally ramify in the ventral acoustic neuropil of the ipsilateral side, now crossed the midline and terminated in the contralateral, unoperated, ventral auditory neuropil. This interneuron is accessible electrophysiologically (Casaday & Hoy 1977) and we may anticipate some interesting findings in the future.

MUTANTS AND THE ANALYSIS OF NEURAL DEVELOPMENT

Many examples throughout this review emphasize the importance of bringing the genetics of experimental animals under control.

Two species of invertebrates have proved to be especially useful in the analysis of neural development using genetic tools: the nematode *Caenorhabditis elegans* and the fruitfly *Drosophila melanogaster*. In other species genetic effects have been demonstrated and isogenic strains produced, but the stocks for extensive genetic manipulations or even the examination of a large variety of defined mutants are not available. A convenient listing of invertebrate mutants is given in King (1975).

One important feature of mutants must be stressed: in a mutant animal every cell is mutant—presynaptic, postsynaptic, nerve, muscle, etc—and one can never assume that a particular effect is due to the expression of the mutation in only a single class of cells. Mosaic animals in which some cells are mutant and others are wild type can be produced by a variety of procedures, especially in *Drosophila* (Hall, Gelbart & Kankel 1976), and this often allows clean experimental designs (for some examples see Palka 1979b). Occasionally special features of a particular mutant may provide alternative ways of controlling for the potentially confounding multiple effects of single mutations, but the problem is ever present and must be explicitly dealt with in every study in which the interaction of cells is at issue.

With this caveat, let us consider some inferences that can be made about the formation of neural pathways from a study of mutants. Albertson et al (1978) have studied mutants of *C. elegans* whose neuron precursors, the blast cells, do not undergo the usual series of cell divisions. Even though they fail to divide, they show approximately the increase in DNA content within their single nucleus that would be expected for producing their progeny by means of normal cell division. The cell shape assumes some of the characteristics expected of the progeny. Hence we infer that (*a*) the proper number of DNA replications and (*b*) expression of the part of the genome that influences cell shape, can both continue even if cytokinesis does not take place.

The principal studies on mutants affecting neural connectivity have been done in *Drosophila* (see the detailed review of Hall & Greenspan 1980). We illustrate them here with current analyses of homeotic mutants. Homeosis is the transformation of one body part into another. Thus in the mutant *Antennapedia* mesothoracic legs form in the place of antennae (e.g. Post-lethwait & Schneiderman 1971). No mosaic experiments have been published yet on this system, but histological preparations give no indication that the brain, especially its characteristically structured antennal lobes, are affected by the mutation. What of the axons of the sensory cells of the homeotic legs, which grow into the brain rather than their normal destination, the mesothoracic neuromere of the fused thoracico-abdominal nerve mass? Will they terminate in the antennal projection areas, will they grow through the brain down into the thorax, or will they wander indiscriminately? The vast majority of homeotic leg axons terminate in the apparently normal antennal centers (Stocker et al 1976). Their endings within these centers are distributed differently than are those of normal antennal axons in wild type flies, and their pattern shows at least some attributes of a leg projection, notably that it is ipsilateral whereas the projection of each antenna is bilateral. The antennae and legs are believed to be serially homologous structures (Postlethwait & Schneiderman 1971) and perhaps that is why they reach areas appropriate to antennal axons. The altered distribution of endings in the antennal lobe suggests that within a given territory of the CNS, the different developmental programs of leg and antennal sensory neurons are being expressed.

Deak (1976) and Stocker (1977) have shown that stimulation of contact chemoreceptors on the homeotic legs elicits the proboscis eversion reflex, a behavior normally elicited by stimulation of thoracic legs but not of antennae. Therefore at least a few of the translocated receptors regularly form functional contacts with neurons in this reflex pathway. The anatomy of neurons producing the reflex is not known, but one possible point of convergence of receptor axons from thoracic and homeotic legs is in the antennal lobes, since Strausfeld (1976) and Ghysen (personal communication) have both reported that in normal flies a few chemoreceptor axons from the legs travel all the way to the antennal lobes. However, stimulation of a single sensory bristle is sufficient to elicit this reflex in other flies (Dethier 1955), so it is impossible to infer from behavior what proportions of "right" and "wrong" connections are made by the homeotic axons; all we know is that in most mutants at least one axon makes functional connections with the reflex pathway.

A case studied in more detail with anatomical methods is a series of closely related mutants affecting the anatomy of the wings on the mesothorax and halteres on the metathorax of *Drosophila* (Palka 1977b, Ghysen 1978, Palka, Lawrence & Hart 1979). A variety of receptor types are found on these appendages, some types being present on both and others only on the wing. Each receptor type on each appendage appears to project to the CNS in a characteristic way, although not all types have been analyzed.

In mutants that form wings instead of halteres (and thus have four wings and no halteres), three classes of receptors respond in three ways to what is effectively a translocation from mesothorax to metathorax of the peripheral wing receptor cell bodies: 1. Certain axons from the posterior (homeotic) wings appear to terminate in the same locations in all three thoracic neuromeres of the CNS as the equivalent axons from the normal wings do, i.e. they overlap the corresponding normal wing projection. 2. Another class of axons forms a duplicate projection pattern in the third neuromere, which resembles that formed by the equivalent normal axons in the second neuromere. 3. A third class from the homeotic wing seems to follow the distribution that would have been followed by the haltere axons that they replaced.

A difference in emphasis in the interpretation of these findings has arisen. Ghysen (1978), focusing on the class 1 axons, has suggested that the CNS contains specific trail markers that the incoming axons follow regardless of the nerve root through which they enter. Since the wing and haltere nerves enter almost at opposite extremes of the fused ganglion, the markers must be readable by the axons when approached from either direction.

Palka, Lawrence & Hart (1979), impressed by the different behaviors of the three classes of axons, emphasize the interplay between ganglionic "substrate" factors and the developmental programs of the receptor cells. Different classes of receptor cells are presumed to acquire different search or interaction programs with the hypothetical ganglionic factors or markers as part of their overall cellular differentiation, so that no single statement such as "axons entering the CNS via an abnormal nerve seek out their normal destinations" applies to all axons. Class 2 homeotic axons, which form a duplicate rather than an overlapping projection with their equivalents from the normal wing, illustrate this point, as do the homeotic leg axons terminating in antennal centers in *Antennapedia* (Stocker et al 1976).

Only one mosaic study on the neurobiology of homeotic flies has been published (Palka, Lawrence & Hart 1979). In this study several lines of evidence counter the assumption of Ghysen (1978) that homeotic mutations affecting the haltere segment leave the CNS unaffected. Projections from clones of mutant tissue in flies, which were otherwise wild type, were compared with projections from homeotic wings in completely mutant flies. They were found to be similar but not identical, suggesting that the mutation not only affects the haltere but also affects the CNS in some as yet unspecified way. This result emphasizes the need for mosaic experiments in studies on the formation of central pathways. For example, the findings of Goodman (1978) on aberrant paths taken by interneurons in certain isogenic strains of locusts were interpreted in terms of changes in the CNS (an alteration in the effectiveness of "epicenters"). However, they could equally well indicate changes in the search programs of the interneurons.

One attempt to address the question of what might constitute the ganglionic "markers" to which incoming axons respond is based on the following argument (Schubiger & Palka 1979 and in preparation): The class 3 axons from homeotic wings follow the path of haltere axons that are no longer present rather than the path of axons from morphologically similar receptors on the wing. Suppose that the recognition of ganglionic markers is accomplished by a process that uses them up or physically occupies them. Then the class 3 axons may follow haltere markers rather than wing markers simply because the latter have been pre-empted by normal wing axons that reached them first or had a greater affinity for them. Eliminating normal wing axons might make the markers available for use or occupancy by homeotic axons. However, elimination of the normal wings of homeotic flies by the mutation wingless, or by surgical removal of the part of the wing imaginal disc destined to form the wing blade, does not deflect homeotic axons onto the wing axon path; rather they continue to follow the haltere path. The hypothesis of pre-empted markers, or competition for a fixed set of markers, thus seems unlikely to account for this situation. It should be pointed out, however, that no adequate mosaic experiment has been done for this class of axons.

Mutants offer the advantage of regularly producing changes in the cell complements and the wiring diagrams of nervous systems without the need for surgical intervention and its accompanying trauma, degeneration, and regeneration. However, considerable caution must be exercised in interpreting these effects, especially if mosaic experiments have not been done.

RESPONSES TO INJURY

The responses of invertebrate neurons to injury are examined here from three main points of view: the fate of severed distal segments, the reaction of glia, and the capacity to achieve specific reinnervation.

Fate of Distal Segments

Distal segments of certain invertebrate motoneurons and interneurons persist for long periods after severance from their cell bodies, a capacity that provides perhaps the most striking contrast between the biology of vertebrate and invertebrate axons. The persistence of distal neuronal segments, often with structural and functional integrity, has attracted attention in Crustacea (Hoy, Bittner & Kennedy 1967, Hoy 1969, Nordlander & Singer 1972, 1973a, Wine 1973, Atwood, Govind & Bittner 1973, Bittner, Ballinger & Larimer 1974, Bittner & Johnson 1974, Bittner & Mann 1976, Bittner & Nitzberg 1975, Krasne & Lee 1977a,b, Meyer & Bittner 1978a,b), insects (Clark 1976a, b, Pipa 1978), annelids (Frank, Jansen & Rinvik 1975, Van Essen & Jansen 1976, Nicholls, Wallace & Adal 1977, Birse & Bittner 1976, 1978), and cephalopods (Wilson 1960).

Survival of distal segments is however not a general phenomenon in these groups. Sensory axons generally degenerate rapidly after they have been severed from the cell body (Lamparter, Akert & Sandri 1967, Boeckh, Sandri & Akert 1970, Pareto 1972, Nordlander & Singer 1973a, Stocker et al 1976). The degeneration of severed sensory axons appears to occur more rapidly in insects than in crustaceans. While cut cricket sensory axons show first signs of degeneration within 4–6 hours, with the aggregation of osmiophilic protein masses at terminals, crustacean sensory axons conduct action potentials for 5–10 days after section (Bittner & Johnson 1974) and can synthesize acetylcholine for up to 14 days (Barker et al 1972). In some cases degeneration of motor and interneurons may be as rapid as that of sensory axons. An explanation of the diversity of response may lie in the evolutionary adaptive strategies of different animals to injury (Bittner 1973).

Some illustrative examples of persistent distal neuron segments follow. Motor axons of the crayfish (*Procambarus clarkii*) claw opener muscle severed in the periphery persist for over 200 days (Hoy, Bittner & Kennedy 1967, Hoy 1969), but for shorter periods at higher ambient temperatures (Norlander & Singer 1973b). Other motor axons in the crayfish, such as those of tonic abdominal flexor neurons, lose severed distal segments somewhat faster, but can remain functional for more than 45 days (Hoy 1969). Distal segments of the medial giant interneurons of crayfish degenerate very slowly and remain competent to conduct impulses and to show synaptic transmission for up to 150 days (Bittner, Ballinger & Larimer 1974, Wine 1973). The diversity of response to injury is emphasized by Meyer & Bittner (1978a,b) who double cut the medial and lateral giant neurons of crayfish, thus isolating segments from perikaryal and from postsynaptic contact. Isolated medial giant axon segments persisted for at least 40 days while those of lateral giant axons degenerated within 7 days. In locusts, distal segments of severed motor axons may remain physiologically active for over 30 days (Rees & Usherwood 1972) and interneurons can persist for at least 10 days (Melamed & Trujillo-Cenóz 1962, Rowell & Dorey 1967, Boulton 1969, Boulton & Rowell 1969).

In the cricket *Teleogryllus oceanicus* axons and arborizations of a thoracic leg motor nerve persist for over 150 days after severance from the soma (Clark, 1976a,b) provided the cut is made within the neuropil. The major arborization of this neuron is contralateral to the soma, and the distal segment retains extensive connections with its presynaptic neighbors and with its glial sheath. The normal pattern of arborization may be preserved for as long as 50 days, after which a slow degeneration proceeds from smaller to larger branches. Some supernumerary sprouts or branches may appear during the first 50 days after the operation. It seems, thus, that the cell can even undergo limited morphogenetic activity in the absence of the nucleus and perikaryal apparatus.

The response to loss of the cell body in this cricket motoneuron is paralleled by that of a sensory interneuron, interneuron A of the crayfish abdominal cord, whose cell body is so positioned in the terminal ganglion that it can be removed (Krasne & Lee 1977a). Postsynaptic reaction to synaptic input from ipsilateral tactile afferents of the tail persisted in more than 50% of operated animals for over 50 days. This neuron, deprived of its own soma, did not establish a substitute connection with any other soma, and was even able to function as a target for regenerating sensory axons (Krasne & Lee 1977b).

In contrast to the longevity of neurons that retain extensive arborizations in the neuropil after severance from the cell body, distal segments of motor axons in the insect do not survive long. The persistence of motor axons in *T. oceanicus* mentioned above (Clark, 1976a,b) requires the central arborization, for isolated distal segments of the same neuron degenerate promptly. Similarly in the cockroach *Periplaneta americana* distal segments of motor nerves degenerate within a few days of isolation from the ganglion (Bodenstein 1957, Guthrie 1967, Jacklet & Cohen 1967) and axons of cockroach giant interneurons degenerate within a few days of isolation from their cell bodies and arborizations (Hess 1960). Ultrastructural manifestations of degeneration are evident within two days of cutting motor nerves that innervate the anlagen of developing dorso-longitudinal flight muscle in the giant silk moth *Antheraea polyphemus* (Nüesch & Stocker 1975).

In annelids, also, distal segments may survive for remarkably long periods. Those of giant axons in the earthworm *Lumbricus terrestris* survive and make specific reconnections, even when several of the ganglia they traverse are removed (Birse & Bittner 1976, 1978). The new connections appear to take the form of electrotonic junctions rather than complete cell fusions. Isolated segments of interneurons in a leech survive after surgical disruption and reestablish specific connections as described below (Frank, Jansen & Rinvik 1975, Carbonetto & Muller 1977).

Mechanisms of Survival

Attention to mechanisms of persistence of distal segments of arthropod neurons flared into prominence with the provocative report by Hoy, Bittner & Kennedy (1967) that proximal segments can fuse face to face with distal segments of the claw opener neuron in the crayfish *Procambarus clarkii*. The strongest evidence for this claim, which is reviewed by Bittner (1977), rests primarily on electrophysiological data. Nordlander & Singer (1972) provided ultrastructural evidence for maintenance of the distal segment through newly formed cell contacts between neurons and glia and among neurons at the wound site, where prolific sprouting occurs from the proximal stump. It may be that electrotonic junctions between the sprouts from the proximal stump and the distal segment maintain its excitor function until a process from the stump establishes functional contact with the muscle. The evidence for this mechanism of regeneration in the crayfish is not strong, but the recent report of interneuron regeneration in a leech (Carbonetto & Muller 1977) describes just such a process. The longitudinally oriented S-cell interneuron of each segment communicates at its anterior and posterior ends with its counterpart in the adjacent segments through electrical synapses. When the axon of the S-cell interneuron is cut it reestablishes electrical continuity (Frank, Jansen & Rinvik 1975) by means of sprouts from the proximal stump, which establish electrical synapses with the distal segment (Carbonetto & Muller 1977). Subsequent growth from the proximal stump finally leads to direct contact with the neighboring S-cell axon, thus bypassing the old distal segment, which has survived about 4 weeks at 16°C. Fernández & Fernandez (1974) reported the proliferation of synaptic junctional specializations at sites of lesions in the leech CNS. In both the leech and the crayfish it seems that junctional fields formed in response to wounding may provide new pathways for interchange of materials as well as ample opportunity for proximal and distal segments of neurons to rediscover each other.

There is much evidence for the view that glial cells play an important role in the maintenance of isolated axon segments. In the crayfish the survival of distal medial giant axon segments was correlated with swelling of the glial sheath (Meyer & Bittner 1978a,b). That is not the whole story, however, for the more rapid degeneration of isolated segments produced by double cuts was taken to imply the need for additional trophic support from pre- or postsynaptic sources. Trophic relations of the medial giant axon segments were explored further by means of autoradiographic studies of tritiated leucine incorporation (Meyer & Bittner 1978b). Protein synthesis in glial sheaths surrounding severed medial giants was twice that of intact control axons, and the grain density over the axoplasm of severed MGA's was four times that of control axons. The long-term survival of severed MGA's thus seems to be based on the transfer of materials from the glial sheath to the axoplasm. Transaxonal transport has been demonstrated from glia to the giant axons of a squid (Lasek, Gainer & Przybylski 1974) while the finding of Nordlander, Masnyi & Singer (1975) that raised levels of protein are transported into severed crustacean axons in comparison with control axons implies the modulation of this process according to physiological state. Freeze fracture studies have revealed transglial channels in crustacean nerve roots (Shivers & Brightman 1976) through which rapid translocation of tracer protein can occur (Shivers 1975). These results, together with the demonstration of junctional changes during degeneration (Shivers 1977) and regeneration (Shivers & Brightman 1977) point to the need for a fuller exploration of the modulation of cell communication during degeneration, distal segment survival, and regeneration.

The capacity for prolonged independent existence of isolated axon segments raises many unanswered questions as to their metabolic competence. Data for the stability of mRNA in an insect (Kafatos & Reich 1968) suggest that the longer axonal survival times must require either new synthesis or retardation of turnover. Krasne & Lee (1977a) have marshalled evidence from vertebrate neurochemistry pointing to the capacity of neurons to modulate synthetic activities and suggest that such mechanisms may come into play in surviving somaless neurons. Whatever the mechanisms, it seems from the wide range of survival times encountered in different species, as well as within a single individual, that the characteristics of each neuron must be seen in an evolutionary adaptive context, with its response attuned to the vulnerability and regenerative program of the particular limb or its various neuromuscular components.

REGENERATION AND NEURAL SPECIFICITY

The capacity of severed neurons to regenerate functional connections with their normal targets or with regenerated body parts has been abundantly demonstrated in most major phyla, and while it tends to be more pervasive in simpler animals it is not uniform within a given grade of organization. Invertebrate sensory neurons arising *de novo* during regeneration form functional connections, as for example in the regeneration of arthropod limbs and annelid segments. Experimental studies demonstrating the specificity of neuronal connections formed during regeneration have emphasized the positive finding that functional connections are restored with remarkable fidelity. However, the number of regenerate neurons that may terminate in functionally irrelevant connections or that fail to connect has remained an open question in most specificity studies because of limitations in anatomical and physiological technique. Lack of sufficiently intimate knowledge of normal circuitry and the difficulty of tracing fibers with adequate resolution have contributed significantly to this problem. Further, estimates of the degree of restoration after regeneration are normally made after the process is complete, with regeneration a *fait accompli,* and seldom take into account the earliest projections of axons that might be provisional or evanescent. With the advent of new methods for axon tracing it should be possible to sharpen up studies of specificity, especially in invertebrate systems where the target neurons are uniquely identifiable. Aspects of neural regeneration in insects have been reviewed in some detail (Edwards & Palka 1976, Nüesch 1968) while properties of crustacean neuromuscular specificity are considered by Bittner (1973) and Govind (1980).

At the tissue level it is a long-held generalization that simpler animals can make more extensive restitution of body parts, including the nervous system. In the coelenterate *Hydra,* a representative of the simplest level of metazoan organization, evidence has accumulated over the years that cells of the nerve net play a decisive role in the regulation of morphogenesis and regeneration. Campbell (1979) reviews reports that chemical destruction of the nervous system eliminates the capacity to regenerate, and Browne & Davis (1978) have most recently presented evidence for a morphogenetic role of the nerve net. These conclusions must be assessed in the light of the demonstration by Campbell and colleagues (reviewed in Campbell 1979) that nerve free *Hydra* produced by colchicine treatment retain the capacity to regulate all aspects of morphogenetic pattern formation. Indeed they can grow and proliferate indefinitely. Reintroduction of interstitial cells by grafting leads to complete restitution of the nerve net, but is not necessary for regeneration of the body.

The capacity of flatworms such as *Planaria* to replace entire ganglia during the regeneration of anterior ends is well known, but is not necessarily general among Platyhelminthes, for polyclads undergo only limited regeneration and require the presence of the brain for repair (Olmsted 1922, Koopowitz, personal communication). Koopowitz, Silver & Rose (1975) obtained evidence that the polyclad *Leptoplana* can establish functional connections between severed parts of nerve trunks, and that alternative pathways, perhaps already existing in the nerve net, may also be recruited following lesions.

In annelids, too, the capacity to regenerate cerebral ganglia and restore behavioral capacities is variable. It has long been known that earthworms replace lost anterior segments (e.g. Herlant-Meewis & Deligne 1965) and regain behavioral responses even though the restitution of brain structure is far from complete. Several species of leech (*Hirudo medicinalis, Haemopus grande,* and *Macrobdella* sp) proved unable to replace CNS cell bodies after varying degrees of ablation, although one species (*Clymnella torquata*) evidently could (Hulsebosch & Bittner 1978).

Some characteristics of neural regeneration in molluscs are exemplified by several recent studies. The pulmonate mollusc *Melampus* regenerates cerebral commisures. After removal of one cerebral ganglion, the remaining segment of the cerebral commissure grows toward the severed end of the labial nerve, which appears to act as a focal area for regrowth of a node that is described from light microscope observations as containing neuropil and some cell bodies of unknown origin (Price 1977). The long known regeneration of gastropod tentacles has recently been shown to require an intact cerebral ganglion. The brain appears to be necessary for regeneration of eye stalk neural components, but not the epithelium, in the land snail *Helix aspersa* (Fernandez & Flores 1978). After excision of the protocerebrum together with the tentacular and optic nerve on one side, the tentacles were amputated on both sides. After 30–40 days a complete epithelium was regenerated on both sides, but no neural structures were found on the side of the brain lesion. On the control side a tentacular ganglion, eye pigmentation, and photoreceptors were present.

Specificity of regeneration at the cell population level has been demonstrated in some Mollusca, notably in the octopus where the capacity to regulate skin color and texture is recovered after denervation (Sanders &

Young 1974). Stereotyped color changes in *Octopus* (Packard & Sanders 1971) are effected by means of differential relaxation and contraction of minute muscles attached to chromatophores. Motor control of the chromatophore system is mediated by neurons with cell bodies in the chromatophore lobe of the brain that reach the skin via the paired pallial nerves. Regional color patterns in the skin are restored 40–60 days after denervation by outgrowth from cut or crushed stumps, with better regeneration from the latter treatment.

The capacity for vigorous replacement of appendages and other body parts by arthropods has provided students of regeneration with an instructive range of experimental systems. Motor and sensory appendages of decapod crustaceans and of orthopteroid insects have been most used in the numerous studies reviewed by Nüesch (1968), Bittner (1973), Edwards & Palka (1976), and Pipa (1978). Our objectives here are to sample some more recent examples of experimental studies that contribute to a more complete picture of the capacity for neuronal specificity in regeneration.

Sensory regeneration of abdominal cerci of the house cricket brings neurons differentiated *de novo* in the integument to the CNS where appropriate terminations are made with identified giant interneurons even after prolonged periods of postembryonic development during which connections are absent (Edwards & Palka 1976).

Some examples of the response of regenerating neurons to surgical alterations of the spatial relationships between center and periphery provide a complementary approach to questions currently being posed with homeotic mutants (see section on mutants). Sensory neurons that normally make ipsilateral central contacts recognize the contralateral counterparts when left and right abdominal cerci of *Acheta* are transposed (Palka & Schubiger 1975). Motoneurons in cockroach thoracic ganglia that had been rotated through 180° in the longitudinal axis after cutting of all roots, connected appropriately with muscles of the limb on the side opposite to their normal target (Bate 1976b). Detailed knowlege of the neurophysiology and anatomy of the cockroach leg enabled Pearson & Bradley (1972) to show that motor regeneration is very highly specific when regeneration occurs at the normal site. Specific recognition of appropriate muscles by serially homologous neurons is possible when a metathoracic leg is transplanted to the neighboring mesothoracic segment (Young 1972). Regenerating motor nerves will, however, make abnormal connections with other muscles when their normal target muscle is absent (Whitington 1977). More recently, Fourtner, Drewes & Holzmann (1978) using this system have shown that specific monosynaptic afferent, as well as efferent, connections are formed after the main metathoracic leg nerve is rerouted to make contact with the stump of the contralateral leg.

The five afferent tail fan roots carrying mechanoreceptor axons to the ipsilateral interneuron A arborizations in the last abdominal ganglion of a crayfish, will enter the stumps of their contralateral counterparts and synapse with the contralateral interneuron A. If the redirected roots are tied to their intact contralateral counterparts they enter the ganglion but do not synapse with the already innervated interneuron A. Nor do they cross the midline to their normal sites of synaptic input, but they do remain viable and if, after a period of 6 weeks, the roots of the side to which the nerve was redirected are severed, the misplaced fibers now form functional contacts with the contralateral interneuron A (Krasne & Lee 1977b). Taken together, these diverse experimental maneuvers indicate that right and left are not distinguished even though laterality is strictly observed in normal development.

Denburg (1975) has proposed a mechanism for specific recognition of muscles by motor axons based on the detection, by means of SDS polyacrylamide gels, of unique protein fractions in each of the coxal depressor muscles he examined. These candidate molecules may be cytoplasmic rather than cell surface components and it is argued that motor neuron outgrowths must form junctional complexes sufficient to transfer specific proteins rather than simply make contact with surface materials. The formation of provisional contacts described by Stocker & Nüesch (1975) perhaps provides ultrastructural evidence for such a mechanism during the establishment of neuromuscular relationships in metamorphosing moths.

A detailed chronology of the response of an insect motor nerve to crushing (Denburg, Seecof & Horridge 1977) provides a comprehensive account of cellular events during motor nerve regeneration. The motor nerve that supplies coxal depressor muscles from the metathoracic ganglion grew at a rate of 0.9 mm/day at 22–23°C from the crushed stump to regenerate a pattern of innervation to the coxal depressor muscles that is identical to the normal. Cobalt filling of numerous neurons demonstrated that misdirected axons reached the muscle surface but they subsequently degenerated. From 20 to 44 days after nerve crush, an average of 3.5 unidentified neurons per ganglion were detected that were not normally associated with the nerve branch in question. The number of such errors declined tenfold, to 0.33 neurons per ganglion, between 45 and 90 days. The general picture is of a profuse outgrowth from proximal stumps beginning 13 days after the crush and leading to a pattern of low specificity before neuromuscular junctions develop. The subsequent high specificity of neuromuscular connections implies a recognition mechanism at the muscle.

Another neural system that has proven to be most important for the analysis of specificity in regeneration at the cellular level is afforded by leech ganglia, in which special features of thoroughly characterized identified neurons have been ingeniously exploited by Baylor & Nicholls (1971), Jansen & Nicholls (1972), Jansen, Muller & Nicholls (1974), Frank, Jansen & Rinvik (1975), Van Essen & Jansen (1976), and Carbonetto & Muller (1977) using in vivo preparations. Nicholls, Wallace & Adal (1977) review the regenerative capacities of individual neurons in the leech. Intracellular recordings showed that central neurons regenerate connections with high precision, although synaptic responses are modified in consistently repeatable ways. Despite uncertainties concerning the nature of apparently inappropriate connections and the multiplicity of sprouting responses at lesion sites, there are clear examples of correct regeneration with specific cells.

The value of the leech preparation is further emphasized by studies of regeneration in ganglia maintained in organ culture (Wallace, Adal & Nicholls 1977). After connectives between ganglia were crushed so that conduction was initially abolished, recovery was monitored anatomically and electrophysiologically. After 5–10 days conduction of impulses through the lesion site was restored and characteristic excitatory and inhibitory synaptic potentials were recorded from identified sensory and motor cells. The regrowth of severed axons was followed by injecting individual sensory neurons with horseradish peroxidase, which demonstrated profuse branching at the site of the lesion by 7 days. One or more of the branches subsequently reached the ganglion beyond the crush while other branches grew back to the ganglion containing the cell body. Sprouting was also observed within the ganglion close to the cell body. After 14 days, stimulation of individual mechanosensory neurons with intracellular electrodes evoked synaptic potentials in the normal target motoneuron in the ganglion beyond the lesion. In this case then, specific connections were reestablished, but extensive supernumerary sprouting and the appearance of synaptic profiles in the region of the lesion raise the possibility that nonspecific or respecified connections were set up in response to the lesion and represent a relaxation of specificity that is unique to damage repair mechanisms. The synaptic profiles that appear in response to the crushing of connectives persist for more than 100 days (Fernández & Fernandez 1974).

The introduction in the leech preparation of the Pronase technique, which makes it possible to destroy individual identified cells (Parnas & Bowling 1977, Bowling, Nicholls & Parnas 1978), opens the way to even more precise analysis of sprouting and synaptic rearrangement.

Studies of identified neurons in gastropods also point the way to a refined analysis of specificity in regeneration (Murphy & Kater 1978). In *Melisoma trivolis,* the salivary neuroeffector complex, a subsystem of the neuron group that mediates feeding behavior, includes an identified pair of electrically coupled cells in the buccal ganglion that send axons via the esophageal nerves to terminate in chemical synapses on salivary gland cells.

The entire complex can be explanted and maintained in a healthy state in the body cavity of a host snail. Severed nerves reinnervated the salivary gland within 5 days and restored normal behavior. Injection of the dye Lucifer yellow showed early and prolific sprouting from the cut stump and also in undamaged areas of the ganglion. Some neurites extended into wrong branches but doubled back on themselves and returned through the ganglion to the correct route. No evidence was found for incorrect innervation either by the regenerating axons or by "foreign" neurons of the salivary glands, although the possibility of rare or transient errors could not be eliminated.

The analysis of neural regeneration has advanced dramatically in level of refinement in the last decade. The emergence of several invertebrate systems as experimental models, each with its own admirable features, serves to establish triangulation points for an accurate charting of the behavior of neurons and glia. The examples selected above suggest that future progress will combine methods for the electrophysiological and anatomical monitoring of single cells with the ultrastructural analysis of membrane interactions and junction formation of various kinds.

ACKNOWLEDGMENTS

We are grateful for the support we have received from the U.S. Public Health Service (NIH Grant 07778 to JSE and JP) and the Commonwealth Fund of New York (HA). We also express our thanks to the many people who sent us reprints, preprints, and unpublished information.

Literature Cited

Albertson, D. G., Sulston, J. E., White, J. G. 1978. Cell cycling and DNA replication in a mutant blocked in cell division in the nematode *Caenorhabditis elegans. Dev. Biol.* 63:165–78

Altman, J. S., Tyrer, N. M. 1974. Insect flight as a system for the study of the development of neuronal connections. In *The Experimental Analysis of Insect Behaviour,* ed. L. Barton Browne, pp. 159–79. Berlin-Heidelberg-New York: Springer

Altman, J. S., Tyrer, N. M. 1977. The locust wing hinge stretch receptors. II. Variation, alternative pathways and "mistakes" in the central arborizations. *J. Comp. Neurol.* 172:431–40

Anderson, D. T. 1972. *Embryology and Phylogeny in Annelids and Arthropods.* London: Pergamon

Anderson, H. 1978a. Postembryonic development of the visual system of the locust *Schistocerca gregaria.* I. Patterns of growth and developmental interactions in the retina and optic lobe. *J. Embryol. Exp. Morphol.* 45:55–83

Anderson, H. 1978b. Postembryonic development of the visual system of the locust, *Schistocerca gregaria.* II. An experimental analysis of the formation of the retina-lamina projection. *J. Embryol. Exp. Morphol.* 46:147–70

Anderson, H., Bacon, J. 1979. Developmental determination of neuronal projection patterns from wind-sensitive hairs in the locust, *Schistocerca gregaria. Dev. Biol.* In press

Atwood, H. L., Govind, C. K., Bittner, G. D. 1973. Ultrastructure of nerve terminals and muscle fibres in denervated crayfish muscle. *Z. Zellforsch. Mikrosk. Anat.* 146:155–65

Babu, K. S. 1975. Postembryonic development of the central nervous system of the spider *Argiope aurantia. J. Morphol.* 146:325–42

Ball, E., Young, D. 1974. Structure and development of the auditory system in the prothoracic leg of the cricket Teleogryllus commodus (Walker). II. Postembryonic development. Z. Zellforsch. Mikrosk. Anat. 147:313–24

Barker, D. L., Herbert, E., Hildebrand, J. G., Kravitz, E. A. 1972. Acetylcholine and lobster sensory neurons. J. Physiol. London 226:205–29

Bate, C. M. 1976a. Pioneer neurons in an insect embryo. Nature 260:54–56

Bate, C. M. 1976b. Nerve growth in cockroaches (Periplaneta americana) with rotated ganglia. Experientia 32:451–52

Bate, C. M. 1976c. Embryogenesis of an insect nervous system. I. A map of the thoracic and abdominal neuroblasts in Locusta migratoria. J. Embryol. Exp. Morphol. 35:107–23

Bate, C. M. 1978. Development of sensory systems in arthropods. In Handbook of Sensory Physiology Vol IX: Development of Sensory Systems, ed. M. Jacobson, pp. 1–53. Berlin-Heidelberg-New York: Springer

Baylor, D. A., Nicholls, J. G. 1971. Patterns of regeneration between individual nerve cells in the central nervous system of the leech. Nature 232:268–69

Benjamin, P. R. 1976. Interganglionic variation in cell body location of snail neurones does not affect synaptic connections or central axonal projections. Nature 260:338–40

Bentley, D. 1975. Single-gene cricket mutants: Effects on behavior, sensilla, sensory neurons and identified interneurons. Science 187:760–64

Bentley, D. 1977. Development of insect nervous system. In Identified Neurons and Behavior of Arthropods, ed. G. Hoyle, pp. 461–81. New York: Plenum

Birse, S. C., Bittner, G. D. 1976. Regeneration of giant axons in earthworms. Brain Res. 113:575–81

Birse, S. C., Bittner, G. D. 1978. Cell to cell specificity of regenerated giant axons in earthworms. Neurosci. Abstr. 4:529

Bittner, G. D. 1973. Degeneration and regeneration in crustacean neuromuscular systems. Am. Zool. 13:379–408

Bittner, G. D. 1977. Trophic interactions of crustacean neurons. In Identified Neurons and Behavior of Arthropods, ed. G. Hoyle, pp. 507–32. New York: Plenum

Bittner, G. D., Ballinger, M. L., Larimer, J. 1974. Crayfish CNS: Minimal degenerative-regenerative changes after lesioning. J. Exp. Zool. 189:13–36

Bittner, G. D., Johnson, A. L. 1974. Degeneration and regeneration in crustacean peripheral nerves. J. Comp. Physiol. 89:1–21

Bittner, G. D., Mann, D. W. 1976. Differential survival of isolated portions of crayfish axons. Cell. Tissue Res. 169: 301–11

Bittner, G. D., Nitzberg, M. 1975. Degeneration of sensory and motor axons in transplanted segments of a crustacean peripheral nerve. J. Neurocytol. 4:7–21

Bodenstein, D. 1957. Studies on nerve regeneration in Periplaneta americana. J. Exp. Zool. 136:89–115

Boeckh, J., Sandri, C., Akert, K. 1970. Sensorische Eingänge und synaptische Verbindungen in Zentralnervensystem von Insekten. Experimentelle Degeneration in der antennalen Sinnesbahn in Oberschlundganglion von Fliegen und Schaben. Z. Zellforsch. Mikrosk. Anat. 103:429–46

Boulton, P. S. 1969. Degeneration and regeneration in the insect nervous system I. Z. Zellforsch. Mikrosk. Anat. 101:98–118

Boulton, P. S., Rowell, C. H. F. 1969. Degeneration and regeneration in the insect central nervous system II. Z. Zellforsch. Mikrosk. Anat. 101:119–34

Bowling, D., Nicholls, J., Parnas, I. 1978. Destruction of a single cell in the central nervous system of the leech as a means of analyzing its connexions and functional role. J. Physiol. London 282:169–80

Brenner, S. 1974. The genetics of Caenorhabditis elegans. Genetics 77:71–94

Browne, C. L., Davis, L. E. 1978. The role of nerve cell density in the regulation of bud production in Hydra. Wilhelm Roux Arch. Dev. Biol. 184:95–108

Burrows, M. 1973. The morphology of an elevator and a depressor motoneuron in the hindwing of a locust. J. Comp. Physiol. 83:165–78

Campbell, R. D. 1979. Development of Hydra lacking interstitial and nerve cells ("Epithelial Hydra"). Symp. Soc. Dev. Biol., 38th. In press

Carbonetto, S., Muller, K. J. 1977. A regenerating neuron in the leech can form an electrical synapse on its severed axon segment. Nature 267:450–52

Carlson, J. R. 1977a. The imaginal ecdyses of the cricket (Teleogryllus oceanicus). I. Organization of motor programs and roles of central and sensory control. J. Comp. Physiol. 115:299–317

Carlson, J. R. 1977b. The imaginal ecdyses of the cricket (Teleogryllus oceanicus) II. The roles of identified motor units. J. Comp. Physiol. 115:319–36

Casaday, G. B., Hoy, R. R. 1977. Auditory interneurons in the cricket Teleogryllus oceanicus: Physiological and anatomical properties. *J. Comp. Physiol.* 121: 1–13

Chiarodo, A. T. 1963. The effects of mesothoracic leg disc extirpation on the postembryonic development of the nervous system of the blowfly *Sarcophaga bullata. J. Exp. Zool.* 153:263–77

Clark, R. 1976a. Structural and functional changes in an identified cricket neuron after separation from the soma. I. Structural changes. *J. Comp. Neurol.* 170:253–66

Clark, R. 1976b. Structural and functional changes in an identified neuron after separation from the soma. II. Functional changes. *J. Comp. Neurol.* 170:267–78

Cloney, R. A. 1978. Ascidian metamorphosis: review and analysis. In *Settlement and Metamorphosis of Marine Invertebrate Larvae,* ed. F. S. Chia, M. Rice, pp. 255–82. Amsterdam: Elsevier/North Holland

Crick, F. H. C. 1971. The scale of pattern formation. *Symp. Soc. Exp. Biol.* 25: 429–38

Dahl, E. 1959. The ontogeny and comparative anatomy of some protocerebral sense organs in Notostracan Phyllopods. *Q. J. Microsc. Sci.* 100:445–62

Davis W. J., Davis, K. B. 1973. Ontogeny of a simple locomotor system: role of the periphery in the development of central nervous circuitry. *Am. Zool.* 13:409–25

Deak, I. I. 1976. Demonstration of sensory neurones in the ectopic cuticle of spineless-aristapedia, a homeotic mutant of *Drosophila. Nature* 260:252–54

Debaisieux, P. 1944. Les yeux de Crustacés. Structure, développement, réactions a l'éclairement. *La Cellule* 50:9–122

Denburg, J. L. 1975. Possible biochemical explanation for specific reformation of synapses between muscle and regenerating motoneurones in cockroach. *Nature* 258:535–37

Denburg, J. L., Seecof, R. L., Horridge, G. A. 1977. The path and rate of growth of regenerating motor axons in the cockroach. *Brain Res.* 125:213–26

Deppe, U., Schierenberg, E., Cole, T., Krieg, C., Schmitt, D., Yoder, B., von Ehrenstein, G. 1978. Cell lineages of the embryo of the nematode *Caenorhabditis elegans. Proc. Natl. Acad. Sci. USA* 75:376–80

Dethier, V. G. 1955. The physiology and histology of the contact chemoreceptors of the blowfly. *Q. Rev. Biol.* 30:348–71

Edwards, J. S. 1969. Postembryonic development and regeneration of the insect nervous system. *Adv. Insect Physiol.* 6:97–137

Edwards, J. S. 1977a. Pathfinding by arthropod sensory nerves. In *Identified Neurons and Behavior of Arthropods,* ed. G. Hoyle, pp. 483–94 New York: Plenum

Edwards, J. S. 1977b. One organism, several brains: Evolution and development of the insect central nervous system. *Ann N. Y. Acad. Sci.* 299:59–71

Edwards, J. S., Chen, S.-W. 1979. Embryonic development of an insect sensory system, the abdominal cerci of *Acheta domesticus. Wilhelm Roux Arch. Dev. Biol.* 186:151–78

Edwards, J. S., Milner, M. J., Chen, S.-W. 1978. Integument and sensory nerve differentiation of Drosophila leg and wing imaginal discs in vitro. *Wilhelm Roux Arch. Dev. Biol.* 185:59–77

Edwards, J. S., Palka, J. 1976. Neural generation and regeneration in insects. In *Simple Networks and Behavior,* ed. J. Fentress, pp. 167–85. Sunderland: Sinauer

Elofsson, R. 1969. The development of the compound eyes of *Penaeus duorarum* (Crustacea: Decapoda) with remarks on the nervous system. *Z. Zellforsch. Mikrosk. Anat.* 97:323–50

Elofsson, R., Dahl, E. 1970. The optic neuropiles and chiasmata of crustacea. *Z. Zellforsch. Mikrosk. Anat.* 107: 343–60

Fernandez, A. D., Flores, D. V. 1978. Studies in the regeneration of the sensory tentacular organs under denervation in *Helix aspersa. Rivista de Microscopia Electronica* 5:186–87

Fernández, J. 1979. Embryonic development of the glossiphonid leech *Theromyzon rude:* Characterization of developmental stages. *Dev. Biol.* In press

Fernández, J., Fernandez, M. S. 1974. Morphological evidence for an experimentally induced synaptic field. *Nature* 251:428–30

Fourtner, C. R., Drewes, C. D., Holzmann, T. W. 1978. Specificity of afferent and efferent regeneration in the cockroach: establishment of a reflex pathway between contralaterally homologous cells. *J. Neurophysiol.* 41:885–95

Frank, E., Jansen, J. K. S., Rinvik, E. 1975. A multisomatic axon in the central nervous system of the leech. *J. Comp. Neurol.* 159:1–14

Ghysen, A. 1978. Sensory neurones recognize defined pathways in *Drosophila* central nervous system. *Nature* 274: 869–72

Goodman, C. S. 1974. Anatomy of locust ocellar interneurons: Constancy and variability. *J. Comp. Physiol.* 95:185–201

Goodman, C. S. 1976. Constancy and uniqueness in a large population of small interneurons. *Science* 193:502–4

Goodman, C. S. 1977. Neuron duplications and deletions in locust clones and clutches. *Science* 197:1384–86

Goodman, C. S. 1978. Isogenic locusts: Genetic variability in the morphology of identified neurons. *J. Comp. Neurol.* 182:681–705

Goodman, C. S., O'Shea, M., McCaman, R., Spitzer, N.C. 1979. Embryonic development of identified neurons: Temporal pattern of morphological and biochemical differentiation. *Science* 204:1219–22

Goodman, C. S., Spitzer, N. C. 1979. Embryonic development of identified neurons: Differentiation from neuroblast to neuron. *Nature* 280:208–14

Govind, C. K. 1980. The development of neural and muscular systems. In *Biology of the Crustacea,* Vol. 3., ed. D. Bliss, M. L. Atwood, D. C. Sandeman. New York: Academic. In preparation

Guthrie, D. M. 1967. The regeneration of motor axons in an insect. *J. Insect Physiol.* 13:1593–1611

Guthrie, D. M. 1975. Regeneration and neural specificity - the contribution of invertebrate studies. In *"Simple" Nervous Systems,* ed., P. N. R. Usherwood, D. R. Newth, pp. 119–65. New York: Crane Russak

Gymer, A., Edwards, J. S. 1967. The development of the insect nervous system. I. An analysis of postembryonic growth in the terminal ganglion of *Acheta domesticus. J. Morphol.* 123:191–97

Hall, J. C., Greenspan, R. J. 1980. Genetic analysis of *Drosophila* neurobiology. *Ann. Rev. Genet.* 13: In press

Hall, J. C., Gelbart, W. M. Kankel, D. R. 1976. Mosaic systems. In *The Genetics and Biology of Drosophila,* ed. M. Ashburner, E. Noviski, 1A:265–314

Hamburger, V. 1962. Specificity in Neurogenesis. *J. Cell. Comp. Physiol.* 60: Suppl. 1, pp. 81–92

Hámori, J., Horridge, G. A. 1966a. The lobster optic lamina. I. General organization. *J. Cell Sci.* 1:249–56

Hámori, J., Horridge, G. A. 1966b. The lobster optic lamina. III. Degeneration of retinula cell endings. *J. Cell Sci.* 1:271–74

Herlant-Meewis, H., Deligne, J. 1965. Influence of the nervous system on regeneration in Annelids. In *Regeneration in Animals and Related Problems* ed. V. Kiortsis, H. A. L. Trampusch, pp. 228–39. Amsterdam: North Holland

Hess, A. 1960. The fine structure of degenerating nerve fibres, their sheaths and their terminations in the central nerve cord of the cockroach (*Periplaneta americana*). *J. Biophys. Biochem. Cytol.* 7:399–44

Heywood, R. B. 1965. Changes occurring in the central nervous system of *Pieris brassicae,* L. (Lepidoptera) during metamorphosis. *J. Insect Physiol.* 11: 413–30

Hildebrand, J. G., Hall, L. M., Osmond, B. C. 1979. Distribution of binding sites for I-125-labeled alpha-bungarotoxin in normal and deafferented antennal lobes of *Manduca sexta. Proc. Nat. Acad. Sci. USA* 76:499–503

Horridge, G. A., Meinertzhagen, I. A. 1970. The accuracy of the patterns of connexions of the first- and second-order neurons of the visual system of *Calliphora. Proc. R. Soc. London Ser. B* 175:69–82

Hoy, R. R. 1969. Degeneration and regeneration in abdominal flexor motor neurons in the crayfish. *J. Exp. Zool.* 172: 219–32

Hoy, R. R., Bittner, G. D., Kennedy, D. 1967. Regeneration in crustacean motor neurons: Evidence for axonal refusion. *Science* 156:251–52

Hoy, R. R., Casaday, G. B., Rollins, S. 1978. Absence of auditory afferents alters the growth pattern of an identified auditory interneuron. *Soc. Neurosci. Abstr.* 4:115

Hulsebosch, C. E., Bittner, G. D. 1978. A comparative study of CNS regenerative abilities. *Neurosci. Abstr.* 4:532

Jacklet, J. W., Cohen, M. J. 1967. Nerve regeneration: correlation of electrical, histological and behavioral events. *Science* 156:1640–43

Jacobson, M. 1978. *Developmental Neurobiology.* New York: Plenum. 2nd ed.

Jansen, J. K. S., Muller, K. J., Nicholls, J. G. 1974. Persistent modification of synaptic interactions between sensory and motor cells following discrete lesions in the central nervous system of the leech. *J. Physiol. London* 242:289–305

Jansen, J. K. S., Nicholls, J. G. 1972. Regeneration and changes in synaptic connections between individual cells in the central nervous system of the leech. *Proc. Natl. Acad. Sci. USA* 69:636–39

Johannson, A. S. 1957. The nervous system of the milkweed bug *Oncopeltus fasciatus* (Dallas) (Heteroptera, Lygaeidae) *Trans. Am. Entomol. Soc.* 83:119–83

Kafatos, F. C., Reich, J. 1968. Stability of differentiation specific and non-specific messenger RNA in insect cells. *Proc. Natl. Acad. Sci. USA* 60:1458–65

Kankel, D. R., Ferrús, A., Garen, S. H., Harte, P. J., Lewis, P. E. 1979. The structure and development of the nervous system. In *The Genetics and Biology of Drosophila*, ed. M. Ashburner, E. Novitski. New York: Academic. In Press

King, R. C., ed. 1975. *Handbook of Genetics, Vol. 3. Invertebrates of Genetic Interest.* New York: Plenum

Koopowitz, H., Silver, D., Rose, G. 1975. Neuronal plasticity and recovery of function in a polyclad flatworm. *Nature* 256:737–38

Krasne, F. B., Lee, S. H. 1977a. Survival of functional synapses on crustacean neurons lacking cell bodies. *Brain Res.* 121:43–57

Krasne, F. B., Lee, S. -H. 1977b. Regenerating afferents establish synapses with a target neuron that lacks its cell body. *Science* 198:517–19

Krieg, C., Cole, T., Deppe, U., Schierenberg, E., Schmitt, D., Yoder, B., von Ehrenstein, G. 1978. The cellular anatomy of embryos of the nematode *Caenorhabditis elegans. Dev. Biol.* 65:193–215

Kriegstein, A. 1977a. Development of the nervous system of *Aplysia californica. Proc. Natl. Acad. Sci. USA* 74:375–78

Kriegstein, A. R. 1977b. Stages in the posthatching development of *Aplysia californica. J. Exp. Zool.* 199:275–88

Kuffler, D. P., Muller, K. J. 1974. The properties and connections of supernumerary sensory and motor nerve cells in the central nervous system of an abnormal leech. *J. Neurobiol.* 5:331–48

Kume, M., Dan, K. 1968. *Invertebrate Embryology.* English trans. J. Dan. HO-LIT, Belgrade

Kunkel, J. G. 1976. Cockroach molting I. Temporal organization of events during the molting cycle of *Blattella germanica. Biol. Bull.* 148:259–73

Kutsch, W. 1974. The influence of the wing sense organs on the flight motor pattern in maturing adult locusts. *J. Comp. Physiol.* 88:413–24

Lamparter, M. E., Akert, K., Sandri, C. 1967. Wallerische Degeneration im Zentralnervensystem der Ameise. Electronmikroskopische Untersuchungen am Prothorakalganglion von *Formica lugubris* Zett. *Schweiz. Arch. Neurol. Psychiatr.* 100:337–54

Lasek, R. J., Gainer, H., Przyblski, R. J. 1974. Transfer of newly synthesised proteins from Schwann cells to giant squid axon. *Proc. Natl. Acad. Sci. USA* 71:1188–92

Letourneau, P. 1975. Cell-to-substratum adhesion and guidance of axonal elongation. *Dev. Biol.* 44:92–101

LoPresti, V., Macagno, E. R., Levinthal, C. 1973. Structure and development of neuronal connections in isogenic organisms: cellular interactions in the development of the optic lamina of *Daphnia. Proc. Natl. Acad. Sci. USA* 70:433–37

LoPresti, V., Macagno, E. R., Levinthal, C. 1974. Structure and development of neuronal connections in isogenic organisms: transient gap junctions between growing optic axons and lamina neuroblasts. *Proc. Natl. Acad. Sci. USA* 71:1098–1102

Macagno, E. R. 1977. Abnormal synaptic connectivity following UV-induced cell death during *Daphnia* development. In *Cell and Tissue Interactions,* ed. J. W. Lash, M. M. Burges. New York: Raven

Macagno, E. R. 1978. Mechanism for the formation of synaptic projections in the arthropod visual system. *Nature* 275: 318–20

Macagno, E. R. 1979. Cellular interactions and pattern formation in the development of the visual system of *Daphnia magna* (Crustacea, Branchiopoda) I. Interactions between embryonic retinular fibers and laminar neurons. *Dev. Biol.* In press

Macagno, E. R., LoPresti, V., Levinthal, C. 1973. Structure and development of neuronal connections in isogenic organisms: Variations and similarities in the optic system of *Daphnia magna. Proc. Natl. Acad. Sci. USA* 70:57–61

Matsumoto, S. G., Murphey, R. K. 1977. Sensory deprivation during development decreases the responsiveness of cricket giant interneurons. *J. Physiol. London* 268:533–48

Matsumoto, S. G., Murphey, R. K. 1978. Sensory deprivation in the cricket central nervous system: evidence for a critical period. *Soc. Neurosci. Abstr.* 4:201

Maxwell, G. D., Hildebrand, J. G. 1978. Neurochemical and anatomical studies of normal and deafferented optic lobes of the moth *Manduca sexta. Soc. Neurosci. Abstr.* 4:120

Meinertzhagen, I. A. 1972. Erroneous projection of retinula axons beneath a dislocation in the retinal equator of Calliphora. *Brain Res.* 41:39–49

Meinertzhagen, I. A. 1973. Development of the compound eye and optic lobes of insects. In *Developmental neurobiology of Arthropods*, ed. D. Young, pp. 51–104. Cambridge Univ. Press

Meinertzhagen, I. A. 1975. The development of neuronal connection patterns in the visual system of insects. In *Cell Patterning. CIBA Found. Symp.* 29 (NS): 265–88. Amsterdam: Elsevier-Excerpta Medica-North-Holland

Meinertzhagen, I. A. 1976. The organization of perpendicular fibre pathways in the insect optic lobe. *Philos. Trans, R. Soc. London Ser. B* 274:555–96

Melamed, J., Trujillo-Cenóz, O. 1962. Electron microscopic observations on the reactional changes occurring in insect fibres after transection. *Z. Zellforsch. Mikrosk. Anat.* 59:851–56

Mellon, D., Stephen, P. J. 1978. Limb morphology and function are transformed by contralateral nerve section in snapping shrimps. *Nature* 272:246–248

Mendenhall, B., Murphey, R. K. 1974. The morphology of cricket giant interneurons. *J. Neurobiol.* 5:565–80

Meyer, M. R., Bittner, G. D. 1978a. Histological studies of trophic dependence in crayfish giant axons. *Brain Res.* 143: 195–211

Meyer, M. R., Bittner, G. D. 1978b. Biochemical studies of trophic dependence in crayfish giant axons. *Brain Res.* 143:213–32

Meyerowitz, E. M., Kankel, D. R. 1978. A genetic analysis of visual system development in *Drosophila melanogaster*. *Dev. Biol.* 62:112–42

Meyer-Rochow, V. B. 1975. Larval and adult eye of the Western Rock Lobster. *Cell Tissue Res.* 162:439–57

Moroff, T. 1912. Cyto-histogenetische studien I. Entwicklung des Facettenauges bei Crustaceen. *Zool. Jahrb. Abt. Anat. Ontog. Tiere* 34:473–558

Mouze, M. 1974. Interactions de l'oeil et du lobe optique au cours de la croissance post-embryonnaire des Insectes odonates. *J. Embryol. Exp. Morphol.* 31: 377–407

Mouze, M. 1978. Rôle des fibres post-rétinniennes dans la croissance du lobe optique de la larve de'*Aeshna cyanea* Müll (Insecte, Odonate). *Wilhelm Roux Arch. Dev. Biol.* 184:325–49

Murphey, R. K. 1977. Development of an identified neuron: morphological effects of blocking presynaptic activity or removing presynaptic neurons. *Soc. Neurosci. Abstr.* 3:186

Murphey, R. K. 1978. Postembryonic development of an identified interneuron in the cricket *Acheta domesticus. Soc. Neurosci. Abstr.* 4:202

Murphey, R. K., Levine, R. K. 1979. Mechanisms responsible for the changes observed in the response properties of partially deafferented insect interneurons. *J. Neurophysiol.* In press

Murphey, R. K., Matsumoto, S. G. 1976. Experience modifies the plastic properties of identified neurons. *Science* 191: 564–66

Murphey, R. K., Matsumoto, S. G., Mendenhall, B. 1976. Recovery from deafferentiation by cricket interneurons after reinnervation by their peripheral field. *J. Comp. Neurol.* 169:335–46

Murphey, R. K., Mendenhall, B., Palka, J., Edwards, J. S. 1975. Deafferentation slows the growth of specific dendrites of identified giant interneurons. *J. Comp. Neurol.* 159:407–18

Murphey, R. K., Palka, J., Hustert, R. 1977. The cercus-to-giant interneuron system of crickets. II. Response characteristics of two giant interneurons. *J. Comp. Physiol.* 119:285–300

Murphy, A. D., Kater, S. B. 1978. Specific reinnervation of a target organ by a pair of identified molluscan neurons. *Brain Res.* 156:322–28

Nässel, D. R. 1976. The retina and retinal projection on the lamina ganglionaris of the crayfish *Pacifastacus leniusculus* (Dana). *J. Comp. Neurol.* 167:341–60

Nicholls, J. G., Wallace, B. Adal, M. 1977. Regeneration of individual neurones in the nervous system of the leech. In *Synapses*, ed. G. A. Cottrell, P. N. R. Usherwood, pp. 249–63. New York: Academic

Nordlander, R. H., Edwards, J. S. 1968. Morphological cell death in the postembryonic development of the insect optic lobes. *Nature* 218:780

Nordlander, R. H., Edwards, J. S. 1969. Postembryonic brain development in the monarch butterfly. II. The optic lobes. *Wilhelm Roux Arch. Dev. Biol.* 163: 197–220

Nordlander, R. H., Edwards, J. S. 1970. Postembryonic brain development in the monarch butterfly. III. Morphogenesis of centres other than the optic lobes. *Wilhelm Roux Arch. Dev. Biol.* 164:247–60

Nordlander, R. H., Masnyi, J. A., Singer, M. 1975. Distribution of ultrastructural tracers in crustacean axons. *J. Comp. Neurol.* 161:499–513

Nordlander, R. H., Singer, M. 1972. Electron microscopy of severed motor fibres in the crayfish. *Z. Zellforsch. Mikrosk. Anat.* 126:157–81

Nordlander, R. H., Singer, M. 1973a. Degeneration and regeneration of severed crayfish fibres: an ultrastructural study. *J. Comp. Neurol.* 152:175–92

Nordlander, R. H., Singer, M. 1973b. Effects of temperature on the ultrastructure of severed crayfish motor axons. *J. Exp. Zool.* 184:289–301

Nordlander, R., Singer, M. 1978. The role of ependyma in regeneration of the spinal cord in the urodele amphibian tail. *J. Comp. Neurol.* 180:349–73

Nüesch, H. 1968. The role of the nervous system in insect morphogenesis and regeneration. *Ann. Rev. Entomol.* 13:27–44

Nüesch, H., Stocker, R. F. 1975. Ultrastructural studies on neuromuscular contacts and the formation of functions in the flight muscle of *Antheraea polyphemus* II. Changes after motor nerve section. *Cell Tissue Res.* 164:331–55

Olmstead, J. M. D. 1922. The role of the nervous system in the regeneration of polyclad Turbellaria. *J. Exp. Zool.* 36:49–56

Packard, A., Sanders, G. D. 1971. Body patterns of *Octopus vulgaris* and maturation of the response to disturbance. *Anim. Behav.* 19:780–90

Palka, J. 1977a. Abnormal neural development in invertebrates. In *Function and Formation of Neural Systems,* ed. G. Stent, pp. 139–59. Berlin: Dahlem Konferenzen

Palka, J. 1977b. Neurobiology of homeotic mutants in *Drosophila. Soc. Neurosci. Abstr.* 3:187

Palka, J. 1979a. Theories of pattern formation in insect neural development. *Adv. Insect Physiol.* 14:256–349

Palka, J. 1979b. Mutants and mosaics, tools in insect developmental neurobiology. *Symp. Soc. Neurosci.* 4:209–27

Palka, J., Edwards, J. S. 1974. The cerci and abdominal giant fibres of the house cricket, *Acheta domesticus.* II. Regeneration and effects of chronic deprivation. *Proc. R. Soc. London Ser. B* 185:105–21

Palka, J., Lawrence, P. A., Hart, H. S. 1979. Neural projection patterns from homeotic tissue of *Drosophila* studied in *bithorax* mutants and mosaics. *Dev. Biol.* 69:549–75

Palka, J., Levine, R., Schubiger, M. 1977. The cercus-to-giant interneuron system of crickets. I. Some attributes of the sensory cells. *J. Comp. Physiol.* 119:267–83

Palka, J., Olberg, R. 1977. The cercus-to-giant interneuron system of crickets. III. Receptive field organization. *J. Comp. Physiol.* 119:301–17

Palka, J., Schubiger, M. 1975. Central connections of receptors on rotated and exchanged cerci of crickets. *Proc. Natl. Acad. Sci. USA* 72:966–69

Panov, A. A. 1960. The structure of the insect brain during successive stages of postembryonic development. III. optic lobes. *Éntomologicheskoe Obozrenie* (Transl.) 39:86–105

Panov, A. A. 1961. The structure of the insect brain at successive stages in postembryonic development. 4. The olfactory centre. *Entomol. Rev.* 40:140–45 (English transl.)

Pareto, A. 1972. Die Zentrale Verteilung der Fühlerafferentz bei Arbeiterinnen der Honigbiene *Apis mellifera* L. *Z. Zellforsch. Mikrosk. Anat.* 131:109–40

Parnas, I., Bowling, D. 1977. Killing of single neurones by intracellular injection of proteolytic enzymes. *Nature* 270:626–28

Peabody, E. B. 1939. Development of the eye of the isopod, *Idotea. J. Morphol.* 64:519–54

Pearson, K. G., Bradley, A. B. 1972. Specific regeneration of excitatory montoneurons to the leg muscles in the cockroach. *Brain Res.* 47:492–96

Pearson, K. G., Goodman, C. S. 1979. Correlation of variability in structure with variability in synaptic connections of an identified interneuron in locusts. *J. Comp. Neurol.* 184:141–65

Pipa, R. L. 1978. Patterns of Neural Reorganisation during the postembryonic development of insects. *Int. Rev. Cytol.* Suppl. 7, pp. 403–38

Pitman, R. M., Tweedle, C. D., Cohen, M. J. 1973. The form of nerve cells: determination by cobalt impregnation. *Intracellular Staining in Neurobiology,* ed. S. B. Kater, G. Nicholson, pp. 83–97. New York: Springer

Postlethwait, J. H., Schneiderman, H. A. 1971. Pattern formation and determination in the antenna of the homeotic mutant *Antennapedia* of *Drosophila melanogaster. Dev. Biol.* 25:606–40

Prestige, M. 1970. Differentiation, degeneration and the role of the periphery: Quantitative considerations. In *The Neurosciences Second Study Program,* ed. F. O. Schmidt, pp. 73–82. New York: Rockefeller Univ. Press

138   ANDERSON, EDWARDS & PALKA

Price, C. H. 1977. Regeneration in the central nervous system of a pulmonate mollusc *Melamplus. Cell Tissue Res.* 180: 529–36

Provine, R. R., Seshan, K. R., Aloe, L. 1976. Formation of cockroach interganglionic connectives: an in vitro analysis. *J. Comp. Neurol.* 165:17–30

Quinn, W. G., Gould, J. L. 1979. Nerves and genes. *Nature* 278:19–23

Rakic, P., Sidman, R. L. 1973. Weaver mutant mouse cerebellum: Defective neuronal migration secondary to abnormality of Bergmann glia. *Proc. Natl. Acad. Sci. USA* 70:240–44

Rees, D., Usherwood, P. N. R. 1972. Fine structure of normal and degenerating motor axons and nerve muscle synapses in the locust, *Schistocerca gregaria. Comp. Biochem. Physiol.* 43:83–101

Rowell, C. H. F., Dorey, A. E. 1967. The number and size of axons in the thoracic connectives of the desert locust *Schistocerca gregaria* (Forsk). *Z. Zellforsch. Mikrosk. Anat.* 83:288–94

Sanders, G. D., Young, J. Z. 1974. Reappearance of specific colour patterns after nerve regeneration in *Octopus. Proc. R. Soc. London Ser. B* 186:1–11

Sanes, J. R., Hildebrand, J. G., Prescott, D. J. 1976. Differentiation of insect sensory neurons in the absence of their normal synaptic targets. *Dev. Biol.* 52:121–27

Sbrenna, G. 1971. Postembryonic growth of the ventral nerve cord in Schistocerca gregaria (Orthoptera, Acrididae). *Bull. Zool.* 38:49–71

Schacher, S., Kandel, E. R., Woolley, R. 1979. Development of neurons in the abdominal ganglion of *Aplysia californica.* I. Axo-somatic synaptic contacts. Dev. Biol. 71:163–75

Schubiger, M., Palka, J. 1979. The haltere-like projection of segmentally translocated wing receptors in *Drosophila* homeotic mutants. *Soc. Neurosci. Abstr.* 5:178

Selverston, A. I., Mulloney, B. 1974. Synaptic and structural analyses of a small neural system. In *Neurosciences, Third Study Program,* ed. F. O. Schmitt, F. G. Worden, pp. 389–95. Cambridge: MIT

Selverston, A. I., Remler, M. P. 1972. Neural geometry and activation of crayfish fast flexor motoneurons. *J. Neurophysiol.* 35:797–814

Shivers, R. R. 1975. Trans-glial channel-facilitated translocation of tracer protein across ventral nerve root sheaths of crayfish. *Brain Res.* 108:47–58

Shivers, R. R. 1977. Formation of junctional complexes at sites of contact of hemo-

cytes with tissue elements in degenerating nerves of the crayfish *Orconectes virilis. Tissue Cell* 9:43–56

Shivers, R. R., Brightman, M. W. 1976. Trans-glial channels in ventral nerve roots of crayfish. *J. Comp. Neurol.* 167: 1–26

Shivers, R. R., Brightman, M. W. 1977. Formation of hemi-desmosomes during regeneration of crayfish nerve root sheath as studied with freeze fracture. *J. Comp. Neurol.* 173:1–22

Spitzer, N. C. 1979. Ion channels in development. *Ann. Rev. Neurosci.* 2:363–97

Stewart, W. W. 1978. Functional connections between cells as revealed by dye-coupling with a highly fluorescent naphthalimide tracer. *Cell* 14:741–59

Stocker, R. F. 1977. Gustatory stimulation of a homeotic mutant appendage, *Antennapedia,* in *Drosophila melanogaster. J. Comp. Physiol.* 115:351–61

Stocker, R. F., Edwards, J. S., Palka, J., Schubiger, G. 1976. Projection of sensory neurons from a homeotic mutant appendage, *Antennapedia,* in *Drosophila melanogaster. Dev. Biol.* 52:210–20

Stocker, R. F., Nüesch, H. 1975. Ultrastructural studies on neuromuscular contacts and the formation of junctions in the flight muscle of *Antheraea polyphemus* (Lep.) 1. Normal adult development. *Cell Tissue Res.* 159:245–66

Stowe, S. 1977. The retina-lamina projection in the crab *Leptograpsus variegatus. Cell Tissue Res.* 185:515–25

Strausfeld, N. J. 1976. *Atlas of an Insect Brain.* Berlin, Heidelberg, New York: Springer

Sulston, J. E. 1976. Post-embryonic development in the ventral nerve cord of *Caenorhabditis elegans. Philos. Trans. R. Soc. London. Ser. B* 275:287–97

Sulston, J. E., Horvitz, H. R. 1977. Postembryonic cell lineages of the nematode, *Caenorhabditis elegans. Dev. Biol.* 56:110–56

Taylor, H. M., Truman, J. W. 1974. Metamorphosis of the abdominal ganglia of the tobacco hornworm, *Manduca sexta. J. Comp. Physiol* 90:367–88

Treistman, S. N., Schwartz, J. H. 1976. Functional constancy in *Aplysia* nervous systems with anomalously duplicated identified neurons. *Brain Res.* 109:607–14

Trujillo-Cenóz, O., Melamed, J. 1973. The development of the retina-lamina complex in muscoid flies. *J. Ultrastruct. Res.* 42:554–81

Truman, J. W. 1976. Development and hormonal release of adult behavior patterns in silkmoth. *J. Comp. Physiol.* 107: 39–48

Truman, J. W., Reiss, S. E. 1976. Dendritic reorganization of an identified motoneuron during metamorphosis of the tobacco hornworm moth. *Science* 192: 447–79

Van Essen, D., Jansen, J. K. S. 1976. Repair of specific neuronal pathways in the leech. *Cold Spring Harbor Symp. Quant. Biol.* 40:495–502

Van Ruiten, T. M., Sprey, T. E. 1974. The ultrastructure of the developing leg disk of *Calliphora erythrocephala. Z. Zellforsch. Mikrosk. Anat.* 147:373–400

Wallace, B. G., Adal, M. N., Nicholls, J. G. 1977. Regeneration of synaptic connections by sensory neurons in leech ganglia maintained in culture. *Proc. R. Soc. London Ser. B* 199:567–85

Ward, S. 1977 Invertebrate neurogenetics. *Ann. Rev. Genet.* 11:415–50

Ward, S., Thomson, N., White, J. G., Brenner, S. 1975. Electron microscopical reconstruction of the anterior sensory anatomy of the nematode *Caenorhabditis elegans. J. Comp. Neurol.* 160: 313–37

Ware, R. W., Clark, D., Crossland, K., Russell, R. L. 1975. The nerve ring of the nematode *Caenorhabditis elegans. J. Comp. Neurol.* 162:71–110

Weisblat, D. A., Sawyer, R. T., Stent, G. S. 1978. Cell lineage analysis by intracellular injection of a tracer enzyme. *Science* 202:1295–98

Wheeler, W. M. 1891. Neuroblasts in the arthropod's embryo. *J. Morphol.* 4: 337–43

Wheeler, W. M. 1893. A contribution to insect embryology. *J. Morphol.* 8:1–160

White, J. G., Albertson, D. G., Anness, M. A. R. 1978. Connectivity changes in a class of motoneurone during the development of a nematode. *Nature* 271: 764–66

White, J. G., Southgate, E., Thomson, J. N., Brenner, S. 1976. The structure of the ventral nerve cord of *Caenorhabditis elegans. Philos. Trans. R. Soc. London. Ser. B* 275:327–48

White, K., Kankel, D. R. 1978. Patterns of cell division and cell movement in the formation of the imaginal nervous system in Drosophila melanogaster. *Dev. Biol.* 65:296–321

Whitington, P. M. 1977. Incorrect connections made by a regenerating cockroach motoneuron. *J. Exp. Zool.* 201:339–44

Wigglesworth, V. B. 1977. Structural changes in the epidermal cells of *Rhodnius* during tracheoli capture. *J. Cell. Sci.* 26:161–74

Wilson, D. M. 1960. Nervous control of movement in cephalopods. *J. Exp. Biol.* 37:57–72

Wine, J. J. 1973. Invertebrate central neurons: Orthograde degeneration and retrograde changes after axotomy. *Exp Neurol.* 38:157–69

Young, D. 1972. Specific re-innervation of limb transplanted between segments in the cockroach *Periplaneta americana. J. Exp. Biol.* 57:305–16

Young, D. 1973. *Developmental Neurobiology of Arthropods,* Cambridge: Cambridge Univ. Press

REFERENCE ADDED IN PROOF   Sanes, J. R., Hildebrand, J. G. 1976. Structure and development of antennae in a moth, *Moduca sexta. Dev. Biol.* 51:282–99

Ann. Rev. Neurosci. 1980. 3:141–67

IONIC CURRENTS
IN MOLLUSCAN SOMA

❖11538

David J. Adams

Department of Physiology & Biophysics, University of Washington
School of Medicine, Seattle, Washington 98195

Stephen J Smith

Department of Physiology & Anatomy, University of California,
Berkeley, California 94720

Stuart H. Thompson

Hopkins Marine Station, Stanford University, Pacific Grove, California 93950

INTRODUCTION

The present understanding of nerve excitability is founded on voltage clamp studies of axonal membranes. It has become clear, however, that the excitability of neuronal somata (cell bodies), dendrites, and terminal regions involves processes substantially different from those evident in axons. Somata, for instance, can exhibit a variety of slow oscillatory firing patterns and long lasting aftereffects of activity that are not observed in axonal preparations. In addition, voltage dependent calcium currents may play a major role in nerve excitability everywhere except the axon, perhaps in connection with the widespread intracellular messenger role of this ion (e.g. excitation–secretion coupling). The giant somata of the gastropod molluscs have so far yielded the most fundamental insights into the mechanisms of the distinctive electrical activity of nonaxonal membrane. It is our purpose here to review these insights.

The giant neurons in ganglia of the gastropods have approximately spherical somata, which range to over 500 μm in diameter, and most have

141

0147-006X/80/0301-0141$01.00

only a single axo-dendritic process. The somata are arrayed accessibly in a cortical layer on the ganglion. A richly varied range of excitability phenomena can be observed in these cells under conditions ranging from largely intact, behaving animals (Willows, Dorsett & Hoyle 1973, Kandel 1976) to internally dialysed, isolated cells (Kostyuk, Krishtal & Pidoplichko 1975b). These phenomena include action potential firing, spike frequency encoding of constant current stimuli, frequency adaptation, spontaneous pacemaking, burst firing, and a variety of spike or tetanic afterpotentials. Oscillations and afterpotentials with characteristic time courses ranging from milliseconds to minutes are readily observable.

The suitability of gastropod somata to voltage clamp was first demonstrated by Hagiwara & Saito (1959). Studies of the ionic basis of giant soma excitation under voltage clamp account for the major advances in our understanding of this system. The recent developments of intracellular ion indicators and internal dialysis techniques are opening a new range of questions concerning molecular mechanisms of ion transport and the relationship between membrane state and intracellular milieu.

This review is concerned with the properties of ionic conductance systems activated directly or indirectly by changes in membrane potential. Ionic conductances sensitive to a variety of neurotransmitters and other chemical agonists are also observed in gastropod somata. Although some of these phenomena may ultimately prove to reflect different aspects or modulations of the same mechanisms discussed here, such agonistic effects are outside the scope of the present work. Agonist effects have been reviewed by Gerschenfeld (1973), Kupfermann (1979), and Barker & Smith (1979).

Soma Voltage Clamp Arrangements

Three general requirements for voltage camp analysis of ionic current are: (a) the ability to impose a specified potential uniformly over a membrane area, (b) the ability to measure current flow through that area, and (c) the ability to change the composition of solutions outside or inside the cell. All voltage clamping arrangements have limitations in each of these regards and the extent to which these requirements are satisfied must be considered carefully in interpreting any voltage clamp observations.

POTENTIAL CONTROL   Control of soma membrane potential is usually accomplished by means of a feedback amplifier and pairs of intracellular and extracellular electrodes: one pair for measuring potential, another for passing current (Cole 1968, Katz & Schwartz 1974). Sometimes a single extracellular electrode is used both for current return and potential reference, but the method is prone to errors due to electrode resistance and polarization. Intracellular electrode contact is made either through micropipettes (Hagiwara & Saito 1959), or through a disrupted membrane area sealed beneath a suction orifice (Kostyuk et al 1975b, Lee, Akaike & Brown 1977). The latter method allows a low resistance contact for both intracellular electrodes but increases the effective series resistance. Newly developed twin orifice and orifice-micropipette methods circumvent this difficulty. These arrangements appear to hold the potential uniform within a few millivolts over the entire soma membrane in most cells (Hagiwara & Saito 1959, Connor & Stevens 1971a; but see Kado 1973). This maintenance of isopotentiality, even during the flow of large ionic currents, must reflect the low internal resistances characteristic of sphere-like soma geometry. The rate at which membrane potential will follow the command signal is limited by preparation series resistance (Hodgkin, Huxley & Katz 1952, Katz & Schwartz 1974, Adams & Gage 1979a). Typical cellular series resistance and membrane capacitance values limit potential settling to exponential time constants on the order of 20 μsec.

CURRENT MEASUREMENT   The presence of an unclamped axo-dendritic process complicates the measurement of soma membrane current. With intact cells, total clamping current consists of significant contributions from both soma and axon. The total current, therefore, has limited usefulness as an indicator of current flow in the well-clamped soma region. Measurement of any rapid somatic conductance change is impossible due to the prolonged (on the order of 10 msec) and nonlinear step charging response of the unclamped axon. Axonal action potentials and other artifacts of poor spatial control complicate interpretation of all clamping current measurements. These problems have been dealt with in two different ways: (a) by patch or focal current measurement at the soma membrane (Frank & Tauc 1964, Neher & Lux 1969, Kado 1973), and (b) by total clamp current measurement after ligation or removal of the axonal process (Connor & Stevens 1971a, Kostyuk, Krishtal & Doroshenko 1974, Connor 1977). With very careful application of either method, capacity transient settling to 0.1% of maximum can be achieved within less than 500 μsec of step onset. At least at low temperatures, this allows resolution of sodium conductance, the fastest known gating process in these neurons (e.g. Adams & Gage 1979a).

SOLUTION CHANGES   Most voltage clamp analysis involves either adding a pharmacological agent or changing the ionic composition at the external surface of the cell. This is often performed by exchanging the external perfusate, since this allows quantitative specification of dosage or concentration, at least in principle. Molluscan somata lend themselves well to this operation, being on the outside of their ganglia, but certain limitations must be recognized. There may be prolonged diffusion delays, and for some substances such as ions, accumulation or depletion effects may displace concentrations near the cell away from bulk perfusate values. Such effects are normally minimized by removing the sheath of connective tissue overlying the ganglion, and can be further reduced by isolated cell techniques. Even in desheathed preparations, however, there is evidence for restricted diffusion access (Eaton 1972, Neher & Lux 1973, Ahmed & Connor 1979). Such restriction probably reflects the highly infolded geometry of the surface membrane and the presence of satellite cells closely associated with the soma (Coggeshall 1967, Graubard 1975). Recently developed techniques for internal dialysis (Kostyuk et al 1975b, Lee et al 1977) and internal ion replacement (Russell, Eaton & Brodwick 1977, Tillotson & Horn 1978) allow alteration and control of the intracellular ionic environment.

Composition of the Total Ionic Current

Corresponding to the diversity of electrical behavior observed in unclamped gastropod somata, the ionic current under voltage clamp is exceedingly complex. After step changes in membrane potential, ionic current relaxations occur on time scales ranging from milliseconds to minutes, and these relaxations may take widely different forms depending on the specific voltage trajectory imposed. The abstraction of order from this diverse phenomenology has proceeded by identification of distinct, relatively simple components of ionic current. The original precedent for this approach is Hodgkin & Huxley's (1952) analysis of axon membrane ionic current.

As is the case for the axon, the most striking features of the voltage clamp current in the soma appear to be due to voltage dependent membrane permeabilities to specific ions. Several major extensions to the axon framework have been necessary, however, to encompass observations of soma currents. First, calcium ions carry a significant fraction of the ionic current in giant somata, while calcium current in squid axons appears to make only a negligible contribution to the total current. Second, the activation kinetics and pharmacology of the soma membrane potassium permeability are far more complex than those observed for the axon potassium permeability. Third, the time dependence of permeability change in voltage clamped somata extends to a range orders of magnitude slower than that encompassed by the squid axon analysis. Finally, there is good evidence that part of the voltage dependence of potassium permeability is actually mediated by intracellular calcium accumulation, while axonal permeability changes appear to be due directly to an action of the electrical field on membrane macromolecules.

The major features of the ionic current in molluscan somata can be summarized by a scheme representing six distinct components as indicated in Table 1. This scheme is a conjunction of components identified by numerous workers over the years upon which we have imposed our own terminology in the hopes of a more coherent presentation. There is no claim that it represents a consensus of opinions even among those workers whose discoveries it purports to encompass. It is adopted mainly as a vehicle for collection and discussion of findings obtained in diverse experimental contexts. The scheme is not exhaustive, since there is no consideration of metabolically driven ionic currents or of currents carried by anions. Neither does it encompass the currents observed with large hyperpolarizations from the resting potential.

Each current component is characterized as an ionic conductance with fixed ion specificity but variable magnitude. The individual components have been distinguished experimentally on the basis of differences in ionic and pharmacological sensitivities with the intent of establishing the simplest possible description of the overall dependence of ionic conductance on voltage and time. It is tempting to suppose that the six distinct current components may correspond to six populations of ionic channel macromolecules, but no such conclusion can be rigorously supported at this time.

The detailed properties of each component listed in the table and the rationale for its separation from total ionic current are discussed in sections on individual components below.

INWARD IONIC CURRENTS

In some molluscan neurons influxes of both Na^+ and Ca^{2+} contribute to the rising phase of the action potential. The soma membrane continues to produce action potentials in either sodium-free or calcium-free solutions but becomes inexcitable if both sodium and calcium ions are removed from the external solution (*Aplysia:* Junge 1967, Geduldig & Junge 1968, Carpenter & Gunn 1970, Wald 1972, *Helix:* Gerasimov et al 1965, Meves 1968).

Table 1 Summary of ionic currents

Symbol	Name	Selectivity
I_{Na}	Sodium current	Na^+
I_{Ca}	Calcium current	Ca^{++}
I_B	Slow inward current	Na^+, Ca^{++}
I_A	Transient potassium current	K^+
I_K	Voltage-activated late current	K^+
I_C	Calcium-activated late current	K^+

Evidence for the independence of sodium and calcium currents is obtained from kinetic, electrochemical, and pharmacological arguments. Differences in the time course and voltage dependence of inward sodium and calcium currents were reported for the *Aplysia* neuron, R_2 (Geduldig & Gruener 1970). Although studies of membrane currents in other cells suggested that the kinetics of Na^+ and Ca^{2+} ion fluxes might be more similar. Newer analyses with better suppression of potassium current have shown that the Na^+ and Ca^{2+} currents differ in gating kinetics and voltage dependence of activation and inactivation. The substitution of impermeant organic cations such as Tris (tris-hydroxy-methylamino methane), choline, or tetramethylammonium, and the effects of varying the external sodium concentration indicate that the earlier, most rapidly rising, inward current recorded in molluscan neurons is carried by sodium ions (Geduldig & Gruener 1970, Kostyuk & Krishtal 1977a, Lee et al 1977, Adams & Gage 1979a). A slower and more sustained inward current is carried by Ca^{2+} ions. It can be blocked selectively by external application of the inorganic cations Ni^{2+}, La^{3+}, Cd^{2+}, Mn^{2+}, Co^{2+}, Mg^{2+} (Geduldig & Gruener 1970, Kostyuk & Krishtal 1977b, Akaike et al 1978b, Adams & Gage 1979a).

Independence of Na and Ca channels is also suggested from the study of gating currents. When all the ionic currents are blocked, subtraction of symmetrical capacitive and leakage currents reveal nonlinear displacement currents. Such displacement currents are believed to be due to intramembranous charge movements involved in opening and closing ion channels in response to changes in the electric field and are called gating currents (see review by Almers 1978). Detection of gating currents associated with activation of Na and Ca channels has been reported in *Aplysia* and *Helix* neurons (Adam & Gage 1976, 1979c, Kostyuk, Krishtal & Pidoplichko 1977). A rapid component is thought to represent Na gating current while a slower component is thought to represent Ca gating current. The amplitudes of gating currents vary with clamp potential in a way that corresponds to the voltage dependence of Na^+ and Ca^{2+} ionic currents.

The Sodium Current, I_{Na}

The activation of sodium current by voltage has been described in *Aplysia* (Adams & Gage 1979b) and in *Helix* neurons (Kostyuk & Krishtal, 1977a). Sodium current in the *Aplysia* neuron, R_{15}, is activated at potentials more positive than –25 mV and is steeply voltage dependent with half maximal conductance occurring at a potential of –8 mV. During maintained depolarization, sodium current activates and then inactivates as in axons. The peak current occurs within a few milliseconds while the time constant of inactivation is 10–20 msec (Adams & Gage 1979a, b, Kostyuk & Krishtal 1977a).

Rates of activation and inactivation are voltage dependent, increasing with depolarization. The time course of Na current in both *Aplysia* and *Helix* can be adequately described by m^3h kinetics where m and h are the Hodgkin and Huxley parameters for activation and inactivation (Adams & Gage 1979b, Kostyuk & Krishtal 1977a). The amplitude and duration of conditioning voltage pulses strongly effect the amplitude of sodium current during a test pulse by affecting the level of inactivation. Unlike sodium current in the squid axon, little if any additional inactivation is removed with conditioning potentials more negative than the resting potential. The voltage dependence of sodium inactivation is similar in *Aplysia* and *Helix* neurons with half-inactivation occurring at about –30 mV (Geduldig & Gruener 1970, Adams & Gage 1979b, Standen 1975, Kostyuk & Krishtal 1977a). Recovery from inactivation is exponential with a time constant of about 30 msec at the resting potential (*Aplysia;* Geduldig & Gruener 1970, Bergmann, Klee & Faber 1974, Adams & Gage 1979b; *Helix:* Kostyuk & Krishtal 1977a).

The measured reversal potential for sodium current is about +50 mV in solutions containing the normal external Na^+ concentration (*Aplysia:* Geduldig & Gruener, 1970; Adams & Gage, 1979a; *Helix:* Kostyuk et al 1975b, Lee et al 1977). Although no comprehensive study of the ion selectivity of the sodium channel has been undertaken, comparison of the relative permeabilities of monovalent cations in R_{15} of *Aplysia* indicates that the sodium channel is permeable to Li^+ ions but relatively impermeable to Cs^+ and K^+ ions as found in squid and frog axonal membranes (Adams & Gage 1979a; see review by Hille 1975). Tetrodotoxin (TTX) selectively blocks ion movements through the channel independent of ionic species (Geduldig & Gruener 1970, Kado 1973, Adams & Gage 1979a). In most gastropod neurons, the sodium current is less sensitive to TTX than in squid axon, and TTX-insensitive channels have been described in *Helix* (Kostyuk et al 1974, Kostyuk & Krishtal 1977a). In some cases, relative insensitivity may be related to the prior use of trypsin to dissociate cells, since this treatment has been shown to inhibit the action of TTX in *Helix* (Lee et al 1977, but see Connor 1977).

The Calcium Current, I_{Ca}

Compelling evidence for a calcium current in gastropod somata comes from the ionic and pharmacological sensitivities of voltage clamp current, and from chemical indicators of the intracellular free calcium level. The gating characteristics of the calcium inward current are quite different from those of the sodium current, beyond the fact that both are activated by membrane depolarization.

MEASUREMENT OF CALCIUM CURRENT   Though the presence of a calcium conductance system is demonstrated readily, separation of the current for quantitative study has proven difficult. Separation by time dependence or voltage dependence is possible only for a limited range of activation conditions, and the diverse effects of calcium outside of its role as charge carrier complicate the interpretation of ion substitution experiments (see section on calcium-activated potassium current below; see also Frankenhaeuser & Hodgkin 1957, Hille, 1968). Direct electrical observation of calcium current at positive potentials is normally thwarted by the activation of much larger currents carried by sodium and potassium ions. The sodium current can be suppressed in several different ways but potassium current poses a less tractable problem, particularly because of evidence that part of the potassium current depends on intracellular calcium. This dependence rules out the procedure of subtracting current after substitution for Ca^{2+} as an assay of calcium current.

This section discusses characteristics of the calcium current when studied after separation by one of several alternative methods:

1. Examination of early inward current in sodium-free medium or in the presence of TTX (Geduldig & Gruener 1970). Calcium current activation occurs somewhat more rapidly than potassium activation, so that early measurements should reflect calcium current characteristics. Separation by this method is only partial and does not allow observation of the true time course of calcium current during maintained depolarizations.

2. Examination of calcium tail currents in axotomized neurons upon repolarization to the potassium equilibrium potential (Connor 1977, Adams & Gage 1979b). This method requires speed and good spatial uniformity of the voltage clamp arrangement, because the tail current is very rapid and any poorly clamped membrane area could result in confusion of calcium and potassium currents.

3. Pharmacological suppression of outward potassium current with high concentrations of tetraethylammonium ions, TEA (Adams & Gage 1979a, Connor 1979). The applicability of this method is limited, however, since TEA does not totally suppress potassium current (see Pharmacology of Potassium Currents).

4. Replacement of intracellular potassium by a relatively impermeant cation, thus eliminating or reducing current through potassium channels (see section on I_K). Two methods have been employed: internal dialysis (Kostyuk et al 1975b, Lee et al 1977) and nystatin-mediated monovalent cation exchange (Tillotson & Horn 1978). A disadvantage of either method is uncertainty about possible effects of the ion replacement procedures on the calcium current itself.

5. Measurement of intracellular free calcium concentrations by a chemical indicator. This provides an alternative to electrical techniques for detection of calcium entry. Both the photoprotein aequorin (Stinnakre & Tauc 1973, Eckert, Tillotson & Ridgway 1977, Lux & Heyer 1977) and the metallochromic dye arsenazo III (Gorman & Thomas 1978, Ahmed & Connor 1979) have been used as indicators of intracellular calcium, but the latter has now been shown to provide both better calcium detection sensitivity and more readily quantifiable results (see Ahmed & Connor 1979). Indicator measurements have the advantage of providing excellent specificity for calcium movement without the use of blocking agents or ionic substitutions. On the other hand, quantitative inferences about the calcium current are restricted by imperfect knowledge of the cellular calcium clearance mechanisms that affect the accumulation monitored by an indicator response.

ION PERMEATION   The reversal (zero-current) potential obtained for the calcium current in neurons exposed to high external TEA concentrations or loaded with Cs^+ ions is between $+60$ and $+70$ mV (Connor 1979, Tillotson & Horn 1978, Adams & Gage 1979a). A major discrepancy exists between this and the value estimated from voltage dependent changes in the intracellular calcium concentration measured with aequorin and arsenazo III, which is about $+130$ mV (Eckert et al 1977, Ahmed & Connor 1979). A similar high value for the reversal potential was obtained after removal of internal K^+ by the dialysis technique (Kostyuk & Krishtal 1977a, Akaike et al 1978b). The indicator and dialyzed cell values are in good agreement with that predicted for a calcium-selective channel. Factors that may contribute to these differences in reversal potential measurements are: (a) the block of potassium current by TEA becomes ineffective at large positive potentials, and (b) the calcium channel is permeable to some extent to the monovalent cations Na^+ and K^+, which move out of the cell at positive voltages. Evidence suggesting that monovalent cations permeate the Ca channel is provided by single channel conductance measurements in the presence of sodium ions.

   Although no comprehensive study of the ion selectivity of the calcium channel has been completed, there is information concerning permeability of some divalent cations. Sr^{2+} and Ba^{2+} have been found to be permeable while no measurable current is obtained on the replacement of Ca^{2+} by Mg^{2+} (Gola et al 1977, Magura 1977, Connor 1977, 1979, Akaike et al 1978b, Adams & Gage 1979b). The selectivity sequence obtained from reversal potential measurements in dialyzed *Helix* neurons is $Ba^{2+} \cong Sr^{2+} > Ca^{2+}$ (Akaike et al 1978).

Calcium currents are blocked by some other external cations in a concentration dependent and reversible manner (Geduldig & Gruener 1970, Kostyuk & Krishtal 1977a, Akaike et al 1978a, Adams & Gage 1979a). The sequence of effectiveness from dose-response curves in *Helix* neurons is $Ni^{2+} > La^{3+} > Cd^{2+} > Co^{2+} > Mg^{2+}$ (Akaike et al 1978b). Organic calcium antagonists such as verapamil and D-600 are effective in blocking calcium current though nonspecifically (Kostyuk & Krishtal 1977a, Akaike et al 1978b, Adams & Gage 1979a). This pharmacology corresponds closely to that of calcium currents in other excitable membranes (see reviews by Baker & Glitsch 1975, Hagiwara 1975). Internal dialysis of *Helix* neurons has shown that calcium current is blocked irreversibly by internal perfusion with fluoride ions and is blocked by a high intracellular free-calcium ion concentration (Kostyuk et al 1975b, Kostyuk & Krishtal 1977b, Akaike et al 1978a).

The relationship between peak calcium current amplitude and external calcium concentration exhibits saturation with calcium concentrations above normal (Standen 1975, Akaike et al 1978b, Adams & Gage 1979a). Furthermore, Akaike et al (1978b) provide evidence for voltage dependent binding of calcium to membrane sites in the channel such that the affinity of the site for calcium is decreased with depolarization. Competitive block of calcium channels by various metal cations and saturation at modest elevations of external calcium ion concentration are consistent with the external site binding model proposed for the Ca channel in barnacle muscle fibers by Hagiwara (1975).

Noise measurements presumed to represent calcium channel gating fluctuations have been obtained by measuring current fluctuations before and after blocking calcium channels. Measurements obtained by this method in internally dialyzed *Helix* neurons provide a single channel conductance for the Ca channel of 0.5 pS or less when Ca^{2+} is the charge carrier and a value of 1.1 pS for Ba^{++} transfer through the channel (Akaike et al 1978a). A theory relating the power spectrum of current fluctuations to the macroscopic gating behavior of the calcium conductance has been proposed by Akaike et al (1978b).

GATING CHARACTERISTICS   A sufficient depolarizing step leads to activation of calcium current, which approaches a maximum with a time constant of 5 to 10 msec (*Helix:* Kostyuk & Krishtal 1977a, Akaike et al 1978b; *Aplysia:* Adams & Gage 1979b). If the depolarization is maintained, the calcium current declines along a relatively prolonged time course. Upon repolarization, calcium current deactivates exponentially with a time constant somewhat faster than the activation process. Some studies have suggested that calcium currents may facilitate or increase during trains of identical depolarizing pulses (Eckert et al 1977, Lux & Heyer 1977). It now appears that the facilitation observed in these studies probably reflects properties of the Ca^{2+} indicator (aequorin) employed rather than the calcium current itself (Ahmed & Connor 1979, Smith & Zucker 1979).

Activation of Ca conductance in *Aplysia* and nudibranch neurons occurs at potentials more positive than those required for activation of sodium conductance (Geduldig & Gruener 1970, Tillotson & Horn 1978, Connor 1977, 1979, Adams & Gage 1976, 1979a), although in *Helix* cells, Ca^{2+} and Na^+ currents first begin to activate at about the same potential (Kostyuk et al 1974, Standen 1975, Kostyuk & Krishtal 1977a, Lee et al 1978). Conductance-voltage curves for marine and terrestrial snails increase sigmoidally with increasing depolarization and maximum calcium conductance is attained at potentials more positive than +30 mV.

The development of inactivation during maintained depolarization has been a point of uncertainty in the published data on Ca current. Many early microelectrode studies reported complete inactivation occurring within a fraction of a second after stepping to positive potentials. Slower or less complete inactivation has been reported where cells have been bathed in high concentrations of TEA (Connor 1979, Adams & Gage 1979b). Furthermore, in dialyzed cells after removal of intracellular potassium ions, a fraction of the calcium current failed to inactivate even at very positive potentials (Kostyuk & Krishtal 1977a, Lee et al 1978). The apparent dependence of calcium channel behavior on procedures designed to block or remove potassium current suggests two possibilities: (*a*) calcium channels become less prone to inactivation in cells treated with TEA and in internally dialyzed cells, or (*b*) suppression of potassium current in such cells reveals normal, incomplete inactivation of calcium channels. Arsenazo III studies suggest that the latter possibility is the case. Intracellular calcium is observed to accumulate progressively for several seconds during prolonged clamp steps, even in intact cells bathed in normal external media (marine: Gorman & Thomas 1978, Ahmed & Connor 1979; terrestrial: S. J. Smith, unpublished). Sustained accumulation strongly suggests that inward calcium current persists for the entire duration of the pulse. Ahmed & Connor (1979) have shown that external addition of TEA has negligible effect on the arsenazo response to depolarization, which suggests that the slow and incomplete inactivation observed in TEA reflects normal channel behavior. Considerable uncertainty remains concerning the presence of inactivation, its presumed kinetics and the amplitude of the effect. These are major questions, especially when the pervasive role of internal Ca^{2+} in nerve cells is considered. Most of the uncertainty stems from the extreme difficulty encountered in isolating Ca current and further complications arising from internal Ca^{2+} accumulation are anticipated. In the following paragraphs, the data on inactivation are collected in the hope of stimulating further research into this issue.

Ca-current inactivation has been described in *Aplysia* (Geduldig & Gruener 1970, Adams & Gage 1979b) and *Helix* neurons (Standen 1975, Kostyuk & Krishtal 1977a, Akaike et al 1978b). Inactivation develops during depolarizing with an approximately exponential time course that depends on the conditioning prepulse potential. Inward calcium currents are reduced by prepulses or at holding potentials more positive than the resting potential with half-inactivation at about -20 mV. The Ca inactivation curve obtained in *Helix* neurons is less steeply voltage dependent than in *Aplysia*. The Ca^{2+} inactivation curve in the *Aplysia* giant neuron, R_2, exhibits an anomalous depression of the calcium current at hyperpolarized holding potentials (Geduldig & Gruener 1970). Studies in *Helix* cells suggested that the apparent depression of Ca current generated from hyperpolarized holding voltages is due to the simultaneous activation of a transient potassium current, A-current (Neher 1971, Standen 1974). Recent investigation of this phenomenon in the *Aplysia* neuron, R_{15}, measuring calcium ionic and gating currents, however, suggests that the Ca channel itself may indeed be inactivated at hyperpolarized potentials (Adams & Gage 1979b, c).

Recovery from Ca-current inactivation, studied with double pulse techniques, exhibits two phases in *Aplysia* neurons (Tillotson & Horn 1978, Adams & Gage 1979b). The first phase has a time constant on the order of milliseconds and the second phase relaxes over several seconds. Only about 50% recovery of Ca current has occurred 1 sec after conditioning activation.

Differences between inward currents carried by Ba^{2+} and by Ca^{2+} ions bear on the mechanism of Ca-current inactivation. Barium appears to carry current through the same channels as calcium, but barium currents exhibit a much slower and less complete inactivation (Magura 1977, Gola et al 1977, Connor 1977, Adams & Gage 1979b). Since Ba^{2+} ions are known to suppress outward potassium currents (see Pharmacology of Outward Current), two interpretations of the Ba^{2+} substitution effect should be considered: (*a*) Ba^{2+} prevents a residual potassium activation, which is otherwise erroneously identified as Ca-current inactivation, or (*b*) substitution of Ba^{2+} for Ca^{2+} actually alters the inactivation characteristics of the calcium conductance system. Both interpretations may be partially correct.

Connor (1979) has shown that Ca^{2+} inward currents after intracellular EGTA injection resemble currents after Ba^{2+} substitution. If one assumes that EGTA has this effect by its familiar action of calcium chelation, this observation implies that a chemically specific effect of intracellular Ca^{2+} accumulation may account for the observed differences between Ca^{2+} and Ba^{2+} inward currents. One such specific effect might be the activation of potassium current. Connor (1979) has suggested that the Ba^{2+} and the internal EGTA effects are evidence in favor of interpretation (*a*) above. Alternatively, intracellular calcium may have a specific blocking or inactivating action on Ca channels not shared by barium, as suggested by Tillotson (1979).

The Slow Inward Current, I_B

A component of inward current characterized by extremely slow decay after depolarizing pulses was originally observed in bursting pacemaker neurons by Gola (1974) and by Eckert & Lux (1975). This slow current may represent an additional conductance system, distinct from I_{Na} and I_{Ca}, which is characteristic of this cell type alone (Arvanitaki & Chalazónitis 1961). Because of its occurrence in bursting cells and the role it is thought to play in producing bursting activity, we will refer to this slow inward current as *B*-current (I_B).

B-current is active at subthreshold voltages within the pacemaker voltage range. For small depolarizations, I_B activates over a time course of seconds and there is a seconds-long tail current at the end of the pulse (Gola 1974, T. Smith et al 1975, Thompson 1976). *B*-current is the smallest of the ionic currents described in molluscan cells. The largest amplitudes of I_B observed are less than one-hundredth the maximum amplitude of I_{Na} or I_{Ca}. It is, nonetheless, readily distinguished from the other inward currents by its slow kinetics of decay at negative holding potentials. Because of the simultaneous activation of other much larger ionic currents, I_B cannot be observed directly in the positive voltage range. The kinetics of activation during depolarization must be inferred from observations of slow tails at potentials near the reversal potential for outward current following depolarizing pulses of various dimensions. An analysis of the voltage dependence and kinetics of I_B based on tail current measurements was reported by Thompson (1976). Activation appears to be approximately exponential in time and the current does not inactivate with prolonged depolarization. The activation time constant is about 2 sec at –60 mV and decreases with depolarization, reaching about 200 msec at 0 mV (Gola 1974, Eckert & Lux 1975, 1976, Thompson 1976).

Several studies have identified an inward current in the pacemaker voltage range from measurements of steady-state I-V curves (Wilson & Wachtel 1974, Eckert & Lux 1975, 1976, T. Smith et al 1975). The inward current studied in this way bears resemblance to the slow current studied by the tail current method, especially in voltage dependence of activation. A note of caution was suggested by Partridge et al (1979), however, who studied very similar steady-state I-V curves in nonbursting cells where I_B was not observed. The kinetics of inward current relaxation in these cells were very much faster than those of I_B observed in bursters. Smith et al (1975) found that the slow inward current was not blocked by TTX in *Otala* but its amplitude depended on external Na^+ concentration. Eckert & Lux (1976) found that the current amplitude was reduced in low Ca^{2+} in *Helix*. It is not presently clear whether I_B represents the sum of different Na^+ and Ca^{2+} currents or a single mechanism with mixed selectivity. Because of its very low amplitude, I_B is difficult to examine free of contamination by other currents. Neither direct measures of reversal potential nor measures of instantaneous current voltage relations seem feasible. Although the voltage dependence of its amplitude and kinetics suggest a voltage sensitive gating mechanism, the characterization of its activation remains incomplete.

B-current is thought to play a significant role in the generation of bursting activity (Gola 1974, T. Smith et al 1975, Thompson 1976). Because of its slow kinetics in the pacemaker voltage range, I_B gives rise to a prominent depolarizing afterpotential after individual spikes and similarly provides a lasting depolarizing drive that leads to the aggregation of spikes into sustained bursts (Thompson & Smith 1976). There have been suggestions that control of neuronal bursting by neurotransmitters or neurohumoral agents may be affected via modulation of the slow inward current (Barker & Smith 1979, Wilson & Wachtel 1978, see also Pellmar & Carpenter 1979).

OUTWARD IONIC CURRENTS

The outward currents seen during voltage clamp steps are thought to be the sum of three separate potassium conductance systems that can occur in varying ratios in different cells. The three potassium ionic currents are designated I_A, I_K, and I_C. I_A is a transient potassium current first studied by Hagiwara & Saito (1959) and subsequently by Connor & Stevens (1971b) and Neher (1971). I_K is a late outward current controlled by a voltage dependent gating mechanism. I_C is a late outward current activated by increases in internal ionized Ca^{2+} concentration (for review see Meech 1978). I_A can readily be separated from the late outward currents, and from inward sodium and calcium currents (at least over part of the range of activation potentials) because it activates at more negative voltages. It can also be readily identified on kinetic grounds since it activates and inactivates more rapidly than late outward current (Kostyuk et al 1975a, Neher 1971, Connor & Stevens 1971c). Furthermore, I_A appears to have no dependence on intracellular calcium ions (Connor 1979). These considerations and the results of pharmacological experiments described below leave little doubt that A-current represents a distinct and separable component of potassium conductance. It is possible to eliminate it from current records simply by studying membrane currents on depolarization from a holding voltage that results in complete inactivation (Connor & Stevens 1971b). Separation of the late outward current into its components (I_K, I_C) is not so straightforward. Because both components activate over similar voltage ranges and there is overlap in kinetics, independent methods must be used to separate them. Pharmacological techniques allow such a separation.

Pharmacology of Potassium Currents

The potassium channel blockers 4-aminopyridine (4-AP) and tetraethylammonium (TEA), and blockers of Ca^{2+} current can be used either alone or in combination to separate and identify I_K and I_C.

Activation of I_C is blocked by procedures that prevent the concentration of intracellular ionized Ca^{2+} from increasing during depolarization. Connor (1979) outlined several effective procedures: (a) replacement of external Ca^{2+} with Mg^{2+}, (b) blockage of Ca^{2+} influx, (c) internal buffering of Ca^{2+} by EGTA injection, (d) replacement of Ca^{2+} by Ba^{2+}.

Method (a) is the least effective, probably because residual Ca^{2+} in membrane infoldings cannot readily exchange with the bathing medium, but the effectiveness of the approach can be improved by addition of a Ca^{2+} blocking ion. Injection of EGTA into cytoplasm blocks the activation of I_C but does not block Ca^{2+} influx. This suppression of I_C is probably due to prevention of the increase in intracellular Ca^{2+} concentration by the buffering action of EGTA. Connor (1979) has shown that 1–2 mM internal EGTA is sufficient to block the activation of I_C in nudibranch neurons.

The potassium current blocker 4-aminophyridine (4-AP) is a particularly effective antagonist of I_A. One-half block occurs at an external concentration of 1.5 mM in nudibranchs (Thompson 1977). The block by external 4-AP is voltage dependent, increasing with time during hyperpolarizing conditioning steps and decreasing with time during depolarization. In contrast to its effects on axonal membrane, 4-AP does not affect the late outward currents at concentrations that completely block I_A (Thompson 1977, Adams & Gage 1979a).

Tetraethylammonium ion also has differential effects on the components of potassium current. Furthermore, its action and effective concentration differ depending on whether TEA is applied externally or internally. Internal TEA blocks I_A and I_K approximately equally (Neher & Lux 1972). External TEA dramatically reduces late outward current but is much less effective in blocking I_A (Connor & Stevens 1971a). A 50% reduction of I_K occurs at 5–12 mM external TEA in *Helix* and at about 8 mM in marine species (Neher & Lux 1972, Thompson 1977). A-current is 50% blocked at a concentration of 20–80 mM in *Helix* and 100 mM in nudibranchs. Even at high concentrations, however, some of the late outward current remains

TEA insensitive (Hagiwara & Saito 1959, Heyer & Lux 1976, Thompson 1977). This relatively TEA-insensitive component has been identified as the calcium-activated potassium current. Even though TEA is not totally selective in its action, the three components of potassium conductance differ in their sensitivity to it. TEA, in fact, has been used to block I_K in order to measure I_C during depolarization (Aldrich et al 1979a). Hermann & Gorman (1979) have reported that potassium currents due to Ca^{2+} microinjection are more sensitive to external TEA than is I_K.

The Transient Potassium Current, I_A

The activation of I_A results in a transient potassium current that peaks and then inactivates at a much slower rate. Inactivation is removed by a conditioning hyperpolarization. Unlike the late outward currents, I_A is activated at subthreshold voltages near the resting potential, and exerts a major influence on excitability in that voltage range: slowing the rate of depolarization in response to a stimulus thereby delaying the action potential, and slowing the rate of repetitive firing in response to maintained stimuli, especially for the first few spikes in a train (Connor & Stevens 1971c). Gola (1974) has suggested there may be a residual level of A-current activation even at voltages negative to the reversal potential for the process (about –65 mV) and that the inactivation gating of I_A may therefore contribute to the inward rectification that is prominent in some cells.

The activation of A-current on depolarization follows a sigmoid rise to a maximum in 10 to 50 ms, which is followed by an exponential inactivation (Connor & Stevens 1971b, Neher 1971). Both activation and inactivation time constants exhibit shallow voltage dependencies, decreasing with depolarization. The inactivation time constant characteristically differs among various identified cells in a species although it is fairly constant for any single identified cell. To ascertain the role played by I_A as a determinant of the voltage trajectory near threshold, it would be interesting to know how differences in kinetics are correlated with differences in repetitive firing to constant current stimulation.

Several detailed analyses of the voltage dependence of I_A have appeared (Connor & Stevens 1971b, c, Neher 1971, Gola 1974, Partridge & Connor 1978, Thompson 1977, Connor 1978). Connor & Stevens (1971c) described the kinetics of I_A with an equation resembling that used by Hodgkin & Huxley (1952) to describe the Na current in squid axon. The conductance of A-current channels was described by the product of activation and inactivation terms whose values are determined by first-order rate equations employing voltage dependent activation and inactivation time constants. The value of the activation term was raised to the fourth power to account for the delay in the rise of I_A in nudibranchs. Neher (1971) used a third power relation to describe the kinetics in *Helix*.

A-current is first seen with depolarization ranging from -60 to -45 mV in the nudibranchs and to -70 mV in *Helix*. Measurement of the voltage where the conductance activates fully is complicated by activation of other currents and has not been clearly determined. Steady-state inactivation is complete near the resting potential and is removed completely at about -100 mV.

The Voltage-Activated Late Potassium Current, I_K

A number of studies describing the properties of total late outward current have appeared (Connor & Stevens 1971a, Neher 1971, Meech & Standen 1975, Adams & Gage 1979a, Leicht et al 1971). Such studies are important because they relate directly to the normal operation of the cell, but they do not permit description of the properties of I_K in isolation. As in other sections of this review, we restrict ourselves to those studies that have attempted to fully isolate I_K from I_C by any of the methods mentioned and we consider data where I_K has been clearly separated from I_A either by subtraction or by using a holding voltage at which A-current is fully inactivated (Heyer & Lux 1976, Kostyuk et al 1975a, Meech & Standen 1975, Eckert & Lux 1977, Thompson 1977, Aldrich et al 1979a).

During prolonged depolarization from the resting potential, K-current rises with a delay to a peak and then declines over several seconds to a nonzero steady-state value. The decline in I_K appears to be due to inactivation of potassium conductance and not due to extracellular accumulation of potassium near the membrane. Although potassium accumulation can occur (Alving 1969, Eaton 1972, Neher & Lux 1973), inactivation is observed under conditions that do not result in a shift in reversal potential (Gola 1974, Heyer & Lux 1976, Aldrich et al 1979a).

Slow inactivation of late outward current has been noted by several authors (Hagiwara et al 1961, Hagiwara & Saito 1959, Connor & Stevens 1971a, Eckert & Lux 1977, Gola 1974, Heyer & Lux 1976, Aldrich et al 1979a, Leicht et al 1971, Kostyuk et al 1975a, Barker & Smith, 1979). Furthermore, progressive frequency dependent decline in outward current amplitude occurs during repetitive depolarization. When voltage pulses are presented at low frequencies (e.g. 1 Hz) the maximum outward current during the second pulse is often less than the current at the end of the preceding pulse, which gives the impression that inactivation continues at about the same rate even during the repolarized interval (Kostyuk et al 1975a, Neher & Lux 1972). Also characteristic of this process is a slowing of the rate of rise of I_K as inactivation progresses toward saturation during repetitive depolarizations. This distinctive inactivation process is seen in many molluscan neural somata but it differs qualitatively from that observed in axonal preparations and from the inactivation of I_A. The process has been called cumulative inactivation (Aldrich et al 1979a).

K-current begins to activate with depolarization to -30 mV and activation increases with depolarization up to positive voltages. The rising phase is sigmoidal, begins after a substantial delay, and is best described by a sum of exponentials rather than by a power of a single exponential (H. Reuter & C. F. Stevens, personal communication). Measurements of the voltage dependence and kinetics of activation of I_K have not been accurately conducted because of complications introduced by the inactivation process.

Inactivation during prolonged depolarization is incomplete and the degree of inactivation depends on voltage and differs among cells. I_K is about one-half inactivated at rest and inactivation is maximal between $+10$ and $+20$ mV. With further depolarization, inactivation begins to decrease again. The steady-state inactivation curve is, therefore, a U-shaped function of voltage (Magura et al 1971, Kostyuk et al 1975a, Heyer & Lux 1976, Eckert & Lux 1977, Aldrich et al 1979a). The time course of inactivation measured from the decline of outward current during a prolonged depolarization can be reasonably fitted by a single exponential function. The time constant is on the order of seconds for *Helix* (Heyer & Lux 1976) and nudibranchs (Aldrich et al 1979a) and is a bell-shaped function of voltage, first decreasing then increasing with depolarization. A more complete picture is obtained when the time course of inactivation is measured with a method that compares the effect of pre-pulses of various durations on the peak amplitude of I_K during a test pulse. When measured by this method the time course is best fitted by the sum of two exponential functions. The slower relaxation corresponds to the decline in current during a long depolarization to the same voltage, while the more rapid relaxation is up to ten times faster and occurs on a time scale similar to the rise time of I_K on depolarization. In dorid neurons, as much as 90% of inactivation occurs by the fast process (Aldrich et al 1979a). Significant inactivation occurs before the time of peak current during depolarization so that in the absence of inactivation the peak current would be much larger.

Recovery from inactivation is extremely slow near the resting potential; it requires 20–30 seconds in *Helix* and as much as 1 min. in *Archidoris* (Kostyuk et al 1975a, Heyer & Lux 1976, Aldrich et al 1979a). The recovery time course is usually measured with a two-pulse procedure, where the persistence of inactivation accrued during the first pulse is measured at different intervals by its effect on a test pulse. For interpulse intervals less than about one second, inactivation increased with interval duration while for greater intervals gradual recovery is observed. The overall recovery time course is, therefore, U-shaped (Heyer & Lux 1976, Eckert & Lux 1977,

Aldrich et al 1979a). Final recovery from inactivation is very slow, at least an order of magnitude slower than the onset of inactivation at the same voltage. Very slow recovery necessitates the use of long (30–60 sec) inter-pulse intervals during voltage clamp experiments to insure return to control conditions. Kostyuk et al (1975a) and Aldrich et al (1979a) have presented models for I_K inactivation. Both models assume a single population of channels that are subject to two kinds of inactivation processes.

Heyer & Lux (1976) and Eckert & Lux (1977) reported voltage dependent inactivation of I_K, but also reported that the distinguishing properties of cumulative inactivation (frequency dependence, greater inactivation for the second pulse in a pair, prolonged recovery) were absent in *Helix* neurons when Co^{2+} was substituted for Ca^{2+}. Concluding that cumulative inactivation was a property of the Ca^{2+} dependent potassium current (I_C), they suggested that I_C could be blocked by intracellular Ca^{2+} ions that accumulate during depolarization (see Plant 1978). Presumably the recovery rate from inactivation would then be determined by the rate of clearance of free Ca^{2+} in the cell. In nudibranch cells, however, cumulative inactivation occurs in Co^{2+} substituted saline (Aldrich et al 1979a). Furthermore, Kostyuk & Krishtal (1977b) observed cumulative inactivation in *Helix* cells dialyzed internally against EGTA solutions, conditions where accumulation of Ca^{2+} should be prevented. Differences between these results seem to be more procedural than actual, and more studies on specifically identified cells will be needed to clarify this difference.

The reversal potential for I_K measured from tail currents is about -65 mV and is similar to I_A reversal. This value contrasts with the ionic equilibrium potential for potassium in *Aplysia* (-75 mV) measured with ion selective electrodes (Kunze, Walker & Brown 1971). The discrepancy can be explained if the channel has a finite permeability to other ions. Ionic selectivity of the channel has been studied in *Helix* neurons from reversal potential measurements and found to be $Tl^+ > K^+ > Rb^+ > Cs^+ > Li^+ > Na^+$. Single channel conductance measured for K-current in the same study using dialyzing voltage clamp and fluctuation analysis was 2–3 pS (H. Reuter & C. F. Stevens, personal communication).

Consideration of its rate of activation suggests that the main function of I_K is to provide outward current to repolarize the membrane during a spike. Because of this, cumulative inactivation can have a significant effect on spike shape, contributing to broadening of the action potential during repetitive firing (Barker & Smith 1979, Aldrich et al 1979b). It is interesting that unlike the curves for the other currents, the voltage dependence of cumulative inactivation is steep at the resting potential so that small changes in subthreshold voltage can lead to significant changes in the level of inactivation of I_K.

The Calcium-Activated Late Potassium Current, I_C

The properties of I_C, the calcium-activated potassium current, in molluscan neurons and in other cell types were reviewed recently (Meech 1978). It seems clear that activation of I_C occurs subsequent to an increase in cytoplasmic free calcium concentration whether the calcium enters through the plasma membrane during depolarization, is released from internal storage sites by photosimulation, or is injected into the cell as a Ca^{2+} salt or EGTA-Ca buffer (Eckert & Tillotson 1978, Thomas & Gorman 1977, Andressen & Brown 1979, Meech 1974, Ahmed & Connor 1979). Calcium influx can be uncoupled from potassium flux by appropriate buffering of internal Ca^{2+} (Connor 1979). The exact relationship between internal free calcium concentration and potassium conductance is not known for neurons (see Simons 1976, for studies on red blood cells), but experiments using injected EGTA-Ca buffers indicate a liminal effective Ca^{2+} concentration of $1-9 \times 10^{-7}$ M (Meech 1974).

C-current has an apparent voltage dependence because of its requirement for internal Ca^{2+} and the fact that I_{Ca} is voltage dependent. The apparent voltage dependence is bell-shaped, rising to a peak with increasing depolarization and then falling off as the reversal potential for Ca current is approached (Meech 1974, Thompson 1977). The more direct question of an intrinsic voltage dependence of I_C gating, which is additional to or synergistic with the Ca^{2+} requirement, has not been addressed.

Meech (1974) reported that Ca^{2+}, Sr^{2+}, and Ba^{2+} ions could activate potassium conductance when injected into *Helix* neurons but Mg^{2+} was not effective. When activated by depolarization the amplitude of I_C is greatly reduced if Ba^{2+} or Sr^{2+} replace external Ca^{2+} (Connor 1979, Eckert & Lux 1977). Sr^{2+} is not as effective in suppressing I_C as Ba^{2+}. From experiments on cells bathed in both Ba^{2+} and Ca^{2+}, Connor (1979) concluded that Ba^{2+} does not act like a blocker of I_C, but more like an ion which fails to activate the conductance mechanism as well as Ca^{2+} does.

The selectivity of the channel for various ionic species is not yet known, but measures of reversal potential suggest that I_C may be somewhat less selective for K^+ ions than I_A or I_K (Meech & Standen 1975). Reversal potentials for I_C, measured from tail currents, are somewhat more positive than those for I_K or I_A. This could reflect a less than perfect potassium selectivity, but the measures are almost certainly contaminated by relaxations of other currents, and because of the long clamp pulses needed to activate significant current there may be shifts due to potassium accumulation. Studies on the ion selectivity of I_C will be important in helping to identify it as a unique component of potassium conductance.

The time course of I_C during depolarization can be observed in cells bathed in saline-containing TEA. C-current increases much more slowly than I_K on depolarization and it continues to increase throughout prolonged pulses (Aldrich et al 1979a). Similar results were reported by Heyer & Lux (1976) and Eckert & Lux (1977) for *Helix* neurons. During repetitive pulses, Heyer & Lux (1976) and Eckert & Lux (1977) reported frequency dependent depression of I_C. This effect was not observed in nudibranch cells (Aldrich et al 1979a). In the protocol of Heyer & Lux (1976) the contribution of open I_C channels is neglected and the incremental increase in I_C during repetitive pulses was noted to decrease. The decrease could result from a number of causes other than I_C inactivation: (*a*) partial inactivation of the Ca^{2+} current (see above), (*b*) approach to saturation of the conductance governing I_C, (*c*) a nonlinear relation between intracellular Ca^{2+} and I_C. Further work is needed before one can decide between the alternatives.

The simplest model for *C*-current activation would be to assume that the onset of I_C on depolarization and its relaxation on repolarization are proportional to the rate of change of internal Ca^{2+} concentration near the membrane. Relaxation of I_C after a pulse can last for more than a minute in nudibranch cells after a brief activating pulse (Thompson 1976, S. Smith 1978, Partridge et al 1979). The relaxation of I_C follows a time course quite similar to the absorbance change of the Ca^{2+} indicator, arsenazo III (Gorman & Thomas 1978). The correspondence is not exact, especially at short times after the end of a pulse, but it does argue for the general validity of the simple model. Models describing the time course of intracellular Ca^{2+} concentration transients have been presented for the case of Ca^{2+} entry through the membrane (Thompson 1977, S. Smith 1978) and for the release of Ca^{2+} from bound stores on photostimulation (Andresson, Brown & Yasui 1979). It is not yet clear whether Ca^{2+} is a direct activator of the channel or if a cofactor is required for activation as in muscle contraction or various enzyme activation processes.

As noted above, the relaxation of I_C after depolarizations or spikes can be extremely prolonged. *C*-current is evidently responsible for several very slow events such as post-burst hyperpolarization and spike frequency adaptation, which have a powerful effect on the integrative function of the neuron. Furthermore, because of the dependence on internal Ca^{2+} concentration, *C*-current is potentially subject to influence by factors such as hormones or synaptic modulators, which modify cellular metabolism.

CONCLUSION

Curves depicting the steady-state voltage dependencies of all of the ionic currents have been collected from the literature and presented in Figure 1. The information comes from a number of sources but an effort has been made to select representative data in order to give the best impression of the alignment of the curves. Inspection of the figure shows the relationship

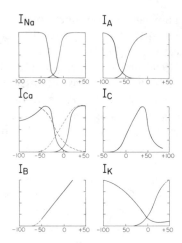

Figure 1 Steady-state voltage dependence of six ionic current components showing the relative magnitude of activation and inactivation variables plotted against a millivolt scale. Note that the voltage scale for I_C is shifted 50 mV to the left. The data used in plotting this figure come from several sources; I_{Na}, Adams & Gage 1979b; I_A, Thompson 1977; I_{Ca}, Adams & Gage 1979b (solid); Akaike et al 1978b (dashed); I_C, Thompson 1977, Smith 1978; I_B, Smith 1978; I_K, Aldrich et al 1979a.

of the currents to each other and to the resting potential and spike threshold. It provides some basis for understanding the range of excitability phenomena in these central neurons when combined with information on the kinetics and relative magnitudes of the currents. Several of the components are activated by depolarizations from rest into the threshold voltage range (I_{Na}, I_{Ca}, I_K). These components have rapid kinetics, relatively large maximum conductances, and are responsible for action potential firing. Other components (I_A, I_B, I_C) are active at more negative voltages, influence the subthreshold voltage behavior of the cell directly and are, therefore, important to the frequency encoding function of the neuron.

The importance of particular properties of the ionic current to electrical excitability can be investigated by mathematical reconstruction procedures. This approach was first used by Hodgkin & Huxley (1952) to demonstrate that the sodium and potassium currents they measured in axons could account for the firing and propagation of action potentials. Reconstruction of excitability behavior is most meaningful when all necessary parameters are determined strictly from voltage clamp data, following the precedent of Hodgkin and Huxley. The first study to account for distinctive features of soma excitability in this way was the analysis by Connor & Stevens (1971c) of repetitive firing in dorid neurons (see also Connor 1978). Subsequent investigations have applied similar procedures to analyze spike frequency adaptation (Partridge & Stevens 1976), bursting pacemaker activity (Smith & Thompson 1975, Gola 1976, Thompson 1976, S. Smith 1978), and effects of drugs or temperature on firing behavior (Williamson & Crill 1976, Partridge & Connor 1978).

A striking feature of the central neurons in molluscan ganglia is the distinctive differences in membrane current patterns encountered in different identified cells. For example, bursting pacemaker cells have B-current, whereas other cell types do not (Thompson & Smith 1976). Aldrich et al (1979a) have documented another source of variability that involves differences in the relative contribution of I_K and I_C to the total outward current among cells. It seems likely that this kind of finding will be extended to the other ionic currents as well, allowing an explanation of the distinguishing features of different cells in terms of the relative contributions of the possible ionic conductance systems.

The calcium current in gastropod somata appears to possess many of the same properties as the calcium current involved in transmitter release at presynaptic terminals of the squid and of frog motoneurons (Katz 1969, Llinás, Steinberg & Walton 1976). The function of calcium entry at the soma, however, is not as well understood. Though the calcium current does contribute to the ability to fire action potentials, it is probably secondary to sodium current in this role. The localization of calcium channels to soma as opposed to axonal regions suggests the likelihood of a specific somatic function (Kado 1973, Junge & Miller 1974, but see Horn 1978). The importance of calcium currents to slow conductance changes involved in pacemaking and long-term variations in excitability has already been noted, but it seems likely that the major significance of the calcium influx may lie in the ability of calcium concentration changes to transmit information within the cytoplasm (see Kretsinger 1977).

ACKNOWLEDGMENTS

This work was supported by NIH grant 1R01 NS14519-01 to S.H.T.; S.J.S. is a Miller Research Fellow and D.J.A. is a recipient of a MDA postdoctoral fellowship. It is a pleasure to acknowledge the assistance of R. Aldrich and P. Bock in preparation of the manuscript and H. Reuter and C. F. Stevens for making unpublished data available to us.

164   ADAMS, SMITH & THOMPSON

Literature Cited

Adams, D. J., Gage, P. W. 1976. Gating currents associated with sodium and calcium current in an *Aplysia* neurone. *Science* 192:783–84

Adams, D. J., Gage, P. W. 1979a. Ionic currents in response to membrane depolarization in an *Aplysia* neurone. *J. Physiol. London* 289:115–42

Adams, D. J., Gage, P. W. 1979b. Characteristics of sodium and calcium conductance changes produced by membrane depolarization in an *Aplysia* neurone. *J. Physiol. London* 289:143–62

Adams, D. J., Gage, P. W. 1979c. Sodium and calcium gating currents in an *Aplysia* neurone. *J. Physiol. London* 291:467–82

Ahmed, Z., Connor, J. A. 1979. Measurement of calcium influx under voltage clamp in molluscan neurones using the metallochronic dye arsenazo III. *J. Physiol. London* 286:61–82

Akaike, N., Fishman, H. M., Lee, K. S., Moore, L. E., Brown, A. M. 1978a. The units of calcium conduction in *Helix* neurones. *Nature* 274:379–82

Akaike, N., Lee, K. S., Brown, A. M. 1978b. The calcium current of *Helix* neuron. *J. Gen. Physiol.* 71:509–31

Aldrich, R. W. Jr., Getting, P. A., Thompson, S. H. 1979a. Inactivation of delayed outward current in molluscan neurone somata. *J. Physiol. London* 291:507–30

Aldrich, R. W. Jr., Getting, P. A., Thompson, S. H. 1979b. Mechanism of frequency-dependent broadening of molluscan neurone soma spikes. *J. Physiol. London* 291:531–44

Almers, W. 1978. Gating currents and charge movements in excitable membranes. *Rev. Physiol. Biochem. Pharmacol.* 82:96–190

Alving, B. O. 1969. Differences between pacemaker and nonpacemaker neurons of *Aplysia* on voltage clamping. *J. Gen. Physiol.* 54:512–31

Andressen, M. C., Brown, A. M. 1979. Photoresponses of a sensitive extraretinal photoreceptor in *Aplysia*. *J. Physiol. London* 287:267–82

Andressen, M. C., Brown, A. M., Yasui, S. 1979. The role of diffusion in the photoresponse of an extraretinal photoreceptor of *Aplysia*. *J. Physiol. London* 287:283–301

Arvanitaki, A., Chalazónitis, N. 1961. Slow waves and associated spiking in nerve cells of *Aplysia*. *Bull. Inst. Oceanogr. Monaco* 58:1

Baker, P. F., Glitsch, H. G. 1975. Voltage-dependent changes in the permeability of nerve membranes to calcium and other divalent cations. *Philos. Trans. R. Soc. London Ser. B* 270:389–409

Barker, J. L., Smith, T. G. 1979. Peptides as neurohormones. In *Society for Neuroscience Symposium*, ed. W. M. Cowan, J. A. Ferrendelli, Bethesda: Soc. Neurosci. 2:340–73

Bergmann, M. C., Klee, M. R., Faber, D. S. 1974. Different sensitivities to ethanol of three early transient voltage clamp currents of *Aplysia* neurons. *Pfleugers Arch.* 348:139–53

Carpenter, D., Gunn, R. 1970. The dependence of pacemaker discharge of *Aplysia* neurons upon Na^+ and Ca^{++}. *J. Cell. Physiol.* 75:121–27

Coggeshall, R. E. 1967. A light and electron microscope study of the abdominal ganglion of *Aplysia* californica. *J. Neurophysiol.* 30:1263–87

Cole, K. S. 1968. *Membranes, Ions and Impulses.* Berkeley: Univ. Calif. Press. 569 pp.

Connor, J. A. 1977. Time course separation of two inward currents in molluscan neurons. *Brain Res.* 119:487–92

Connor, J. A. 1978. Slow repetitive activity from fast conductance changes in neurons. *Fed. Proc.* 37:2139–45

Connor, J. A. 1979. Calcium current in molluscan neurones: measurement under conditions which maximize its visibility. *J. Physiol. London* 286:41–60

Connor, J. A., Stevens, C. F. 1971a. Inward and delayed outward membrane currents in isolated neural somata under voltage clamp. *J. Physiol. London* 213:1–20

Connor, J. A., Stevens, C. F. 1971b. Voltage clamp studies of a transient outward current in gastropod neural somata. *J. Physiol. London* 213:21–30

Connor, J. A., Stevens, C. F. 1971c. Prediction of repetitive firing behaviour from voltage clamp data on an isolated neurone soma. *J. Physiol. London* 213:31–53

Eaton, D. 1972. Potassium ion accumulation near a pace-making cell of *Aplysia*. *J. Physiol. London* 224:421–40

Eckert, R., Lux, H. D. 1975. A non-inactivating inward current recorded during small depolarizing voltage steps in snail pacemaker neurons. *Brain Res.* 83:486–89

Eckert, R., Lux, H. D. 1976. A voltage-sensitive persistent calcium conductance in neuronal somata of *Helix*. *J. Physiol. London* 254:129–51

Eckert, R., Lux, H. D. 1977. Calcium-dependent depression of a late outward current in snail neurons. *Science* 197:472–75

Eckert, R., Tillotson, D. 1978. Potassium activation associated with intraneuronal free calcium. *Science* 200:437–39

Eckert, R., Tillotson, D., Ridgway, E. B. 1977. Voltage-dependent facilitation of Ca^{2+} entry in voltage-clamped, aequorin-injected molluscan neurons. *Proc. Natl. Acad. Sci. USA* 74:1748–52

Frank, K., Tauc, L. 1964. Voltage-clamp studies of molluscan neuron membrane properties. In *The Cellular Function of Membrane Transport*, ed. J. F. Hoffmann, pp. 113–35. Englewood Cliffs, NJ: Prentice Hall.

Frankenhaeuser, B., Hodgkin, A. L. 1957. The action of calcium on the electrical properties of squid axons. *J. Physiol. London* 137:218–43

Geduldig, D., Gruener, R. 1970. Voltage clamp of the *Aplysia* giant neurons: Early sodium and calcium currents. *J. Physiol. London* 211:217–44

Geduldig, D., Junge, D. 1968. Sodium and calcium components of action potentials in the *Aplysia* giant neurone. *J. Physiol. London* 199:347

Gerasimov, V. D., Kostyuk, P. G., Maiskii, V. A. 1965. The influence of divalent cations on the electrical characteristics of membranes of giant neurones. *Biofizika* 10:447–53

Gerschenfeld, H. M. 1973. Chemical transmission in invertebrate central nervous sytems and neuromuscular junctions. *Physiol. Rev.* 53:1–119

Gola, M. 1974. Neurones a ondes-salves des mollusques, variations cycliques lentes des conductances ioniques. *Pfleugers Arch.* 352:17–36

Gola, M. 1976. Electrical properties of bursting pacemaker neurones. In *Neurobiology of Invertebrates*, ed. J. Salanki, pp. 381–423. New York: Plenum

Gola, M., Ducreux, C., Chagneux, H. 1977. Ionic mechanism of slow potential wave production in barium-treated *Aplysia* neurons. *J. Physiol. Paris* 73:407–40

Gorman, A. L. F., Thomas, M. V. 1978. Changes in the intracellular concentration of free calcium ions in a pacemaker neurone, measured with metallochomic indicator dye arsenazo III. *J. Physiol. London* 275:357–76

Graubard, K. 1975. Voltage attenuation within *Aplysia* neurons: The effect of branching pattern. *Brain Res.* 88:325–32

Hagiwara, S. 1975. Ca-dependent action potential. In *Membranes—A Series of Advances*, ed. G. Eisenmann, 3:359–82. New York: Dekker

Hagiwara, S., Kusano, K., Saito, N. 1961. Membrane changes of *Onchidium* nerve cell in potassium-rich media. *J. Physiol. London* 155:470–89

Hagiwara, S., Saito, N. 1959. Voltage-current relations in nerve cell membrane of *Onchidium verruculatum*. *J. Physiol. London* 148:161–179

Hermann, A., Gorman, A. L. F. 1979. External and internal effects of tetraethylammonium on voltage-dependent and Ca-dependent K^+ currents components in molluscan pacemaker neurons. *Neurosci. Lett.* 12:87–92

Heyer, C. B., Lux, H. D. 1976. Control of the delayed outward potassium currents in bursting pacemaker neurones of the snail, *Helix pomatia*. *J. Physiol. London* 262:349–82

Hille, B. 1968. Charges and potentials at the nerve surface: divalent ions and pH. *J. Gen. Physiol.* 51:221–36

Hille, B. 1975. Ion selectivity of Na and K channels of nerve membranes. See Hagiwara 1975, pp. 255–323

Hodgkin, A. L., Huxley, A. F. 1952. A quantitative description of membrane current and its application to conduction and excitation in nerve. *J. Physiol. London* 117:500–44

Hodgkin, A. L., Huxley, A. F., Katz, B. 1952. Measurement of current-voltage relations in the membrane of the giant axon of *Loligo*. *J. Physiol. London* 116:424–48

Horn, R. 1978. Propagating calcium spikes in an axon of *Aplysia*. *J. Physiol. London* 281:513–34

Junge, D. 1967. Multi-ionic action potentials in molluscan giant neurones. *Nature* 215:546–48

Junge, D., Miller, J. 1974. Different spike mechanisms in axon and soma of molluscan neuron. *Nature* 252:155–56

Kado, R. T. 1973. *Aplysia* giant cell: soma-axon voltage clamp current differences. *Science* 182:843–45

Kandel, E. R. 1976. *Cellular Basis of Behavior*. San Francisco: Freeman. 727 pp.

Katz, B. 1969. *The Release of Neural Transmitter Substances*. Liverpool Univ. Press

Katz, G. M., Schwartz, T. L. 1974. Temporal control of voltage-clamped membranes: an examination of principles. *J. Memb. Biol.* 17:275–91

Kostyuk, P. G., Krishtal, O. A., Shakhovalov, Y. A. 1977a. Separation of sodium and calcium currents in the somatic membrane of mollusk neurones. *J. Physiol. London* 270:545–68

Kostyuk, P. G., Krishtal, O. A. 1977b. Effects of calcium and calcium-chelating agents on the inward and outward current in the membrane of the mollusc neurones. *J. Physiol. London* 270: 569–80

Kostyuk, P. G., Krishtal, O. A., Doroshenko, P. A. 1974. Calcium currents in snail neurones. I. Identification of calcium current. *Pfleugers Arch.* 348: 83–93

Kostyuk, P. G., Krishtal, O. A., Doroshenko, P. A. 1975a. Outward currents in isolated snail neurones—I. Inactivation Kinetics. *Comp. Biochem. Physiol.* 51C:359–63

Kostyuk, P. G., Krishtal, O. A., Pidoplichko, V. I. 1975b. Effects of internal fluoride and phosphate on membrane currents during intracellular dialysis of nerve cells. *Nature* 257:691–93

Kostyuk, P. G., Krishtal, O. A., Pidoplichko, V. I. 1977. Asymmetric displacement currents in nerve cell membrane and effect of internal fluoride. *Nature* 267:70–72

Kretsinger, R. H. 1977. Evolution of the informational role of calcium in eukaryotes. In *Calcium Binding Protein and Calcium Function,* ed. R. H. Wasserman, pp. 63–72. Amsterdam: Elsevier

Kunze, D. L., Walker, J. L., Brown, H. M. 1971. Potassium and chloride activities in identifiable *Aplysia* neurons. *Fed. Proc.* 30:255

Kupfermann, I. 1979. Modulatory actions of neurotransmitters. *Ann. Rev. Neurosci.* 2:447–65

Lee, K. S., Akaike, N., Brown, A. M. 1977. Trypsin inhibits the action of tetrodotoxin in neurones. *Nature* 265: 751–53

Lee, K. S., Akaike, N., Brown, A. M. 1978. Properties of internally perfused, voltage-clamped, isolated nerve cell bodies. *J. Gen. Physiol.* 71:489–507

Leicht, R., Meves, H., Wellhöner, H. H. 1971. Slow changes of membrane permeability in giant neurones of *Helix pomatia. Pfleugers Arch.* 323:63–79

Llinás, R., Steinberg, I. Z., Walton, K. 1976. Presynaptic calcium currents and their relation to synaptic transmission: voltage clamp study in squid giant synapse and theoretical model of the calcium gate. *Proc. Natl. Acad. Sci. USA* 73:2918–22

Lux, H. D., Heyer, C. B. 1977. An aequorin study of a facilitating calcium current in bursting pacemaker neurons of *Helix. Neuroscience* 2:585–92

Magura, I. S. 1977. Long-lasting inward current in snail neurons in barium solutions in voltage-clamp conditions. *J. Memb. Biol.* 35:239–56

Magura, I. S., Krishtal, O. A., Valeyev, A. G. 1971. Behaviour of delayed current under long-duration voltage clamp in snail neurones. *Comp. Biochem. Physiol. A.* 40:715–22

Meech, R. W. 1974. The sensitivity of *Helix aspersa* neurones to injected calcium ions. *J. Physiol. London* 237:259

Meech, R. W. 1978. Calcium-dependent potassium activation in nervous tissues. *Ann. Rev. Biophys. Bioeng.* 7:1–18

Meech, R. W., Standen, N. B. 1975. Potassium activation in *Helix aspersa* neurones under voltage clamp; a component mediated by calcium influx. *J. Physiol. London* 249:211–39

Meves, H. 1968. The ionic requirements for the production of action potentials in *Helix pomatia* neurones. *Pfleugers Arch. Gesamte Physiol. Menschen Tiere* 204:215–41

Neher, E. 1971. Two fast transient current components during voltage clamp on snail neurons. *J. Gen. Physiol.* 58:36–53

Neher, E., Lux, H. D. 1969. Voltage clamp on *Helix pomatia* neuronal membrane: current measurement over limited area of the soma surface. *Pfleugers Arch.* 311:272–77

Neher, E., Lux, H. D. 1972. Differential action of TEA^+ on two K^+ current components of a molluscan neurone. *Pfleugers Arch.* 336:87–100

Neher, E., Lux, H. D. 1973. Rapid changes of potassium concentration at the outer surface of exposed single neurons during membrane current flow. *J. Gen. Physiol.* 61:385–99

Partridge, L. D., Connor, J. A. 1978. A mechanism for minimizing temperature effects on repetitive firing frequency. *Am. J. Physiol.* 234:155–61

Partridge, L. D., Stevens, C. F. 1976. A mechanism for spike frequency adaptation. *J. Physiol. London* 256:315

Partridge, L. D., Thompson, S. H., Smith, S. J., Connor, J. A. 1979. Current-voltage relationships of repetitively firing neurons. *Brain Res.* 164:69–79

Pellmar, T. C., Carpenter, D. O. 1979. Voltage dependent calcium current induced by serotonin. *Nature* 277:483–84

Plant, R. E. 1978. The effects of calcium^{++} on bursting neurons: A modeling study. *Biophys. J.* 21:217–36

Russell, J. M., Eaton, D. C., Brodwick, M. S. 1977. Effects of nystatin on membrane conductance and internal ion activities in *Aplysia* neurons. *J. Membr. Biol.* 37:137–56

Simons, T. J. B. 1976. Calcium-dependent potassium exchange in human red cell ghosts. *J. Physiol. London* 256:227–44

Smith, S. J. 1978. *The mechanism of bursting pacemaker activity in neurons of the mollusc Tritonia diomedia*. PhD thesis. Univ. Wash. Seattle. 101 pp.

Smith, S. J., Thompson, S. H. 1975. Prediction of burst waveforms from voltage clamp measurements in bursting pacemaker neurons. *Neurosci. Abst.* 1:611

Smith, S. J., Zucker, R. S. 1979. Aequorin response facilitation and intracellular calcium accumulation in mulluscan neurons. *J. Physiol. London.* In press

Smith, T. G. Jr., Barker, J. L., Gainer, H. 1975. Requirements for bursting pacemaker potential activity in molluscan neurons. *Nature* 253:450

Standen, N. B. 1974. Properties of a calcium channel in snail neurones. *Nature* 250:340–42

Standen, N. B. 1975. Voltage-clamp studies of the calcium inward current in an identified snail neurone: comparison with the sodium inward current. *J. Physiol. London* 249:253–68

Stinnakre, J., Tauc, L. 1973. Calcium influx in active *Aplysia* neurones detected by injected aequorin. *Nature New Biol.* 242:113–15

Thomas, M. V., Gorman, A. L. F. 1977. Internal calcium changes in a bursting pacemaker neuron measured with arsenazo III. *Science* 196:531

Thompson, S. H. 1976. *Membrane currents underlying bursting in molluscan pacemaker neurons*. PhD thesis. Univ. Wash., Seattle. 124 pp.

Thompson, S. H. 1977. Three pharmacologically distinct potassium channels in molluscan neurones. *J. Physiol. London* 265:465–488

Thompson, S. H., Smith, S. J. 1976. Depolarizing afterpotentials and burst production in molluscan pacemaker neurons. *J. Neurophysiol.* 39:153–61

Tillotson, D. 1979. Inactivation of Ca conductance dependent on entry of Ca ions in molluscan neurons. *Proc. Natl. Acad. Sci. USA* 76:1497–500

Tillotson, D., Horn, R. 1978. Inactivation without facilitation of calcium conductance in caesium-loaded neurons of *Aplysia. Nature* 273:312–14

Wald, F. 1972. Ionic differences between somatic and axonal action potentials in snail giant neurones. *J. Physiol. London* 220:267–81

Williamson, T. C., Crill, W. E. 1976. Voltage clamp analysis of pentylenetetrazol effects upon excitability in mulluscan neurons. *Brain Res.* 116:217–29

Willows, A. O. D., Dorsett, D. A., Hoyle, G. 1973. The neuronal basis of behavior in *Tritonia*. I. Functional organization of the central nervous system. *J. Neurobiol.* 4:207

Wilson, W. A., Wachtel, H. 1974. Negative resistance characteristic essential for the maintenance of slow oscillations in bursting neurons. *Science* 186:932

Wilson, W. A., Wachtel, H. 1978. Prolonged inhibition in burst firing neurons: Synaptic inactivation of the slow regenerative inward current. *Science* 202:722–75

Ann. Rev. Neurosci. 1980. 3:169–87

CHEMICAL NEUROTOXINS ❖11539
AS DENERVATION TOOLS
IN NEUROBIOLOGY

G. Jonsson

Department of Histology, Karolinska Institutet, S-104 01 Stockholm, Sweden

INTRODUCTION

The discovery by Tranzer & Thoenen (1967) that the noradrenaline (NA) analogue 6-hydroxydopamine (6-OH-DA; 2,4,5-trihydroxyphenylethylamine) can induce selective degeneration of sympathetic adrenergic nerves led to the introduction of a new concept into neurobiology, namely *chemical denervation* of a given neuron type. The rationale for this site- or target-directed neurotoxic action is that 6-OH-DA is a generally cytotoxic agent that is selectively accumulated by catecholamine (CA) neurons, thereby restricting the degenerative action of the compound to the structures that accumulate it. This principle was later extended to 5-hydroxytryptamine (5-HT; serotonin) neurons which are acted on by 5,6- and 5,7-dihydroxytryptamine (5,6-HT; 5,7-HT) in the same way (Baumgarten et al 1971, 1973, Baumgarten & Lachenmayer 1972). The discovery of these neurotoxins opened new opportunities and they have over the years been extensively used as denervation tools for morphological, biochemical, physiological, pharmacological, and behavioral investigations to elucidate various aspects of monoamine neurotransmitter functions (see Malmfors & Thoenen 1971, Jonsson et al 1975, Sachs & Jonsson 1975a, Kostrzewa & Jacobowitz 1974, Baumgarten & Björklund 1976, Baumgarten et al 1977, Jacoby & Lytle 1978). This development has also led to much greater awareness of potentially cytotoxic agents and a number of structurally different compounds have been found to possess more or less pronounced neurotoxic actions on

169

monoamine neurons, all with the common denominator of being selectively or preferentially accumulated by these neurons. Some of these are drugs employed in humans where their neurotoxic actions are of practical concern. Another type of neurotoxin that has received great attention in recent years is the so-called "excitotoxic" amino acids, of which kainic acid is the most powerful. These agents also appear to have a site-directed neurotoxic specificity (receptors) that is much broader and more general with respect to neuron type than that observed for the monoamine neurotoxins (see symposium volume edited by McGeer et al 1978c). The chemical neurotoxins have provided neurobiologists with unique tools to investigate basic neurobiological mechanisms and have created possibilities for developing animal models for various human diseases. Chemical neurotoxins have also provided information on degeneration and regeneration processes as well as plasticity phenomena; this information may be of importance for future attempts to prevent degeneration and to promote nerve growth and functional recovery after a lesion. The purpose of the present review is to review briefly the mode of action, specificity, and fields of applicability of the available chemical neurotoxins.

CATECHOLAMINE NEUROTOXINS

The CA neurotoxin that was first discovered and remains the best characterized is 6-OH-DA (Figure 1), still the most useful of the compounds known to possess a neurodegenerative action on CA neurons. Convincing evidence originating from several laboratories has shown that there are two features of fundamental importance for the degenerative action of 6-OH-DA on CA neurons. (*a*) 6-OH-DA is efficiently taken up and accumulated by neurons that have a membrane transport mechanism for catecholamines, thus accounting for the specificity of the 6-OH-DA action. (*b*) 6-OH-DA has a low red-ox potential, and is very susceptible to nonenzymatic oxidation, a property associated with the cytotoxic action of 6-OH-DA (see Sachs & Jonsson 1975a).

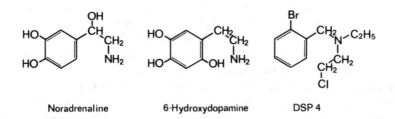

| Noradrenaline | 6-Hydroxydopamine | DSP 4 |

Figure 1   The structures of noradrenaline and the catecholamine neurotoxins 6-hydroxydopamine and DSP 4.

Effects of 6-OH-DA on CA Neurons

In order to induce degeneration 6-OH-DA has to reach a critical concentration in CA neurons. This threshold concentration has been calculated from both in vivo and in vitro experiments to be about 50–100 mM (average concentration in the cytoplasm; Jonsson & Sachs 1971, Jonsson 1976). Consistent with this, it has been found that 6-OH-DA acts very much in an "all-or-none" fashion and causes complete destruction of CA uptake-storage mechanisms and functional properties in a proportion of nerves that depends on the dose of 6-OH-DA administered (Jonsson & Sachs 1972). Although 6-OH-DA can be taken up and stored in the amine storage granules (vesicles), this does not seem to be a prerequisite for the neurodegenerative effects (Jonsson & Sachs 1970). This implies that the critical intraneuronal concentration of 6-OH-DA is related to that reached in the extragranular space, an idea that is also supported by pharmacological experiments (Jonsson et al 1972). The neurotoxic action of 6-OH-DA is very rapid: the whole CA store is depleted within 1/2–1 hr after 6-OH-DA administration. 6-OH-DA can act as a "false transmitter," but only after administration of very low doses that cause limited degeneration (Thoenen & Tranzer 1968, Jonsson 1976). So far there is little evidence that 6-OH-DA causes transient damage that is not associated with degeneration. A word of caution might be appropriate at this point, since an apparent nondegenerative dose of 6-OH-DA may produce transient damage that does not readily show up in the tests used. The axonal membrane functions are damaged at a very early stage in the 6-OH-DA induced degeneration process, as reflected by a loss of the CA membrane pump functions (Jonsson & Sachs 1970, Thoenen & Tranzer 1968) and loss of ability to generate and conduct action potentials (Haeusler 1971, Furness et al 1970). The latter effect has been ascribed to a 6-OH-DA induced permanent depolarization associated with increased Ca^{++} permeability. Ultrastructural signs of degeneration are seen 1–4 hr after 6-OH-DA administration that are very similar to those seen after surgical axotomy (Furness et al 1970, Hökfelt et al 1972) although they appear much faster due to a direct action of 6-OH-DA on the nerve terminals. Both pre- and postsynaptic supersensitivity similar to that seen after surgical denervation develop after 6-OH-DA, as monitored with physiological, behavioral, and biochemical techniques. Concomitant with CA depletion, enzymes responsible for the transmitter biosynthesis also disappear (Uretsky & Iversen 1970).

Molecular Mechanisms of 6-OH-DA Neurotoxicity

The molecular mechanism(s) of the neurodegenerative action of 6-OH-DA is still in question in spite of considerable effort to solve this important problem. There is strong evidence that the cytotoxic actions are associated with the ease of autoxidation of 6-OH-DA; this, however, is a complex process with simultaneous formation of a number of very reactive and potentially cytotoxic products, including quinones (Saner & Thoenen 1971), H_2O_2 (Heikkila & Cohen 1971), 5,6-dihydroxyindole (Blank et al 1972), the superoxide radical (Heikkila & Cohen 1973), the hydroxy radical (Cohen & Heikkila 1974), and singlet oxygen (Heikkila & Cabbat 1977). It has so far not been possible to assign any of these molecular species (see Figure 2) a causal role in initiating the degenerative action of 6-OH-DA. There are at present two molecular theories proposed to explain the cytotoxic effects, both based on the fact that 6-OH-DA is very easily oxidized. One of the theories postulates that H_2O_2 and radicals formed are the causative agents (Heikkila & Cohen 1975), whereas the other theory assumes that the quinoid-like oxidation products of 6-OH-DA act as alkylating agents (Saner & Thoenen 1971). Both processes could result in alterations of the structural and functional properties of important cellular proteins and lipids, thereby causing irreversible damage. There is experimental evidence to support both mechanisms of 6-OH-DA cytotoxicity, although the relative importance of the two is unclear. The evidence supporting a cytotoxic action of superoxide and hydroxy radicals originates from studies using

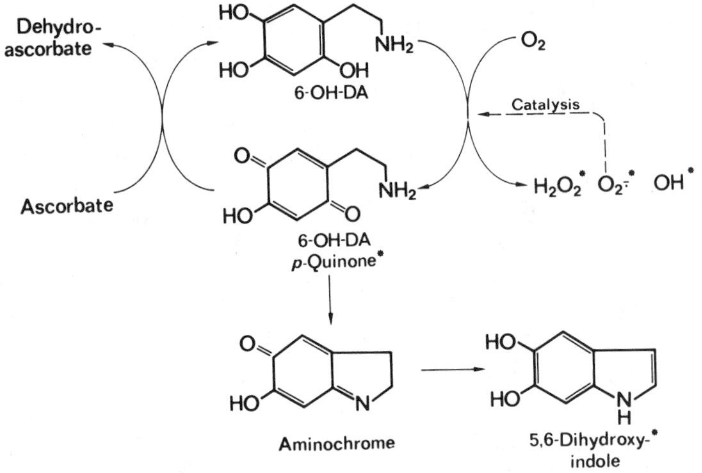

Figure 2 Reaction scheme for the autoxidation of 6-OH-DA with formation of 6-OH-DA p-quinone, H_2O_2, superoxide (O_2^-), and hydroxy (OH·) radicals. The superoxide radical catalyzes the oxidation of 6-OH-DA. The p-quinone can be reduced back to 6-OH-DA by ascorbate or cyclize to aminochrome, which can be reduced to form 5,6-dihydroxyindole. This latter compound is considered to be an intermediate in the formation of various melanins. The potentially cytotoxic compounds are denoted with *.

radical trapping agents (Heikkila & Cohen 1975, Cohen et al 1976b). The data indicate that the radicals play a role in the over-all neurotoxicity of 6-OH-DA partly via a regulatory role in the autoxidation of 6-OH-DA and partly via a direct cytotoxic effect. In this context it is of interest to mention that intraneuronal CA levels can control the degenerative action of 6-OH-DA, probably because CA can trap radicals (Sachs et al 1975). Increased intraneuronal NA thus can protect against the degenerative actions of 6-OH-DA. Data on the cytotoxic role of H_2O_2 formed from 6-OH-DA is so far inconclusive. Although it has been shown that H_2O_2 damages the uptake mechanism for CA neurons in vitro, polarographic studies have demonstrated that the amount of H_2O_2 actually present during air oxidation of 6-OH-DA is quite small. Since generated H_2O_2 further oxidizes 6-OH-DA, however, (Liang et al 1975) H_2O_2 could participate in over-all neurotoxicity by promoting 6-OH-DA oxidation.

Concerning the alkylation theory for cytotoxic effects initially proposed by Saner & Thoenen (1971), it is known that strong electrophiles such as p-quinone and aminochromes are generated upon autoxidation of 6-OH-DA, and that these are capable of rapidly reacting with cellular nucleophiles to form covalent linkages. Subsequent studies (Rotman et al 1976, Jonsson 1976) have confirmed that oxidation products of 6-OH-DA are covalently bound to proteins both in vitro and in vivo. Furthermore, it has been shown that a substantial portion of the 6-OH-DA taken up and accumulated in the adrenergic nerves is covalently bound to neuronal structures and that the time course of this binding parallels the neurodegenerative effects of 6-OH-DA (Jonsson 1976). Also, the covalently bound fraction of 6-OH-DA is markedly increased in situations when the neurotoxic action of 6-OH-DA is potentiated and, conversely, considerably reduced when adrenergic nerves are protected from undergoing degeneration. The correlation between cytotoxicity and the amount of covalently bound oxidation products is compatible with the view that binding may play an important role in the irreversible damage and neurodegeneration produced by 6-OH-DA. There is no direct information on the exact chemical nature of the covalent binding of 6-OH-DA oxidation products to neuronal elements, since attempts to identify the products formed in vivo have not yet been reported. However, evidence obtained from model experiments using substituted 6-OH-DA derivatives, make it conceivable that the alkylation reaction in vivo is a nucleophilic reaction and that the preferred sites of interaction of 6-OH-DA oxidation products are at free sulfhydryl (-SH) groups of neuronal molecules (Borchardt 1975, Borchardt et al 1975, 1977, Creveling et al 1975, Rotman et al 1976). The data available indicate that a nucleophilic attack occurs at one of the electronegative centers of the oxidation products (position 2 of the p-quinone or 4 of the aminochrome). At present it is not known whether one of the initial oxidation products, the p-quinone, or other reactive intermediates, e.g. the semi-quinone or aminochromes, is the active compound that binds covalently to the nucleophiles.

For the sake of completeness, it should also be mentioned that two other contributory mechanisms for 6-OH-DA neurotoxicity have been suggested, each associated with effects of 6-OH-DA on energy metabolism. It has been observed that oxygen consumption during the autoxidation of 6-OH-DA is so rapid that hypoxia may occur within the neuron (Heikkila & Cohen 1972, Liang et al 1975). Furthermore, 6-OH-DA has been shown to be a very potent uncoupler of oxidative phosphorylation (Wagner & Trendelenburg 1971).

Considering the accumulated data on the molecular mechanism of the cytotoxic action of 6-OH-DA, one is apt to conclude that the question of the mechanism of neurotoxicity is still unsettled. In attempts to resolve this, Graham et al (1978) recently investigated whether the cytoxicity of 6-OH-DA and a number of related polyphenols reflected their potential for autoxidation or the sulfhydryl reactivity of their quinone products. These authors concluded from their studies that 6-OH-DA kills cells mainly through production of H_2O_2 and radicals, although the sulfhydryl reactivity of quinones probably makes some contribution. Borchardt et al (1977) investigated the neurotoxic effects of a number of methylated 6-OH-DA analogues on central NA and DA neurons in order to clarify the same problem. They found that 6-OH-DA derivatives capable of generating both reactive quinones and free radicals were more toxic than those derivatives that generated only hydrogen peroxide and radicals. It thus appears that the neurodegeneration produced by 6-OH-DA results from a series of complex interactions involving both electrophiles (6-OH-DA quinone, aminochromes) and the oxygen oxidation products. It has been suggested that radicals might facilitate the reactivity of the 6-OH-DA quinones through denaturation of proteins and concomitant exposure of nucleophilic sites (Creveling et al 1975).

Specificity

From studies on the mode of action of 6-OH-DA it has become clear that the neurotoxic potency of 6-OH-DA can be regulated and modified at many levels, which may account for the considerable heterogeneity in response to 6-OH-DA in different central and peripheral CA neuron systems (see Kostrzewa & Jacobowitz 1974, Sachs & Jonsson 1975a). In general it has been observed that the nerve terminals appear to be the most sensitive, the axons less so, and the cell bodies the structures least sensitive to the neurotoxic action of 6-OH-DA. These differences can at least partly be related to differences in surface-volume relationships. Among the various central CA neuron systems, the locus coeruleus NA system (nerve terminals) appears to be the one most sensitive to 6-OH-DA, whereas the adrenaline neuron systems are most resistant (Jonsson et al 1976). The DA systems range from being relatively sensitive (the nigro-striatal system) to almost completely resistant (the tubero-infundibular DA system) to 6-OH-DA. Since the route of 6-OH-DA administration (systemic, intraventricular, intracisternal, or intracerebral) and the injection technique are crucial factors for the outcome, it is almost impossible to give general guidelines to the sensitivity of various CA systems. The appropriate 6-OH-DA dose, injection volume, time and route of administration vary depending on which CA system or systems the investigator wants to denervate (see Jonsson et al 1975). In spite of all these precautions the extent of the lesion may not be satisfactory and unwanted nonspecific cytotoxic effects may occur. When seeking a strategy to optimize lesioning conditions, one may find it worthwhile to consider the various factors affecting the neurotoxic potency of 6-OH-DA (Table 1). In order to facilitate the discussion, the factors have been distinguished as "extraneuronal" and "neuronal." Although several of these factors may be regarded as constant, it is clear that some can be manipulated experimentally in an advantageous way. Considering that cytotoxicity is associated with the oxidation of 6-OH-DA, which can occur both intra- and extraneuronally, leading to specific and nonspecific effects, respectively, it is obvious that one should use as low a dose of 6-OH-DA as possible and endeavor to keep the "red-ox" equilibrium 6-OH-DA $\rightleftharpoons$ p-quinone to the left extraneuronally, and to the right intraneuronally. One way to keep 6-OH-DA intact and to prolong its "survival" in the extraneuronal compartment is to administer a reducing agent (e.g. ascorbic acid) with 6-OH-DA. Ascorbic acid has been shown to potentiate the neurotoxic potency of 6-OH-DA (Jonsson et al 1974). The use of ascorbic acid may, however, be a double-edged approach, especially if one uses high concentrations, since there are several reports showing a general cytotoxic effect of ascorbic acid (see Waddington & Crow 1979). As to manipulating "intraneuronal" factors, reduction of CA stores with the tyrosine hydroxylase inhibitor, α-methyl-p tyrosine has a potentiating effect, most probably due to reducing the

Table 1   Factors affecting the neurotoxic potency of 6-OH-DA

Extraneuronal	Neuronal
a. Diffusion conditions	a. Membrane uptake mechanism
b. MAO, COMT	b. Surface-volume relationship
c. 'Red-ox' behavior of the extraneuronal milieu	c. Catecholamines
	d. MAO
	e. 'Red-ox' behavior of the intraneuronal milieu

inhibitory effect of CA on the rate of oxidation of 6-OH-DA (Sachs et al 1975). MAO inhibition is also a way of increasing the neurotoxic potency and specificity of 6-OH-DA, as it will reduce the breakdown of 6-OH-DA by MAO catabolism, especially intraneuronally. It is important, however, to administer the MAO inhibitor shortly before 6-OH-DA in order to eliminate the counteracting effect of increased intraneuronal CA levels, which occur relatively soon after MAO inhibition. COMT inhibition can also potentiate the neurotoxic action of 6-OH-DA.

From the discussion above it is clear that the neurotoxic potency of 6-OH-DA is related to an interplay of many factors, some of which are not easy to control. It is therefore perhaps not astonishing that varying and contradictory results have been reported from various laboratories, with respect to specific versus nonspecific effects. This has led some investigators to conclude that 6-OH-DA given by local intracerebral injection does not produce lesions that are more selective than those seen after electrolytic or mechanical or chemical treatment (Poirer et al 1972, Butcher et al 1975). From all the information accumulated over the years showing that it is possible to obtain specific CA denervations both from a structural and a biochemical, as well as a functional viewpoint, it is obvious that such an opinion is too extreme. In spite of the imperfections inherent in use of 6-OH-DA, it is clear that 6-OH-DA allows more complete denervation of CA neuron systems with less nonspecific damage than is possible by any other currently available technique. It is, however, equally obvious that the blind acceptance that all damage produced by 6-OH-DA is restricted exclusively to CA neurons is objectionable. Therefore it has to be emphasized that the specificity of the action has to be evaluated in all experiments by morphological, biochemical, and physiological indices. Appropriate controls are also important especially when 6-OH-DA is applied by local micro-injection. This statement is of course equally valid for all chemical neurotoxins used at present as denervation tools.

Another specificity problem can arise when 6-OH-DA is used with the intention of denervating selectively NA or DA neurons in a CNS region that contains both neuron types. One approach available for selectively denervating DA neurons is to use a potent NA uptake inhibitor such as desipramine or protriptyline in combination with 6-OH-DA. Pretreatment with the uptake inhibitor will protect the NA neurons, but not the DA neurons, from undergoing degeneration. This strategy can be used in most situations. So far it has not been possible to use this approach for the selective denervation of NA, because a potent DA inhibitor that does not affect the uptake mechanism of NA neurons is lacking. The recently discovered neurotoxin DSP 4 (see below), which has been observed to affect NA neurons selectively may prove to be the neurotoxin of choice to overcome this problem.

It should be mentioned in this context that specific uptake inhibitors can be used as controls for the nonspecific effects of 6-OH-DA (see Lidbrink & Jonsson 1975).

Other Compounds and Drugs with CA Neurotoxic Properties

A large number of 6-OH-DA derivatives and related polyphenols have been synthesized, but none of them have been found to be superior to 6-OH-DA as a denervation tool for CA neurons. Of these, 6-OH-DOPA and 6-amino-DA have been used to a certain extent (see Sachs & Jonsson 1975a, Kostrzewa & Jacobowitz 1974); 6-OH-DOPA has the advantage of passing the blood-brain barrier and can thus be used for CNS studies employing systemic injections. However, relatively large doses are needed, risking unwanted nonspecific toxic effects. 6-amino-DA is also more generally toxic than 6-OH-DA (Jonsson & Sachs 1973). Recently a new compound has been discovered, N-(2-chloroethyl)-N-ethyl-bromobenzylamine (DSP 4, see Figure 1), which is structurally different from 6-OH-DA and has been shown to be capable of producing an acute and selective degeneration of both central and peripheral NA nerve terminals, leaving DA, adrenaline and serotonin neurons apparently unaffected (Jonsson et al 1979). Biochemical and fluorescence histochemical studies have shown that the effects of DSP 4 on NA neurons are in many respects similar to those seen with 6-OH-DA. DSP 4 easily passes the blood-brain barrier and after systemic injection of the compound the neurotoxic effects are exerted preferentially on NA nerve terminal projections originating in the locus coeruleus. Since all the effects of DSP 4 on NA neurons can be completely counteracted by pretreatment with the NA uptake blocker desipramine, as with 6-OH-DA, DSP 4 must act directly on the NA neurons. Whether DSP 4 is accumulated intraneuronally in the same way as is 6-OH-DA, and exerts its cytotoxic action there, or primarily attacks the NA uptake sites at the axonal membrane is unclear. Very little is known of the molecular mechanism of action of DSP 4, but it is conceivable that this unstable and highly reactive compound produces its neurotoxic effects via an alkylation reaction. Since DSP 4 easily passes the blood-brain barrier, in contrast to 6-OH-DA, and appears to be selective for NA neurons, it may prove to be a useful denervation tool for studies on various aspects of NA transmitter function in the CNS.

The introduction of the selective neurotoxins has led to the realization that other drugs may also have neurotoxic actions on CA neurons. Thus, prolonged administration of the adrenergic blocking agent guanethidine, used for many years in the treatment of hypertension, has been found to cause a very marked and selective degeneration of sympathetic adrenergic neurons (Jensen-Holm & Juul 1971, Burnstock et al 1971, Heath et al 1972). The short adrenergic neurons innervating the vas deferens appear to be most sensitive to guanethidine (Bittiger et al 1977). The selectivity of the effect of guanethidine is related to its accumulation in adrenergic neurons by the catecholamine uptake mechanism, whereas the cytotoxic effects have been related to the ability of guanethidine to inhibit oxidative phosphorylation. Guanethidine may have some advantages over 6-OH-DA for producing sympathectomy, especially in developing animals (Johnson et al 1975).

Chronic amphetamine treatment has been reported to have a selective neurotoxic effect on DA neurons in the caudate nucleus (Ellison et al 1978). Whether this is due to continuous stimulation of DA nerve terminals or due to formation of toxic metabolites is not known. In this context it might be of interest to mention that it has recently been reported that the condensation products of adrenaline and acetaldehyde (tetrahydroisoquinoline alkaloids) can induce selective degeneration of sympathetic adrenergic nerves similar to that caused by 6-OH-DA, although less extensive (Azevedo & Osswald 1977). Since these alkaloids may be formed in vivo during ethanol intoxication (see Cohen 1976), it is possible that the observed effects may bear on some of the phenomena related to acute and chronic alcoholism. It is obvious that there are a number of structurally unrelated compounds that can induce a selective or preferential degeneration of CA neurons. With increasing knowledge of their mode of action, new neurotoxins of value to the neurobiologist can hopefully be developed.

Pathological Implications of CA Neurotoxins

Soon after the discovery of the unique neurotoxic actions of 6-OH-DA, this compound was implicated as a possible etiological factor for the pathogenesis of schizophrenia (see Stein & Wise 1971). Although very little, if any, evidence has been put forward to substantiate this view, it has become clear with increasing knowledge of the molecular mechanisms involved in cytotoxicity of 6-OH-DA, that the endogenous CA are not chemically inert substances and that they may undergo oxidation reactions with formation of potentially cytotoxic molecular species (Adams 1972, Tse et al 1976), which, apart from being possible factors in the pathogenesis of certain diseases, may also affect the turnover of proteins in monoamine neurons. It is of interest to note that, as with 6-OH-DA, a portion of exogenously administered DA seems to be covalently bound in adrenergic nerves; the fraction of DA bound, however, is quantitatively much smaller than that seen after 6-OH-DA (Jonsson 1976). These findings are consistent with results reported by Maguire et al (1974) who postulated that oxidative reactions of NA are involved in binding of ^{3}H-NA to particulate cell fractions. This raises the question whether the CA neurotransmitters are potentially toxic to their own nerve structures where they are locally stored in very large amounts. It is thus possible that, under certain conditions of an altered intraneuronal milieu, increased aberrant oxidations of CA may be involved in the genesis of pathological conditions where CA neurons are considered to be malfunctioning, e.g. Parkinson's disease, schizophrenia, and manic-depressive illness. Along similar lines, Cohen et al (1976a) have proposed a theory for oxidative damage of nigrostriatal DA neurons in Parkinson's disease.

SEROTONIN NEUROTOXINS

The discovery of the unique properties of 6-OH-DA as a selective CA neurotoxin stimulated the search for compounds with analogous effects on central 5-HT neurons. This led to the demonstration by Baumgarten and co-workers of the neurotoxic actions of 5,6-HT and 5,7-HT on central 5-HT neurons (Baumgarten et al 1971, 1973, Baumgarten & Lachenmayer 1972). These findings were soon confirmed by several laboratories and considerable information on the use of these neurotoxins for studies on 5-HT neurons and various aspects of their transmitter function have now been published (see Jacoby & Lytle 1978, Baumgarten & Björklund 1976, Baumgarten et al 1977). Although 5,6-HT has been shown to have a rather selective neurodegenerative action on 5-HT neurons, with limited effects on CA neurons, it is clear that this compound has serious drawbacks as a selective denervation tool because it causes a substantial degree of unspecific damage to non-monoamine-containing nerve cells. 5,7-HT is a better tool than 5,6-HT because it has a higher neurotoxic potency for 5-HT neurons and more limited nonspecific cytotoxic effects for non-monamine neurons. Although local microinjection of 5,7-HT into the CNS causes greater nonspecific damage than injection of the solvent alone, the lesion is small compared with the large unspecific lesions produced by electrolysis (Lorens et al 1976). 5,7-HT possesses more neurotoxic action on NA neurons than 5,6-HT, but this limitation can be circumvented by pretreatment with disipramine. Thus, 5,7-HT can be used to produce rather selective denervation of 5-HT neurons (Sachs & Jonsson 1975b, Björklund et al 1975) and is the neurotoxin of choice for functional analysis of central 5-HT neurons (Daly et al 1974) and for experimental work on 5-HT neurons during development (Sachs & Jonsson 1975b).

The data available indicate that the molecular mechanisms of action of the 5-HT neurotoxins are very similar in principle to those of 6-OH-DA. The relative importance for neurotoxicity of oxygen oxidation products versus covalent binding of indolequinone oxidation products is also an unresolved question here. Effects of 5-HT neurotoxins on oxidative phosphorylation have also been implicated (see Cohen & Heikkila 1978, Creveling & Rotman 1978). There are some interesting differences between

5,7-HT and 6-OH-DA as to the action on NA neurons, however, and, in addition, differences in the action of 5,7-HT on NA and 5-HT neurons. First, the neurotoxic action of 6-OH-DA appears to be dependent on its rapid spontaneous autoxidation, whereas 5,7-HT does not react as readily with oxygen (Creveling et al 1975). Second, although the neurodegenerative action of 6-OH-DA is potentiated by inhibition of MAO, the corresponding neurotoxicity of 5,7-HT is abolished by MAO inhibition in NA neurons but not in 5-HT neurons (Baumgarten et al 1975, Breese & Cooper 1975, Creveling et al, 1975). Therefore it appears that the molecular mechanisms of action of 5,7-HT and 6-OH-DA are fundamentally different at least in NA neurons. On the other hand, it has been reported that hydroxy radical scavenging agents protect peripheral adrenergic neurons against destruction by both 6-OH-DA and 5,7-HT suggesting a similarity in mechanisms of action for these two neurotoxins (Allis & Cohen 1977, Cohen & Heikkila 1978). Further studies will be needed to clarify these questions.

Another class of compounds that exert a selective neurotoxic action that is primarily restricted to the 5-HT neurons are the halogenated amphetamines. The first evidence for a possible neurotoxic action of this type of compound was the finding that p- or 4-chloroamphetamine (pCAM, Figure 3) produced a long-lasting (as long as 4 months) depletion of 5-HT in the brain (Sanders-Bush et al 1972). Subsequently, it has been shown that 4-bromoamphetamine (Fuller et al 1975) and fenfluramine (Harvey & McMaster 1975, Sanders-Bush et al 1975) also produce long-term and apparently irreversible decreases in the 5-HT content of the brain. The depletion of 5-HT produced by these drugs can be blocked by 5-HT uptake inhibitors, suggesting an action directly on and a possible requirement for transport into 5-HT neurons. Very little is known of the molecular mechanisms for the neurotoxic actions of these compounds. It has been reported that the halogenated amphetamines appear to produce a more selective neurotoxic action on the B9-serotonin cell group than on the B7- and B8-cell body groups (Harvey et al 1975). This is difficult to reconcile, however, with the observations that pCAM appear to cause very marked degenerative effects on 5-HT nerve terminal projections originating in all the B7–B9 cell groups (Köhler et al 1978). Biochemical and fluorescence histochemical data available suggest that pCAM mainly affects the nerve

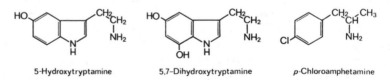

5-Hydroxytryptamine   5,7-Dihydroxytryptamine   p-Chloroamphetamine

Figure 3   The structures of 5-hydroxytryptamine and the serotonin neurotoxins 5,7-dihydroxytryptamine and p-chloroamphetamine.

terminal projections of the ascending but not the descending 5-HT systems, without detectable effects in the mesencephalon. pCAM also seems to have a very weak effect on 5-HT nerve terminals in the hypothalamus, which is known to be innervated by fibers originating both from B7 and B8 and from the pontine B5 and B6 cell groups. The reason for the variable sensitivity of 5-HT terminals in different brain regions is unknown. In view of this it is obvious that pCAM has serious limitations and is not generally applicable as a denervation tool for studies on 5-HT neurons, although the halogenated amphetamines in certain situations may have advantages compared with 5,6-HT and 5,7-HT (see Fuller 1978).

KAINIC ACID AND OTHER EXCITOTOXINS

It has been known for some years that glutamate (see Figure 4), a putative excitatory neurotransmitter, and certain of its structural analogues have striking neurotoxic properties in the CNS (Olney et al 1971). The common feature of these compounds is their neuroexcitatory action on most CNS neurons (Curtis & Watkins 1963, Curtis et al 1972) and the neurotoxic potency of the glutamate analogues has been shown to correlate well with their neuroexcitatory activity (Olney et al 1971, Schwarcz et al 1978). The most powerful of the compounds is kainic acid (Figure 4). The proposed theory for the neurotoxicity of these compounds is that the neuronal damage follows upon prolonged depolarization induced by the excitotoxins; the neurons are excited to death. This action is considered to be mediated via specific receptors, possibly glutamate receptors, although the exact receptor type is not known at present (see McGeer et al 1978a). The subsequent discovery that local intracerebral injection of kainic acid leads to a rapid degeneration of neurons with cell bodies near the site of infusion and spares the axons of passage and afferent nerve terminals (Coyle & Schwarcz 1976, McGeer & McGeer 1976) has established the excitotoxins as very useful denervation tools (see McGeer et al 1978c, Coyle 1978). The information available on kainic acid and related compounds has opened new possibilities for the development of animal models of diseases that involve general loss of neurons in specific brain loci (e.g. Huntington's chorea). In addition, kainic acid has been useful in the exploration and analysis of complex anatomical circuitries and in the localization of various biochemical markers to specific types of structures within the lesioned area. The actions of kainic acid and other excitotoxins appear to be mediated by receptors localized to neuronal perikarya and dendrites. Because most of the neurons are considered to have these receptors (see Coyle 1978), the applicability of the excitotoxins as denervation tools is much broader and general with respect to neuron types compared with the monoamine neurotoxins. However, the use of kainic acid as a selective neurodegenerative agent is compli-

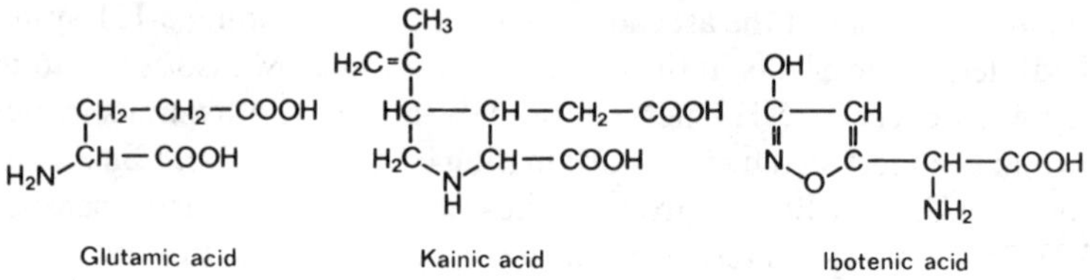

Figure 4 The structures of the excitotoxins glutamic acid, kainic acid, and ibotenic acid.

cated by several factors. Although its action is relatively nonselective with respect to neuron type, some variability in the neurotoxic effect of kainic acid occurs, because the vulnerability of neurons appears to depend on their synaptic input. It has thus been observed that kainate-induced degeneration of neostriatal neurons is dependent on an intact cortico-striatal glutamatergic pathway (McGeer et al 1978b).These data also suggest that kainic acid does not act directly on glutamate receptors. The neuronal vulnerability has been found to differ markedly, between and within various regions in the CNS and the doses of kainic acid required for the destruction of very resistant neurons often approach lethal doses of the compound (see McGeer et al 1978a, Schwarcz et al 1979). Brain maturity is also an important factor for the neurotoxic action of kainic acid, as it has been found that striatal neurons are almost insensitive to kainate at seven days after birth and that their vulnerability increases with age (Campochiaro & Coyle 1978). Another drawback of kainic acid is its rapid diffusion rate that often leads to undesirable degeneration of highly sensitive cells distant from the injection site (see Schwarcz et al 1979). Thus the extent and specificity of kainate lesions must be analyzed carefully. This is even more emphasized by a recent study showing that although kainic acid has its greatest degenerative action on neuronal perikarya, a significant amount of damage to axons of passage may also occur after a local intracerebral injection (Mason & Fibiger 1979). These authors emphasized that it will be necessary to show, in a quantitative fashion, that for every new brain area into which kainic acid is injected, it has indeed spared all fibers of passage.

Another valuable denervation tool is ibotenic acid (Figure 4), a conformationally restricted glutamic acid analogue that has been found to produce a type of lesion similar to those seen after intracerebral injection of kainic acid (Schwarcz et al 1979). Ibotenic acid has the advantage of being less generally toxic and of producing more discrete lesions compared with kainic acid and may, therefore, in many situations be the neurotoxin of choice. In spite of the limitations and the unclear mode of action, it is clear that kainic acid and other excitotoxins are useful tools for the neurobiologist for ablating neuronal cell populations in the brain.

CONCLUSIONS

The introduction into neurobiology of the principle of using selective chemical neurotoxins as denervation tools has opened up broad avenues to investigate important neuronal mechanisms of basic nature. There are at present two principal types of neurotoxins available both of which have a selective site-directed neurotoxic action, namely the monoamine neurotoxins (6-OH-DA, 5,6-HT, 5,7-HT) and the excitotoxins (kainic acid, ibotenic acid). The neurotoxic action is, however, mediated via recognition sites, which have different localizations and are associated with different properties. While the monoamine neurotoxins destroy only a given neuron type, kainic acid and other excitotoxins affect most neurons. However, when injected locally in the CNS, the excitotoxins damage the neuronal perikarya and spare afferent nerve terminals and axons of passage in the injected area. Although it is clear that both types of neurotoxins can produce rather selective denervations when properly used, it is also equally obvious that they do not possess absolute specificity. Experience has shown that the blind acceptance of the degenerative action of a used neurotoxin as being restricted exclusively to structures in accord with the theory is unwarranted. The careful evaluation of the specificity and extent of the lesion is therefore a very important aspect of using neurotoxins as denervation tools. With increasing knowledge of the mode of action of the various neurotoxins, it has become evident that compounds in the CNS that normally serve as neurotransmitters are potentially cytotoxic. The neuropathological implications of this are most intriguing, but require further studies in order to understand better their significance for pathological states. Finally, in view of the rapidly growing number of neurotoxins available for the neurobiologist and with increasing knowledge as to their mode of action at the molecular level, one can foresee future development of new, valuable neurotoxins to be used as denervation tools using specific transport systems and receptors as recognition sites, which we know of today but for which we have no neurotoxins.

ACKNOWLEDGMENTS

Part of the studies reviewed in this paper have been supported by the Swedish MRC (04X-2295). The author wishes to thank Mrs. Birgit Frideen for preparation of the manuscript.

Literature Cited

Adams, R. N. 1972. Stein and Wise theory of schizophrenia: A possible mechanism for 6-hydroxydopamine formation *in vivo. Behav. Biol.* 7:861–66

Allis, B., Cohen, G. 1977. The neurotoxicity of 5,7-dihydroxytryptamine in the mouse atrium: protection by 1-phenyl-3-(2-thioazolyl)-2-thiourea and by ethanol. *Eur. J. Pharmacol.* 43:269–72

Azevedo, I., Osswald, W. 1977. Adrenergic nerve degeneration induced by condensation products of adrenaline and acetaldehyde. *Naunyn-Schmiedebergs Arch. Pharmakol.* 300:139–44

Baumgarten, H. G., Björklund, A. 1976. Neurotoxic indoleamines and monoamine neurons. *Ann. Rev. Pharmacol. Toxicol.* 16:101–11

Baumgarten, H. G., Björklund, A., Lachenmayer, L., Nobin, A. 1973. Evaluation of the effects of 5,7-dihydroxytryptamine on serotonin and catecholamine neurons in the rat CNS. *Acta Physiol. Scand.* Suppl. 391, pp. 1–19

Baumgarten, H. G., Björklund, A., Lachenmayer, L., Nobin, A., Stenevi, U. 1971. Long-lasting selective depletion of brain serotinin by 5,6-dihydroxytryptamine. *Acta Physiol. Scand.* Suppl. 373, p. 1

Baumgarten, H. G., Björklund, A., Nobin, A.,Rosengren, E., Schlossberger, H. G. 1975. Neurotoxicity of hydroxylated tryptamines: structure-activity relationships. *Acta Physiol. Scand.* Suppl. 429, pp. 7–27

Baumgarten, H. G., Lachenmayer, L. 1972, 5,7-dihydroxytryptamine: Improvement in chemical lesioning of indoleamine neurons in mammalian brain. *Z. Zellforsch. Mikrosk.* 135:399–414

Baumgarten, H. G., Lachenmayer, L., Björklund, A. 1977. In *Methods in Psychobiology*, ed. R. D. Myers, 3:47–98. New York: Academic

Bittiger, H., Maître, L., Krinke, G., Schnider, K., Hess, R. 1977. A study of long-term effects of guanethidine on peripheral noradrenergic neurons of the rat. *Toxicology* 8:63–78

Björklund, A., Baumgarten, H. G., Rensch, A. 1975. 5,7-Dihydroxytryptamine: improvement of its selectivity for serotonin neurons in the CNS by pretreatment with desipramine. *J. Neurochem.* 24: 833–35

Blank, C. L., Kissinger, P. T., Adams, R. N. 1972. 5,6-Dihydroxyindole formation from oxidized 6-hydroxydopamine. *Eur. J. Pharmacol.* 19:391–94

Borchardt, R. T. 1975. Affinity labelling of catechol-0-methyltransferase by the oxidation products of 6-hydroxydopamine. *Mol. Pharmacol.* 11:436–45

Borchardt, R. T., Burgess, S. K., Reid, J. R., Liang, Y. O., Adams, R. N. 1977. Effects of 2- and/or 5-methylated analogues of 6-hydroxydopamine on norepinephrine and dopamine containing neurons. *Mol. Pharmacol.* 13:805–18

Borchardt, R. T., Smissman, E. E., Nerland, D., Reid, J. R. 1975. Catechol 0-Methyltransferase. 7. Affinity labelling with the oxidation products of 6-aminodopamine. *J. Med. Chem.* 19: 30–37

Breese, G. R., Cooper, B. R. 1975. Behavioral and biochemical interactions of 5,7-dihydroxytryptamine with various drugs when administered intracisternally to adult and developing rats. *Brain Res.* 98:517–27

Burnstock, G., Evans, B., Gannon, B. J., Heath, J. W., James, V. 1971. A new method of destroying adrenergic nerves in adult animals using guanethidine. *Br. J. Pharmacol.* 43:295–301

Butcher, L. L., Hodge, G. K., Schaeffer, J. C. 1975. In *Chemical Tools in Catecholamine Research,* ed. G. Jonsson, T. Malmfors, Ch. Sachs, 1:83–90. Amsterdam: North-Holland

Campochiaro, P., Coyle, J. T. 1978. Ontogenetic development of kainate neurotoxicity: Correlations with glutamatergic innervation. *Proc. Natl. Acad. Sci. USA* 75:2025–29

Cohen, G. 1976. Alkaloid products in the metabolism of alcohol and biogenic amines. *Biochem. Pharmacol.* 25: 1123–28

Cohen, G., Dembiec, D., Mytilineou, C., Heikkila, R. E. 1976a. In *Advances in Parkinsonism,* ed. W. Birkmayer, O. Hornykiewicz, pp. 251–57. Basel: Editiones Roche

Cohen, G., Heikkila, R. E. 1974. The generation of hydrogen peroxide, superoxide radical and hydroxyl radical by 6-hydroxydopamine, dialuric acid and related cytotoxic agents. *J. Biol. Chem.* 249:2447–52

Cohen, G., Heikkila, R. E. 1978. In *Serotonin Neurotoxins,* ed. J. H. Jacoby, L. O. Lytle, 305:74–84. New York: New York Acad. Sci.

Cohen, G., Heikkila, R. E., Allis, B., Cabbat, F., Dembiec, D., MacNamee, D., Mytilineou, C., Winston, B. 1976b. Destruction of sympathetic nerve terminals by 6-hydroxydopamine: Protection by 1-phenyl-3-(2-thiazolyl)-2-thiourea, diethyldithiocarbamate, methimazole, cysteamine, ethanol, and *n*-butanol. *J. Pharmacol. Exp. Ther.* 199:336–52

Coyle, J. T. 1978. Neuronal mapping with kainic acid. *Trends Neur.* 1:132–35

Coyle, J. T., Schwarcz, R. 1976. Lesion of striatal neurons with kainic acid provides a model for Huntington's Chorea. *Nature* 263:244–46

Creveling, C. R., Rotman, A. 1978. In *Serotonin Neurotoxins*, ed. J. H. Jacoby, L. D. Lytle, 305:57–73. New York: New York Acad. Sci.

Creveling, C. R., Rotman, A., Daly, J. W. 1975. see Butcher et al 1975, pp. 23–32.

Curtis, D. R., Duggan, D., Felix, D., Johnston, G. A. R., Tebecis, A. K., Watkins, J. C. 1972. Excitation of mammalian central neurons by acidic amino acids. *Brain Res.* 41:283–301

Curtis, D. R., Watkins, J. C. 1963. Acidic amino acids with strong excitatory actions on mammalian neurons. *J. Physiol.* 166:1–14

Daly, J., Fuxe, K., Jonsson, G. 1974. 5,7-Dihydroxytryptamine as a tool for the morphological and functional analysis of central 5-hydroxytryptamine neurons. *Res. Commun. Chem. Pathol. Pharmacol.* 1:175–87

Ellison, G., Eison, M. S., Huberman, H. S., Daniel, F. 1978. Long-term changes in dopaminergic innervation of caudate nucleus after continuous amphetamine administration. *Science* 201:276–78

Fuller, R. W. 1978. In *Serotonin Neurotoxins*, ed. J. H. Jacoby, L. D. Lytle, 305:178–84. New York: New York Acad. Sci.

Fuller, R. W., Baker, J. C., Perry, K. W., Molloy, B. B. 1975. Comparison of 4-chloro-, 4-bromo- and 4-fluoroamphetamine in rats: drug levels in brain and effects on brain serotonin metabolism. *Neuropharmacology* 14:739–46

Furness, J. B., Campbell, G. R., Gillard, S. M., Malmfors, T., Cobb, J. L. S., Burnstock, G. 1970. Cellular studies of sympathetic denervation produced by 6-hydroxydopamine in the vas deferns. *J. Pharmacol. Exp. Ther.* 174:111–23

Graham, D. G., Tiffany, S. M., Bell, W. R. Jr., Gutknecht, W. F. 1978. Autoxidation versus covalent binding of quinones as the mechanism of toxicity of dopamine, 6-hydroxydopamine and related compounds toward C1300 neuroblastoma cells *in vitro. Mol. Pharmacol.* 14:644–53

Haeusler, G. 1971. Early pre- and postjunctional effects of 6-hydroxydopamine. *J. Pharmacol. Exp. Ther.* 178:49–62

Harvey, J. A., McMaster, S. E. 1975. Fenfluramine: evidence for a neurotoxic action on midbrain and a long-term depletion of serotonin. *Psychopharmacol. Comm.* 1:217–28

Harvey, J. A., McMaster, S. E., Yunger, L. M. 1975. p-Chloroamphetamine: selective neurotoxic action in the brain. *Science* 187:841–43

Heath, J. W., Evans, B. K., Gannon, B. J., Burnstock, G., James, V. B. 1972. Degeneration of adrenergic neurones following guanethidine treatment: an ultrastructural study. *Virchows Arch. B.* 11:182–97

Heikkila, R. E., Cabbat, F. S. 1977. Chemiluminescence from 6-hydroxydopamine Involvement of hydrogen peroxide, the superoxide and the hydroxyl radical; A potential role for singlet oxygen. *Res. Commun. Chem. Pathol. Pharmacol.* 17:649–62

Heikkila, R., Cohen, G. 1971. Inhibition of biogenic amine uptake by hydrogen peroxide: A mechanism for toxic effects of 6-hydroxydopamine. *Science* 172:1257–58

Heikkila, R., Cohen, G. 1972. Further studies on the generation of hydrogen peroxide by 6-hydroxydopamine. *Mol. Pharmacol.* 8:241–48

Heikkila, R. E., Cohen, G. 1973. 6-Hydroxydopamine: evidence for superoxide radical as an oxidative intermediate. *Science* 181:456–57

Heikkila, R. E., Cohen, G. 1975. See Butcher et al 1975, pp. 7–14.

Hökfelt, T., Jonsson, G., Sachs, Ch. 1972. Fine structure and fluorescence morphology of adrenergic nerves after 6-hydroxydopamine *in vivo* and *in vitro. Z. Zellforsch. Mikrosk.* 131:529–43

Jacoby, J. H., Lytle, L. D. eds. 1978. *Serotonin Neurotoxins, Ann. NY Acad. Sci.* vol. 305. New York: N.Y. Acad. Sci.

Jensen-Holm, J., Juul, P. 1971. Ultrastructural changes in the rat superior ganglion following prolonged guanethidine administration. *Acta Pharmacol. Toxicol.* 30:308–20

Johnson, E. M., Cantor, E., Douglas, J. R. 1975. Biochemical and functional evaluation of the sympathectomy produced by the administration of guanethidine to newborn rats. *J. Pharmacol. Exp. Ther.* 193:503–12

Jonsson, G. 1976. Studies on the mechanisms of 6-hydroxydopamine cytotoxicity. *Med. Biol.* 54:406–20

Jonsson, G., Hallman, H., Ponzio, F., Ross, S. 1979. A new neurotoxic compound for central and peripheral noradrenaline neurons. *J. Neurochem.* In press

Jonsson, G., Fuxe, K., Hökfelt, T., Goldstein, M. 1976. Resistance of central phenylethanolamine-N-methyl transferase containing neurons to 6-hydroxydopamine. *Med. Biol.* 54:421–26

Jonsson, G., Lohmander, S., Sachs, Ch. 1974. 6-Hydroxydopamine induced degeneration of noradrenaline neurons in the scorbutic guinea-pig. *Biochem. Pharmacol.* 23:2585–93

Jonsson, G., Malmfors, T., Sachs, Ch. 1972. Effects of drugs on the 6-hydroxydopamine induced degeneration of adrenergic nerves. *Res. Commun. Chem. Pathol. Pharmacol.* 3:543–56

Jonsson, G., Malmfors, T., Sachs, Ch., eds. 1975. *Chemical Tools in Catecholamine Research* I. Amsterdam: North Holland

Jonsson, G., Sachs, Ch. 1970. Effects of 6-hydroxydopamine on the uptake and storage of noradrenaline in sympathetic adrenergic neurons. *Eur. J. Pharmacol.* 9:141–55

Jonsson, G., Sachs, Ch. 1971. Uptake and accumulation of ^{3}H-6-hydroxydopamine in adrenergic nerves. *Eur. J. Pharmacol.* 16:55–62

Jonsson, G., Sachs, Ch. 1972. Degenerative and nondegenerative effects of 6-hydroxydopamine on adrenegic nerves. *J. Pharmacol. Exp. Ther.* 180:625–35

Jonsson, G., Sachs, Ch. 1973. 6-Aminodopamine induced degeneration of catecholamine neurons. *J. Neurochem.* 21:117–24

Köhler, C., Ross, S. B., Srebro, B., Ögren, S. O. 1978. In *Serotonin Neurotoxins. Ann. NY Acad. Sci.* 305:645–63

Kostrzewa, R. M., Jacobowitz, D. M. 1974. Pharmacological actions of 6-hydroxydopamine. *Pharmacol. Rev.* 26:199–288

Liang, Y-O., Wightman, R. M., Plotsky, P., Adams, R. N. 1975. See Butcher et al 1975, pp. 15–22

Lidbrink, P., Jonsson, G. 1975. On the specificity of 6-hydroxydopamine induced degeneration of central noradrenaline neurons after intracerebral injection. *Neurosci. Lett.* 1:35–39

Lorens, S. A., Guldberg, H. C., Hole, K., Köhler, C., Srebro, B. 1976. Activity, avoidance learning and regional 5-hydroxytryptamine following intrabrain stem 5,7-dihydroxytryptamine and electrolytic midbrain raphe lesions in the rat. *Brain Res.* 108:97–113

Maguire, M. E., Goldmann, P. H., Gilman, A. G. 1974. The reaction of ^{3}H-norepinephrine with particulate fractions of cells response to catecholamine. *Mol. Pharmacol.* 10:563–81

Malmfors, T., Thoenen, H., eds. 1971. *6-Hydroxydopamine and Catecholamine Neurons.* Amsterdam: North Holland

Mason, S. T., Fibiger, H. C. 1979. On the specificity of kainic acid. *Science* 204:1339–41

McGeer, E. G., McGeer, P. L. 1976. Duplication of biochemical changes of Huntington's chorea by intrastriatal injection of glutamic and kainic acid. *Nature* 263:517–19

McGeer, P. L., McGeer, E. G., Hattori, T. 1978a. In *Kainic Acid as a Tool in Neurobiology,* ed. E. G. McGeer, J. W. Olney, P. L. McGeer, pp. 123–38. New York: Raven

McGeer, E. G., McGeer, P. L., Singh, K. 1978b. Kainate-induced degeneration of neostriatal neurons: Dependancy upon corticostriatal tract. *Brain Res.* 139:381–83

McGeer, E. G., Olney, J. W., McGeer, P. L., eds. 1978c. *Kainic Acid as a Tool in Neurobiology.* New York: Raven

Olney, J. W., Ho, O. L., Rhee, V. 1971. Cytotoxic effects of acidic and sulphur-containing amino acids on the infant mouse central nervous system. *Exp. Brain Res.* 14:61–76

Poirer, L. J., Langelier, P., Roberge, A., Boucher, R., Kitsikis, A. 1972. Nonspecific histopathological changes induced by the intracerebral injection of 6-hydroxydopamine (6-OH-DA). *J. Neurol. Sci.* 16:401–16

Rotman, A., Daly, J. W., Creveling, C. R. 1976. Oxygen dependent reaction of 6-hydroxydopamine, 5,6-dihydroxytryptamine and related compounds with proteins *in vitro.* A model for cytotoxicity. *Mol. Pharmacol.* 12:887–99

Sachs, Ch., Jonsson, G. 1975a. Mechanisms of action of 6-hydroxydopamine. *Biochem. Pharmacol.* 24:1–8

Sachs, Ch., Jonsson, G. 1975b. 5,7-dihydroxytryptamine induced changes in the postnatal development of central 5-hydroxytryptamine neurons. *Med. Biol.* 53:156–64

Sachs, Ch., Jonsson, G., Heikkila, R., Cohen, G. 1975. Control of the neurotoxicity of 6-hydroxydopamine by intraneuronal noradrenaline in rat iris. *Acta Physiol. Scand.* 93:345–51

Sanders-Bush, E., Bushing, J. A., Sulser, F. 1972. Long-term effects of p-chloroamphetamine on tryptophan hydroxylase activity and on the levels of 5-hydroxytryptamine and 5-hydroxyindole acetic acid in brain. *Eur. J. Pharmacol.* 20:385–88

Sanders-Bush, E., Bushing, J. A., Sulser, F. J. 1975. Long-term effect of *p*-chloroamphetamine and related drugs on central serotonergic mechanisms. *J. Pharmacol. Exp. Ther.* 192:33–41

Saner, A., Thoenen, H. 1971. Model experiments on the molecular mechanism of action of 6-hydroxydopamine. *Mol. Pharmacol.* 7:147–57

Schwarcz, R., Hökfelt, T., Fuxe, K., Jonsson, G., Goldstein, M., Terenius, L. 1979. Ibotenic acid induced neuronal degeneration: A morphological and neurochemical study. *Prog. Brain Res.* 37: In press

Schwarcz, R., Scholz, D., Coyle, J. T. 1978. Structure-activity relations for the neurotoxicity of kainic acid derivatives and glutamate analogues. *Neuropharmacol.* 17:145–51

Stein, L., Wise, C. D. 1971. Possible etiology of schizophrenia: Progressive damage to the noradrenergic reward system by 6-hydroxydopamine. *Science* 171: 1032–36

Thoenen, H., Tranzer, J. P. 1968. Chemical sympathectomy by selective destruction of adrenergic nerve endings with 6-hydroxydopamine. *Naunyn-Schmiedebergs Arch. Exp. Pathol. Pharmakol.* 261:271–88

Tranzer, J. P., Thoenen, H. 1967. Ultramorphologische Veränderungen der sympatischen Nervenendigungen der Katze nach Vorbehandling mit 5- and 6- Hydroxy-Dopamin. *Naunyn-Schmiedebergs Arch. Exp. Pathol. Pharmakol.* 257:343

Tse, D. C. S., McCreery, R. L., Adams, R. N. 1976. Potential oxidative pathways of brain catecholamines. *J. Med. Chem.* 19:37–40

Uretsky, N. J., Iversen, L. L. 1970. Effects of 6-hydroxydopamine on catecholamine containing neurons in rat brain. *J. Neurochem.* 17:269–78

Waddington, J. L., Crow, T. J. 1979. Drug-induced rotational behaviour following unilateral intracerebral injection of saline-ascorbate solution: neurotoxicity of ascorbic acid and monoamine-independent circling. *Brain Res.* 161: 371–76

Wagner, K., Trendelenburg, U. 1971. Effect of 6-hydroxydopamine on oxidative phosphorylation and on monoamine oxidase activity. *Naunyn-Schmiedebergs Arch. Pharmakol.* 269:112–16

Ann. Rev. Neurosci. 1980. 3:189–226

VISUAL-MOTOR FUNCTION OF THE PRIMATE SUPERIOR COLLICULUS[1]

❖11540

Robert H. Wurtz and Joanne E. Albano

Laboratory of Sensorimotor Research, National Eye Institute,
and Laboratory of Neurobiology, National Institute of Mental Health,
Bethesda, Maryland 20205

INTRODUCTION

One of the major functions of the brain, and probably the original function of all nervous systems, is to produce movement in response to sensory stimulation. In recent years one case of such sensory-motor function, the visual initiation of eye movements, has been studied extensively. Much of this work has centered on the most obvious candidate for visual-motor guidance, the superior colliculus. There has long not been any doubt that the superior colliculus is involved in vision and eye movement—the structure receives direct projections from the retina and over a century ago stimulation of the colliculus was shown to produce eye movements (Adamük 1870). But it has only recently been possible to study the relation of single cell activity within this structure to both movement and vision.

This cellular approach, along with new anatomical, physiological, and behavioral methods, has enabled the investigation of the machinery whereby the visual stimulus initiates motor movement. It is these advances that we concentrate on in this review. We emphasize the primate superior colliculus because most of the work relating to movement has been done in the rhesus monkey, *Macaca mulatta*. At the same time we draw on work done on the cat and tree shrew to supplement areas as yet unexplored in the monkey, although many interesting facets of work on these and other species are not considered. Other recent reviews have considered many of these other aspects (Sprague et al 1970, Gordon 1975, Sprague 1972,

---

[1]The US Government has the right to retain a nonexclusive, royalty-free license in and to any copyright covering this paper.

189

Sprague et al 1972, Sprague 1975, Sparks & Pollack 1977, Goldberg & Robinson 1978).

The organization of this review follows the recent suggestions that the seven alternating fiber and cell layers of the superior colliculus can be functionally divided into two parts (Figure 1): 1. the superficial layers, consisting of stratum zonale, stratum griseum superficiale, and stratum opticum, and 2. the deep layers, consisting of stratum griseum and album intermediale, and stratum griseum and album profundum. This division is based upon behavioral, anatomical, and electrophysiological studies in several mammals.

The subdivision of the superior colliculus was first suggested by behavioral and anatomical studies in the tree shrew (Casagrande et al 1972, Harting et al 1973, Casagrande & Diamond 1974). When lesions were restricted to the superficial layers, tree shrews had deficits in the ability to perform a visual discrimination; larger lesions extending into the deep layers resulted in an additional failure to orient to stationary or moving objects. Harting et al (1973) found that the efferent connections of those layers also differ; the superficial layers that receive direct retinal input project to "visual," i.e. dorsal and ventral lateral geniculate and pulvinar, thalamic nuclei. In contrast, the deep layers project to thalamic nuclei that are not generally considered to be visual centers and to subthalamic and lower brainstem nuclei that are generally considered to be motor areas or reticular formation. Harting et al (1973) interpreted these results as evidence that the colliculus of the tree shrew consists of two divisions: a superficial division, concerned with visual processing, and a deeper division, concerned with orienting movements of the head and eyes in response to stimuli.

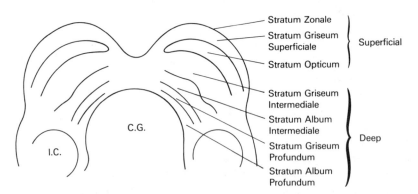

Figure 1 Alternating fiber and cell layers of monkey superior colliculus. A drawing of a coronal section through the colliculus is shown. The seven layers indicated on the right are divided into three layers designated as the superficial division and four layers designated as the deep division. I. C., inferior colliculus; C. G., central gray.

More recently, Edwards (in press) has argued that in the cat there is a complete anatomical segregation of the superficial layers from the deeper layers. First, he pointed out that the morphological features of cells found in the two zones are distinct; cell profiles in the superficial layers are similar to those in other primary sensory nuclei while cell profiles in deeper layers are like those found in other areas of the brainstem reticular core. Second, the afferent and efferent connections are consistent with a division into a sensory part and a reticular-like part. Third, receptive field properties indicate that the superficial cells are like other cells of the primary sensory pathways having limited fields while deep cells have diffuse and often multimodal sensory fields. Finally, repeated attempts to demonstrate connections between these layers have failed and so in the absence of unequivocal evidence Edwards was led to consider that such connections do not exist.

Electrophysiological experiments in the behaving monkey also emphasize the differences between the superficial and deep layers. As we shall see, even in the behaving animal the cells of the superficial layers are primarily visual while cells in the deeper layers discharge in relation to eye movements, although they may also have sensory responses. Moreover, activity between the parts is not necessarily coordinated; visual activity in the superficial layers does not necessarily lead to movement activity in deeper layers and, conversely, movement activity in deeper layers does not require visual activity in the superficial layers (Mohler & Wurtz 1976, Sparks et al 1977, Mays & Sparks 1979).

This review has four parts. In the first part we describe the visual responses of cells in the superficial layers, and in the second part the activity of cells in the intermediate and deep layers in relation to movement. In these two sections our primary goal is to describe the cellular elements within the superior colliculus based on the available physiological and anatomical observations. In the third part we attempt to derive a picture of intracollicular organization. Finally, in the fourth part we consider the functional relationship of the superior colliculus to current views of the oculomotor system. These last two sections concentrate on the functional organization of the colliculus and of necessity rely more on hypotheses than on verified observations.

VISUAL CELLS IN SUPERFICIAL LAYERS

The approach of the electrode tip to the surface of the superior colliculus is signaled by a roar of neuronal activity that is elicited by spots of light in the visual field. Superficial cells respond briskly to visual stimuli and are not fussy about stimulus size, shape, direction, or speed of movement. Some superficial cells also have a more subtle component to their response that is related to rapid eye movements (saccades) made by the animal. In this first section we examine properties of cells responding to passive visual stimulation, the behavioral modulation of the visual responses that occurs in some visual cells, and the afferent and efferent organization of the superficial layers.

Visual Responses

HORIZONTAL AND VERTICAL ORGANIZATION   The contralateral visual field is mapped in an orderly way across the superficial layers of the primate superior colliculus (Cynader & Berman 1972). The central visual field is represented at the anterior pole and lateral margin, while the peripheral visual field is represented at the posterior pole. The upper fields are represented medially, the lower fields laterally. The visual field representation in the primate differs from other mammals in two ways (Lane et al 1971, Cynader & Berman 1972, Kaas et al 1974). First, there is minimal ipsilateral representation of the visual field so that the vertical meridian representation occurs at the anterolateral margin of the colliculus. Second, the magnification factor of the central representation is larger than in most mammals; the central 10° is expanded to include more than 30% of the surface of the colliculus.

This horizontal map is produced by input from the two retinae. The projection of the contralateral hemi-retina includes the entire colliculus while the projection of the ipsilateral hemi-retina is represented only in the anterior portion of the colliculus leaving a monocular representation of the crescent at the posterior pole (Hubel et al 1975).

With increasing depth in the superficial layers, cells have increasingly larger visual receptive fields (Humphrey 1968, Cynader & Berman 1972, Goldberg & Wurtz 1972a, Moors 1978). Cells near the surface have central activating regions with diameters smaller than a degree while cells encountered more ventrally (as well as those with more peripheral receptive fields) have central activating regions often larger than 10° across.

RESPONSE TO STATIONARY STIMULI   The visual receptive fields of superficial layer cells consist of a central activating region surrounded by a zone capable of suppressing the response of the center (Humphrey 1968, Schiller & Koerner 1971, Cynader & Berman 1972, Goldberg & Wurtz 1972a, Updyke 1974, Marrocco & Li 1977, Moors 1978). The majority of these cells respond transiently to the onset and/or offset of spots of light with a latency of about 40–80 msec. There is a gradient of response strength across the activating center; responses to small flashes of light are greater toward the central portion of the field and are weaker toward the periphery (Goldberg & Wurtz 1972a). The activating center exhibits the property of internal summation, that is, a progressively greater number of spikes per burst are elicited as a function of increasing stimulus diameter. The summation effect has two components: there is an increase in response, reaching a maximum at some stimulus diameter smaller than the activating center, and the response latency is also reduced with larger stimuli (Moors 1978). When the diameter of the spot exceeds the boundaries of the central activating region and infringes upon the surround, the center response is suppressed. The relative strength of the center-surround interaction is variable; some surrounds are sufficiently weak as to allow diffuse flashes of light to activate the cells. We should emphasize that stimulation of the suppressive surround of colliculus cells alone evokes no cell discharge (Schiller & Koerner 1971, Goldberg & Wurtz 1972a, Cynader & Berman 1972), whereas the antagonistic surrounds of some retinal ganglion cells and lateral geniculate neurons can be independently excited by annuli (e.g. Dreher et al 1976). We therefore suggest that the term suppressive surround be used to describe receptive fields of superior colliculus cells to distinguish them from the antagonistic surrounds found in other parts of the visual pathways.

RESPONSE TO MOVING STIMULI   Most cells of the superficial layers also respond to moving stimuli. In contrast to the cat (e.g. Sterling & Wickelgren 1969), few sampled cells in the monkey are selective for the direction of movement. For the few directionally sensitive cells, tuning curves are broad, extending over 180° or more (Goldberg & Wurtz 1972a). Cells are also nonselective for stimulus velocities: the majority respond well to low stimulus velocities (as slow as 5°/sec), and many cells respond also to high stimulus velocities sometimes equivalent to saccadic velocities (600–900°/sec) (Goldberg & Wurtz 1972a, Robinson & Wurtz 1976, Marrocco & Li 1977, Moors 1978).

Three types of response patterns to moving stimuli have been described (Marrocco & Li 1977). One type consists of two sets of leading and lagging edge responses resulting from the passage of the stimulus edge through the initial and final receptive field borders. A second response pattern consists of a single set of leading and lagging edge responses occurring only at the initial receptive field border; an inhibitory effect resulting from the initial discharge eliminates the leading and lagging edge response at the second receptive field border. A third type of pattern is a sustained response throughout the receptive field elicited by a small moving stimulus. This response type is more commonly encountered in the lower half of the superficial layers. There is possibly a fourth kind of visually responsive cell termed the "jerk" detector, but these cells are either quite deep in the superficial layers or in the intermediate layers and have only been seen in paralyzed preparations (Schiller & Koerner 1971, Cynader & Berman 1972).

Behavioral Modulation of Visual Responses

The use of awake behaving monkeys has allowed examination of the visual responses of these cells when the animal actively responds to the visual stimuli. Two types of modulation of the visual response have been found: one that leads to an enhanced visual response when the monkey uses the visual stimulus as a target for a saccade and another that reduces the response of the cell to stimuli falling on the retina during a saccadic eye movement.

VISUAL ENHANCEMENT   Goldberg & Wurtz (1972b) noticed that for some cells the visual response was more vigorous when a trained monkey was required to make a rapid or saccadic eye movement from a fixation point to a visual stimulus lying within the central activating region. The enhancement took the form of either a more vigorous on-response to the visual stimulus or a prolonged on-response or both. The enhanced on-response was not evident on initial trials when the monkey was first required to make a saccade to the visual stimulus, but it occurred on subsequent trials when the monkey could expect to make a saccade to the receptive field stimulus. Since the enhancement of the visual response occurred before the monkey actually made the saccade, the receptive field stimulus was equivalent in both fixation and saccade trials; only the behavioral significance of the stimulus changed. The enhancement was found more frequently in the more ventral cells of the superficial layers.

The visual enhancement is spatially specific, that is, the visual response of the cell is enhanced only when a saccade is made near the central activating region of a cell. Saccades made to stimuli remote from the receptive field fail to produce an enhanced response. The visual enhancement is therefore not a generalized arousal or alerting effect.

The enhancement phenomena is temporally specific, that is, visually enhanced responses are transient effects temporally related to the beginning of the saccade (Wurtz & Mohler 1976a). By presenting the receptive field stimulus at varying times before and after the initiation of the saccade, response enhancement could be seen in visual responses occurring as early as 200 msec before saccade onset and also in visual responses occurring after the end of the saccade. The enhancement does not occur when the monkey responds to a stimulus but does not make a saccade to it.

Goldberg & Wurtz (1972b) suggested that the enhancement of visual responses at the cellular level might be a correlate of the selective attention seen at the behavioral level. The subsequent experiments showing the spatial and temporal relation of visual enhancement to eye movement suggest that the enhancement should be regarded either as only one type of selective attention or as part of the process specifically related to the initiation of

saccades (Wurtz & Mohler 1976a). The subsequent finding of a more generalized enhancement effect in parietal cortex (Robinson et al 1978, Bushnell
et al 1978) reinforces the view that the enhancement in the colliculus is
related specifically to saccades. The visual enhancement effect has been
reviewed more extensively elsewhere (Wurtz et al, in press).

The enhancement effect is also different from the modification of the
visual response of a cell due to a remote stimulus, which has been demonstrated in the cat (Rizzolatti et al 1974). This remote effect (at least in the
monkey) can result from interaction between two stimuli rather than from
the monkey's use of the stimulus (Richmond & Wurtz 1978).

VISUAL SUPPRESSION   The second type of modulation found in some
visual cells produces a suppression of their discharge rate following saccades (Goldberg & Wurtz 1972a). In these behavioral experiments, two
paradigms were used. The monkey was first required to maintain fixation
while a visual stimulus was swept at saccadic velocities across the receptive
field. The monkey was then required to make a saccade to a second fixation
point while a visual stimulus was appropriately positioned in the visual field
so that the effect of the saccade was to sweep the visual stimulus across the
receptive field. Robinson & Wurtz (1976) found that in this second condition a suppression of cell activity occurs that is appropriately timed and
powerful enough to eliminate the response of the cell to visual stimuli even
if the stimuli are 1 to 2 log units above background. The suppression occurs
with saccades in many directions and is only slightly influenced by saccadic
amplitude. The effect of the suppression is to reduce sensitivity to all visual
input during an eye movement.

This reduction of sensitivity could not result from any visual input since
the suppression persists in total darkness (Goldberg & Wurtz 1972a). Nor
could it result from peripheral sensory feedback such as proprioception
since Richmond & Wurtz (1977) found that the suppression persisted in the
collicular cells even when the eye was paralyzed by a retrobulbar block.
They concluded that the suppression, and by inference the reduced sensitivity to visual stimulation, was a result of a corollary discharge to saccadic eye movement impinging on these visually related cells in the colliculus.
Some cells show both enhancement of response to a relevant stimulus before
a saccade as well as reduced sensitivity to an irrelevant stimulus during the
saccade. These cells showing either type of modulation will be referred to
as *modulated visual cells* in contrast to other *unmodulated visual cells.*

Afferent Influences to Superficial Layers

The afferent influences that contribute to the visual responses of superficial
layer cells and to their modulation are summarized in Table 1. The three

Table 1 Connections of the superficial layers

Afferent connections	Efferent connections
Retina	Thalamus:
	Dorsal lateral geniculate
Visual cortex:	(magnocellular, interlaminar)
Striate & Prestriate	Pregeniculate
	(Ventral lateral geniculate)
Other cortical areas:	Inferior pulvinar
Frontal eye fields	
Midbrain:	Midbrain:
*Parabigeminal	Parabigeminal
*Pretectum	Pretectum

anatomically identified sources are the retina, several cortical areas, and two midbrain nuclei. Projections from the midbrain are shown with asterisks since these have not yet been demonstrated in the primate.

RETINOTECTAL Axons of retinal ganglion cells enter the superior colliculus bilaterally to terminate chiefly within the upper portion of stratum griseum superficiale (Hendrickson et al 1970, Tigges & Tigges 1970, Lund 1972a,b, Tigges & O'Steen 1974, Hubel et al 1975). As seen with autoradiography, the termination zone is particularly dense in the upper 200 μm except in the foveal region where retinotectal projection is sparse (Hubel et al 1975). In the parafoveal region receiving binocular input, terminals from the ipsilateral and contralateral eye appear as patches. Patches of contralateral eye inputs are concentrated in the upper 50–125 μm while patches from the ipsilateral eye extend somewhat more ventrally. A similar organization has been seen in the cat (Graybiel 1975). These anatomical observations may account for the electrophysiological observation that within the superficial layers there are intercalated areas of ipsilateral and contralateral eye dominance (Hubel et al 1975, Marrocco & Li 1977). All cells within 0.8 mm of the surface could be consistently driven by optic chiasm stimulation, which suggests a direct (but not necessarily monosynaptic) connection from retina to colliculus. More ventral cells in stratum opticum and the upper portion of the intermediate layers are also driven but with longer latencies (Marrocco & Li 1977). For many cells the short latency responses persist after ablation of striate cortex (Marrocco 1978).

Of the three major categories of retinal ganglion cells (i.e. *X, Y, W*) only two have been shown to project to the colliculus (de Monasterio & Gouras 1975, Schiller & Malpeli 1977, de Monasterio 1978a,b). These groups are both nonspectrally selective and have been termed 1. broad-band cells and 2. "rarely encountered cells." Primate broad-band cells are similar to Y-cells found in the cat retina (see Rodieck 1979 for review). They respond with a transient burst to the onset or offset of stationary spots of light, have antagonistic on- or off-surrounds, and are rapidly conducting. The second group projecting to the colliculus is the "rarely encountered" group (also called atypical ganglion cells). This second group is characterized by slowly conducting axons and low spontaneous activity. Some cells have on-off centers with suppressive surrounds, others have either on- or off-centers and may lack surrounds altogether. It seems likely that these cells represent a heterogeneous group of primate W-like cells and may be the major source of retinal input to the primate superior colliculus (de Monasterio 1978a). This notion is consistent with the similarities in visual receptive field structure of rarely encountered retinal ganglion cells and superficial layer colliculus cells. Both groups of cells have on-off centers, suppressive surrounds, and lack spectral sensitivity. The generally larger sizes of colliculus activating centers compared to the receptive field centers of retinal ganglion cells suggest a substantial convergence of retinal inputs in the colliculus.

CORTICOTECTAL   The projection of striate cortex to the superficial layers is well established in primate and numerous other species (Garey et al 1968, Wilson & Toyne 1970, Abplanalp 1970, Harting & Noback 1971, Kadoya et al 1971, Sterling 1971, Lund 1972b, Kawamura et al 1974, Robson & Hall 1975, Finlay et al 1976, Powell 1976). The projection of striate cortex arises from layer V and terminates for the most part ipsilaterally throughout the superficial layers (Lund et al 1975, Powell 1976). Slight degeneration is sometimes also found in the intermediate layers.

The receptive field properties of corticotectal cells are more specific for stimulus features than colliculus cells, yet they are less specific than nontectally projecting cells of striate cortex (Finlay et al 1976). When compared to cortical cells as a whole, corticotectal cells have somewhat larger receptive fields, have weak orientation specificity, and tend to have a weak directional preference for moving stimuli. They are binocular with complex-type receptive fields and high spontaneous activity.

Surprisingly, removal of the striate input, either by ablation or by cooling, has little or no effect upon the visual responses of cells in the superficial layers of the primate colliculus (Schiller et al 1974). Observed changes are limited to "patchy" receptive fields having subareas of on- or off-responses and an increase in monocularly dominated receptive fields. The visual responses of cells in the anterior colliculus representing the central visual field, which appears to receive a relatively weak input from the retina, are not especially affected by interruption of striate-tectal inputs. These results taken together with the lack of similarity in the receptive field organization of corticotectal cells suggest that striate cortex does not make a significant contribution to the passive visual properties in the superficial layers of the superior colliculus. Nor is striate cortex likely to be an important source of input to modulated cells since the enhancement effect seen in striate cortex is not spatially selective as it is in the colliculus (Wurtz 1969, Wurtz & Mohler 1976b) and the suppression effect in striate cortex is at best weak and is often not seen (Wurtz 1969, Bartlett & Doty 1974, Duffy & Burchfiel 1975). The contribution of striate cortex to the activity in the superficial layers remains a puzzle in the primate.

The frontal eye fields (area 8) project to the superficial layers and stratum opticum, and also to the dorsal part of intermediate layers (Kuypers & Lawrence 1967, Astruc 1971, Kunzle & Akert 1974, Kunzle et al 1976). Nearly half of the sampled neurons in a restricted part of area 8 respond to visual stimuli, and their receptive field properties are similar to those of the superficial layers (Mohler et al 1973, Wurtz & Mohler 1976b). They respond transiently to small flashes of light with a latency of 70–110 msec, in general are not selective for stimulus orientation or direction of movement, and show enhancement similar to that seen in the colliculus. Unlike colliculus cells, their receptive fields are commonly 20° or more in diameter and may extend 10–20° into the ipsilateral visual field. While the exact contribution of these cells is unknown, it may be that visual neurons in the frontal eye fields provide facilitative input to colliculus cells showing visual enhancement.

In the cat, other cortical visual areas project to the superficial layers (Kawamura et al 1974). In general, projections of cortical areas more remote from area 17 are to increasingly deeper colliculus layers. In the monkey, regions of prestriate cortex have been shown to project to the superficial layers (Kuypers & Lawrence 1967, Benevento & Davis 1977), but the organization of connections within the superficial layers is unknown.

Reciprocal and Efferent Influences of the Superficial Layers

In keeping with the visual character of afferent connections to the superficial layers, the efferent connections also distribute to known visual nuclei of the thalamus and have reciprocal connections with two other visual structures found in the midbrain (Table 1).

THALAMUS The superficial layers project mainly to three visual thalamic nuclei: the dorsal lateral geniculate, the pregeniculate nucleus (ventral lateral geniculate), and the inferior pulvinar (Mathers 1971, Harting et al 1973, Benevento & Fallon 1975, Trojanowski & Jacobson 1975, Graham 1977, Harting et al 1978). In the tree shrew the cells giving rise to these projections are stratified within the stratum griseum superficiale; cells projecting to the dorsal and ventral lateral geniculate arise primarily from the upper portion while cells projecting to pulvinar are located primarily in the lower half (Albano et al 1979). The nature of the tectal input to these visual structures has not been studied. However, some cells in the pregeniculate nucleus are modulated by saccadic eye movements in complete darkness (Büttner & Fuchs 1973). Projections to nonvisual thalamic nuclei have been described in monkey (Benevento & Fallon 1975) but have not been confirmed in other species (Harting et al 1973, Graham 1977). These nonvisual projections were found using anterograde degeneration and were only evident in cases where the lesions invaded the anterior colliculus where damage to fibers of passage and the pretectum is particularly likely.

MIDBRAIN   In addition to these visual thalamic projections, the superficial layers receive from and send connections to two visual midbrain structures, the pretectal complex and the parabigeminal complex. In cat, the superficial layers receive connections from two pretectal nuclei: the nucleus of the optic tract and the posterior pretectal nucleus (Edwards et al 1979). Several other studies in cat, tree shrew, and monkey have reported reciprocal connections from the superficial layers back to the pretectum but do not agree upon which specific nuclei receive these connections (Harting et al 1973, Berman 1977, Benevento et al 1977, Graham 1977, Itoh 1977). Thus, these connections may be reciprocal only in a general sense since it is not yet clear whether the specific nuclei that project to the colliculus also receive the collicular projections.

In cat, the parabigeminal nucleus, a small midbrain nucleus lying in the lateral tegmental area adjoining the superior colliculus, also sends and receives connections to the superficial layers (Harting et al 1973, Benevento & Fallon 1975, Graham 1977, Harting 1977, Graybiel 1978a, Edwards et al 1979, Henkel & Edwards 1978). The responses of visual cells of the superficial layers and the parabigeminal cells are strikingly similar (Sherk 1978). Like cat colliculus cells, parabigeminal cells are frequently direction selective but are not selective for stimulus size or speed of movement. Receptive fields consist of a central activating region and a surround with diameters corresponding to the range encountered in the colliculus. The reciprocal connection of the parabigeminal nucleus led Graybiel (1978a) to consider the parabigeminal nucleus as a "satellite system" possibly modulating the visual responses in the superficial layers.

Function of Superficial Layers

What role the cells in the superficial layers play in behavior has not been investigated in the monkey. The prominent ascending projections through the thalamus suggest that the visual processing seen in the superior colliculus could contribute to the form discrimination functions usually attributed to cerebral cortex. While this may be true in other mammals (for example, tree shrew, see Casagrande & Diamond 1974; cat, see Berlucchi et al 1972, Sprague et al 1977, Tunkel & Berkley 1977), the role in form discrimination in primates appears to be slight (Rosvold et al 1958, Anderson & Symmes 1969, Butter 1974a,b) or simply not yet revealed with the tasks that have been tried. Human psychophysical studies suggest that a saccade to a part of the visual field may alter detection thresholds in that part of the field (Singer et al 1977); this type of perceptual modulation may be related to ascending effects of the colliculus in general and of visual enhancement in particular. The superficial layers also might have critical effects on activity in the deep layers, which we consider later.

MOVEMENT CELLS IN THE DEEP LAYERS

As a microelectrode passes from superficial layers into the deep layers of the superior colliculus, a dramatic change in the relation of collicular activity to visual-motor behavior occurs. The cells now discharge in close temporal relation to saccades in the dark as well as in the light. In this section we summarize the types of these cells, their relation to eye movements, and anatomical connections of the strata intermediale and profundum.

Properties of Cells Related to Movement

Cells related to movement discharge before saccadic eye movements made to visual targets. Most cells also discharge before saccades made spontaneously in the light or the dark (Wurtz & Goldberg 1971, Schiller & Koerner 1971, Wurtz & Goldberg 1972a) and before the quick phase of optokinetic (Schiller & Koerner 1971) and vestibular nystagmus (Schiller & Koerner 1971, Wurtz & Goldberg 1972a). These cells were first studied in the monkey (Schiller & Koerner 1971, Wurtz & Goldberg 1971), although such cells had been reported previously in the cat superior colliculus (Straschill & Hoffmann 1969). In the monkey there is considerable agreement on most of the characteristics of these cells (Schiller & Koerner 1971, Wurtz & Goldberg 1971, Schiller 1972, Schiller & Stryker 1972, Wurtz & Goldberg 1972a,c, Robinson & Jarvis 1974, Wurtz & Mohler 1974, Sparks 1975, Mohler & Wurtz 1976, Sparks et al 1976, Sparks & Pollack 1977, Sparks et al 1977, Sparks 1978, Wurtz 1978, Mays & Sparks 1979) so that we cite references in this section only when specific or conflicting findings are involved.

Cells related to movement have three salient characteristics: 1. They discharge before saccadic eye movements, usually leading the onset of eye movement by about 50 to nearly 150 msec. 2. The cells only discharge before an eye movement to one area of the visual field, an area referred to as the movement field of the cell (Wurtz & Goldberg 1972a) in analogy to the visual field of visual neurons. 3. The movement fields are organized in retinotopic coordinates rather than in spatial coordinates. That is, if the movement field of a cell is located 20° to the right as the monkey looks straight ahead, the cell will discharge before a 20° rightward saccade. The cell will also discharge in a similar manner before a 20° rightward saccade starting from any other orbital position. We consider each of these characteristics in turn.

TIMING OF DISCHARGE   The time of onset of movement cell discharge varies with depth within the colliculus (Mohler & Wurtz 1976). The most dorsal movement cells, found at the junction of stratum opicum and stratum griseum intermediale, show the shortest lead time before onset of saccades, usually about 40–50 msec. The burst of activity is superimposed on a low background rate. Other cells deeper in the colliculus have an earlier but more gradual onset in their rate of discharge before a saccade, which is superimposed on a higher background rate of discharge. The increase in discharge rate of these cells begins at least 100–150 msec before the onset of the saccade. The slow rate of onset of the early discharge and the typically high background rate of these cells probably account for several larger estimates of their lead before saccades (200–300 msec, see Wurtz & Goldberg 1972a; 70–500 msec, see Schiller & Koerner 1971). Variation in the lead time among these movement cells suggests a continuum, but further work might well reveal a series of discrete cell types. For ease of reference, we refer to the cells discharging close to the onset of the saccade as short-lead cells and to the other cells as long-lead cells.

Another distinction between cells related to movement is based on the presence or absence of a burst of action potentials preceding the eye movement (Sparks et al 1976, Sparks et al 1977). Saccade-related burst cells have a relatively discrete burst of high frequency discharge beginning about 20 msec before saccade onset even though the earliest increase in activity may occur 80–100 msec before a saccade. Other cells show a gradual buildup of activity starting 80–100 msec before the saccade but have no such burst of activity. The depth of the cell types has not been determined. It seems clear, however, that the short-lead cells are all burst-type cells; the long-lead cells may or may not be burst-type cells. This overlap of the short-lead type and the burst-type neuron will be important since these cells are probably output cells of the colliculus, and we consider them again later.

The relation of the cell discharge to the end of the eye movement also varies, occasionally ending with the end of the saccade, but frequently continuing beyond the end of the saccade. In general, the deeper the cell within the colliculus, the longer the discharge continues after the end of the saccade (Mohler & Wurtz 1976), but no systematic study has been devoted to this relationship.

MOVEMENT FIELDS   A movement field is determined by having the monkey saccade to one point in the visual field on a series of trials, then to other points, until the discharge of the cell preceding a series of points is obtained. While the cell discharge may increase before saccades to a large area of the visual field, there is a gradient in the vigor of response across the field. Saccades to points in the center of the movement field are preceded by a more vigorous discharge than saccades to the edge of the movement field. The gradient within a movement field declines more sharply towards the fovea (Sparks et al 1976) as does the gradient of the visual receptive fields in the superficial layers (Goldberg & Wurtz 1972a). The onset of the discharge has also been reported to start earlier in the central area of the movement field in some cases (Sparks et al 1976) but not in others (Mohler & Wurtz 1976).

HORIZONTAL AND VERTICAL ORGANIZATION   Within the superior colliculus there is an orderly horizontal organization of movement fields. This map of the movement fields is in register with the retinotopic map of the superficial layers, but the detailed topography of this movement map is even less well known than that of the visual map. A difference between the visual and movement fields is that while visual fields have not been shown to cross the vertical meridian into the ipsilateral field (Cynader & Berman 1972), movement fields definitely cross into the ipsilateral field especially when the movement fields include the vertical meridian. The size of the movement fields varies with location on the retinotopic map. Fields near the fovea are smaller in overall size (5° across) and have steeper gradients of cell discharge to given points within the field while fields further from the fovea are larger (20–30° across) with rather gradual response gradients.

Neither the duration nor the vigor of cell discharge is related to location of the movement field in the visual field (Sparks et al 1976). Thus a cell with a movement field 5° from the fovea discharges just as vigorously as a cell with a movement field 20° away. This is in contrast to cells in the pontine reticular formation whose duration of discharge varies with saccade amplitude. The organizational principle in the colliculus is based on which cell discharges, not how much it discharges.

Movement cells are first encountered at the junction of the superficial and intermediate gray layers. As movement cells are located farther within the deep layers the size of the movement fields gradually increases (Mohler & Wurtz 1976). The movement fields of the most dorsal cells (at the dorsal border of stratum griseum intermediale) are about the same size as the visual receptive fields of the superficial layer cells just above. Movement field size increases with depth until the field can easily cover a quadrant of the visual field. Exactly how deep within the colliculus movement cells are found has surprisingly never been determined.

Visual-Motor Integration

SENSORY RESPONSES   Many of these cells related to movement also respond to visual stimulation as first reported by Schiller & Koerner (1971), and this sensory response is all that is evident in the paralyzed monkey (Cynader & Berman 1972, Updyke 1974). In the awake monkey the visual response is very slight compared to the movement related discharge of such cells, often being only a few spikes even to an optimal stimulus. The visual receptive fields of these cells always overlap the movement fields but are seldom coterminous with them. These visual responses probably do not result from the visual activity in the superficial layers since following ablation or cooling of striate cortex these cells of the deep layers lose their visual responses (Schiller et al 1974). In addition, the enhanced visual response, prominent in the deeper parts of the superficial layers, has not been found in these deep layer cells (Goldberg & Wurtz 1972b, Mohler & Wurtz 1976).

Another visually related movement cell is the "quasi-visual cell" reported by Sparks et al (1977), and Mays & Sparks (1979). These cells have a fixed latency response to the onset of a visual target and continue to discharge until a saccade of appropriate amplitude and direction occurs. Because the quasi-visual cells continue to discharge even after the saccade target has been removed, Mays & Sparks (1979) point out that these cells "hold" the information of eye position error. Since these cells begin their discharge long before the onset of a saccade, usually without a discrete burst of activity, we think they could be regarded as a new variety of long-lead cell. Like long-lead cells, quasi-visual cells are usually below the first movement cells encountered in a microelectrode penetration and like some long-lead cells, quasi-visual cells have visual responses.

Some cells in the deep layers of the colliculus also respond to auditory and somatosensory stimuli (Updyke 1974) and this has been studied most extensively in the cat and mouse (Wickelgren 1971, Stein & Arigbede 1972, Gordon 1973, Stein et al 1976b, Drager & Hubel 1975, 1976). The relation of such multisensory input has not been studied in awake animals, but since eye movements can obviously be initiated by other than visual stimuli, the possible function of these inputs is intriguing.

VISUALLY TRIGGERED MOVEMENT CELLS   The visually triggered movement cells identified by Mohler & Wurtz (1976) discharge before saccades into the movement field of that cell but only when the saccade is triggered by a visual target. Spontaneous eye movements in light or dark are not accompanied by such a discharge as would be the case with the movement cells considered thus far. These visually triggered movement cells are located at the dorsal border of the stratum griseum intermediale and are identical to the short-lead neurons except for the added requirement of a visual trigger. These cells usually also show a slight response to visual stimulation. But it must be emphasized that this visual response is not necessary for the "gating" of such cells since the movement part of its discharge can be elicited without a visual response. Similar cells have recently been reported by Mays & Sparks (1979).

Collicular Relation to Saccade Metrics

STIMULATION   In the awake monkey, Robinson (1972) and Schiller & Stryker (1972) (Stryker & Schiller 1975) found that stimulating the superior colliculus produced results paralleling those found with single cell recording. Stimulation produced saccades whose amplitude was related to the point stimulated, even with suprathreshold currents. The shortest latency for a saccade following stimulation was 20 msec—a time equal to the lead time of burst neurons before a saccade. Thresholds for eliciting saccades were lowest in the intermediate layers.

Schiller & Stryker (1972) also compared the direction and amplitude of saccades evoked by micro-stimulation at one point in the colliculus with the movement fields of cells recorded at the same point. After stimulation with the same microelectrode the eyes moved to the same part of the visual field represented by the movement fields of adjacent cells. The map of eye movement directions corresponded to the retinotopic map found in superficial layers (Cynader & Berman 1972). This retinotopically organized movement map in the primate may differ from areas of the cat colliculus where stimulation has been reported to be related to the position of the eye in the orbit (Straschill & Rieger 1973, Roucoux & Crommelinck 1976, Crommelinck et al 1977) although others have not found this (Stryker & Blakemore 1972, Stein et al 1976a, Harris 1979).

EYE-HEAD   In monkeys free to move their heads, eye and head movements usually occur together (Bizzi et al 1971) so that the movement cells in the superior colliculus might be related not only to saccades but also to attempted head movement. Robinson & Jarvis (1974) allowed the monkey to make horizontal head movements as well as eye movements but found that the colliculus cells still discharged in close temporal relation to eye movement, not to head movement. By stimulating the colliculus, Stryker & Schiller (1975) found that eye movements were elicited with short latency, with fixed threshold, and with the same amplitude regardless of the initial position of the eye in the orbit. Head movements following stimulation had variable latency ($>$ 90 msec), variable threshold, and usually occurred when the eye had deviated to one side of the orbit. Both studies reach the conclusion that the monkey superior colliculus does not determine the occurrence of head movement. This may be species dependent since stimulation and recording experiments in the cat colliculus suggest that this structure might be related to head movement (Harris 1979, Roucoux et al, in press).

If a head movement occurs along with an eye movement, the amplitude of the eye movement is shortened primarily by feedback from the vestibular system, called the vestibular ocular reflex (Bizzi et al 1971). The shift in gaze (eye plus head movement) remains the same regardless of the occurrence of the head movement. When reduction of saccade amplitude occurs, the discharge of collicular movement-related cells in the colliculus is appropriate for the saccade that would have occurred if the head had not moved (Robinson & Jarvis 1974). The discharge of the movement cells can therefore be regarded as related to the retinal error, which leads to a shift in gaze regardless of whether the error is reduced by a saccade or not.

SACCADE METRICS   Another indication that the colliculus cells are not necessarily tied to the metrics of the executed saccade is the observation that a movement cell that discharged before 40° long saccades also discharged when the monkey executed two serial 20° saccades (Mohler & Wurtz 1976). The cell did not discharge preceding ordinary 20° saccades since such a saccade did not fall into the movement field of the cell. The cell discharged in relation to the size of the error, not necessarily the amplitude of the saccades accomplishing the movement.

Mohler & Wurtz (1976) concluded that the discharge of the movement cells (including short-lead cells) could be decoupled from the saccade since they observed cells in a monkey that discharged before saccades and continued to discharge at the appropriate time even when the monkey stopped making the saccades. Sparks (1978), using a more elaborate paradigm, rarely observed such decoupling in the burst neurons and concluded that these neurons are more tightly coupled to saccades than Mohler & Wurtz (1976) thought. Unfortunately, no one has compared systematically the tightness of coupling with the type of cell, long-lead or short-lead. The short and consistent interval between the discharge of short-lead cells and the onset of the saccade is probably the best argument that short-lead cells are the output neurons of the superior colliculus; decoupling, like the occurrence of head movements along with saccades or the execution of two saccades instead of one, may occur between the colliculus and the site of saccade generation.

SMOOTH PURSUIT   Those movement cells that discharge before saccades to targets outside the foveal region do not discharge before smooth pursuit eye movements except before the catch-up saccades during pursuit movements (Wurtz & Goldberg 1972a). Cells within the foveal region of the superior colliculus do discharge during tracking eye movements, but it seems likely that this discharge relates to the movement of the target stimulus rather than the eye movement itself (Schiller & Koerner 1971, Goldberg & Wurtz 1971). The relation of colliculus cells to the slow phase of nystagmus or vergence eye movements has not been investigated.

Afferent and Efferent Connections

The characteristics of the movement cells and their influence on other brain areas is dependent upon the afferent and efferent connections of the deep layers summarized in Table 2. Quite unlike the simpler patterns of connections of the superficial layers, the number of structures projecting to the deep layers is overwhelming; subcortical projections involve nearly all regions of the brainstem and corticotectal projections include regions from prefrontal to occipital cortex. With the exception of certain sensory areas of cerebral cortex, these connections do not generally involve structures associated with primary sensory pathways.

AFFERENT CONNECTIONS   For purposes of simplification, the afferent connections may be organized into seven groups (Table 2). Many connections are indicated with asterisks since most of the data is derived from cat and few of these connections have been established in the primate.

Cortical afferents to the deep layers arise from numerous areas including striate, prestriate, auditory, and somesthetic cortex, parietal and temporal cortical areas, prefrontal cortex and frontal eye fields (Kuypers & Lawrence 1967, Garey et al 1968, Paula-Barbosa & Sousa-Pinto 1973, Kawamura et al 1974, Kunzle & Akert 1974, Kunzle et al 1976, Goldman & Nauta 1976, Benevento & Davis 1977, Jones & Wise 1977, Wise & Jones 1977, Hartmann-von Monakow et al 1979). These corticotectal projections are diffuse and do not terminate within a single lamina although there is some evidence of stratification since striate and prestriate areas tend to terminate more dorsally than more remote cortical areas (Kawamura et al 1974). Actually the projections from striate and prestriate regions have two terminal foci, one in the superficial layers, which appears discrete, and a second in the

Table 2 Connections of the intermediate and deep layers

Afferent connections	Efferent connections
Cortex	Ascending
Striate & Prestriate cortex	*Subthalamus & thalamus*
Auditory, Somesthetic, Motor cortex	Zona incerta
Regions of Parietal, Temporal cortex	Fields of forel
Prefrontal cortex	Reticular, Limitans n.
Frontal eye fields	Reunions n.
Diencephalon	Intralaminar n.
	(Parafascicular, Centromedian)
* Zona incerta	Mediodorsal n. (rim)
* Reticular n.	Suprageniculate
* Pregeniculate	Medial geniculate (magnocellular)
(Ventral lateral geniculate)	*Midbrain*
Pretectum	Anterior, Posterior pretectal n.
* n. Posterior commissure	n. Posterior commissure
* Anterior, Posterior pretectal n.	
* n. Optic tract	Descending ipsilateral
Midbrain	(Tectopontine/Tectobulbar)
* Cuneiform, Subcuneiform	*Midbrain*
* Substantia nigra (pars reticularis)	Parabigeminal, Peri-parabigeminal
* Parabigeminal, Peri-parabigeminal	Paralemniscal
* Paralemniscal	Subcuneiform, Cuneiform
* n. Brachium inferior colliculus	Inferior colliculus
* External n. inferior colliculus	External capsule
* Pericentral n.	*Pons*
* n. Sagulum	Reticularis tegmenti pontis
* Locus coeruleus	Reticularis pontis oralis
* Raphe dorsalis	Dorsolateral pontine n.
* Lateral parabrachial n.	Facial motor n.
Pons and medulla	Descending contralateral
* n. Pontis caudalis, oralis	(Tectospinal/Predorsal Bundle)
* Reticularis tegmenti pontis	*Pons*
* Ventral n. lateral lemniscus	Reticularis tegmenti pontis
* Dorsomedial periolivary n.	Reticularis pontis oralis, caudalis
* Medial n. trapezoid body (medial)	Abducens, perioculomotor regions
* Sensory, Spinal trigeminal	Facial n.
Cerebellum	*Medulla*
* n. Gigantocellularis	Subnucleus B Medial accessory n.
* n. Paragigantocellularis lateralis	inferior olive
* Medial vestibular n.	Raphé
* Perihypoglossal	*Cervical spinal cord*
* Cuneate, Gracile n.	Commissural pathway
Cervical spinal cord	
Lateral cervical nucleus	

deep layers, which appears diffuse. Areas 19, 21, 7 and Clare-Bishop project diffusely and only to the deeper layers (Kawamura et al 1974); projections from other cortical areas have a similar pattern (Kuypers & Lawrence 1967, Garey et al 1968, Goldman & Nauta 1976).

The subcortical afferents to the intermediate and deep layers of the cat have been surveyed recently by Edwards and his colleagues (1979). They found over 40 sources of afferent input, which are also summarized in Table 2. These nuclei include reticular nuclei of the thalamus, midbrain, pons and medulla, several pretectal nuclei, the deep cerebellar nuclei and inputs from the sensory and spinal trigeminal, the cuneate and gracile nucleus and cervical spinal cord (Grofová et al 1978, Edwards et al 1979). In addition, the intermediate and deep layers receive a "patchy" projection from substantia nigra (Graybiel 1978b).

EFFERENT CONNECTIONS   Efferent fibers of the intermediate and deep tectal pathways are organized in Table 2 into four pathways: ascending, descending ipsilateral, descending contralateral, and commissural. These pathways have been established in a diverse group of species including the primate (Harting et al 1973, Benevento & Fallon 1975, Kuypers & Maisky 1975, Graham 1977, Edwards 1977, Edwards & Henkel 1978, Edwards et al 1979, Kawamura & Hashikawa 1978, Holcombe & Hall 1978). The ascending pathway takes a route through the stratum opticum and the brachium of the superior colliculus to connect with regions of the thalamus and subthalamus such as the medial and inferior pulvinar, the magnocellular division of the medial geniculate, fields of Forel, certain nuclei of the intralaminar complex, zona incerta, and the reticular nucleus of the thalamus (Harting et al 1973, Benevento & Fallon 1975, Graham 1977).

The target structures of the ipsilateral descending pathway are primarily located in the midbrain and pons (Harting et al 1973, Graham 1977, Harting 1977, Baleydier & Magnin 1979). In the midbrain region, fibers distribute to the parabigeminal-lateral tegmental area, the mesencephalic reticular formation (nucleus cuneiformis and subcuneiformis) and the external nucleus of the inferior colliculus. In the pons, target structures involve the pontine reticular formation (nucleus reticularis pontis) and the so-called dorsolateral pontine nuclei (Harting et al 1973, Frankfurter et al 1976, Graham 1977, Harting 1977). The descending contralateral pathway, also called the tectospinal tract or the predorsal bundle, has more extensive regions of termination including regions of the pontine and medullary reticular formation, regions near and possibly within the abducens nucleus, subnucleus B of the medial accessory nuclei of the inferior olive, and cervical spinal cord (Harting et al 1973, Frankfurter et al 1976, Harting 1977, Castiglioni et al 1978, Weber et al 1978).

Now that we have emphasized the number and diversity of these efferent connections, we can begin to notice several patterns. First, several of the structures listed in Table 2 project to the cerebellum and may therefore provide pathways linking the superior colliculus with visual-oculomotor regions of the cerebellar cortex (Frankfurter et al 1976, Harting 1977, Weber et al 1978). These nuclei include the dorsolateral pontine nuclei, regions of the pontine reticular formation, and subnucleus B of the medial accessory nuclei of the inferior olive. Second, other projections involve structures concerned with movements involved in orienting responses. In cat, the descending colliculus connections from paralemniscal nuclei to the facial nucleus have been implicated in the control of pinnae movements, and in tree shrew, section of the predorsal bundle results in a loss of visual orienting responses (Henkel & Edwards 1978, Raczkowski et al 1976). Third, these layers project to regions in and about the abducens and oculomotor nuclei (Harting 1977, Edwards & Henkel 1978). This pathway provides for disynaptic, sometimes monosynaptic, excitation directly to oculomotor neurons in addition to less direct pathways through the pontine reticular formation (Grantyn & Grantyn 1976). Finally, many of the same structures we have discussed also project back to the superior colliculus.

Whether there is a laminar organization of cells that give rise to specific efferent projections is not yet certain. In cat, cells involved in the commissural pathway are located primarily in the stratum griseum intermediale (Edwards 1977, Magalhaes-Castro et al 1978). Axons coarsing in the predorsal bundle appear to arise from cell bodies located in the intermediate layer in squirrel (Holcombe & Hall 1978) and in the deep layers in galago (Raczkowski & Diamond 1978). Projections to the paralemiscal region arise from the deep layers while the projections to the parabigeminal nucleus arise from cells in the intermediate layers (Henkel & Edwards 1976). Projections to perioculomotor areas of the abducens arise primarily from the intermediate layer (Edwards & Henkel 1978). There is also disagreement concerning the laminar distribution of cells of origin that project to brainstem reticular regions. Graham (1977), using autoradiographic technique, found the greatest amount of transport in cases with the deepest injections. However, studies using retrograde tracing techniques find that after brainstem injections of reticular nuclei and spinal cord most labeled cells are located in both the intermediate and deep layers (Kuypers & Maisky 1975, Hashikawa & Kawamura 1977, Kawamura & Hashikawa 1978, Castiglioni et al 1978). The only tentative conclusion that may be made at this point is that each layer appears to give rise to somewhat different efferent connections. We suspect that future anatomical studies will reveal that each colliculus target structure will have its own laminar pattern such as has been already revealed in the superficial layers.

INTRACOLLICULAR ORGANIZATION

We have thus far considered five elements within the superior colliculus. Two elements, modulated and unmodulated visual cells, are found within the superficial layers and three elements, visually-triggered, short-lead, and long-lead movement cells, are found within the deep layers. Anatomical studies have provided no clues concerning the functional connections between these cells. Electrophysiological studies indicate, however, a temporal order of cell discharge to the onset of a visual target and the onset of a saccade, which provides a clue about the sequence of processing occurring within the superficial and deeper layers. The timing of discharge of each of the five elements is summarized in Figure 2A, the hypothetical relationships between cells derived from this temporal order is illustrated in Figure 2B, and the possible output signal conveyed by each cell type is shown in Figure 2C. This list of cell types, drawing of possible connections, and identification of possible functions is made explicit not to imply certainty but to make specific the hypotheses to be tested and modified.

Superficial Layers

UNMODULATED VISUAL CELLS   Based on the latency of response to a visual stimulus, the temporal sequence of cell discharge in the superficial layers is from top down. As we have previously noted, unmodulated visual cells tend to have shorter response latencies and also tend to be located more dorsally in the superficial layers. Unmodulated visual cells could be regarded as conveying a retinal error signal; that is, the population of cells that discharge in the retinotopically organized map can indicate the difference between where the monkey is currently looking and where a saccade target is located. It appears that these cells derive their visual properties primarily from the retina and so may be considered as neurons of the primary visual pathways.

MODULATED VISUAL CELLS   The discharge of the modulated visual cells that show enhancement could convey a selected retinal error signal; one visual error signal selected from many to be used as a target for a saccadic eye movement. Since the visually enhanced response incorporates information that an eye movement is about to occur, it can hardly be regarded as an early stage in the decision to make a visually guided eye movement; it is more accurately viewed as the result of this process. One function of the enhancement may be to prolong the visual response which then more easily overlaps in time with the discharge of the movement-related cells in the deep layers.

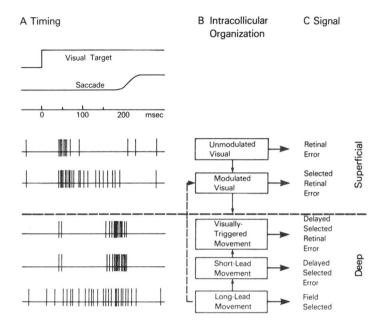

Figure 2   Temporal sequence of cell discharge in monkey superior colliculus (*A*), hypothetical intracollicular organization based on this sequence (*B*), and possible signal conveyed by
each of these elements.

The drawing at the top of *A* indicates the onset of a visual target on the top line with the
subsequent saccade from a fixation point to that target 200 msec later shown on the second
line. Subsequent lines show schematic representations of the discharge of each type of cell in
relation to stimulus onset and the subsequent saccade. The dashed horizontal line indicates
the division between the superficial and intermediate-deep layers. The discharge of the visually
triggered movement cell is similar to that of the short-lead cell but the cell would not show
such a discharge if the saccade were not to a visual target. Two examples of the discharge of
a long-lead cell are shown; the lower line illustrates a case when increase in cell discharge leads
the onset of the saccade target in a series of trials in which the monkey can anticipate the
temporal sequence.

B shows a hypothetical sequence of these cells. Small arrows between boxes emphasize that
the intracollicular connections drawn are based on timing of discharge, not known anatomical
connections. No afferents are shown; arrows on the right emphasize the possibility of outputs
from every layer even though only certain outputs (visually triggered movement cells and
short-lead movement cells) are emphasized in the text. It is obvious that the temporal sequence
shown only eliminates some alternatives but does not require the connections shown. The
connections drawn are derived from the temporal relation of cell discharge to visually initiated
saccades but other nonsaccadic functions of the colliculus might produce different temporal
relations implying different intracollicular connections.

C shows for each variety of cell the type of signal it could convey to the rest of the nervous
system since the superior colliculus must be part of an extended system for the initiation of
saccades.

The function of the suppressive input to some of these active cells might be to ensure that no new retinal error signal is generated during a saccade; the retinal error signal would not only be selected but it would usually not be updated until the saccade is over. The source of the suppression seen in many of the active cells in the superficial layers is unknown.

Deep Layers

LONG-LEAD CELLS   In contrast to the top down sequence shown for the superficial layers and in traditional notions of collicular function, the order of the onset of the saccade-related discharge in the intermediate and deep layers is from the bottom up (Mohler & Wurtz 1976). Long-lead movement cells deep in the colliculus start to discharge earlier, before the onset of saccades, than the more dorsal cells. In fact, the discharge of long-lead cells occurs as early as any other demonstrated saccade-related neurons in the brain. For example, long-lead movement cells in the colliculus precede a saccade by 150 msec or more while long-lead burst neurons in the pontine reticular formation may lead the eye movement by about 100–125 msec (Fuchs & Luschei 1972, Keller 1974).

The discharge of these long-lead cells could be regarded as a selected movement field signal since their discharge indicates that one part of the field has been selected as opposed to another. This selection is not necessarily dependent upon visual information present at the moment of saccade onset since the discharge of these cells can anticipate the visual signal to make a saccade and can occur before saccades made in total darkness. One wonders whether the slow and irregular start of the cell may reflect a tentativeness of the selection and whether the large movement fields of these cells reflect a lack of spatial specificity in the selection process. Certainly, the movement fields of these cells taken individually would seem too large to accurately define a selected eye movement. This selected movement field signal presumably arrives at the colliculus rather than being generated by it, but the source of this signal may be any of the myriad possibilities listed in Table 2.

SHORT-LEAD CELLS   The transition from long-lead to short-lead cells involves both a temporal change, trimming the prolonged discharge on a high background rate to a final burst on a near silent background, and a spatial change, narrowing the size of the movement field from broad to narrow. This latter step might result if short-lead movement fields are derived from the overlap of subjacent fields in a way comparable to that suggested by McIlwain (1976) for visual fields. Alternatively, these fields might be spatially refined by an extracollicular input. Because of the smaller size of their movement fields, short-lead movement cells are most closely related to the amplitude of the saccade. For this reason and because the discrete and vigorous saccade-related discharge of these cells precedes the eye movement by the same latency (20 msec) as a saccade elicited by stimulation, these cells appear to be a likely output element of the colliculus.

The discharge of the short-lead movement cells may be considered as a delayed error signal—a delayed signal because the discharge occurs just before the saccade rather than after the visual stimulus, and an error signal because their discharge is associated with a particular saccadic amplitude that reduces the error between where the eye is now and where it will soon be.

VISUALLY TRIGGERED CELLS   The visually triggered movement cell is shown in Figure 2 as successive stage in the processing sequence because it incorporates both movement-related activity and activity resulting from the appearance of a visual target. The visual target for these cells is a necessary condition for the movement-related discharge to occur: the movement discharge may be viewed as "gated" by the presence of the visual stimulus. This gating effect [or partial gating in some cells (Mohler & Wurtz 1976, Mays & Sparks 1979)] is independent of any discrete visual response the cell may have to the visual stimulus. Thus these cells suggest a sequence of processing that involves not only an upward flow of movement activity within deep cells, but also a downward flow of visual processing from the superficial layers. Since the visually triggered movement cells are encountered in the most dorsal part of the movement layers, an input from visual cells could be quite direct, perhaps involving dendro-dendritic connections. Although there is no anatomical evidence on this point, Golgi material indicates considerable overlap in the distribution of superficial and intermediate layer cells (Valverde 1973, Langer & Lund 1974, Tokunaga & Otani 1976) and electron micrographs show that dendro-dendritic connections are frequent (Sterling 1971). Alternatively this connection may be indirect and may involve an extracollicular source such as the frontal eye fields and other extra-striate visual areas. The close functional relationship between the visually triggered movement cells and the active cells with enhanced responses is emphasized by the difficulty in distinguishing in some cases one cell type from the other (Wurtz & Mohler 1976a).

Visually triggered cells might be regarded as transmitting a delayed retinal error signal since these cells add the requirement of a visual gating signal to the delayed error signal. Like the short-lead cells, these cells probably also represent an important collicular output to the oculomotor system.

A second possible connection that crosses the boundaries of the two divisions of the colliculus is also suggested by the temporal sequence of discharge. This connection might provide the input producing the visual enhancement effect in the active visual cells and is shown as a dashed line in Figure 2B from the long-lead movement cells to the active visual cells. Long-lead movement cells discharge during the time that a stimulus target for a saccade still falls on the receptive field of a cell and, when these long-lead cells show an anticipatory response during a series of trials to the same target, their discharge would overlap even the onset of the visual stimulus. Since there is no evidence that the deep cells connect to the superficial cells directly (Edwards, in press), this connection might be indirect, perhaps via the parabigeminal-lateral tegmental region. Alternatively, the input to the superficial layers could arise from the same source that activates the long-lead cells.

This summary indicates how our thinking about the organization of the superior colliculus has come full circle. The original ideas of collicular organization centered on a transition from visual activity in the superficial layers to movement activity in the intermediate and deep layers (Schiller & Koerner 1971, Robinson 1972, Wurtz & Goldberg 1972b, Sprague 1975). This view was made compelling by the registration between the visual and movement maps, but the subsequent view emphasizing the separation of the colliculus into superficial and deeper layers (summarized in the introduction) left little reason for the visual-motor convergence. The present emphasis on the convergent flow of visual and motor processing provides a new rationale for the registration but maintains the fundamental separation between superficial and deeper layers.

RELATION TO OCULOMOTOR SYSTEM

In the previous section we identified five cell elements within the colliculus. We now consider the relation of these elements to the oculomotor system and specifically to visually guided eye movements. Initiation of visually guided saccades must depend on at least three events: the selection of the visual target, the determination of retinal error, and the triggering of the saccade.

Selection

Target selection within the visual field is usually taken for granted in analyses of oculomotor control since reference is made to *the* target or *the* retinal error, but even in rarified laboratory experiments more than one visual target is usually present, and some type of selection process must precede the initiation of the saccade. While it would be premature to argue tnat the selection process is complete within the colliculus, several cell types already considered are appropriate for this function: long-lead cells show an early discharge related to one part of the visual field; visual responses are enhanced before saccades to one area of the field; visually triggered movement cells discharge only before saccades to a visual target in one part of the visual field. This spatial selection function is essentially the shift of visual attention view applied specifically to eye movements (Goldberg & Wurtz 1972b, Wurtz & Goldberg 1972c), which subsequent experiments have shown is an appropriate restriction (Wurtz & Mohler 1976a).

Retinal Error

This factor, essentially the visual guidance of the saccade, has been the factor most thoroughly considered by control theory models of saccade generation (see Robinson 1973 for review). The basis of these control theory models is that the eye musculature requires a pulse and a step of activity in the oculomotor neurons to move the eye to a new position and hold it there. Neural elements that could provide this input to the oculomotor neurons have been identified in the paramedian pontine reticular formation. These are the medium-lead burst cells, which discharge in a vigorous burst of activity 15 msec prior to the saccade, and the tonic cells, which have a discharge rate related to eye position. Thus both components of the oculomotor discharge are coded in the temporal domain: the pulse, a transient change in discharge frequency, and a step, a steady state change in frequency. Any scheme concerning the connection of the superior colliculus to the brainstem occulomotor system must therefore involve a translation from "which" cell is firing to the "frequency" of cell discharge. This translation is frequently referred to as a spatial to temporal transformation. The result of this transformation is that larger saccade amplitudes are represented by higher frequencies of cell discharge.

How this transformation comes about is unknown, but we think that there may be two mechanisms by which the colliculus translates its neuronal discharge, coded by location, to an oculomotor discharge, coded in frequency. First, the spatial to temporal transformation may result from the density of connections from different parts of the movement map. In cat, Precht et al (1974) found that the thresholds for producing EPSP's in the cat abducens nucleus was higher from stimulating electrodes in the anterior colliculus than in the posterior colliculus. Later, Edwards & Henkel (1978) found that the number of cells labeled by HRP injection of the pontine oculomotor area of the cat was sparse in the intermediate layers of the anterior colliculus and more dense in the posterior colliculus. They suggested that the greater density of neurons in the posterior colliculus might be the anatomical mechanisms by which the posterior colliculus exerts a stronger effect upon the abducens nucleus and thereby produces saccades of greater amplitude. If we assume that synaptic drive is related to discharge frequency in brainstem oculomotor neurons, then the result of activating more posterior regions of the colliculus, which represent more peripheral regions of the visual field, is to produce a greater frequency of discharge. However, the evidence for this mechanism for the spatial to temporal transformation is not suggested by the experiments of Raybourn & Keller (1977), who stimulated the superior colliculus and found responses in the eye movement-related cells of the pontine reticular formation, but did not find any lower threshold for the stimulation effects in the anterior than in the posterior colliculus (for long-lead burst cells). Instead, they found that monosynaptic connections were more common in the anterior than in the posterior colliculus.

A second mechanism which may contribute to this spatial to temporal transformation is the size of the movement fields within the colliculus. In general, the greater the amplitude of the saccade, the larger the movement field. Larger movement fields imply that any given saccade will have more cells discharging before the saccade and therefore greater synaptic drive to the brainstem oculomotor areas. Again, larger saccades would produce greater synaptic drive, which would lead to higher frequency in oculomotor areas.

An implication of this type of transformation is that any damage to the colliculus should produce a shortening of saccades, never a lengthening. Following partial ablation of the colliculus, saccades do fall short of the target as indicated by an increased frequency of small corrective saccades that move the eye farther away from the original starting position (Mohler & Wurtz 1977). This shortening of saccades has been studied only in the central 25° of the visual field. The hypothesis suggests that the damage would be more striking in the periphery.

Another transformation of the error signal that would be required by a recent model of the oculomotor system (Robinson 1975, Zee et al 1976) is a shift from a retinotopic coordinate system to a spatial coordinate system. This new model of the oculomotor system has several conceptual advantages for producing the appropriate burst of neural activity required to initiate a saccade. One advantage of the model is that since the error signal is in spatial coordinates, this signal would be available to guide head, hand, and body movements as well as eye movements, a particularly perspicacious parsimony. The disadvantage as far as analysis of the colliculus is concerned is that the model requires the error signal to be in spatial coordinates at least a hundred msecs before the onset of the saccade. As we have seen, the movement-related colliculus cells that discharge as late as 20 msec before saccade onset are in a retinotopic coordinate system, not a spatial one. At this point one can neither expect cells with position-related signals to be found, and preliminary reports have suggested that such signals might exist (Sparks et al 1977, Peck & Schlag-Rey 1978), or one can modify the model. The modification needed to fit the characteristics of the cells described would require a shift into spatial coordinates much later in the processing sequence, preferably just before the control of burst duration. The visual cells or enhanced visual cells in the colliculus could then provide the retinal error signal and, more important, the visually triggered movement cells could provide a delayed selected retinal error signal. The cellular activity within the colliculus would then be incorporated into a logical visual-motor sequence instead of being excluded as in the model formulated by Robinson and his colleagues (Robinson 1975, Zee et al 1976).

Trigger

It is necessary for the oculomotor system to know when a saccade should be initiated as well as where it should be directed, hence recent models of the oculomotor system incorporate a trigger in the model for initiation of the saccade (Robinson 1975, Zee et al 1976). This trigger closes a switch that enables the burst generator. The cells that might correspond to this switch have been identified as the pause cells (Keller 1974) lying near the midline of the pons; when they pause, it is suggested that they release the medium-lead burst neurons from inhibition. Initiation of this pause, and therefore the saccade, could be another function of the output cells of the superior colliculus, one of triggering saccades, not necessarily one of guiding them.

Tests of Oculomotor Relations

While the superior colliculus could logically be related to the oculomotor system in any of the ways described, that it is so related remains to be established. The close relationship of the superior colliculus to the initiation of saccadic eye movements demonstrated by electrical stimulation has already been considered, but electrophysiological and ablation experiments have also evaluated this relationship.

The best established output from the superior colliculus to the oculomotor system is to the pontine reticular formation. "Burst" type movement cells are most frequently activated antidromically by stimulation of the pontine area (Keller 1980), which is consistent with the emphasis on the short-lead movement cells as important output cells of the colliculus, as "burst" cells are quite similar (see Figure 2). The relation of the types of pontine cells to the colliculus has been investigated (Raybourn & Keller 1977) by stimulating the superior colliculus and recording the response of pontine neurons. Stimulation of the colliculus had the most direct and powerful effect on the pontine long-lead burst neurons and medium-lead burst cells. Thus, if the output of the colliculus contributes to the frequency of discharge related to the amplitude of saccades, it must do so indirectly through the long-lead burst cells. Collicular stimulation also produced short latency excitation in pause neurons, which suggests that the colliculus contributes to the background discharge which, when removed, may be a trigger for saccadic eye movements.

One would expect ablation of the superior colliculus to be devastating to the generation of saccadic eye movements but this has not been the case. One earlier report did suggest difficulty in moving the eyes following ablation of the monkey superior colliculus (Denny-Brown 1962) but another study found no such deficit (Pasik et al 1966). When eye movements were recorded (Wurtz & Goldberg 1972b, Mohler & Wurtz 1977), the deficit in saccades following large but incomplete unilateral collicular ablation was clear but circumscribed. The monkeys with ablations continued to make visually guided saccades into the visual field related to the ablated superior colliculus, but they tended to do so with longer (by 50–250 msec) latencies to initiate saccades, and they showed an increased frequency of small corrective saccades. The longer latency to make saccades might result from a deficit in any of the three necessary processes from the use of a less efficient pathway for selecting the saccadic target, determining retinal error, or triggering the saccade.

The lack of a substantial visual guidance deficit must indicate either that the colliculus does not perform the stimulus selection and retinal error processes or that it performs them in parallel with another source. The other source of this visual input must be dependent upon the striate cortex since ablation of this area along with the colliculus eliminates visually guided saccades (Mohler & Wurtz 1977). On the other hand, ablation of striate cortex also leaves the monkey able to make saccades to a visual target once his recovery is sufficient for him to detect the target, which indicates that the colliculus can indeed function without cortex. The remaining pathway for this visual guidance may involve the frontal eye fields since stimulation of this area elicits saccades even following collicular ablation (Schiller 1977). The ability of humans to make visually guided saccades and pointing movements even with extensive occipital brain damage (and a lack of perception of the target) has been reported (Pöppel et al 1973, Perenin & Jeannerod 1975, Weiskrantz et al 1974).

The frequency of spontaneous saccades is also reduced following ablation of the colliculus in monkey (Rosvold et al 1958, Denny-Brown 1962, Pasik et al 1966, Albano & Wurtz 1978) and in man (Heywood & Ratcliff 1975). This deficit in spontaneous saccades might indicate a decoupling of the ability of a stimulus to initiate a saccade. This is consistent with the reduction in distractibility during a fixation task as judged by the number of saccades made to a suddenly appearing peripheral stimulus in the monkey (Albano & Wurtz 1978). An analogous reduction in distractibility was reported earlier for the rat (Goodale & Murison 1975). These deficits might well be related to the stimulus selection or trigger functions of the colliculus we have already discussed.

Additional ablation studies in monkeys have found deficits in reaching tasks, particularly in the peripheral visual field (Keating 1974, 1976, Butter 1974a,b, Butter et al 1978) and more complex tasks (Rosvold et al 1958, Anderson & Symmes 1969, Kurtz 1977, Latto 1978), but whether these deficits relate to the initiation of saccades is unknown.

CONCLUSION

We have concentrated on one function of the superior colliculus, the initiation of saccadic eye movements, because the evidence on this particular function allows us to begin to outline how sensory information might lead to movement. Rather than a direct visual to motor transition within the colliculus, we find several steps, which are part of a sequence required in executing visually guided saccadic eye movements. The colliculus in turn is part of a more extended system, which probably includes the cerebral cortex and certainly connects to the pontine areas related to saccadic eye movement.

Saccadic eye movements, moreover, are only one part of a more general mechanism of gaze shift or orientation that involves head and body as well as eyes. Other animals rely more heavily on head and body movements than do primates and the organization of the superior colliculus or optic tectum in these animals may be related to such differences in behavior. For example, the cat has a tendency to use its head proportionately more than its eyes for orienting and the superior colliculus in the cat is more clearly related to head movements than is the colliculus in the monkey. The primate superior colliculus may represent a specialization of a more general function —that of orientation.

ACKNOWLEDGMENTS

It is a pleasure to thank Nita Hight and Norma Craig for their dedicated assistance in the preparation of this manuscript. We also thank Lance Optican for valuable discussions on models of the oculomotor system.

Literature Cited

Abplanalp, P. 1970. Some subcortical connections of the visual system in tree shrews and squirrels. *Brain Behav. Evol.* 3:155–68

Adamük, E. 1870. Über die innervation der augenbewegungen. *Zentr. Med. Wiss.* 8:65

Albano, J. E., Norton, T. T., Hall, W. C. 1979. Laminar origin of projections from the superficial layers of the superior colliculus in the tree shrew. *Tupaia glis. Brain Res.* In press

Albano, J. E., Wurtz, R. H. 1978. Modification of the pattern of saccadic eye movements following ablation of monkey superior colliculus. *Neurosci. Abstr.* 8:161

Anderson, K. V., Symmes, D. 1969. The superior colliculus and higher visual functions in the monkey. *Brain Res.* 13:37–52

Astruc, J. 1971. Corticofugal connections of area 8 (frontal eye field) in *Macaca mulatta. Brain Res.* 33:241–56

Baleydier, C., Magnin, M. 1979. Afferent and efferent connections of the parabigeminal nucleus in cat revealed by retrograde axonal transport of horseradish peroxidase. *Brain Res.* 161:187–98.

Bartlett, J. R., Doty, R. W. Sr. 1974. Influence of mesencephalic stimulation on unit activity in striate cortex of squirrel monkeys. *J. Neurophysiol.* 37:642–52

Benevento, L. A., Davis, B. 1977. Topographical projections of prestriate cortex to pulvinar nuclei in the macaque monkey. An autoradiographic study. *Exp. Brain Res.* 30:405–24

Benevento, L. A., Fallon, J. H. 1975. The ascending projections of the superior colliculus in the rhesus monkey (*Macaca mulatta*) *J. Comp. Neurol.* 160:339–62

Benevento, L. A., Rezak, M., Santos-Anderson, R. 1977. An autoradiographic study of the projections of the pretectum in the rhesus monkey (*Macaca mulatta*): Evidence for sensorimotor links to the thalamus and oculomotor nuclei. *Brain Res.* 127:197–218

Berlucchi, G., Sprague, J. M., Levy, J. , Di Berardino, A. C. 1972. Pretectum and superior colliculus in visually guided behavior and in flux and form discrimination in the cat. *J. Comp. Physiol. Psychol.* 78:123–72

Berman, N. 1977. Connections of the pretectum in the cat. *J. Comp. Neurol.* 174:227–54.

Bizzi, E., Kalil, R. E., Tagliasco, V. 1971. Eye-head coordination in monkeys: Evidence of centrally patterned organization. *Science* 173:452–54

Bushnell, M. C., Robinson, D. L., Goldberg, M. E. 1978. Dissociation of movement and attention: neuronal correlates in posterior parietal cortex. *Neurosci. Abstr.* 4:621

Butter, C. M. 1974a. Effect of superior colliculus, striate and prestriate lesions on visual sampling in rhesus monkeys. *J. Comp. Physiol. Psychol.* 87:905–17

Butter, C. M. 1974b. Visual discrimination impairments in rhesus monkeys with combined lesions of striate cortex and superior colliculus. *J. Comp. Physiol. Psychol.* 87:918–29

Butter, C. M., Weinstein, C., Bender, D. B., Gross, C. G. 1978. Localization and detection of visual stimuli following superior colliculus lesions in rhesus monkeys. *Brain Res.* 156:33–49

Büttner, U., Fuchs, A. F. 1973. Influence of saccadic eye movements on unit activity in simian lateral geniculate and pregeniculate nuclei. *J. Neurophysiol.* 36:127–41

Casagrande, V. A., Diamond, I. T. 1974. Ablation study of the superior colliculus in the tree shrew (*Tupaia glis*). *J. Comp. Neurol.* 156:207–38

Casagrande, V. A., Harting, J. K., Hall, W. C., Diamond, I. T., Martin, G. F. 1972. Superior colliculus of the tree shrew: a structural and functional subdivision into superficial and deep layers. *Science* 177:444–47

Castigloni, A. J., Gallaway, M. C., Coulter, J. D. 1978. Spinal projections from the midbrain in monkey. *J. Comp. Neurol.* 178:329–45

Crommelinck, M., Guitton, D., Roucoux, A. 1977. In *Control of Gaze by Brain Stem Neurons, Developments in Neuroscience,* ed. R. Baker, A. Berthoz, 1:425–35. Amsterdam: Elsevier. 514 pp.

Cynader, M., Berman, N. 1972. Receptive-field organization of monkey superior colliculus. *J. Neurophysiol.* 35:187–201

de Monasterio, F. M. 1978a. Properties of concentrically organized X and Y ganglion cells of macaque retina. *J. Neurophysiol.* 41:1394–1417

de Monasterio, F. M. 1978b. Properties of ganglion cells with atypical receptive field organization in retina of macaques. *J. Neurophysiol.* 41:1435–49

de Monasterio, F. M., Gouras, P. 1975. Functional properties of ganglion cells of the rhesus monkey retina. *J. Physiol. London* 251:167–95

Denny-Brown, D. 1962. The midbrain and motor integration. *Proc. R. Soc. Med.* 55:527–38

Dräger, U. C., Hubel, D. H. 1975. Responses to visual stimulation and relationship between visual, auditory, and somatosensory inputs in mouse superior colliculus. *J. Neurophysiol.* 38:690–713

Dräger, U. C., Hubel, D. H. 1976. Topography of visual and somatosensory projections to mouse superior colliculus. *J. Neurophysiol.* 39:91–101

Dreher, B., Fukada, Y., Rodieck, R. W. 1976. Identification, classification and anatomical segregation of cells with X-like and Y-like properties in the lateral geniculate nucleus of old-world primates. *J. Physiol. London* 258:433–52

Duffy, F. H., Burchfiel, J. L. 1975. Eye movement-related inhibition of primate visual neurons. *Brain Res.* 89:121–32

Edwards, S. B. 1977. The commissural projection of the superior colliculus in the cat. *J. Comp. Neurol.* 173:23–40

Edwards, S. B. 1980. The deep cell layers of the superior colliculus: Their reticular characteristics and structural organization. In *The Reticular Formation Revisited: Specifying Functions for a Nonspecific System.* ed. A. Hobson, M. Brazier. New York: Raven. In press

Edwards, S. B., Ginsburg, C. L., Henkel, C. K., Stein, B. E. 1979. Sources of subcortical projections to the superior colliculus in the cat. *J. Comp. Neurol.* 184:309–30

Edwards, S. B., Henkel, C. K. 1978. Superior colliculus connections with the extraocular motor nuclei in the cat. *J. Comp. Neurol.* 179:451–68.

Finlay, B. L., Schiller, P. H., Volman, S. F. 1976. Quantitative studies of single cell properties in monkey striate cortex. IV. Corticotectal cells. *J. Neurophysiol.* 39: 1352–61

Frankfurter, A., Weber, J. T., Royce, G. J., Strominger, N. L., Harting, J. K. 1976. An autoradiographic analysis of the tecto-olivary projection in primates. *Brain Res.* 118:245–57

Fuchs, A. F., Luschei, E. S. 1972. Unit activity in the brainstem related to eye movement. *Bibl. Ophthalmol.* 82:17–27

Garey, L. J., Jones, E. G., Powell, T. P. S. 1968. Interrelationships of striate and extrastriate cortex with the primary relay sites of the visual pathway. *J. Neurol. Neurosurg. Psychiatry* 31:135–57

Goldberg, M. E., Robinson, D. L. 1978. Visual system: Superior colliculus. In *Handbook of Behavioral Neurobiology,*

ed. R. B. Masterton, 1:119–64. New York: Plenum

Goldberg, M. E., Wurtz, R. H. 1971. Response of single cells in monkey superior colliculus during pursuit eye movement and stationary fixation. *Neurology* 21:435

Goldberg, M. E., Wurtz, R. H. 1972a. Activity of superior colliculus in behaving monkey. I. Visual receptive fields of single neurons. *J. Neurophysiol.* 35:542–59.

Goldberg, M. E., Wurtz, R. H. 1972b. Activity of superior colliculus in behaving monkey. II. Effect of attention on neuronal responses. *J. Neurophysiol.* 35: 560–74.

Goldman, P. S., Nauta, W. J. H. 1976. Autoradiographic demonstration of a projection from prefrontal association cortex to the superior colliculus in the rhesus monkey. *Brain Res.* 116:145–49

Goodale, M. A., Murison, R. C. C. 1975. The effects of lesions of the superior colliculus on locomotor orientation and the orienting reflex in the rat. *Brain Res.* 88:243–61

Gordon, B. 1973. Receptive fields in deep layers of cat superior colliculus. *J. Neurophysiol.* 36:157–78

Gordon, B. 1975. Superior colliculus: Structure, physiology, and possible connections. In *Physiology,* ed. C. C. Hunt, Ser. 1, 3:185–231. London: Butterworths

Graham, J. 1977. An autoradiographic study of the efferent connections of the superior colliculus in the cat. *J. Comp. Neurol.* 173:629–54

Grantyn, A. A., Grantyn, R. 1976. Synaptic actions on tectofugal pathways on abducens motoneurons in the cat. *Brain Res.* 105:269–85

Graybiel, A. M. 1975. Anatomical organization of retinotectal afferents in the cat: An autoradiographic study. *Brain Res.* 96:1–23

Graybiel, A. M. 1978a. A satellite system of the superior colliculus: the parabigeminal nucleus and its projections to the superficial collicular layers. *Brain Res.* 145:365–74

Graybiel, A. M. 1978b. A stereometric pattern of distribution of acetylthiocholinesterase in deep layers of the superior colliculus. *Nature* 272:539–41

Grofová, I., Ottersen, O. P., Rinvik, E. 1978. Mesencephalic and diencephalic afferents to the superior colliculus and periaqueductal gray substance demonstrated by retrograde axonal transport of horseradish peroxidase in the cat. *Brain Res.* 146:205–20

Harris, L. R. 1979. *The superior colliculus and movements of the eyes and head in cat. J. Physiol.* In press

Harting, J. K. 1977. Descending pathways from the superior colliculus: an autoradiographic analysis in the rhesus monkey (*Macaca mulatta*). *J. Comp. Neurol.* 173:583–612

Harting, J. K., Casagrande, V. A., Weber, J. T. 1978. The projection of the primate superior colliculus upon the dorsal lateral geniculate nucleus: autoradiographic demonstration of interlaminar distribution of tectogeniculate axons. *Brain Res.* 150:593–99

Harting, J. K., Hall, W. C., Diamond, I. T., Martin, G. F. 1973. Anterograde degeneration study of the superior colliculus in *Tupaia glis:* Evidence for a subdivision between superficial and deep layers. *J. Comp. Neurol.* 148:361–86

Harting, J. K., Noback, C. R. 1971. Subcortical projections from the visual cortex of the tree shrew (*Tupaia glis*). *Brain Res.* 25:21–33

Hartmann-von Monakow, K., Akert, K., Kunzle, H. 1979. Projections of precentral and premotor cortex to the red nucleus and other midbrain areas in *Macaca fascicularis. Exp. Brain Res.* 34:91–105

Hashikawa, T., Kawamura, K. 1977. Identification of cells of origin of tectopontine fibers in the cat superior colliculus: An experimental study with the horseradish peroxidase method. *Brain Res.* 130:65–79

Hendrickson, A., Wilson, M. E., Toyne, M. J. 1970. The distribution of optic nerve fibers in *Macaca mulatta. Brain Res.* 23:425–27

Henkel, C. K., Edwards, S. B. 1976. Laminar differences in some uncrossed projections of the superior colliculus to the midbrain. *Neurosci. Abstr.* 2:1117

Henkel, C. K., Edwards, S. B. 1978. The superior colliculus control of pinna movements in the cat: possible anatomical connection. *J. Comp. Neurol.* 182:763–76.

Heywood, S., Ratcliff, G. 1975. In *Basic Mechanisms of Ocular Motility and their Clinical Implications.* ed. G. Lennerstrand, P. Bach-y-Rita, pp. 561–64. New York: Pergamon. 584 pp.

Holcombe, V., Hall, W. C. 1978. Laminar origins of two descending pathways from the superior colliculus in the grey squirrel. (*Sciurus carolinensis*). *Neurosci. Abstr.* 4:632

Hubel, D. H., LeVay, S., Wiesel, T. N. 1975. Mode of termination of retinotectal fibers in Macaque monkey: An autoradiographic study. *Brain Res.* 96:25–40

Humphrey, N. K. 1968. Responses to visual stimuli of units in the superior colliculus of rats and monkeys. *Exp. Neurol.* 20:312–40

Itoh, K. 1977. Efferent projection of the pretectum in the cat. *Exp. Brain Res.* 30:89–105

Jones, E. G., Wise, S. P. 1977. Size, laminar and columnar distribution of efferent cells in the sensory-motor cortex of monkeys. *J. Comp. Neurol.* 175:391–438

Kaas, J. H., Harting, J. K., Guillery, R. W. 1974. Representation of the complete retina in the contralateral superior colliculus of some mammals. *Brain Res.* 65:343–46

Kadoya, S., Massopust, L. C. Jr., Wolin, L. R. 1971. Striate cortex-superior colliculus projections in squirrel monkey *Exp. Neurol.* 32:98–110

Kawamura, K., Hashikawa, T. 1978. Cell bodies of origin of reticular projections from the superior colliculus in the cat: an experimental study with the use of horseradish peroxidase as a tracer. *J. Comp. Neurol.* 182:1–16

Kawamura, S., Sprague, J. M., Niimi, K. 1974. Corticofugal projections from the visual cortices to the thalamus, pretectum and superior colliculus in the cat. *J. Comp. Neurol.* 158:339–62

Keating, E. G. 1974. Impaired oreintation after primate tectal lesions. *Brain Res.* 67:538–41

Keating, E. G. 1976. Effects of tectal lesions on peripheral vision in the monkey. *Brain Res.* 104:316–20

Keller, E. L. 1974. Participation of medial pontine reticular formation in eye movement generation in monkey. *J. Neurophysiol.* 37:316–32

Keller, E. L. 1980. Colliculoreticular organization in the oculomotor system. In *Progress in Brain Res.* Amsterdam: Elsevier. In press

Kunzle, H., Akert, K. 1974. Efferent connections of cortical area 8 (frontal eye field) *Macaca fasicularis* a reinvestigation using the autoradiographic techniques. *J. Comp. Neurol.* 173:147–64

Kunzle, H., Akert, K., Wurtz, R. H. 1976. Projection of area 8 (frontal eye field) to superior colliculus in the monkey. An autoradiographic study. *Brain Res.* 117:487–92

Kurtz, D. 1977. *Eye movements of monkeys with superior colliculus lesions during visual discrimination performance.* PhD thesis. Univ. Mich.

Kuypers, H. G. J. M., Lawrence, D. G. 1967. Cortical projections to the red nucleus and the brain stem in the rhesus monkey. *Brain Res.* 4:151–88

Kuypers, H. G. J. M., Maisky, V. A. 1975. Retrograde axonal transport of horseradish peroxidase from spinal cord to brainstem cell groups in the cat. *Neurosci. Lett.* 1:9–14

Lane, R. H., Allman, J. M., Kaas, J. H. 1971. Representation of the visual field in the superior colliculus of the grey squirrel (*Sciurus carolinensis*) and tree shrew (*Tupaia glis*). *Brain Res.* 26:277–92

Langer, T. P., Lund, R. D. 1974. The upper layers of the superior colliculus of the rat: A Golgi study. *J. Comp. Neurol.* 158:405–34

Latto, R. 1978. The effects of bilateral frontal eye-field, posterior parietal or superior collicular lesions on visual search in the rhesus monkey. *Brain Res.* 146:35–50

Lund, J. S., Lund, R. D., Hendrickson, A. E., Bunt, A. H., Fuchs, A. F. 1975. The origin of efferent pathways from the primary visual cortex, area 17, of the macaque monkey as shown by retrograde transport of horseradish peroxidase. *J. Comp. Neurol.* 164:287–303

Lund, R. D. 1972a. Synaptic patterns in the superficial layers of the superior colliculus of the monkey, *Macaca mulatta. Exp. Brain Res.* 15:194–211

Lund, R. D. 1972b. Anatomic studies on the superior colliculus. *Invest. Ophthalmol.* 11:434–44

Magalhaes-Castro, H. H., deLima, A. D., Saraiva, P. E. S., Magalhaes-Castro, B. 1978. Horseradish peroxidase labeling of cat tectotectal cells. *Brain Res.* 148:1–13

Marrocco, R. T. 1978. Conduction velocities of afferent input to superior colliculus in normal and dicorticate monkey. *Brain Res.* 140:155–58

Marrocco, R. T., Li, R. H. 1977. Monkey superior colliculus: properties of single cells and their afferent inputs. *J. Neurophysiol.* 40:844–60

Mathers, L. H. 1971. Tectal projection to the posterior thalamus of the squirrel monkey. *Brain Res.* 35:295–98

Mays, L. E., Sparks, D. L. 1979. Dissociation of visual and saccade-related responses in superior collicular neurons. *J. Neurophysiol.* In press

McIlwain, J. T. 1976. Large receptive fields

and spatial transformations in visual system. *Int. Rev. Physiol.* 2:223–48

Mohler, C. W., Goldberg, M. E., Wurtz, R. H. 1973. Visual receptive fields of frontal eye field neurons. *Brain Res.* 61:385–89

Mohler, C. W., Wurtz, R. H. 1976. Organization of monkey superior colliculus: intermediate layer cells discharging before eye movement. *J. Neurophysiol.* 39:722–44

Mohler, C. W., Wurtz, R. H. 1977. Role of striate cortex and superior colliculus in visual guidance of saccadic eye movements in monkeys. *J. Neurophysiol.* 40:74–94

Moors, J. 1978. Single unit responses to moving and stationary flashing stimuli in the superior colliculus of the rhesus monkey (*Macaca mulatta*). PhD thesis. Univ. Nijmegen, Nijmegen, Netherlands

Pasik, T., Pasik, P., Bender, M. B. 1966. The superior colliculi and eye movements. *Arch. Neurol.* 15:420–36

Paula-Barbosa, M. M., Sousa-Pinto, A. 1973. Auditory cortical projections to the superior colliculus in the cat. *Brain Res.* 50:47–61

Peck, C. K., Schlag-Rey, M. 1978. Visual-motor properties of units in the superior colliculus of the alert cat. *Neurosci. Abstr.* 4:640

Perenin, M. T., Jeannerod, M. 1975. Residual vision in cortically blind hemifields. *Neuropsychologia.* 13:1–7

Pöppel, E., Held, R., Frost, D. 1973. Residual visual function after brain wounds involving the central visual pathways in man. *Nature* 243:295–96

Powell, T. P. S. 1976. Bilateral cortico tectal projection from visual cortex in the cat. *Nature* 260:526–27

Precht, W., Schwindt, P. C., Magherini, P. C. 1974. Tectal influences on cat ocular motoneurons. *Brain Res.* 82:27–40

Raczkowski, D., Casagrande, V. A., Diamond, I. T. 1976. Visual neglect in the tree shrew after interruption of the descending projections of the deep superior colliculus. *Exp. Neurol.* 50:14–29

Raczkowski, D., Diamond, I. T. 1978. Cells of origin of several efferent pathways from superior colliculus in *Galago senegalensis. Brain Res.* 146:351–57

Raybourn, M. S., Keller, E. L. 1977. Colliculoreticular organization in primate oculomotor system. *J. Neurophysiol.* 40:861–78

Richmond, B. J., Wurtz, R. H. 1977. Visual responses during saccadic eye movement: a corollary discharge to superior colliculus. *Neurosci. Abstr.* 3:574

Richmond, B. J., Wurtz, R. H. 1978. Visual masking by remote stimuli in monkey superior colliculus neurons. *Neurosci. Abstr.* 4:642

Rizzolatti, G., Camarda, R., Grupp, L. A., Pisa, M. 1974. Inhibitory effect of remote visual stimuli on visual responses of cat superior colliculus: spatial and temporal factors. *J. Neurophysiol.* 37: 1262–75

Robinson, D. A. 1972. Eye movements evoked by collicular stimulation in the alert monkey. *Vision Res.* 12:1795–808

Robinson, D. A. 1973. Models of the saccadic eye movement control system. *Kybernetik* 14:71–83

Robinson, D. A. 1975. See Heywood & Ratcliff 1975, pp. 337–74

Robinson, D. L., Goldberg, M. E., Stanton, G. B. 1978. Parietal association cortex in the primate: sensory mechanisms and behavioral modulations. *J. Neurophysiol.* 41:910–32

Robinson, D. L., Jarvis, C. D. 1974. Superior colliculus neurons studied during head and eye movements of the behaving monkey. *J. Neurophysiol.* 37:533–40

Robinson, D. L., Wurtz, R. H. 1976. Use of extraretinal signal by monkey superior colliculus neurons to distinguish real from self-induced stimulus movement. *J. Neurophysiol.* 39:852–70

Robson, J. A., Hall, W. C. 1975. Connections of layer VI in striate cortex of the grey squirrel. *Brain Res.* 93:133–39

Rodieck, R. W. 1979. Visual pathways. *Ann. Rev. Neurosci.* 2:193–225

Rosvold, H. E., Mishkin, M., Szwarcbart, M. K. 1958. Effects of subcortical lesions in monkeys on visual discrimination and single alternation performance. *J. Comp. Physiol. Psychol.* 51:437–44

Roucoux, A., Crommelinck, M. 1976. Eye movements evoked by superior colliculus stimulation in the alert cat. *Brain Res.* 106:349–63

Roucoux, A., Crommelinck, M., Meulders, M. 1980. Visual fixation: A collicular reflex? In *Progress in Brain Research.* Amsterdam: Elsevier. In press

Schiller, P. H. 1972. The role of the monkey superior colliculus in eye movement and vision. *Invest. Opthalmol.* 11: 451–60

Schiller, P. H. 1977. The effect of superior colliculus ablation on saccades elicited by cortical stimulation. *Brain Res.* 122:154–56

Schiller, P. H., Koerner, F. 1971. Discharge characteristics of single units in supe-

rior colliculus of the alert rhesus monkey. *J. Neurophysiol.* 34:920–36

Schiller, P. H., Malpeli, J. G. 1977. Properties and tectal projections of monkey retinal ganglion cells. *J. Neurophysiol.* 40:428–45

Schiller, P. H., Stryker, M. 1972. Single-unit recording and stimulation in superior colliculus of the alert rhesus monkey. *J. Neurophysiol.* 35:915–24

Schiller, P. H., Stryker, M., Cynader, M., Berman, N. 1974. Response characteristics of single cells in the monkey colliculus following ablation or cooling of visual cortex. *J. Neurophysiol.* 37: 181–94

Sherk, H. 1978. Visual response properties and visual field topography in the cat's parabigeminal nucleus. *Brain Res.* 145:375–79

Singer, W., Zihl, J., Pöppel, E. 1977. Subcortical control of visual thresholds in humans: Evidence for modality specific and retinotopically organized mechanisms of selective attention. *Exp. Brain Res.* 29:173–90

Sparks, D. L. 1975. Response properties of eye movement-related neurons in the monkey superior collisulus. *Brain Res.* 90:147–52

Sparks, D. L. 1978. Functional properties of neurons in the monkey superior colliculus: coupling of neuronal activity and saccade onset. *Brain Res.* 156:1–16

Sparks, D. L., Holland, R., Guthrie, B. L. 1976. Size and distribution of movement fields in the monkey superior colliculus. *Brain Res.* 113:21–34

Sparks, D. L., Mays, L. E., Pollack, J. G. 1977. Saccade-related unit activity in the monkey superior colliculus. In *Control of Gaze by Brainstem Neurons,* ed. R. Baker, A. Berthoz, 437–44. Amsterdam: Elsevier. 514 pp.

Sparks, D. L., Pollack, J. G. 1977. The neural control of eye movements: The role of the superior colliculus. In *Eye Movements,* ed. B. A. Brooks, F. J. Bajandas. New York: Plenum

Sprague, J. M. 1972. The superior colliculus and pretectum in visual behavior. *Invest. Ophthalmol.* 11:473–82

Sprague, J. M. 1975. Mammalian tectum: Intrinsic organization, afferent inputs, and integrative mechanisms. Anatomical substrate. *Neurosci. Res. Program Bull.* 13:204–13

Sprague, J. M., Berlucchi, G., Di Berardino, A. 1970. The superior colliculus and pretectum in visual by guided behavior and visual discrimination in the cat. *Brain Behav. Evol.* 3:285–94

Sprague, J. M., Berlucchi, G., Rizzolatti, G. 1972. The role of the superior colliculus and pretectum in vision and visually guided behavior. In *Handbook of Sensory Physiol.* ed. R. Jung, Vol. VII/B: 27–101. Berlin: Springer. 738 pp.

Sprague, J. M., Levy, J., Di Berardino, A., Berlucchi, G. 1977. Visual cortical areas mediating form discrimination in the cat. *J. Comp. Neurol.* 172:441–88

Stein, B. E., Arigbede, M. O. 1972. Unimodal and multimodal response properties of neurons in the cat's superior colliculus. *Exp. Neurol.* 36:179–96

Stein, B. E., Goldberg, S. J., Clamann, H. P. 1976a. The control of eye movements by the superior colliculus in the alert cat. *Brain Res.* 118:469–74

Stein, B. E., Magalhaes-Castro, B., Kruger, L. 1976b. Relationship between visual and tactile representations in cat superior colliculus. *J. Neurophysiol.* 39: 401–19

Sterling, P. 1971. Receptive fields and synaptic organization of the superficial gray layer of the cat. *Brain Res.* Suppl. 3:309–28

Sterling, P., Wickelgren, B. G. 1969. Visual receptive fields in the superior colliculus of the cat. *J. Neurophysiol.* 32:1–15

Straschill, M., Hoffmann, K.-P. 1969. Functional aspects of localization in the cat's tectum opticum. *Brain Res.* 13:274–83

Straschill, M., Rieger, P. 1973. Eye movements evoked by focal stimulation of cat's superior colliculus. *Brain Res.* 59:221–27

Stryker, M., Blakemore, C. 1972. Saccadic and disjunctive eye movements in cats. *Vision Res.* 12:2005–13

Stryker, M. P., Schiller, P. H. 1975. Eye and head movements evoked by electrical stimulation of monkey superior colliculus. *Exp. Brain Res.* 23:103–12

Tigges, J., O'Steen, W. K. 1974. Termination of retino-fugal fibers in squirrel monkey: a reinvestigation using autoradiographic methods. *Brain Res.* 79:489–95

Tigges, M., Tigges, J. 1970. The retinofugal fibers and their terminal nuclei in *Galago crassicaudatus* (primates). *J. Comp. Neurol.* 138:87–102

Tokunaga, A., Otani, K. 1976. Dendritic patterns of neurons in the rat superior colliculus. *Exp. Neurol.* 52:189–205

Trojanowski, J. Q., Jacobson, S. 1975. Peroxidase labeled subcortical afferents to pulvinar in rhesus monkey. *Brain Res.* 97:144–50

Tunkel, J. E., Berkley, M. A. 1977. The role of the superior colliculus in vision: visual form discrimination in cats with superior colliculus ablations. *J. Comp. Neurol.* 176:575–87

Updyke, B. V. 1974. Characteristics of unit responses in superior colliculus of the cebus monkey. *J. Neurophysiol.* 37:896–909

Valverde, R. 1973. The neuropilin superficial layers of the superior colliculus of the mouse. A correlated Golgi and electron microscopic study. *Z. Anat. Entwicklungsgesch.* 142:117–47

Weber, J. T., Partlow, G. D., Harting, J. K. 1978. The projection of the superior colliculus upon the inferior olivary complex of the cat: an autoradiographic and horseradish peroxidase study. *Brain Res.* 144:369–77

Weiskrantz, L., Warrington, E. K., Sanders, M. D., Marshall, J. 1974. Visual capacity in the hemianopic field following a restricted occipital ablation. *Brain* 97:709–28

Wickelgren, B. G. 1971. Superior colliculus: some receptive field properties of bimodally responsive cells. *Science* 173: 69–72

Wilson, M. E., Toyne, M. J. 1970. Retinotectal and cortico-tectal projections in *Macaca mulatta. Brain Res.* 24:395–406

Wise, S. P., Jones, E. G. 1977. Cells of origin and terminal distribution of descending projections of the rat somatic sensory cortex. *J. Comp. Neurol.* 175:129–58

Wurtz, R. H. 1969. Response of striate cortex neurons to stimuli during rapid eye movements in the monkey. *J. Neurophysiol.* 32:975–86.

Wurtz, R. H. 1978. The primate superior colliculus and visually guided eye movements. In *Recent Advances in Primatology Vol. I. Behaviour,* ed. D. J. Chivers, J. Herbert, pp. 643–53. New York: Academic. 980 pp.

Wurtz, R. H., Goldberg, M. E. 1971. Superior colliculus cell responses related to eye movements in awake monkeys. *Science* 171:82–84

Wurtz, R. H., Goldberg, M. E. 1972a. Activity of superior colliculus in behaving monkey. III. Cells discharging before eye movements. *J. Neurophysiol.* 35: 575–86

Wurtz, R. H., Goldberg, M. E. 1972b. Activity of superior colliculus in behaving monkey. IV. Effects of lesions on eye movements. *J. Neurophysiol.* 35:587–96

Wurtz, R. H., Goldberg, M. E. 1972c. The primate superior colliculus and the shift of visual attention. *Invest. Ophthalmol.* 11:441–50

Wurtz, R. H., Mohler, C. W. 1974. Selection of visual targets for the initiation of saccadic eye movements. *Brain Res.* 71:209–14

Wurtz, R. H., Mohler, C. W. 1976a. Organization of monkey superior colliculus: Enhanced visual response of superficial layers cells. *J. Neurophysiol.* 39:745–65

Wurtz, R. H., Mohler, C. W. 1976b. Enhancement of visual response in monkey striate cortex and frontal eye fields. *J. Neurophysiol.* 39:766–72

Wurtz, R. H., Goldberg, M. E., Robinson, D. L. 1980. Behavioral modulation of visual responses in the monkey: A neurophysiological approach to attention. In *Progress in Physiological Psychology and Psychobiology.* New York: Academic. In press

Zee, D. S., Optican, L. M., Cook, J. D., Robinson, D. A., Engel, W. K. 1976. Slow saccades in spinocerebellar degeneration. *Arch. Neurol.* 33:243–51

Ann. Rev. Neurosci. 1980. 3:227–68
Copyright © 1980 by Annual Reviews Inc. All rights reserved

SUBSTANCE P AS A TRANSMITTER CANDIDATE[1]

❖11541

R. A. Nicoll

Departments of Pharmacology and Physiology, University of California, San Francisco, California 94143

C. Schenker and S. E. Leeman

Department of Physiology, Harvard Medical School, Boston, Massachusetts 02115

INTRODUCTION

Substance P is a peptide which was first described in 1931 (von Euler & Gaddum 1931) in extracts of brain and intestine, but which was not purified to homogeneity until 1970 (Chang & Leeman 1970). The isolation of substance P was accomplished subsequent to the discovery of a sialogogic peptide in hypothalamic extracts (Leeman & Hammerschlag 1967), which was shortly thereafter characterized as substance P. The name substance P (for preparation) had been used in the laboratory of origin to designate the active agent in a particular preparation of tissue extracts. This nondescript term entered the literature in 1934 and has persisted (Gaddum & Schild 1934). The amino acid sequence of substance P (SP), H-ARG-PRO-LYS-PRO-GLN-GLN-PHE-PHE-GLY-LEU-MET-NH$_2$, was established in 1971 (Chang et al 1971) and shortly thereafter synthetic peptide was pre-

[1] Abbreviations used: ACh, acetylcholine; AChE, acetylcholinesterase; CAT, choline acetyltransferase; CNS, central nervous system; DA, dopamine; 5,6-DHT, 5,6-dihydroxytryptamine; 5,7-DHT, 5,7-dihydroxytryptamine; EPSP, excitatory postsynaptic potential; GAD, glutamic acid decarboxylase; 5-HT, 5-hydroxytryptamine; mepp, miniature endplate potential; NE, norepinephrine; RIA, radioimmunoassay SP, substance P; SPLI, substance P-like immunoreactivity.

227

0147-006X/80/0301-0227$01.00

pared (Tregear et al 1971), permitting the development of precise methods for biochemical, histochemical, physiological, and pharmacological studies of the peptide.

In addition to the hypotensive and smooth muscle-contracting activity that led to its discovery (von Euler & Gaddum 1931), SP has many pharmacological effects (see Skrabanek & Powell 1977) that may be of functional significance. One likely physiological role for SP is that of a neurotransmitter. Indeed, SP is widely but selectively distributed in both the central and peripheral nervous systems, is found in fiber tracts and nerve endings, is released upon depolarization, and can alter the activity of some neurons when applied in their vicinity.

In this review we examine the evidence that SP may be a transmitter in three *substance P systems:* the nociceptive primary afferents, the striato-nigral tract, and the habenulo-interpeduncular tract. Other areas where SP may play such a role are only briefly considered. In addition, we discuss several topics which are both sufficiently well documented and interesting enough to merit separate treatment, namely (*a*) the coexistence of SP and 5-HT in some neurons, (*b*) the functional interaction of the postulated SP receptor with the calcium channel in salivary gland cells, (*c*) the interaction of SP with some nicotinic receptors, and (*d*) the association of SP with blood vessels. For convenience we have summarized in Table 1 the distribution of SP in the nervous system of various species.

Our discussion draws primarily on the literature published after 1971. Pernow (1953, 1963) and Lembeck & Zetler (1962, 1971) have reviewed earlier work. For a discussion of general pharmacological, endocrine, cardiovascular, and behavioral effects of SP we refer to the following reviews: Skrabanek & Powell (1977), the most complete source of references, listing and briefly describing virtually all publications dealing with SP that have appeared before the end of 1977; von Euler & Pernow (1977), a collection of research papers presented at the 37th Nobel Symposium and covering a wide range of topics on the chemistry and pharmacology of SP; and Mroz & Leeman (1977), a rather complete general review discussing the chemistry, distribution, and pharmacology of SP as well as assay methods.

A Few Caveats to Consider

Following and evaluating the interdisciplinary literature dealing with peptides in the nervous system presents obvious problems. For the sake of the nonspecialist we have, therefore, selected a few areas in which intrinsic difficulties in interpretation are easily overlooked.

IMMUNOCHEMISTRY   Immunological methods have been extremely important in determining the amounts of a particular peptide present in a tissue sample by the use of radioimmunoassy (RIA) and for localizing the peptide to specific neuronal structures by immunohistochemical techniques. The immunological methods are critically dependent on the species of anti-sera used which may cross-react with compounds structurally related to the immunogen and hence may give rise to *false positives*. To control for this an antiserum is generally characterized in terms of its cross-reactivity with known peptides and their analogs. Furthermore, partial identity of im-munoreactive tissue extracts with synthetic peptide can be checked by independent means (e.g. chromatography, electrophoresis). However, even if these controls are carried out, one cannot exclude cross-reactivity with some unknown compound, particularly in immunohistochemical studies. This problem is further discussed by Mroz & Leeman (1977) and by Ljung-dahl et al (1978a).

Immunochemical methods can also give rise to *false negatives*. Antigenic sites may be lost during fixation and antibodies may not penetrate to all antigenic sites. The indirect immunofluorescence technique of Coons (see Coons 1958) is sometimes not sufficiently sensitive to visualize very low concentrations of antigen in areas where immunoreactivity has been demon-strated by RIA. Axons and cell bodies seem to be particularly difficult to detect with some antisera (Ljungdahl et al 1978a), perhaps because they do not cross-react with a possible precursor. The peroxidase-antiperoxidase method of Sternberger (see Sternberger et al 1970) may provide greater sensitivity and has the advantage of producing an electron dense reaction product, thus permitting visualization in the electron microscope.

In the case of SP, oxidation of the methionine residue slowly occurs when the peptide is kept in dilute solution (Floor & Leeman, in press). Since the relative immunoreactivity of SP-sulfoxide, as compared to SP, is different for different antisera, this may introduce errors in measurement of tissue levels of SP. Addition of 2-mercapto-ethanol to the tissue extraction and resuspension medium largely prevents such oxidation. To what extent SP is oxidized during tissue preparation for histochemistry is not known.

In summary, depending on the method and antisera used, immunochemi-cal procedures may give rise both to false positives and false negatives. On the other hand, cross-reactivity with a precursor or metabolite is of obvious usefulness for biosynthetic or metabolic studies. In this review we have used the term *substance P-like immunoreactivity* (SPLI) when referring to values obtained by immunochemical procedures.

ELECTROPHYSIOLOGY/IONTOPHORESIS

Specificty of the effect   Changes in membrane potential or in neuronal activity subsequent to iontophoresis of a peptide can be nonspecific. An observed response could be due to a nonspecific membrane effect by the peptide (as opposed to an action mediated via a specific receptor) or to a simple effect of current or pH. (Peptides are usually ejected as cations at an acidic pH.) Because the transport number of most peptides is quite low, relatively large currents may be required to obtain effects and thus current and pH artifacts become more of a problem. The technique of pressure ejection is, therefore, gaining popularity in studies on the neuronal action of peptides. Blockade of a peptide response with a specific antagonist is probably the most meaningful test of specificity. However, no such antagonist is known for substance P.

Apparent time-course of the response   The time course of the response observed upon iontophoresis of a peptide depends on the time it takes for the peptide to reach the target membrane, as well as the time course of the physiological response. The first component is largely determined by the time it takes to eject the peptide from the electrode, which in turn depends on the transport number of the peptide and the amount and duration of retaining current used (which electrophoreses the peptide up the electrode, away from the tip). It is possible to determine the relative contributions of these two components (electrophoretic vs physiological) on the apparent time course of the response by calibrating the micropipettes used. The extent to which a given retaining current delays the ejection of SP from micropipettes was specifically investigated by Guyenet et al (1979). As expected, a larger retaining current led to a greater delay in ejecting SP from the micropipettes, as determined by RIA. Similarly, when tested on locus coeruleus neurons in vivo, iontophoretic ejection of SP, following a large retaining current (13 nA), gave a sluggish increase in firing rate whereas following a weak retaining current (0.4 nA) the SP-induced rise in activity was brisk (5 sec lag) and no different from that induced by ACh. The magnitude of the response was also increased.

Apparent mode of action   A given response (excitatory or inhibitory) to an applied substance can be due to a direct effect on the neuron from which the recording is made, or it can be mediated via local interneurons or equivalent structures. If the response is mediated by an inhibitory interneuron, the direct membrane action (excitatory or inhibitory) of the applied substance will result in the opposite (inhibitory or excitatory) overall effect. Such indirect effects have been clearly established for glutamate in the olfactory bulb (Nicoll 1971) and enkephalin in the hippocampus (Zieglgänsberger et al 1979). The use of Co^{2+} or Mg^{2+} to block local transmission can reveal such indirect effects, but this test assumes that the substance is not acting via a Ca^{2+} channel. If one knows the identity of a suspected intervening transmitter, specific antagonists can be used to block the indirect effect.

Comparing the membrane effects of a transmitter candidate with those of natural stimulation presents an additional set of problems. Natural stimulation may be heterogeneous in terms of input neurons and their conductance changes. Selective stimulation of a homogeneous population of input neurons, however, may not be possible. Furthermore, if the neuron has extended dendrites, it may not be possible to accurately record dendritic synaptic conductance changes from the soma.

In short, it is exceedingly difficult to determine the direct action of a peptide (or other compound) in the CNS and to satisfy the criterion of mimicry. On the other hand, electrophysiological studies in the peripheral nervous system and in tissue culture have been particularly rewarding.

ELECTRONMICROGRAPHS Electronmicrographs represent a slice in time of normally dynamic processes. This is particularly relevant to the appearance of synaptic boutons that discharge vesicles during activity. Specifically, differences in vesicle size, number, and immunoreactivity in synaptic boutons may reflect differences in their past histories or intrinsic differences between the neurons to which they belong. It is difficult to distinguish between these two possibilities without an extensive study directly aimed at the problem (e.g. see Basbaum & Heuser 1979).

THE DISTRIBUTION OF SUBSTANCE P

Table 1 summarizes the distribution of SPLI in the CNS of various species, as determined by RIA. Small amounts of SPLI have also been measured in various peripheral organs of several species (Nilsson & Brodin 1977). Immunohistochemical data is difficult to tabulate or describe without doing injustice to the often exquisite pictures. We therefore refer directly to the original papers. The distribution, determined by Hökfelt's group, of immunofluorescent SP in the rat central nervous system is summarized in Ljungdahl et al (1978a,b), which also contains an extensive list of references; that determined by the Cambridge Medical Research Council group is summarized in Cuello & Kanazawa (1978). SP immunofluorescence has been described in the mammalian central nervous system (Nilsson et al 1974, Hökfelt et al 1976b, 1977c, 1978b,c) in primary afferents (Nilsson et al 1974, Hökfelt et al 1975, 1976a, 1977c,d, Schultzberg et al 1978b), and in the periphery (Hökfelt et al 1977b, 1978d, Schultzberg et al 1978a). Hökfelt et al (1978a) compare the relative distributions of various peptides in the nervous system of the rat.

Table 1 The distribution of substance P in the central nervous system[a]

Region	Rat pmol/10 mg wet weight[b]	Rat pmol/mg protein[c]	Human pmol/mg protein[d]	Human pmol/10 mg wet weight[e]	Pigeon pmol/10 mg wet weight[f]
Somatosensory system					
Dorsal root ganglia	0.6				
Dorsal horn	9.4				10.1
Trigeminal nucleus	12.1				9.0
Dorsal column	1.1				
Dorsal column nucleus	1.5				
Thalamic nucleus	0.2				
Somatosensory cortex	0.2				
Visual system					
Retina					< 0.1
Optic nerve					< 0.1
Lateral geniculate body	0.7	0.9			
Superior colliculus		1.8			
Visual cortex	0.2				
Basal ganglia					
Striatum		0.9			3.3[g]
Caudate nucleus	2.2		3.7		
Putamen			3.3		
Globus pallidus	2.9		18.0		1.2[h]
Substantia nigra	15.1				7.5[i]
Pars compacta		2.9	47.2	4.1	
Par reticularis		11.38	47.9	7.4	
Pars lateralis		3.0			
Subthalamic nucleus	2.0				
Hypothalamus		2.1	5.2		2.1
Medial	5.5				
Middle	4.5				
Lateral	4.3				
Medial preoptic nucleus		4.4			
Lateral preoptic nucleus		3.3			
Periventricular nucleus		3.3			
Suprachiasmatic nucleus		1.6			
Supraoptic nucleus		1.6			
Anterior hypothalamic nucleus		3.2			
Paraventricular nucleus		3.1			
Arcuate nucleus		2.5			
Ventromedial nucleus		2.5			
Dorsomedial nucleus		3.5			
Perifornical nucleus		2.9			
Ventral premammilary nucleus		3.3			
Dorsal premammilary nucleus		1.7			
Posterior hypothalamic nucleus		2.8			
Medial forebrain bundle, anterior		3.1			
Medial forebrain bundle, posterior		2.3			
Median eminence		1.0			
Posterior pituitary		0.5[j]			

Table 1 *(Continued)*

Region	Rat pmol/10 mg wet weight[b]	Rat pmol/mg protein[c]	Human pmol/mg protein[d]	Human pmol/10 mg wet weight[e]	Pigeon pmol/10 mg wet weight[f]
Limbic system					
Olfactory bulb	0.5	0.2			0.7
Olfactory tubercle	2.6				
Olfactory cortex	0.4				
Amygdala	3.3	3.4			
Hippocampus	0.3				
Habenula	3.3				
Interpeduncular nucleus	5.2	5.9			8.5
Septum	3.5	1.2			
Dorsal septal nucleus		2.8			
Lateral septal nucleus		3.6			
Interstitial nucleus of stria terminals		3.3			
Nucleus accumbens	2.4				4.3
Mammilary body	1.8				
Anterior thalamic nucleus	1.9				
Other regions					
Locus coeruleus					5.3
Red nucleus		1.3			
Medial geniculate		0.8			
Inferior colliculus		1.2			
Central gray		2.9			
Cerebellum	< 0.1	< 0.1	0.2		< 0.1
Pineal	< 0.1[j]		0.5		
Ventral horn					1.1
Anterior pituitary	0.2[j]				

[a] As determined by RIA; standard errors have been omitted and values rounded off.

[b] From Kanazawa & Jessel (1976); original values in ng/g wet weight have been converted to pmol/10 mg wet weight using a molecular weight of SP to facilitate comparison with other studies; value/10 mg wet weight is on the same scale as value/mg protein since nervous tissue is about 10% protein.

[c] From Brownstein et al (1976).

[d] From Gale et al (1978).

[e] From Kanazawa et al (1977a); original values in ng/g.

[f] From Reubi & Jessel (1978); original values in pmol/g.

[g] Paleostriatum augmentatum, a possible counterpart of the mammalian striatum.

[h] Paleostriatum primitivum.

[i] Nucleus tegmenti pedunculo-ponti.

[j] M. H. Fernstrom and S. E. Leeman, unpublished.

EVIDENCE FOR SUBSTANCE P AS A TRANSMITTER

Our concepts of chemical transmission, as derived from the study of the neuromuscular junction, may need to be expanded when applied to the central nervous system. On morphological grounds, all combinations of interaction between axons, dendrites, and cell bodies have been postulated (see Sheperd 1974). Furthermore, a released compound may act on several target cells in its vicinity [Beaudet & Descarries (1978) discuss this possibility for monoamines in the cerebral cortex] or it may act upon its neuron of origin, as suggested for the dendritically released DA in the substantia nigra (Groves et al 1975). In each of these cases the site of action of the released compound will be determined by the location of specific receptors relative to the location and efficiency of some inactivating system. In our discussion of SP as a neurotransmitter, we will include any or all of the above possibilities in addition to a simple presynaptic-to-postsynaptic action. Such diversity of action is in fact suggested by the ultrastructural appearance of boutons with SPLI in the spinal cord (Barber et al 1979), as discussed below.

Since the biosynthesis of SP has not yet been characterized and its mode of degradation has not been described at a site relevant to transmission, the evidence supporting a role of SP in neurotransmission is limited to presence, release, and mimicry.

Substance P in Nociceptive Primary Afferent Fibers

The most convincing evidence for a role of SP in neurotransmission comes from studies on the sensory C fibers (and possibly A delta fibers) which convey noxious information. The search for the sensory transmitter(s) has a long and interesting history. Dale, for instance, maintained an active interest in the discovery of the transmitter in sensory fibers, as illustrated in a letter to Eccles in 1953, written at the age of 78. After reading a preprint of the initial work of Eccles, Fatt & Koketsu (cf. 1954) demonstrating the cholinergic nature of the transmission from motoneuron axon collaterals onto Renshaw cells, Dale wrote:

> It is extremely satisfactory to have the direct evidence of a cholinergic transmission from the ending of the collateral of a cholinergic axon, which, as you say, could have been predicted. I myself emphasize, in 1934, in a *Nothnagel Lecture*[2], which I gave in Vienna, the fact that the chemical function appeared to be a function, not merely of the nerve ending, but of the whole neurone, and speculated at the time concerning the possibility, that the identification of the peripheral transmitter of the so-called "antidromic vasodilatation" might give a clue to the transmitter, at the other, central synaptic ending, of what appeared to be normally an afferent nerve fibre.
>
> A number of people seem now to be taking up this clue, and trying to identify the peripheral transmitter of the antidromic vasodilation, in the hope of identifying one of the missing central transmitters. I suspect, however, that they are overlooking the fact, that Gasser and Hinsey[3] showed, years ago, that this antidromic vasodilation was to small, so-called "C" fibres in the dorsal roots, which certainly would not be concerned in the monosynaptic reflex effects on large motoneurons with which you have been largely concerned.

[2]Although the formation of "Dale's Principle" first appears in the Nothnagel Lecture (1935a), it is the Dixon Lecture (1935b), which was given a month later, that is usually quoted on this topic.

[3]Dale is referring to the paper of Hinsey & Gasser (1930).

PRESENCE AND LOCALIZATION   There were two major laboratories that were "taking up this clue," that of Hellauer & Umrath (1948) and that of Lembeck (1953). Hellauer & Umrath (1948) measured noncholinergic vasodilator activity in extracts of dorsal and ventral roots and found that considerably more activity was present in dorsal than ventral roots. They proposed that this vasodilator activity represented the action of the excitatory transmitter of sensory fibers. Lembeck (1953) obtained similar results and, in addition, found that the extract caused gut contraction. These properties were identical to those of SP, which had been extracted earlier from gut and brain by von Euler & Gaddum (1931) and found to be a peptide (von Euler, 1936). Thus, Lembeck proposed that SP might be the sensory transmitter.

These studies were extended by Otsuka and collaborators who isolated a peptide from bovine dorsal roots that had pharmacological properties similar to SP, but also was found to depolarize motoneurons of the isolated frog spinal cord (Otsuka et al 1972). This was an important finding because until this time no clearly defined CNS effect of SP extracts had been detected. The dorsal root peptide was subsequently shown (Takahashi et al 1974) to have precisely the same pharmacological, chemical, enzymatic, and immunological properties as the undecapeptide isolated by Chang & Leeman (1970) from hypothalamus and identified by them as SP. Using bioassay, Takahashi & Otsuka (1975) found that SP was highly concentrated in the dorsal horn of the spinal cord and that following dorsal root ligation or section, the concentration of SP in the dorsal horn and the segment of the dorsal root nearest the cord fell markedly, whereas the concentration in the segment nearest the cell bodies increased manyfold. Compared to the changes in the dorsal horn, only a slight fall in SP was detected in the ventral horn after dorsal root section. These results suggest that SP is synthesized in the cell body and subsequently transported to the dorsal horn of the spinal cord. Interestingly, years earlier, Holton (1959) had shown with similar ligation experiments that there was transport of SP into the peripheral branch of sensory neurons.

Immunohistochemical studies have shown that SPLI-positive fibers form a dense plexus in the substantia gelatinosa of the spinal cord (Hökfelt et al 1975, Chan-Palay & Palay 1977, Ljungdahl et al 1978a, Cuello & Kanazawa 1978, Barber et al 1979) and that this immunoreactivity is markedly decreased by dorsal root ligation (Hökfelt et al 1977c) or dorsal rhizotomy (Barber et al 1979). Only about 20% of the cells in the dorsal root ganglia of the rat are SPLI-positive (Hökfelt et al 1975, 1976a). A similar percentage has been found for spinal ganglia maintained in culture (Schultzberg et al 1978b). Although it is possible that negative cells may also contain slight, low concentrations of SPLI, the fact that a portion of the unlabeled cells contain another peptide, somatostatin (Hökfelt et al

1976a), suggests that there is a distinct population of SP-containing neurons. These cells have small somas and small, probably unmyelinated or thinly myelinated axons (Hökfelt et al 1977c). This discovery, in addition to the presence of free fibers with SPLI in skin, raises the possibility that SP may be localized in fibers involved in nociception. This association is made stronger by immunohistochemistry done on tooth pulp afferents, which are considered to mediate only the sensation of pain in humans (Anderson et al 1970). Olgart et al (1977b) found SPLI-positive fibers in teeth, which disappeared after sectioning the inferior alveolar nerve, thus supporting their sensory role. At the light microscopic level, it was not possible to establish with certainty whether these fibers were unmyelinated or contained a thin myelin sheath.

Further support for the association of SP with pain sensory fibers comes from studies on the distribution of SP in the trigeminal nucleus. It is generally accepted that the nucleus caudalis of the trigeminal nuclear complex receives nociceptive input (Dubner et al 1976). Immunohistochemical studies have revealed that the highest concentration of SPLI-positive fibers is located in this part of the nucleus (Cuello & Kanazawa, 1978) and that most of this immunoreactivity disappears after section of the trigeminal nerve (Cuello et al 1978a).

Peripheral nerve section is known to produce a marked cell loss in spinal and trigeminal ganglia and a degeneration of central axon branches of primary sensory neurons (Knyihár & Csillik 1976, Gobel & Binck 1977). This cell loss appears to be largely limited to small neurons (Aldskogius & Arvidsson 1978). Recently Jessell et al (1979b) found that following peripheral nerve section, there was a marked fall in SP content measured with radioimmunoassay in the spinal cord, with a time course similar to that of neuronal degeneration. There was no change in the concentration of GAD or CAT activity in the spinal cord after such lesions.

Studies on the action of capsaicin, a derivative of homovanillic acid, also link SPLI-positive afferents to a nociceptive function and a possible role in the axon reflex. Administered acutely, capsaicin produces intense pain and neurogenic plasma extravasation, whereas chronic administration of capsaicin renders animals insensitive to painful chemogenic stimuli (Jancsó 1968).

In newborn rats, capsaicin leads to a selective degeneration of chemosensitive primary afferents (Jancsó et al 1977) and to an irreversible depletion of SPLI in those areas containing primary afferents but not in other regions of the CNS (Gamse et al 1979a). In older rats capsaicin does not lead to a degeneration of primary afferents (Joö et al 1969), but results nevertheless in a decrease in SPLI in the substantia gelatinosa (Jessell et al 1978b). This decrease in SPLI is reversible (Gamse et al 1979a) and may be related to the ability of capsaicin to induce (Ca^{2+} dependent) release of SPLI from spinal cord in vitro (Theriault et al 1979, Gamse et al 1979c) and in vivo (Jessell et al 1979a). [Interestingly, capsaicin does not release SPLI from slices of hypothalamus or substantia nigra (Gamse et al 1979c).] In the skin, capsaicin inhibits the plasma extravasation that can normally be induced by antidromic stimulation of sensory neurons and decreases the levels of SPLI. Because intra-arterial infusion of SP induces plasma extravasation in the corresponding skin area, SP may play a role in the axon reflex (i.e. Dale's antidromic vasodilatation) (Gamse et al 1979a).

The localization of SPLI in synaptic terminals has also been examined at the ultrastructural level using the peroxidase-antiperoxidase method (Chan-Palay & Palay 1977, Cuello et al 1977, Hökfelt et al 1977c, Pickel et al 1977, Pelletier et al 1977, Barber et al 1979). The most detailed study is by Barber et al (1979), who found immunoreactivity in synaptic terminals containing both small agranular and large granular vesicles. This immunoreactivity appeared to be present over the granular vesicles and in the terminal cytoplasm associated with the exterior surfaces of small, agranular synaptic vesicles (Chan-Palay & Palay 1977). The presence of SPLI in terminals that contain two populations of vesicles raises the interesting possibility that these terminals may contain another neurotransmitter in addition to SP. Alternatively, the smaller vesicles may be a consequence of membrane retrieval subsequent to exocytosis of the large granular ones and hence, may be empty. Because the SP immunoreactivity was not always concentrated at the "active sites" of the synaptic junctions, and the terminals had features resembling neuroendocrine terminals, Barber et al (1979) suggested that SP might be released from nonsynaptic sites as well as from conventional synaptic junctions. SP might also play a role in presynaptic inhibition (or excitation?), as SPLI-positive axo-axonic synapses exist in the spinal cord (Barber et al 1979). This notion is supported by the observation that SP does have a direct depolarizing action on primary afferent fibers (Nicoll 1976).

RELEASE   Otsuka & Konishi (1976) have studied the release of SP from the isolated spinal cord of the newborn rat and found that repetitive stimulation of the dorsal root or exposure of the cord to a Krebs solution containing 55 mM K^+ increased the efflux of SPLI. The release of SPLI was abolished in a calcium deficient and magnesium rich Krebs solution. They also found that increasing the stimulus duration, which would activate the smaller fibers, markedly increased the amount of SPLI released. However, it is unclear in these experiments whether release of SP only occurred from unmyelinated C fibers. A calcium dependent release has also been found in slices of trigeminal nucleus (Jessell & Iversen 1977) and in cultures of dissociated dorsal root ganglion cells (Mudge et al 1979). Release of SPLI from the mammalian spinal cord has also been demonstrated in vivo by perfusion of the spinal subarachnoid space (Jessell et al 1979a). Local perfusion of potassium or capsaicin released SPLI; this release was reduced by cobalt. Sciatic nerve stimulation also released SPLI, but only at stimulus intensities which recruited A delta and C fibers. This finding provides strong physiological support for the presence of SP in, and its release from, small diameter fibers. If SP is the mediator of antidromic vasodilation, as might be predicted from its presence in small cutaneous fibers (Hökfelt et al 1977c) and its known action on blood vessels, then it should be possible to demonstrate its release from peripheral axons. Attempts thus far have failed to show such release (Burcher et al 1977), but preliminary results from tooth pulp afferents (Olgart et al 1977a) suggest that antidromic stimulation can elicit SP release. It would be of interest to know if this release is calcium dependent, as sensory terminals do contain small numbers of vesicles (Iggo 1974).

MIMICRY   In a random sampling of dorsal horn neurons, Henry et al (1975) found that the iontophoretic application of SP excited approximately half of the neurons tested. Compared to glutamate, the excitations were slow in onset and considerably outlasted the application. It was argued that the slow time course of action seemed incompatible with SP being the main excitatory transmitter released from primary afferents, and it was suggested that it might modulate activity over a longer period of time (however, cf Guyenet et al 1979). Subsequent studies on dorsal horn neurons have attempted to correlate SP sensitivity to a specific sensory modality. Henry (1976) found that only neurons classified as nociceptive were excited by SP, but that only about half of these units responded. These findings strengthened the idea that SP is involved in the transmission of nociceptive information. As pointed out by Henry, the failure of some cells to respond might be due to the low rate of release of SP from the pipette, or might occur because the neuron to which SP was applied did not receive monosynaptic input from primary afferents. It is perhaps surprising that such a correlation was found, as it is based on the assumption that only those cells that receive an SP input are sensitive to SP; yet, in a previous study on the cuneate nucleus (Krnjević & Morris 1974), with only moderate levels of SP immunofluorescence (Ljungdahl et al 1978a), approximately half of the neurons were excited by SP. Nevertheless, the association of SP sensitivity and activation by nociceptive stimuli has also been made by Randić & Miletić (1977) in the dorsal horn of the spinal cord and in the trigeminal nucleus caudalis in which the neurons were activated by tooth pulp stimulation (Andersen et al 1978). It has also been reported that following dorsal root

section, neurons in the dorsal horn become supersensitive to an SP analogue, an eledoisin-related peptide (Wright & Roberts 1978). These results
are compatible either with increased responsiveness of the neurons or with
greater access of the peptide to receptor sites following de-afferentation.
Because the major site of termination of spinal unmyelinated primary afferents is in the substantia gelatinosa on the dendrites of secondary neurons,
one might expect SP to be more effective when applied at this site. Duggan
et al (1979) has recorded from lamina IV and V neurons while applying SP,
either by iontophoresis or pressure ejection in the substantia gelatinosa.
However, when SP was applied at those sites at which enkephalin was
effective in blocking nociceptive input (see below), it had a weak action. One
explanation for these results is that enkephalin can act at some distance
from the nerve terminals, and indeed autoradiographic studies on localization of opiate receptors support this idea (Atweh & Kuhar 1977).

A precise comparison of the mechanism of action of SP and of the
transmitter released from nociceptive primary afferents is extremely difficult
for a number of reasons. First, the input to neurons in the dorsal horn is
thought to occur mainly in the substantia gelatinosa, on the dendrites, and
at a considerable distance from the cell body. This anatomical arrangement
makes it very difficult to determine whether the primary afferent transmitter
alters the conductance of the neuronal membrane and, if so, to determine
the reversal potential of the transmitter action. Also, the slow conduction
velocity of unmyelinated afferents makes it difficult to conclude conclusively
that the neuron under investigation is being activated monosynaptically by
afferent fibers. Zieglgänsberger & Tulloch (1979) have recorded the effect
of iontophoretically applied SP to dorsal horn neurons with intracellular
recording. They found that SP causes a reversible depolarization that parallels the increase in firing rate observed with extracellular recording. They
were unable to detect any change in resistance during the SP depolarization.
However, the SP was applied 100–160 μm from the recording site, and it
is possible that a small change in resistance might not be detected with such
a separation between the site of application and recording.

The evidence outlined above provides very strong support for the idea
that SP is involved in synaptic transmission from primary afferents that
mediate nociception. The slow time course of SP action is not a particularly
negative finding because the time course of transmitter action from nociceptive afferents is not well characterized, and the low transport number of SP
(Guyenet et al 1979) can result in delayed release from microelectrodes.
Other evidence that would strengthen the transmitter role of SP at nociceptive afferents is as follows. It would be desirable to show a selective release
of SP in the spinal cord during noxious stimulation. However, this would
still not conclusively show that the release is directly from primary afferents

and not from SP-containing interneurons. Such a differentiation would be difficult to make, although it is clear from cell culture studies (Mudge et al 1979) that SP can indeed be released from afferent neurons. The ionic mechanism underlying the action of SP and the transmitter released from nociceptive afferents is unclear, and the technical problems associated with determining the mechanism will be difficult to surmount. The development of potent and selective SP antagonists would greatly advance our understanding of the physiological role of SP. The synthesis of analogues of SP, as has been done so successfully for angiotensin II (cf Khosla et al 1974), might be a fruitful approach.

In their original work on the *dorsal root peptide*, Otsuka and collaborators made the observation that this peptide depolarized motoneurons, and thus it was logical to conclude that it might be present in those primary afferents that end monosynaptically on motoneurons, i.e. the large 1a afferents. Further support for this idea came from the observation that Lioresal blocked the 1a EPSP on motoneurons and the action of SP (Saito et al 1975). Although Lioresal appears to have some specificity toward the action of SP (Otsuka & Yanagisawa 1979), it is not entirely clear whether the specificity is sufficiently good to permit one to conclude that the block of 1a EPSPs is due to an antagonism of SP. The histochemical studies do not entirely exclude the existence of SP in 1a afferents, but strongly suggest that it is not present.

In the context of SP sensitivity of motoneurons, it is of interest that immunohistochemical studies indicate that at least two SP-containing systems, in addition to primary afferents, exist in the spinal cord. In colchicine-treated rats, numerous SP-positive cells were observed in the dorsal horn, immediately lateral to the dorsal horn, and in the dorsal part of the ventral horn (Hökfelt et al 1977d, Ljungdahl et al 1978a). In addition, the ventral horn contains a moderate density of SP-positive fibers (Hökfelt et al 1977c, Ljungdahl et al 1978a). These fibers remain after dorsal root section but disappear following cord transection, indicating that they represent a descending projection to motor nuclei. As mentioned above, SP exerts a depolarizing action on motoneurons but considerable controversy exists concerning the mechanism underlying this depolarization. Bath application of SP to the spinal cord of the newborn rat (Otsuka 1978) and to the frog spinal cord (Nicoll 1976, 1978) causes a depolarization associated with an increased conductance of the motoneuron membrane. The effect in the frog can be seen in preparations bathed in magnesium to block indirect synaptic effects. In the cat, iontophoretically applied SP has been reported to produce either no change in conductance (Zieglgänsberger & Tulloch 1979) or a decrease in conductance (Krnjević 1977). The basis for these conflicting results is not entirely clear, but species differences and differences in the technique of application might provide an explanation.

ASSOCIATION OF OPIATE ACTION WITH NOCICEPTIVE AFFERENTS
As the association of SP with nociceptive afferents has emerged, it has become increasingly clear that the action of opiates and enkephalin-containing systems are also intimately associated with these afferents. In the dorsal horn, met-enkephalin-positive cell bodies and nerve terminals are in close proximity to SP-positive cell bodies and nerve terminals; lesion experiments suggest that these neurons are interneurons (Hökfelt et al 1977d). Results from a number of studies suggest that opiates and enkephalin act on the primary afferent terminals and fibers. LaMotte et al (1976) have found that opiate receptors are highly concentrated in the dorsal horn of the spinal cord and that dorsal root section results in a marked decrease in binding in this region. Jessell et al (1979b) also found a decrease in opiate receptor binding following sciatic nerve and dorsal root section. Although it is conceivable that transneuronal degeneration might explain these results, the findings suggest that opiate receptors are present on afferent fibers. Autoradiographic localization of opiate receptor sites, utilizing selective binding of a potent opiate antagonist [^{3}H] diprenorphine, demonstrates a narrow band of silver grains in the substantia gelatinosa of the spinal cord (Atweh & Kuhar 1977). Opiate receptor binding was high only in those sensory nuclei that receive C fiber innervation, which suggests that some opiate receptors are associated with the C fiber terminals. The study of Hiller et al (1978) provides direct evidence that opiate receptors are located on primary afferent fibers. Opiate receptor binding was localized primarily in the neuritic outgrowth from sensory dorsal root ganglion cells grown in culture.

The presence of opiate receptors, presumably synthesized in the cell body of sensory neurons, on the intramedullary portion of primary afferents, as for example in the vagus (Atweh & Kuhar 1977), raises the question as to whether these receptors are preferentially transported centrally, or whether they might also be transported into the peripheral axon. There is some evidence that this may indeed be the case for GABA receptors because it can be shown pharmacologically that GABA receptors are present both in the central (Nicoll & Alger 1979) and peripheral portion of sensory fibers (Brown & Marsh 1978). To examine this possibility for opiate receptors, binding studies have been performed on the vagal trunk, which contains a high proportion of unmyelinated C-fiber afferents that are thought to possess opiate receptors on their central terminals (Atweh & Kuhar 1977). The vagus nerve does have some binding sites, whereas the sciatic nerve was devoid of binding sites (M. J. Kuhar, personal communication). It will be of interest to determine whether these opiate receptors in the vagus nerve are localized on SP fibers. About 10% of the vagal fibers contain SPLI and are mostly unmyelinated (Gamse et al 1979b). Furthermore, cell bodies with SPLI are found in the nodose ganglion, the sensory ganglion of the vagus nerve (Lundberg et al 1978).

A number of pharmacological studies have found an association of opiate action with input from small afferent fibers and/or with SP. In a well controlled study, Duggan et al (1976) have found that iontophoretically applied opiates and enkephalin in the substantia gelatinosa block the activation of lamina IV and V neurons by noxious, but not innocuous, stimulation, and that the effect is reversed by small systemic doses of the opiate antagonist naloxone. Although these experiments do not entirely exclude a postsynaptic site of action, for instance on an interneuron in the substantia gelatinosa, they are most easily explained by a selective presynaptic block of transmitter release from nociceptive afferents. A similar differential effect of enkephalin on responses of nucleus caudalis neurons in the trigeminal nucleus to tooth pulp stimulation was found by Andersen et al (1978). In spinal cord explants with attached sensory dorsal root ganglia, sensory-evoked synaptic activity in the dorsal horn region, but not in the ventral horn region, is blocked by opiates and opioid peptides, and this effect is prevented by naloxone (Crain et al 1977, 1978). These results do not distinguish between pre- or postsynaptic actions, but they do indicate that only the effects of certain primary afferents are blocked as a consequence of opiate receptor activation. MacDonald & Nelson (1978) have simultaneously recorded intracellularly from synaptically coupled dorsal root ganglion cells and spinal neurons and have tested the effect of iontophoretically applied etorphine, a potent opiate agonist, on the EPSP evoked by ganglion cell stimulation. Using the coefficient of variation of the EPSP to determine quantal content and quantal size, they found that etorphine had no effect on the size of the quanta but reduced the quantal content. This result indicates that opiate receptor activation depressed the EPSP solely by reducing the release of excitatory transmitter. Interestingly, all the cells they examined gave a similar result suggesting either that (a) they were very fortunate in selecting sensory afferents that have opiate receptors or (b) that all of the sensory afferents in their cultures possess opiate receptors, a situation that appears not to be the case in the culture experiments of Crain et al (1977, 1978). Jessell & Iversen (1977) found that the potassium-induced release of SP from slices of the trigeminal nucleus was reduced by opiates and opioid peptides; this block of release was prevented by naloxone. Although the precise site of action of opiates could not be determined in this preparation, these experiments clearly establish a functional association between opiates and substance P-containing neurons. Mudge et al (1979) also showed that enkephalin can inhibit release of substance P from cultured neurons; because sensory neurons were the only cell types present, enkephalin must have acted directly on the sensory neurons.

Two likely mechanisms by which opiates might inhibit transmitter release are either (*a*) that they cause a conductance increase so that the action potential does not effectively depolarize the terminal membrane or (*b*) that they inhibit the calcium influx which is necessary for transmitter release.

Picrotoxin-sensitive presynaptic inhibition by GABA is associated, at least temporally, with a depolarization of primary afferents (PAD) and there is evidence that an increase in chloride conductance may be the mechanism of both presynaptic inhibition and PAD (Nicoll & Alger 1979). Whether the depolarization itself contributes to the presynaptic inhibition is unclear, but it is likely of secondary importance because at the crustacean neuromuscular junction, presynaptic inhibition is associated with a conductance increase that results in a hyperpolarization of the motor nerve terminal (Kawai & Niwa 1977).

In isolated spinal cord preparations, enkephalin has been found to have a direct hyperpolarizing action which is blocked by naloxone; in these experiments enkephalin had little effect on ventral root potentials elicited by dorsal root stimulation (Evans & Hill 1978, R. A. Nicoll, unpublished observation). Sastry (1978b) has found that iontophoretically applied enkephalin decreases the excitability of $A\delta$ afferent terminals of the cat, which is consistent with a hyperpolarizing action. Such an action could block transmitter release if the hyperpolarization blocked impulse invasion into the nerve terminals or if the conductance increase reduced the size of the action potential in the terminal.

In the trigeminal nucleus, tooth pulp afferents are depolarized by stimulating the periaqueductal grey matter, a site that elicits a naloxone reversible analgesia. However, naloxone fails to alter this depolarization (Hu et al 1978). These results suggest that naloxone reversible analgesia is not mediated by a depolarization of the afferents.

Mudge et al (1979) suggest that enkephalin may inhibit substance P release from cultured sensory neurons inhibiting Ca^{2+} influx. Using intracellular electrodes, they find that enkephalin can decrease the duration of Ca^{2+}-action potentials recorded in the cell soma. The effect is antagonized by naloxone. It is likely that the calcium channels present in the cell soma membrane share common properties with calcium channels that activate transmitter release from the processes. These channels might, therefore, serve as models for those at the inaccessible release sites (Baker 1972). In these cultured sensory neurons, enkephalin did not alter the resting membrane properties of the neurons; there was no change in membrane potential or conductance. Rather, enkephalin inhibited a voltage-sensitive conductance.

As reviewed by Fields & Basbaum (1978), descending neuronal pathways that are in part serotonergic can suppress transmission of pain signals at the level of the spinal cord. The finding that serotonin, GABA, and norepinephrine can inhibit the release of SP from cultures of dorsal root ganglion cells (Mudge et al 1979) and decrease the calcium component of the action potential in sensory neurons (Dunlap & Fischbach 1978) raises the possibility that a variety of neurotransmitter systems may be capable of suppressing pain transmission by a presynaptic mechanism.

The results discussed above indicate that a very intimate relationship exists between nociceptive afferents (which in all probability utilize SP as a neurotransmitter) and endogenous opiate systems. Important questions for future research include 1. Do enkephalin terminals, in fact, make axo-axonic synapses on SP-containing afferent terminals? 2. What is the precise mechanism underlying the presynaptic action of opiates? 3. How are the presumed enkephalinergic presynaptic inhibitory pathways activated and can this pathway be demonstrated electrophysiologically?

Substance P in the Striatonigral Tract

The reciprocal striatonigral-nigrostriatal system is part of the extra-pyramidal motor system and is involved in sensory-motor integration. Its clinical importance is emphasized by two major diseases directly linked to defects in this system: Parkinson's disease (paralysis agitans), characterized by tremor, rigidity, and hypokinesis associated with degeneration of nigral and pallidal neurons (Calne 1970); and Huntington's chorea, characterized by involuntary muscular contractions apparently related to degeneration of striatal, pallidal, and cortical neurons (Shoulson & Chase 1975, Lange et al 1976). It has become clear that the striatonigral tract, in addition to the GAD-positive neurons (Kim et al 1971, Fonnum et al 1974, Precht & Yoshida 1971), contains the majority of the SP fibers that give rise to the dense plexus of SPLI in the substantia nigra. Because, in addition, SP is released from nigral tissue upon stimulation, and because it has an excitatory effect upon the dopaminergic nigrostriatal neurons, SP is very likely a transmitter in the substantia nigra, possibly acting directly upon the dopaminergic neurons.

PRESENCE AND LOCALIZATION   The substantia nigra contains the highest level of SPLI of any microdissected brain region (Table 1). Within the nigra of a number of species, several laboratories have found, using RIA, that SPLI is more concentrated in the pars reticularis than in the pars compacta (Brownstein et al 1976, Gauchy et al 1979, Kanazawa et al 1977a;

Table 1), although other laboratories have found little difference (Jessell et al 1978a, Gale et al 1978). The reason for this discrepancy is not clear. SP immunofluorescence studies (Hökfelt et al 1977c, Ljungdahl et al 1978a,b, see also Cuello & Kanazawa 1978) reveal an extremely dense network of fluorescent fibers interspersed with occasional nonfluorescent cell bodies that spans the entire pars reticularis. The SP immunofluorescence decreases from pars reticularis to pars compacta except in the rostral part of the substantia nigra. In the pars compacta, a considerably less dense plexus containing SPLI surrounds the many nonfluorescent cell bodies. As is evident from adjacent sections, many of these cell bodies are tyrosine hydroxylase(TH)-positive and hence are most likely the DA-containing nigrostriatal neurons (Ljungdahl et al 1978b).

The dendrites of these neurons extend far into the pars reticularis (Björklund & Lindvall 1975) and comparison of adjacent sections stained for SP and TH, respectively (Ljungdahl et al 1978b), clearly demonstrates the overlap of the dendritic field of the amine-containing neurons and the plexus of the SPLI-positive fibers. Direct synaptic contact, however, remains to be demonstrated at the electron microscope level.

Subcellularly, most of the SPLI is found in the synaptosomal fraction (Duffy et al 1975). No electron micrographs of SP immunoreactive terminals in the nigra have appeared in the literature. Hökfelt et al (1977c) report, however, that SPLI is confined to "nerve endings containing both small agranular and large granular vesicles" when examined under the electron microscope. The reaction product obtained by the peroxidase-antiperoxidase method is reportedly located in the cytoplasm or over the large granular vesicles.

Lesion studies indicate that the majority of the neurons giving rise to the SPLI in the rat substantia nigra have their cell bodies in the anterior striatum and hence are part of the striato-nigral tract (Mroz et al 1977a,b, Brownstein et al 1977, Hong et al 1977, Gale et al 1977, Kanazawa et al 1977b, Jessell et al 1978a). Substance P immunofluorescent cell bodies have indeed been observed in the anterior striatum in the rat (Kanazawa et al 1977b, Ljungdahl et al 1978a), and SP immunofluorescent fibers are seen in the caudal portion of the internal capsule (Ljungdahl et al 1978a,b, Jessell et al 1978a) and in the adjacent medial forebrain bundle (Ljungdahl et al 1978a). Whether there are substance P-containing pallido-nigral neurons as well is still a matter of debate. The problem stems from the fact that the efferents from the anterior striatum pass caudo-medio-ventrally through the globus pallidus on their way to the substantia nigra (e.g. see Fox & Rafols 1976). Brownstein et al (1977) concluded that the anterior striatum constitutes the major afferent input to the substantia nigra pars reticularis and probably the sole afferent input of anterior origin, because lesions immediately rostral to the substantia nigra (Mroz et al 1977a) produced no greater fall in nigral SPLI than did lesions separating the anterior striatum from the globus pallidus. On the other hand, Kanazawa et al (1977b) found that the fall in nigral SPLI (about 90%) caused by electrolytic lesions of the globus pallidus exceeded the decrease (about 70%) caused by even the largest anterior striatal lesion. In addition, they detected some large SP immunofluorescent cell bodies in the globus pallidus, although Ljungdahl et al (1978a) were unable to confirm this finding in colchicine-treated rats. The lesion experiments of Kanazawa et al (1977b) are somewhat ambiguous because, as they point out, their pallidal lesions "generally involved other structures such as parts of the striatum and the internal capsule."

In addition to SPLI, the striatonigral tract also contains GABA. The two systems have different but overlapping distributions. Whereas most of the SPLI-containing striatonigral neurons originate in the anterior striatum, the GABA-or GAD-positive afferents to the nigra seem to have their cell bodies located more caudally, as determined from lesion studies in the rat (Brownstein et al 1977, Gale et al 1977, Jessell et al 1978a). Whether any neurons in the overlap zone contain both substance P and GABA has not yet been investigated.

The neuronal population in the striatum consists largely (over 95%) of medium sized, spiny neurons, with a minor fraction of large aspiny ones (Kemp & Powell 1971, Fox et al 1972/73a,b, Lange et al 1976). Initial Golgi studies had suggested that only the large aspiny neurons project out of the striatum (Fox et al 1972/73b). From more recent evidence it has become clear, however (see Fox & Rafols 1976, Graybiel & Ragsdale 1979, Kitai et al 1976), that it is the medium sized spiny neurons that constitute the striatal efferents. It can therefore be expected that the SP (and GABA) neurons belong to this latter class. Indeed, as outlined by their im- munofluorescence, the cell bodies with SPLI in the striatum appear small to medium sized (Ljungdahl et al 1978a).

In Huntington's chorea, the smaller striatal neurons and some pallidal neurons degenerate (Lange et al 1976). This cell loss is accompanied by a decrease in SPLI in the substantia nigra (Kanazawa et al 1977a, 1979, Gale et al 1978) and in the internal part of the globus pallidus (Kanazawa et al 1979).

RELEASE   It has been clearly established, qualitatively, that depolariza- tion of nigral tissue leads to a Ca^{2+} dependent release of SPLI (Schenker et al 1976, Jessell 1978). In superfused slices of substantia nigra, either K^+ or veratridine are effective depolarizing agents. SPLI efflux is approxi- mately linearly related to the external K^+ concentration over the range of 15–60 mM K and, at constant K^+, to the external Ca^{2+} concentration over a range of 0.1–3 mM (with concomitant decrease of Mg^{2+}) (Jessell 1978). Sequential K^+ pulses in superfused synaptosomes induce progressively less Ca^{2+} dependent SPLI release, which suggests that the readily releasable pool of SPLI is of limited size or that there is inactivation at some step in the depolarization-secretion process (Schenker et al 1976). This is consistent with the observation that in the continued presence of Ca^{2+} and depolarizing amounts of K^+, the SPLI release rises to an early peak and then declines back toward the basal level (see von Euler & Pernow, 1977, p. 216).

Although these studies were designed to optimize the stimulation-induced release and demonstrate the voltage and Ca^{2+} dependence of the release process, they give no quantitative information about the amount, the time course, and the conditions under which SP is released in the intact functioning nigra. It also remains to be determined whether there normally is a spontaneous release of SP, whether the release is quantal, whether release occurs via exocytosis, and whether a single action potential is sufficient or whether a train of spikes is required to induce release over baseline.

Indirect evidence in support of spontaneous release of SP in vivo comes from studies done by Glowinski's group in the cat (Chéramy et al 1978a, Michelot et al 1979). Infusion of anti-SP gamma-globulins into the substantia nigra by means of a push-pull cannula led to a decrease of the spontaneous ^3H-DA release in the ipsilateral caudate nucleus, monitored simultaneously as an index of the activity of the nigrostriatal DA neurons. Control experiments done by infusing nonspecific gammaglobulins were without effect. Some of these experiments will be further discussed below. Whether this inferred spontaneous release of SP also occurs in the intact animal or represents an artifact due to local tissue damage will be difficult to determine because completely noninvasive techniques for measuring release of a compound in the brain are not yet available.

Stimulation of nigral tissue induces the release not only of SPLI but also of DA (Geffen et al 1976, Nieoullon et al 1977a), of GABA (Reubi et al 1977) and of 5-HT (Reubi & Emson 1978). These transmitter candidates affect each other's release from slices or in vivo in a complex way (Reubi et al 1978, Jessell 1978, Chéramy et al 1978b, Michelot et al 1979, Starr 1979), probably reflecting their functional interactions in the intact nigra but unfortunately giving few clues about the local circuitry. GABA (10–50 μM) inhibits the K^+-induced release of SPLI from superfused nigral slices but has no effect on the basal SPLI efflux (Jessell 1978). This inhibitory action of GABA is abolished in the presence of picrotoxin (50 μM), indicating that the effect is mediated via specific GABA receptors. It would be interesting to determine whether these GABA receptors are located on SP axon terminals or on some intervening structure. This question could possibly be resolved by repeating these experiments using synaptosomes and high superfusion rates. The release of SPLI in the nigra seems to be tonically regulated by GABA because high concentrattions (100 μM) of picrotoxin alone induce SPLI release over baseline (Jessell 1978). DA (50 μM), on the other hand, does not affect SP release (Reubi et al 1977).

MIMICRY   The cell bodies in the substantia nigra are topographically arranged according to their respective projection to the striatum, the thalamus, and the tectum (Faull & Mehler 1978). Because the terminal field of the SPLI-containing striatonigral neurons covers the entire substantia nigra (see Ljungdahl et al 1978a,b), these neurons can be expected, a priori, to innervate all three nigral efferent projections in addition to possible intrinsic neurons. Whether this is in fact the case has not yet been investigated. Because a transmitter candidate must mimic the action of the natural transmitter, it must first be determined which synaptic interaction SP must mimic. This will prove to be extremely difficult in view of the complexity of the substantia nigra. It is not surprising, then, that only a few studies have examined the effects of SP on nigral neurons.

Davies & Dray (1976) recorded extracellularly from spontaneously active neurons in the nigra and tested the effect of iontophoretic application of SP on firing rate. SP induced a small increase in firing rate in 23/34 neurons (eight neurons did not respond to SP and three were depressed). All of these neurons were also excited by glutamate and ACh and inhibited by GABA. The location of the micropipette was identified histochemically but Davies & Dray fail to indicate where the responsive neurons were located. Dray mentions, however, in a lecture (Dray & Straughan 1976) that neurons in the compacta as well as in the reticularis responded to SP. This would make it likely that at least some dopaminergic nigrostriatal neurons are excited by SP, either directly or indirectly. The paper by Walker et al (1976) is sometimes quoted as evidence that SP may have an excitatory effect on nigral neurons. They indeed observed an increase in firing rate upon iontophoresis of SP in a dilute solution at pH 5–6. At that pH, SP is clearly a cation. Yet Walter et al obtained positive results only with "anodal current." If by "anodal current" they mean ejection of cations, their results confirmed those obtained by Davies & Dray (1976); but if their "anodal current" ejected anions, then their observations are unrelated to any action of SP.

The activity of the dopaminergic nigrostriatal neurons can also be estimated from ^{3}H-DA release in the caudate nucleus by means of push-pull cannulas (Nieoullon et al 1977b). Such an approach has the advantage of monitoring chemically identified neurons in the anesthetized cat in vivo. Unilateral infusion of SP into the substantia nigra via a push-pull cannula leads to an increase in ^{3}H-DA release in the ipsilateral caudate (Chéramy et al 1977, Michelot et al 1979), whereas infusion of anti-SP gammaglobulins decreases the ^{3}H-DA release in the ipsilateral caudate (Chéramy et al 1978a, Michelot et al 1979). These observations strongly suggest that SP increases the firing rate of dopaminergic neurons and seems to do so tonically in the anesthetized cat in vivo. Whether SP acts directly upon these neurons or whether its action is mediated via the local circuitry remains to be determined. It might be interesting, for example, to repeat the above experiments during local transmission block.

Nieoullon et al (1977c) had previously made the puzzling observation that in the anesthetized cat in vivo, an increase in ^{3}H-DA release in the caudate is accompanied by a decrease in ^{3}H-DA release in the ipsilateral nigra; at the same time the converse changes in ^{3}H-DA release occur in the contralateral nigrostriatal system (Nieoullon et al 1977c). Similarly, whereas unilateral nigral infusion of SP increases ^{3}H-DA release in the ipsilateral caudate, it inhibits ^{3}H-DA release in the ipsilateral nigra in the cat in vivo (Michelot et al 1979); nigral application of anti-SP gammaglobulins have the opposite effect (Chéramy et al 1978a). [Interestingly, application of 10^{-5} M SP to superfused nigral slices increases ^{3}H-DA release (Reubi et al 1978).] On the other hand, unilateral nigral application of SP does not affect the contralateral nigrostriatal system in vivo (Michelot et al 1979). It is neither known in what way SP achieves this uncoupling of the two nigrostriatal systems, nor by what pathway they are normally coupled and reciprocally controlled.

How can the overall excitatory effect of SP on nigral neurons be reconciled with the overall inhibitory response observed in nigral neurons during striatal stimulation (Yoshida & Precht 1971, Dray et al 1976)? Striatal electrical stimulation can be expected to activate a heterogeneous population of neurons, many of which are the striatonigral GABA neurons. Indeed, the inhibitory field potential recorded in the nigra upon stimulation of the caudate is blocked by picrotoxin (Precht & Yoshida 1971) and by bicuculline (Dray et al 1976). It is therefore possible that an excitatory response may be masked by the inhibition produced by the striatonigral GABA Neurons. This is consistent with the observation that inhibition is sometimes preceded by brief excitation (Dray et al 1976, Preston et al 1978, S. T. Kitai, personal communication).

Substance P in the Habenulo-Interpeduncular Tract

PRESENCE AND LOCALIZATION   SP immunofluorescent cell bodies in the medial habenula (Hökfelt et al 1975, Cuello et al 1978b, Cuello & Kanazawa 1978, Ljungdahl et al 1978a,b) give rise to fibers that can be followed all along the fasciculus retroflexus (Ljungdahl et al 1978a). The SP immunofluorescence becomes particularly strong at the ventral surface of the brain before the fibers enter the lateral nucleus. The very high levels of SPLI detected in the interpeduncular nucleus by RIA in the rat (Brownstein et al 1976, Kanazawa & Jessell 1976; Table 1) are concentrated in a dense band of immunofluorescent fibers in the ventrolateral part of the nucleus and only occasional patches in the central part of the nucleus (Hökfelt et al 1977c, Ljungdahl et al 1978a,b, Cuello et al 1978b). Lesion of the medial habenula(e) leads to a loss of SPLI in the interpeduncular nucleus, as determined by RIA (Mroz et al 1976, 1977a, Hong et al 1976, Cuello et al 1978b) or by immunofluorescence (Hökfelt et al 1977c, Cuello et al 1978b). Transsection of the fasciculus retroflexus similarly depletes the interpeduncular nucleus of SPLI, and SPLI accumulates proximal to the cut (Emson et al 1977, Cuello et al 1978b). The loss of SPLI after these procedures is not complete, which is consistent with the presence of SP immunofluorescent cell bodies observed in the interpenduncular nucleus (Ljungdahl et al 1978a, Cuello & Kanazawa 1978, Cuello et al 1978b). It is not known whether these SPLI-positive neurons are intrinsic to the nucleus. They do not, however, project to the habenulae since no loss in SPLI is observed in these nuclei following destruction of the interpenduncular nucleus and the ventral tegmental area (Cuello et al 1978b).

In addition to the SPLI neurons, the habenulo-interpenduncular tract also contains cholinergic neurons (Kataoka et al 1973, Léránth et al 1975, Emson et al 1977, Cuello et al 1978b) and these two populations appear to have distinct origins. Separation of the medial from the lateral habenula leads to (*a*) a decrease in CAT activity and AChE staining in the medial habenula and (*b*) a decrease in SPLI in the lateral habenula and the interpenduncular/ventral tegmental area (Cuello et al 1978b). Since no change in SPLI was observed in the medial habenula, or any decrease in AChE staining or CAT activity in the lateral habenula, the majority of SP and cholinergic fibers must have a separate origin. Although the cell bodies of these SPLI neurons are in the medial habenula, those of the cholinergic neurons are most likely in the lateral habenula. It is also likely that they reciprocally project axon collaterals or axons into the lateral and medial habenula, respectively. However, these studies do not resolve whether all of these nuerons project to the interpenduncular nucleus and whether a separate population of habenulo-interpenduncular neurons contain both SP and ACh.

RELEASE   Whether SP is released from the interpenduncular nucleus upon stimulation has, to our knowledge, not yet been investigated.

MIMICRY   Sastry (1978a) has randomly sampled the firing rate of spontaneously active neurons in the interpeduncular nucleus and tested their response to habenular stimulation and to iontophoretically applied SP and ACh in anesthetized rats. Most cells responded to habenular stimulation (92/98), to SP (83/98), and to ACh (82/98). However, whereas the response to habenular stimulation was either excitation, inhibition, or inhibition followed by excitation (as determined from the discharge rates), the response to SP and ACh was almost invariably excitation. Using intracellular electrodes, Ogata (1979) has recorded from cells in slices of the interpeduncular nucleus an SP-induced depolarization, which persists after blockade of synaptic transmission by a low calcium medium. These results suggest that both the SP and ACh pathways have an excitatory action on interpeduncular neurons. In addition, habenular stimulation seems to activate some unidentified inhibitory neurons, possibly local interneurons. The ACh innervation clearly predominates because the excitatory effect of habenular stimulation was almost completely inhibited by doses of atropine that completely blocked ACh responses, but had no effect on SP responses. The residual atropine resistant excitation could conceivably be due to release of endogenous SP. In some neurons SP and ACh applied together gave a potentiated response (not seen with SP + glutamate or ACh + glutamate). Sastry (1978a) suggested that this potentiation could reflect the simultaneous activation of the paired *crest synapses* found in the interpeduncular nucleus (see Lenn 1976). It will be interesting to investigate this proposition with immunohistochemical methods.

Because of the relatively simple, synaptic connections (predominantly axodendritic, a few axosomatic, see Lenn 1976) and the possibility of blocking the muscarinic response, the interpeduncular nucleus may prove a suitable place to compare the excitatory effects of SP and ACh in the same appropriately chosen neurons by intracellular recording. Perfused slices could be used to determine the ionic requirements of the responses.

Substance P in Other Areas of the Central Nervous System

It is not unreasonable to expect that SP will have a transmitter function in all areas where terminals containing SPLI are present. Such areas are quite numerous (see Ljungdahl et al 1978a,b) but, except for the SP systems described above, the cell bodies of these terminals have not been localized and release of SPLI has only been demonstrated in the hypothalamus (Iversen et al 1976). However, iontophoresed SP increases the firing rate of neurons in the cuneate nucleus (Krnjević & Morris 1974), the locus coeruleus (Guyenet & Aghajanian 1977, Guyenet et al 1979), and the amygdala (Le Gal La Salle & Ben-Ari 1977). All of these areas contain at least moderately dense SP immunofluorescent fibers which could be terminals, whereas the amygdala exhibits an extremely dense SPLI-positive plexus. If SP release can be demonstrated in these regions and if the action of SP can be shown to be direct, then a general pattern emerges that SP in the CNS is an excitatory transmitter. Supporting evidence for a direct effect of SP in the locus coeruleus comes from the observation that SPLI-positive terminals make axo-dendritic synapses there (Pickel et al 1979).

Substance P and Synaptic Transmission in Autonomic Ganglia

Recent findings suggest that SP may play a role in autonomic transmission. Immunohistochemical studies (Hökfelt et al 1977b) have shown that fibers in several sympathetic ganglia contain SP. The inferior mesenteric ganglion of the guinea pig is particularly rich in these fibers which surround principal ganglion cells in a basket-like manner. A very dense network of SP-positive nerve terminals also occurs around cell bodies in the myenteric plexus (Hökfelt et al 1977b). No SPLI-positive cell bodies were found in any ganglia and thus the origin of these fibers remains unknown. It is possible that some are derived from primary sensory neurons. However, the presence of SP-positive fibers in organotypic tissue culture of intestine, in which all extrinsic innervation is presumably removed, indicates that SP neurons are present in the gut (Schultzberg et al 1978a). Interestingly, both inferior mesenteric ganglion cells and myenteric plexus neurons receive slow noncholinergic excitatory synaptic innervation. In inferior mesenteric ganglia the slow potential is associated with a fall in membrane resistance in the majority of cells (Neild 1978). Dun & Karczmar (1979) found that SP (10–100 nM) caused a depolarization in these cells that was unaffected by cholinolytic drugs or by blockade of synaptic transmission with a low Ca^{2+}/high Mg^{2+} solution. In the majority of cells this response was also associated with a fall in membrane resistance. Of particular importance was the finding that the slow synaptic potential was abolished when ganglion cells were desensitized by large doses of SP. The slow synaptic potential in myenteric plexus neurons (Katayama et al 1979) is associated with a marked increase in membrane resistance. Similarly, SP elicits a slow depolarization with an increase in membrane resistance. Experiments in which the concentration of extracellular potassium was altered suggest that SP acts by decreasing resting potassium permeability. Grafe et al (1979) have questioned the involvement of SP in the slow synaptic potential of myenteric plexis neurons because the serotonin antagonist, methysergide, blocks the synaptic potential but not the action of SP. Thus they favor serotonin as the transmitter for the slow synaptic potentials in this ganglion. However, one cannot rule out that with different stimulus parameters SP may contribute to the slow potentials.

These physiological studies of autonomic ganglia provide pharmacological evidence that SP may mediate some forms of non-cholinergic slow synaptic excitation. Of particular interest is the finding that the excitatory action of SP in different types of ganglion cells is generated by distinctly different ionic mechanisms. Furthermore, these studies emphasize the value of simple neuronal systems in elucidating the cellular actions of SP.

SUBSTANCE P AND 5-HYDROXYTRYPTAMINE COEXIST IN SOME MEDULLARY NEURONS

The presence of two or more putative transmitters in the same mature neuron has been documented in invertebrates (Kerkut et al 1967, Brownstein et al 1974, but see also Osborne 1977) and in the mammalian peripheral (Hökfelt et al 1977a) and central nervous system (Chan-Palay et al 1978, Hökfelt et al 1978b). In several nuclei of the lower medulla oblongata in the rat, there are neurons that contain either 5-HT or SPLI, or both 5-HT and SPLI (Chan-Palay et al 1978, Hökfelt et al 1978b). The presence of both 5-HT and DOPA decarboxylase immunofluorescence in the SPLI-containing neurons (Hökfelt et al 1978b) argues against the possibility that the 5-HT derives solely from active uptake; rather, it would seem that there is a distinct population of neurons that can synthesize both 5-HT and SP. All searches for an active uptake system for SP have been negative (Iversen et al 1976, Segawa et al 1976, 1977, A. W. Mudge and S. E. Leeman, unpublished).

Where do these "5-HT + SP" neurons project? A serotonergic raphé-spinal pathway has been described (Dahlström & Fuxe 1965) and, because the SPLI in the ventral horn is of supraspinal origin (Hökfelt et al 1977c), it is possible that the "5-HT + SP" neurons are a component of the raphé-spinal pathway. Hökfelt et al (1978b) and Björklund et al (1979), Singer et al (1979) have examined this possibility. Selective destruction of serotonergic neurons was obtained by means of neurotoxins. Intracysternal or intraventricular administration of 5,6- or 5,7-dihydroxytryptamine (DHT) into rats pretreated with protryptiline to prevent destruction of catecholamine neurons (Hökfelt et al 1978b) resulted in the virtual disappearance of 5-HT immunofluorescence in both the dorsal and ventral horns and a decrease in the SP immunofluorescence in the ventral horn. Radioimmunoassay data indicate that there is a slight decrease in SPLI in the dorsal horn as well (Hökfelt et al 1978b, Björklund et al 1979), but by far the most severe depletion of SPLI (–93%) was seen in the lumbar ventral horn (Björklund et al 1979). These results suggest that the majority of SP neurons projecting to the ventral horn and a smaller population of SP neurons projecting to the dorsal horn contain both 5-HT and SP in their terminals. However, the possibility can not be entirely excluded that destruction of the 5-HT neurons with the neurotoxins may have damaged some SP neurons nonspecifically or may have led to SPLI depletion by a transsynaptic mechanism. The question can only be settled definitively at the ultrastructural level. Multiple action of 5,6-DHT is perhaps suggested by the interesting observation (Björklund et al 1979) that the SPLI levels fail to recover in the spinal cord even after 20 months. By that time, 5,6-DHT treated 5-HT neurons regenerate some axons that reach all levels of the cord (see Björk-lund & Stenevi 1979). The SPLI levels in the raphé, on the other hand, did recover to control values by 20 months after the 5,6-DHT treatment (Björklund et al 1979). If it can be verified that a large component of the raphé-spinal projection consists of "5-HT + SP" neurons, then it is unlikely that these neurons represent a transitional state from "5-HT-only" to "SP-only" or vice versa, as transitional neurons would be expected to constitute only a small fraction of the total neuronal population (e.g. see Patterson 1978).

Particularly intriguing is the possibility of a descending "5-HT + SP" projection to the dorsal horn. SP is thought to be the transmitter of some nociceptive primary afferents (as discussed above), whereas 5-HT probably mediates the descending inhibition of nociceptive transmission in the dorsal horn (see Fields & Basbaum 1978). Speculations concerning such an arrangement include the possibility of centrally induced *peripheral pain,* in addition to simple potentiation or depression of incoming signals.

Aside from the synaptic organization of the proposed "5-HT + SP" terminals in the spinal cord, the most interesting question concerns the control of release of the two transmitter candidates. In particular, if both are released, are they always released simultaneously and at the same site or can their release be controlled separately? It is interesting, in this context, that reserpine treatment depletes the ventral horn only of 5-HT while the SPLI levels remain unchanged (Hökfelt et al 1978b). Whether this effect reflects a differential action of reserpine on the release (and hence storage or neuronal origin) of the two compounds, or simply indicates a selective inhibition of the 5-HT reuptake mechanism merits more specific investigation.

It is likely that more instances will be found in which two transmitter candidates are present in the same neuron. It will be important to demonstrate whether such neurons synthesize both compounds, or whether active uptake accounts for the presence of the second transmitter. The two possibilities have different functional implications.

THE SUBSTANCE P RECEPTOR IN SALIVARY GLANDS

SP was isolated on the basis of its sialogogic activity (Chang & Leeman 1970), which is not mediated via the adrenergic or cholinergic input to the salivary glands (Leeman & Hammerschlag 1967). It is likely that salivary glands are innervated by SP fibers. SP immunofluorescence has been reported in the vicinity of secretory structures (Hökfelt et al 1977c). Also, the SPLI levels in the rat submaxillary gland decrease upon section of the chorda tympani (Robinson et al 1979). The precise location of the SP terminals and their function (sensory or effector) in the glands are not known, however. Direct application of SP to rat parotid gland slices induces K^+ and amylase secretion (Rudich & Butcher 1976). The action of SP in this system is very similar to that of α-adrenergic and muscarinic agonists, but is not mediated by any of these receptors. In particular, all three agonists are more potent secretagogues for K^+ than for amylase, and the K^+ release is Ca^{2+} dependent.

Putney's group has recently accumulated considerable evidence supporting the hypothesis that in rat parotid acinar cells, three distinct receptor systems (α-adrenergic, muscarinic, and SP) regulate the K^+ permeability by interacting with a single population of Ca^{2+} channels (Putney 1977, 1978, Marier et al 1978, Putney et al 1978, see also Putney 1979). They had shown previously that the agonist-induced increase in K^+ permeability, as measured by $^{86}Rb^+$ release from parotid acinar slices, is biphasic: an initial transient surge in secretion is followed by a steady, slowly declining phase (Putney 1976, 1978). The initial transient can occur in the absence of external Ca^{2+}, whereas the sustained phase requires Ca^{2+} in the medium. In the absence of external Ca^{2+} and in the presence of EGTA, the initial transient is a one-shot event: once it has been induced by any one of the three agonists, carbachol, phenylephrine or SP, it can no longer be elicited by any of the other two agonists (Putney 1977). That the three receptors are known to be distinct (Putney 1976, Rudich & Butcher 1976), suggests that all three receptors affect a common site. This cross-receptor inactivation can be prevented if Ca^{2+} is added to the medium before exposing the tissue to the second agonist. Then a second transient can be elicited. The initial transient, therefore, has the features of a capacitative event and Putney (1977) suggested that it might be mediated by the release of bound Ca^{2+} from a site inaccessible to EGTA, but accessible to all three receptors.

Further experiments tested the possibility that the three receptors might also regulate the sustained phase of $^{86}Rb^+$ release by acting at a common site. Because a Ca^{2+} ionophore, in the presence of external Ca^{2+}, can mimic agonist induced K^+ release (Selinger et al 1974) and because the sustained release is blocked by cobalt (Marier et al 1978), it is likely that this phase is triggered by Ca^{2+} influx. Indeed, $^{45}Ca^{2+}$ uptake into dispersed parotid acinar cells is stimulated by carbachol, epinephrine, isoproterenol, and SP (Putney et al 1978). Furthermore, the carbachol and SP-stimulated $^{45}Ca^{2+}$ uptake is blocked by cobalt (cobalt block of adrenergic stimulation was not tested.) If the three receptors acted on different Ca^{2+} channels, then their ability to induce the sustained phase of ^{86}Rb release should be additive. Marier et al (1978) found, however, that in the presence of a submaximal concentration of external Ca^{2+}, supramaximal concentrations of any two agonists were not additive. This strongly suggests that all three receptors act on the same population of Ca^{2+} channels and individual cells possess all three receptors.

These studies represent the first clear description linking the postulated SP receptor with an identified ionic channel. They further demonstrate that the SP-stimulated increase in Ca^{2+} permeability is cobalt sensitive. Finally, they suggest that in parotid acinar cells a limiting number of these Ca^{2+} channels are shared by three distinct receptors. What still need to be demonstrated are the specific, saturable binding sites for SP in these cells.

Preliminary data from Putney's group (Wheeler et al 1979, J. W. Putney Jr., personal communication) indicates that specific binding of SP and analogs in dispersed parotid acinar cells can be demonstrated by the inhibition of ^{125}I-physalaemin binding that they cause. (Physalaemin is an SP-related peptide of amphibian origin and has a higher affinity for salivary gland tissue than SP.) ^{125}I-physalaemin binding is rapid, reversible, and saturable with an apparent K_m of 1.4 nM. Maximum specific binding appears to be very low (1.67 fmol /mg protein) and is therefore difficult to compare to the higher values (95.7 fmol/mg protein) obtained by Nakata et al (1978) for rabbit brain membranes. From Scatchard analysis it was estimated that there are only about 200 ^{125}I-physalaemin receptors per parotid acinar cell. The apparently excellent agreement between binding and pharmacological potency (^{86}Rb release) of SP and other peptides suggest, however, that these few binding sites may very well represent the physiologically relevant SP receptors. These studies are somewhat preliminary and should be extended to a more complete pharmacological characterization of the SP receptor by means of additional SP fragments and analogs. Also, it would be very useful to carry out ^{125}I-physalaemin binding studies with neuronal tissue for comparison. [Several laboratories have been unsuccessful in repeating Nakata et al's (1978) binding studies (S. H. Snyder, personal communication).]

An interesting feature of the receptors regulating Ca^{2+} permeability in salivary gland cells is the ligand-induced, Ca^{2+} independent turnover of phosphatidyl inositol (Oron et al 1975, Jones & Michell 1975) that they mediate. Preliminary studies suggest that SP also promotes phosphatidyl inositol turnover in parotid acinar cells (Jones & Michell 1978b). This is analogous to a vast number of other instances in various tissues in which a physiological response (e.g. secretion, exocytosis, contraction) seems to depend on Ca^{2+} entry and in which agonist binding leads to phosphatidyl inositol turnover in the absence of external Ca^{2+} (for a review see Michell 1975, Jones & Michell 1978a). Michell (1975) has argued that phosphatidyl inositol breakdown, which perhaps precedes ^{32}P incorporation, may be involved in the coupling of receptor occupation to Ca^{2+} channel opening. This proposition is speculative and somewhat controversial, but merits being tested by direct experiments.

It is an open question whether the characteristics of neuronal SP receptors are the same as those of the SP receptor(s) in salivary glands. By analogy with the nicotinic receptor, which appears to be different in neurons and in skeletal muscle (see Morley et al 1979 for a review), a similar situation may exist for SP receptors. Furthermore, just as there are nicotinic and muscarinic receptors for ACh, SP may have a set of receptors possibly associated with different ionic channels. Hence, a priori extrapolation seems unwise. But Putney's studies do suggest that it is worth investigating whether SP can induce Ca^{2+} currents in neurons as well.

THE INTERACTION OF SUBSTANCE P WITH SOME NICOTINIC RECEPTORS

SP depresses the activity induced by ACh in Renshaw cells (Krnjević & Lekić 1977, Belcher & Ryall 1977, Ryall & Belcher 1977) and in cultured adrenal chromaffin cells (Livett et al 1979).

Renshaw cells, which are innervated by motoneuron collaterals, can be identified by their characteristic high frequency discharge upon ventral root stimulation or upon iontophoresis of ACh. During iontophoresis of SP the nicotinic ACh-induced excitation is markedly reduced or completely abolished in virtually all Renshaw cells tested in the cat (Krnjević & Lekić 1977, Belcher & Ryall 1977, Ryall & Belcher 1977). Spontaneous firing or firing in response to ventral root stimulation is depressed to a lesser extent, possibly because the peptide does not diffuse to all activated synapses. Since excitation by glutamate, aspartate, DL-homocysteate, or a muscarinic agonist remains unaltered in the presence of SP, the SP-induced depression most likely occurs at the level of the nicotinic receptor.

A similar curare-like action of SP has been observed in cultured bovine adrenal cells (Livett et al 1979). Activity of these cells can be estimated from the stimulus-induced, Ca^{2+} dependent release of ^3H-NE into the medium. SP inhibits the ACh-or nicotine-induced release of ^3H-NE in a dose related manner. Neither the basal efflux nor the K^+-induced release of ^3H-NE are affected by SP. Because these cells do not respond to muscarinic agonists, these results are most easily explained by an interaction of SP with the nicotinic receptor-channel complex. The precise site and mechanism of action of SP remains to be determined. Rephrasing the suggestions made by Belcher & Ryall (1977), SP may prevent channel opening, block open channels, or promote channel closing. It could do so by competing with ACh or by acting independently. Cultured adrenal chromaffin cells seem to be a suitable system for further characterization of the interaction of SP with the nicotinic receptor.

In contrast to these cases of postsynaptic activity at a cholinergic synapse, the action of SP at the frog neuromuscular junction seems to be primarily presynaptic (Steinacker 1977). The transient decrease in miniature endplate potential (mepp) amplitude caused by high (10^{-4} M) concentrations of SP does, however, raise the possibility that SP may also intereact with the nicotinic receptor in muscle; the high dose required may reflect a diffusional barrier. Alternatively, the receptors may be effectively different (see Morley et al 1979); perhaps physalaemin would be a more suitable ligand to test in the frog. A similar SP-induced decrease in mepp amplitude has also been observed at the Mauthner fiber-giant fiber synapse in the hatchetfish where the transmitter is also thought to be ACh (Steinacker & Highstein 1976).

Is the curare-like effect of SP of physiological relevance? Little is known about the synaptic arrangement of SP-containing terminals in the ventral horn, but appreciable amounts of SPLI are clearly present there (Takahashi & Otsuka 1975), and SPLI fibers have been described in the vicinity of motoneurons (Barber et al 1979). Small amounts of SPLI are also present in the rat adrenal gland as well as in the hatchetfish brain stem (A. W. Mudge and S. E. Leeman, unpublished). It is therefore conceivable that the interaction of SP with these nicotinic receptors may be more than just a pharmacological curiosity.

SUBSTANCE P IN ASSOCIATION WITH BLOOD VESSELS

The presence of SP-positive terminals in the walls of arteries and veins or in contact with capillaries suggests that SP plays a role in regulating blood flow in some tissues, whereas at other sites it is secreted into the circulation and may function as a hormone (Sundler et al 1977, Chan-Palay & Palay 1977, Barber et al 1979, Hökfelt et al 1977c). Hökfelt et al (1977c) report SP fibers in association with cerebral and cutaneous blood vessels. A dense plexus of SPLI-positive terminals around hypophyseal portal vessels of the primate median eminence suggest a role for SP in regulating anterior pituitary secretion (Hökfelt et al 1978c). Barber et al (1979) found that although SP fibers form an extensive network around blood vessels in the dorsal horn of the spinal cord, the fibers are in contact with the perivascular astrocytic processes rather than directly with the endothelial cells. The function of this arrangement is not clear, but Barber et al (1979) discuss some possible implications. The finding that endothelial cells can readily degrade SP (Johnson & Erdös 1977) may pose a problem for the passage of SP into the circulation. Chan-Palay & Palay (1977) indicate transport of membrane-bound SPLI granules through endothelial cells.

SPLI has been detected in the plasma of several species including man (Nilsson et al 1975, Skrabanek et al 1977). Extraction of bovine plasma (Leeman & Carraway 1977) or cat plasma (Gamse et al 1978) yields immunoreactive material that comigrates with synthetic SP in several chromatographic systems. Cannon et al (1979) find differences in the size and stability of endogenous and exogenous SP and suggest that in plasma, SP circulates in association with a carrier of large molecular weight. Their experiments are also consistent with the presence of a circulating precursor that cross-reacts with their antibody. In the cat, the intestine appears to be an important source of SP in plasma (Gamse et al 1978), which may be derived from the SPLI-containing endocrine-like cells located in intestinal mucosa (Heitz et al 1976). The intestinal origin of plasma SPLI again suggests a hormonal role for SP that is perhaps related to some aspect of gastrointestinal function. Of course the SPLI in plasma may partly or merely represent SPLI that has been released and is en route to excretion.

CONCLUSION

In this review we have focused on a few particularly well characterized systems in which SP might function as a neurotransmitter. Because the biosynthesis and inactivation of SP have not been characterized, the evidence for its role as a neurotransmitter is limited to presence, release, and mimicry. Probably the single most important advance in our understanding of SP, and for that matter, other peptides, has been in the category of presence. The development of extremely sensitive and specific immunoassays and immunohistochemical techniques has permitted a detailed description of the neuronal systems containing SP. In primary afferents, localization has been extended to the ultrastructural level. In addition, in this system studies on release have been very successful as it has been possible to unambiguously demonstrate a calcium dependent release of SP from primary afferents. In the central nervous system, the most difficult criterion to satisfy has been that of mimicry. Virtually all neurons that are affected by SP are excited; where examined, this excitation is associated with a depolarization. However, it has not been possible to compare rigorously the membrane effects of SP with the synaptic potentials that are presumed to be mediated by SP. Some of the technical problems that have prevented this comparison may be overcome with the use of slice preparations. The interpeduncular nucleus provides a favorable site for such an approach. On the other hand, in some autonomic ganglia SP has been shown to mimic the membrane effects of slow synaptic potentials, although a detailed analysis of the ionic mechanism and pharmacology are still required. The lack of a selective SP antagonist is a serious handicap in establishing the membrane effects and, hence, in evaluating the neurotransmitter role of SP. In addition, receptor binding studies have had limited success. The preliminary results using I^{125}-physalaemin in salivary gland tissue are promising, and it will be of interest to see if this approach can be used in the CNS.

Some results with SP may require an expansion of our concept of chemical transmission. The ultrastructural studies on primary afferents reveal terminals that have features resembling those of neuroendocrine cells. The intimate association of SP terminals with blood vessels and perivascular astrocytes is also consistent with a neuroendocrine function. The interaction of SP with some nicotinic responses suggests that it might alter synaptic transmission by directly interacting with the receptors of other neurotransmitters. Finally, the clear demonstration of the coexistence of SP and 5-HT in the same neurons raises interesting questions about the physiology of these neurons.

Research on SP has exploded during the past few years and, although the evidence for a classical neurotransmitter role is now very strong, recent findings indicate that it would be unwise to limit our thinking to such a narrow point of view.

Literature Cited

Aldskogius, H., Arvidsson, J. 1978. Nerve cell degeneration and death in the trigeminal ganglion of the adult rat following peripheral nerve transection. *J. Neurocytol.* 7:229–50

Andersen, R. K., Lund, J. P., Puil, E. 1978. Enkephalin and substance P effects related to trigeminal pain. *Can. J. Physiol. Pharmacol.* 56:216–22

Anderson, D. J., Hannan, A. C., Matthews, B. 1970. Sensory mechanisms in mammalian teeth and their supporting structures. *Physiol. Rev.* 50:171–95

Atweh, S. F., Kuhar, M. J. 1977. Autoradiographic localization of opiate receptors in rat brain. 1. Spinal cord and lower medulla. *Brain Res.* 124:53–67

Baker, P. F. 1972. Transport and metabolism of calcium ions in nerve. *Proc. Biophys. Mol. Biol.* 24:177–223

Barber, R. P., Vaughn, J. E., Slemmon, J. R., Salvaterra, P. M., Roberts, E., Leeman, S. E. 1979. The origin, distribution, and synaptic relationships of substance P axons in rat spinal cord. *J. Comp. Neurol.* 184:331–51

Basbaum, C. B., Heuser, J. E. 1979. Morphological studies of stimulated adrenergic axon varicosities in the mouse vas deferens. *J. Cell Biol.* 80:310–25

Beaudet, A., Descarries, L. 1978. The monosynaptic innervation of rat cerebral cortex: Synaptic and non-synaptic axon terminals. *Neuroscience* 3:851–60

Belcher, G., Ryall, R. W. 1977. Substance P and Renshaw cells: A new concept of inhibitory synaptic interactions. *J. Physiol. London* 272:105–19

Björklund, A., Emson, P. C., Gilbert, R. F. T., Skagerberg, G. 1979. Further evidence for the possible co-existence of 5-hydroxytryptamine and substance P in medullary raphe neurons of rat brain. *Br. J. Pharmacol.* In press

Björklund, A., Lindvall, O. 1975. Dopamine in dendrites of substantia nigra neurons: Suggestions for a role in dendritic terminals. *Brain Res.* 83:531–37

Björklund, A., Stenevi, U. 1979. Regeneration of monoaminergic and cholinergic neurons in the mammalian central nervous system. *Physiol. Rev.* 59(1):62–100

Brown, D. A., Marsh, S. 1978. Axonal GABA-receptors in mammalian peripheral nerve trunks. *Brain Res.* 156:187–91

Brownstein, M. J., Mroz, E. A., Kizer, J. S., Palkovits, M., Leeman, S. E. 1976. Regional distribution of substance P in the brain of the rat. *Brain Res.* 116:299–305

Brownstein, M. J., Mroz, E. A., Tappaz, M. L., Leeman, S. E. 1977. On the origin of substance P and glutamic acid decarboxylase (GAD) in the substantia nigra. *Brain Res.* 135:315–23

Brownstein, M. J., Saavedra, J. M., Axelrod, J., Zeman, G. H., Carpenter, D. O. 1974. Coexistence of several putative neurotransmitters in single identified neurons of Aplysia. *Proc. Natl. Acad. Sci. USA* 71:4662–65

Burcher, E., Nilsson, G., Änggård, A., Rosell, S. 1977. An attempt to demonstrate release of substance P from the tongue, skin, and nose of dog and cat. In *Substance P,* ed. U.S. von Euler, B. Pernow,

37:183–85. New York: Raven. 344 pp.

Calne, D. B. 1970. *Parkinsonism: Physiology, Pharmacology, and Treatment.* London: Arnold Baltimore: William & Wilkins 136 pp.

Cannon, D., Skrabanek, P., Powell, D. 1979. Difference in behavior between synthetic and endogenous substance P in human plasma. *Naunyn-Schmiedebergs Arch. Pharmacol.* 307:251–55

Chang, M. M., Leeman, S. E. 1970. Isolation of a sialogogic peptide from bovine hypothalamic tissue and its characterization as substance P. *J. Biol. Chem.* 245:4784–90

Chang, M. M., Leeman, S. E., Niall, H. D. 1971. Amino-acid sequence of substance P. *Nature New Biol.* 232:86–87

Chan-Palay, V., Palay, S. L. 1977. Ultrastructural identification of SP cells and their processes in rat sensory ganglia and their terminals in the spinal cord by immunocytochemistry. *Proc. Natl. Acad. Sci. USA* 74:4050–54

Chan-Palay, V., Jonsson, G., Palay, S. L. 1978. Serotonin and substance P coexist in neurons of the rat's central nervous system *Proc. Natl. Acad. Sci. USA* 75:1582–86

Chéramy, A., Michelot, R., Leviel, V., Nieoullon, A., Glowinski, J., Kerdelhue, B. 1978a. Effect of the immunoneutralization of substance P in the cat substantia nigra on the release of Dopamine from dendrites and terminals of Dopaminergic neurons. *Brain Res.* 155:404–8

Chéramy, A., Nieoullon, A., Glowinski, J. 1978b. GABAergic processes involved in the control of dopamine release from nigrostriatal dopaminergic neurons in the cat. *Eur. J. Pharmacol.* 48:281–95

Chéramy, A., Nieoullon, A., Michelot, R., Glowinski, J. 1977. Effects of intranigral application of Dopamine and substance P on the in vivo release of newly synthesized ^{3}H-Dopamine in the ipsilateral caudate nucleus of the cat. *Neurosci. Lett.* 4:105–9

Coons, A. H. 1958. Fluorescent antibody methods. In *General Cytochemical Methods,* ed. J. F. Danielli, 1:399–422. New York: Academic. 471 pp.

Crain, S. M., Crain, B., Peterson, E. R., Simon, E. J. 1978. Selective depression by opioid peptides of sensory-evoked dorsal-horn network responses in organized spinal cord cultures. *Brain Res.* 157:196–201

Crain, S. M., Peterson, E. R., Crain, B., Simon, E. J. 1977. Selective opiate depression of sensory-evoked synaptic networks in dorsal horn regions of spinal cord cultures. *Brain Res.* 133:162–66

Cuello, A. C., Del Fiacco, M., Paxinos, G. 1978a. The central and peripheral ends of the substance P-containing sensory neurones in the rat trigeminal system. *Brain Res.* 152:499–509

Cuello, A. C., Emson, P. C., Paxinos, G., Jessell, T. 1978b. Substance P containing and cholinergic projections from the habenula. *Brain Res.* 149:413–29

Cuello, A. C., Jessell, T. M., Kanazawa, I., Iversen, L. L. 1977. Substance P: Localization in synaptic vesicles in rat central nervous system. *J. Neurochem.* 29:747–51

Cuello, A. C., Kanazawa, I. 1978. The distribution of substance P immunoreactive fibers in the rat central nervous system. *J. Comp. Neurol.* 178:129–56

Dahlström, A., Fuxe, K. 1965. Evidence for the existence of monoamine neurons in the central nervous system-II Experimentally induced changes in the intraneuronal amine levels of bulbospinal neuron system. *Acta Physiol. Scand.* 64: Suppl. 247, pp. 5–36

Dale, H. H. 1935a. Reizubertragung durch chemische Mittel im peripheren Nervensystem. Samm. der von der Nothnagel-Stiftung veranstalteten Vortrage 4. In: *Adventures in Physiology: A Selection of Scientific Papers.* London: Pergamon, 1953. 652 pp.

Dale, H. H. 1935b. Pharmacology and nerve endings. *Proc. R. Soc. London* 28: 319–32

Davies, J., Dray, A. 1976. Substance P in the substantia nigra. *Brain Res.* 107:623–27

Dray, A., Gonye, T. J., Oakley, N. R. 1976. Caudate stimulation and substantia nigra activity in the rat. *J. Physiol.* 259:825–49

Dray, A., Straughan, D. W. 1976. Synaptic mechanisms in the substantia nigra. *J. Pharm. Pharmac.* 28:400–5

Dubner, R., Gobel, S., Price, D. D. 1976. Peripheral and central trigeminal "pain" pathways. In *Advances in Pain Research and Therapy,* ed. J. J. Bonica, D. Albe-Fessard, pp. 137–48. New York: Raven. 1012 pp.

Duffy, M. J., Mulhall, D., Powell, D. 1975. Subcellular distribution of substance P in bovine hypothalamus and substantia nigra. *J. Neurochem.* 25:305–7

Duggan, A. W., Hall, J. G., Headley, P. M. 1976. Morphine, enkephalin, and the substantia gelatinosa. *Nature* 264: 456–58

Duggan, A. W., Griersmith, B. T., Headley, P. M., Hall, J. G. 1979. Lack of effect by substance P at sites in the substantia gelatinosa where met-enkephalin reduces the transmission of nociceptive impulses. *Neurosci. Lett.* 12:313–17

Dun, N. J., Karczmar, A. G. 1979. Action of substance P on sympathetic neurons. *Neuropharmacology* 18:215–18

Dunlap, K., Fischbach, G. D. 1978. Neurotransmitters decrease the calcium component of sensory neurone action potentials. *Nature* 276:837–39

Eccles, J. C., Fatt, P., Koketsu, K. 1954. Cholinergic and inhibitory synapses in a pathway from motor-axon collaterals to motoneurons. *J. Physiol. London* 126: 524–62

Emson, P. C., Cuello, A. C., Paxinos, G., Jessell, T., Iversen, L. L. 1977. The origin of substance P and acetylcholine projections to the ventral tegmental area and interpenducular nucleus in the rat. *Acta Physiol. Scand. Suppl.* 452: 43–46

Evans, R. H., Hill, R. G. 1978. Effects of excitatory and inhibitory peptides on isolated spinal cord preparations. In *Iontophoresis and Transmitter Mechanisms in the Mammalian CNS,* ed. R. W. Ryall, J. S. Kelly, pp. 101–3. Amsterdam: Elsevier. 494 pp.

Faull, R. L. M., Mehler, W. R. 1978. The cells of origin of nigrotectal, nigrothalamic and nigrostriatal projections in the rat. *Neuroscience* 3:989–1002

Fields, H. L., Basbaum, A. I. 1978. Brainstem control of spinal pain-transmission neurons. *Ann. Rev. Physiol.* 40:217–48

Floor, E., Leeman, S. E. 1979. Substance P-sulfoxide: Separation from substance P by high pressure liquid chromotography, biological and immunological activities, and chemical reduction. *Anal. Biochem.* In press

Fonnum, F., Grofova, I., Rinvik, E., Storm-Mathisen, J., Walberg, F. 1974. Origin and distribution of glutamate decarboxylase in substantia nigra of the cat. *Brain Res.* 71:77–92

Fox, C. A., Andrade, A., Hillman, D. E., Schwyn, R. C. 1972–1973a. The spiny neurons in the primate striatum: A golgi and electron microscopic study. *J. Hirnforsch.* 13:181–201

Fox, C. A., Andrade, A. N., Schwyn, R. C., Rafols, J. A. 1972–1973b. The aspiny neurons and the glia in the primate striatum: A golgi and electron microscopic study. *J. Hirnforsch.* 13:341–62

Fox, C. A., Rafols, J. A. 1976. The striatal efferents in the globus pallidus and in the substantia nigra. In *The Basal*

Ganglia, ed. M. D. Yahr, pp. 37–55. New York: Raven. 474 pp.

Gaddum, J. H., Schild, H. 1934. Depressor substance in extracts of intestine. *J. Physiol.* 83:1–14

Gale, J. S., Bird, E. D., Spokes, E. G., Iversen, L. L., Jessell, T. M. 1978. Human brain substance P: Distribution in controls and Huntington's chorea. *J. Neurochem.* 30:633–34

Gale, K., Hong, J. S., Guidotti, A. 1977. Presence of substance P and GABA in separate striatonigral neurons. *Brain Res.* 136:371–75

Gamse, R., Holzer, P., Lembeck, F. 1979a. Decrease of substance P in primary afferent neurons and impairment of neurogenic plasma extravasation by capsaicin. *Brit. J. Pharmacol.* In press

Gamse, R., Lembeck, F., Cuello, A. C. 1979b. Substance P in the Vagus Nerve. *Naunyn-Schmiedebergs Arch. Pharmakol.* 306:37–44

Gamse, R., Molnar, A., Lembeck, F. 1979c. Substance P release from spinal cord slices by capsaicin. *Life Sci.* 25:629–36

Gamse, R., Mroz, E. A., Leeman, S. E., Lembeck, F. 1978. The intestine as source of immunoreactive substance P in plasma of the cat. *Naunyn-Schmiedebergs Arch. Pharmakol.* 305:17–21

Gauchy, C., Beaujouan, J. C., Besson, M. J., Kerdelhue, B., Glowinski, J., Michelot, R. 1979. Topographical distribution of substance P in the cat substantia nigra. *Neurosci. Lett.* In press

Geffen, L. B., Jessell, T. M., Cuello, A. C., Iversen, L. L. 1976. Release of Dopamine from dendrites in rat substantia nigra. *Nature* 260:258–60

Gobel, S., Binck, J. M. 1977. Degenerative changes in primary trigeminal axons and in neurons in nucleus caudalis following tooth pulp extirpations in the cat. *Brain Res.* 132:347–54

Grafe, P., Mayer, C. J., Wood, J. D. 1979. Evidence that substance P does not mediate slow synaptic excitation within the myenteric plexus. *Nature.* 279:720–21

Graybiel, A. M., Ragsdale, C. W. Jr., 1979. Development and chemical specificity of neurons. *Prog. Brain Res.* 51. In press

Groves, P. M., Wilson, C. J., Young, S. J., Rebec, G. V. 1975. Self-inhibition by Dopaminergic neurons. *Science* 190: 522–29

Guyenet, P. G., Aghajanian, G. K. 1977. Excitation of neurons in the locus coeruleus by substance P and related peptides. *Brain Res.* 136(1):178–84

Guyenet, P. G., Mroz, E. A., Aghajanian, G. K., Leeman, S. E. 1979. Delayed iontophoretic ejection of substance P from glass micropipettes: Correlation with time-course of neuronal excitation in vivo. *Neuropharmacology* 18:553–58

Heitz, P., Polak, J. M., Timson, C. M., Pearse, A. G. E. 1976. Enterochromaffin cells as the endocrine source of gastrointestinal substance P. *Histochemistry* 49:343–47

Hellauer, H. F., Umrath, K. 1948. Uber die Aktionssubstanz der sensiblen Nerven. *Pflugers Arch.* 249:619–30

Henry, J. L. 1976. Effects of substance P on functionally identified units in cat spinal cord. *Brain Res.* 114:439–51

Henry, J. L., Krnjević, K., Morris, M. E. 1975. Substance P and spinal neurones. *Can. J. Physiol. Pharmacol.* 53:423–32

Hiller, J. M., Simon, E. J., Crain, S. M., Peterson, E. R. 1978. Opiate receptors in cultures of fetal mouse dorsal root ganglia (DRG) and spinal cord: Predominance in DRG neurites. *Brain Res.* 145:396–400

Hinsey, J. C., Gasser, H. S. 1930. The component of the dorsal root mediating vasodilatation and Sherrington contracture. *Am. J. Physiol.* 92:679–89

Hökfelt, T., Elde, R., Johansson, O., Luft, R., Nilsson, G., Arimura, A. 1976a. Immunohistochemical evidence for separate populations of somatostatin-containing and substance P-containing primary afferent neurons in the rat. *Neuroscience* 1:131–36

Hökfelt, T., Elde, R., Johansson, O., Ljungdahl, Å., Schultzberg, M., Fuxe, K., Goldstein, M., Nilsson, G., Pernow, B., Terenius, L., Ganten, D., Jeffcoate, S. L., Rehfeld, J., Said, S. 1978a. Distribution of peptide-containing neurons. In *Psychopharmacology: A Generation of Progress,* ed. M. A. Lipton, A. DiMascio, K. F. Killam, pp. 39–65. New York: Raven. 1731 pp.

Hökfelt, T., Elfvin, L. G., Elde, R., Schultzberg, M., Goldstein, M., Luft, R. 1977a. Occurrence of somatostatin-like immunoreactivity in some peripheral sympathetic noradrenergic neurons. *Proc. Natl. Acad. Sci. USA* 74:3587–91

Hökfelt, T., Elfvin, L. G., Schultzberg, M., Goldstein, M., Nilsson, G. 1977b. On the occurrence of substance P-containing fibers in sympathetic ganglia: Immunohistochemical evidence. *Brain Res.* 132:29–41

Hökfelt, T., Johansson, O., Kellerth, J. O., Ljungdahl, Å., Nilsson, G., Nygårds, A., Pernow, B. 1977c. Immunohisto-chemical distribution of substance P. In *Substance P,* ed. U.S. von Euler, B. Pernow, pp. 117–45. New York: Raven. 344 pp.

Hökfelt, T., Kellerth, J. O., Nilsson, G., Pernow, B. 1975. Substance P: Localization in the central nervous system and in some primary sensory neurons. *Science* 190:889–90

Hökfelt, T., Ljungdahl, Å., Steinbusch, H., Verhofstad, A., Nilsson, G., Brodin, E., Pernow, B., Goldstein, M. 1978b. Immunohistochemical evidence of substance P-like immunoreactivity in some 5-hydroxytryptamine-containing neurons in the rat central nervous system. *Neuroscience* 3:517–38

Hökfelt, T., Ljungdahl, Å., Terenius, L., Elde, R., Nilsson, G. 1977d. Immunohistochemical analysis of peptide pathways possibly related to pain and analgesia: Enkephalin and substance P. *Proc. Natl. Acad. Sci. USA* 74:3081–85

Hökfelt, T., Meyerson, B., Nilsson, G., Pernow, B., Sachs, C. 1976b. Immunochemical evidence for SP-containing nerve endings in the human cortex. *Brain Res.* 104:181–86

Hökfelt, T., Pernow, B., Nilsson, G., Wetterberg, L., Goldstein, M., Jeffcoate, S. L. 1978c. Dense plexus of substance P immunoreactive nerve terminals in eminentia medialis of the primate hypothalamus. *Proc. Natl. Acad. Sci. USA* 75:1013–15

Hökfelt, T., Schultzberg, M., Elde, R., Nilsson, G., Terenius, L., Said, S., Goldstein, M. 1978d. Peptide neurons in peripheral tissues including the urinary tract: Immunohistochemical studies. *Acta Pharmacol. Toxicol.* 43:79–89

Holton, P. 1959. Further observations on substance P in degenerating nerve. *J. Physiol. London* 149:35

Hong, J. S., Costa, E., Yang, H-Y.T. 1976. Effects of habenular lesions on the substance P content of various brain regions. *Brain Res.* 118:523–25

Hong, J. S., Yang, H-Y.T., Racagni, G., Costa, E. 1977. Projections of substance P containing neurons from neostriatum to substantia nigra. *Brain Res.* 122:541–44

Hu, J. W., Dostrovsky, J. O., Sessle, B. J. 1978. Primary afferent depolarization of tooth pulp afferents is not affected by naloxone. *Nature* 276:283–84

Iggo, A. 1974. Cutaneous reception. In *The Peripheral Nervous System,* ed. J. I. Hubbard. pp. 397–404. New York: Plenum. 530 pp.

Iversen, L. L., Jessell, T., Kanazawa, I. 1976. Release and metabolism of substance P in rat hypothalamus. *Nature* 264:81–83

Jancsó, N. 1968. Desensitization with capsaicin and related acylamides as a tool for studying the function of pain receptors. *Proc. 3rd Int. Pharmacol. São, Paulo. In Pharmacology and Pain,* 9:33–55. Oxford: Pergamon

Jancsó, N., Kiraly, E., Jancsó-Gabor, A. 1977. Pharmacologically induced selective degeneration of chemosensitive primary sensory neurones. *Nature* 270: 741–43

Jessell, T., Tsunoo, A., Kanazawa, I., Otsuka, M. 1979b. Substance P: Depletion in the dorsal horn of rat spinal cord after section of the peripheral processes of primary sensory neurons. *Brain Res.* 168:247–60

Jessell, T. M. 1978. Substance P release from the rat substantia nigra. *Brain Res.* 151:469–78

Jessell, T. M., Emson, P. C., Paxinos, G., Cuello, A. C. 1978a. Topographic projections of substance P and GABA pathways in the striato-pallido-nigral system: A biochemical and immunohistochemical study. *Brain Res.* 152: 487–98

Jessell, T. M., Iversen, L. L. 1977. Opiate analgesics inhibit substance P release from rat trigeminal nucleus. *Nature* 268:549–51

Jessell, T. M., Iversen, L. L., Cuello, A. C. 1978b. Capsaicin-induced depletion of substance P from primary sensory neurones. *Brain Res.* 152:183–88

Jessell, T. M., Mudge, A. W., Leeman, S. E., Yaksh, T. L. 1979a. Release of substance P and somatostatin, in vivo, from primary afferent terminals in mammalian spinal cord. *Soc. for Neurosci. Abstr.* 5:611

Johnson, A. R., Erdös, E. G. 1977. Inactivation of SP by cultured human endothelial cells. In *Substance P,* ed. U.S. von Euler, B. Pernow, pp. 253–60. New York: Raven. 344 pp.

Jones, L. M., Michell, R. H. 1975. The relationship of calcium to receptor-controlled stimulation of phosphatidyl inositol turnover. Effects of acetylcholine, adrenaline calcium ions, cinchocaine, and a bivalent cation ionophore on rat parotid gland fragments. *Biochem. J.* 148:479–85

Jones, L. M., Michell, R. H. 1978a. Stimulus-response coupling at α-adrenergic receptors. *Biochem. Soc. Trans.* 6: 673–88

Jones, L. M., Michell, R. H. 1978b. Enhanced phosphatidylinositol breakdown as a calcium-independent response to rat parotid fragments to substance P. *Biochem. Soc. Trans.* 6: 1035–37

Joó, F., Szolcesányi, J., Jancsó-Gabor, A. 1969. Mitochondrial alterations in the spinal ganglion cells of the rat accompanying the long lasting sensory disturbance induced by capsaicin. *Life Sci.* 8:621–26

Kanazawa, I., Bird, E. D., Gale, J. S., Iversen, L. L., Jessell, T. M., Muramoto, O., Spokes, E. G., Sutoo, D. 1979. Substance P: Decrease in substantia nigra and globus pallidus of Huntington's chorea. In *2nd International Huntington's Disease Symposium,* ed. T. N. Chase, N. S. Wexler, A. Barbeau, New York: Raven. In press

Kanazawa, I., Bird, E., O'Connell, R., Powell, D. 1977a. Evidence for the decrease of substance P content in the substantia nigra of Huntington's Chorea. *Brain Res.* 120:387–92

Kanazawa, I., Emson, P. C., Cuello, A. C. 1977b. Evidence for the existence of substance P-containing fibres in striato-nigral and pallidonigral pathways in rat brain. *Brain Res.* 119:447–53

Kanazawa, I., Jessell, T. 1976. Post mortem changes and regional distribution of substance P in the rat and mouse nervous system. *Brain Res.* 117:362–67

Kataoka, K., Nakamura, Y., Hassler, R. 1973. Habenulo-interpeduncular tract: A possible cholinergic neuron in rat brain. *Brain Res.* 62:264–67

Katayama, Y., North, R. A., Williams, J. J. 1979. The action of substance P on neurones of the myenteric plexus of the guinea-pig small intestine. *Proc. R. Soc. London Ser. B* In press

Kawai, N., Niwa, A. 1977. Hyperpolarization of the excitatory nerve terminals by inhibitory nerve stimulation in lobster. *Brain Res.* 137:365–68

Kemp, J. M., Powell, T. P. S. 1971. The structure of the caudate nucleus of the cat, light and electron microscopy. *Philos. Trans. R. Soc. London Ser. B* 262:383–401

Kerkut, G. A., Sedden, C. B., Walker, R. J. 1967. Uptake of DOPA and 5-hydroxytryptophane by monoamine-forming neurons in the brain of *Helix Aspersa. Comp. Biochem. Physiol.* 23:159–62

Khosla, M. C., Smeby, R. R., Bumpus, F. M. 1974. Structure-activity relationship in angiotensin II analogs. In *Angiotensin,* ed. I. H. Page, F. Bumpus, pp. 126–61. New York: Springer. 584 pp.

Kim, J. S., Bak, I. J., Hassler, R., Okada, Y. 1971. Role of γ-aminobutyric acid (GABA) in the extrapyramidal motor system. 2. Some evidence for the existence of a type of GABA-rich strionigral neurons. *Exp. Brain Res.* 14:95–104

Kitai, S. T., Kocsis, J. D., Preston, R. J., Sugimori, M. 1976. Monosynaptic imputs to caudate neurons identified by intracellular injection of horseradish peroxidase. *Brain Res.* 109:601–6

Knyihár, E., Csillik, B. 1976. Effect of peripheral axotomy on the fine structure and histochemistry of the Rolando substance: Degenerative atrophy of central processes of pseudounipolar cells. *Exp. Brain Res.* 26:73–87

Krnjević, K. 1977. Effects of substance P on central neurons in cats. In *Substance P*, ed. U.S. von Euler, B. Pernow, pp. 217–30. New York: Raven. 344 pp.

Krnjević, K., Lekić, D. 1977. Substance P selectively blocks excitation of Renshaw cell by acetylcholine. *Can. J. Physiol. Pharmacol.* 55:958–61

Krnjević, K., Morris, M. E. 1974. An excitatory action of substance P on cuneate neurones. *Can. J. Physiol. Pharmacol.* 52:736–44

LaMotte, C., Pert, C. B., Snyder, S. H. 1976. Opiate receptor binding in primate spinal cord: Distribution and changes after dorsal root section. *Brain Res.* 112:407–12

Lange, H., Thörner, G., Hopf, A., Schröder, K. F. 1976. Morphometric studies of the neuropathological changes in choreatic diseases. *J. Neurol. Sci.* 28:401–25

Leeman, S. E., Carraway, R. E. 1977. The discovery of a sialogogic peptide in bovine hypothalamic extracts: Its isolation, characterization as substance P, structure and synthesis. In *Substance P*, ed. U.S. von Euler, B. Pernow, pp. 5–13. New York: Raven. 344 pp.

Leeman, S. E., Hammerschlag, R. 1967. Stimulation of salivary secretion by a factor extracted from hypothalamic tissue. *Endocrinology* 81:803–10

Le Gal La Salle, G., Ben-Ari, Y. 1977. Microiontophoretic effects of substance P on neurons of the medial amygdala and putamen of the rat. *Brain Res.* 135:174–79

Lembeck, F. 1953. Zur Frage der zentralen Übertragung afferenter Impulse. III. Mitteilung, Das Vorkommen und die Bedeutung der Substanz P in den dor-salen Wurzeln des Ruckenmarks. *Naunyn-Schmiedebergs Arch. Pharmakol.* 219:197–213

Lembeck, F., Zetler, G. 1962. Substance P: A polypeptide of possible physiological significance, especially within the nervous system. *Int. Rev. Neurobiol.* 4:159–215

Lembeck, F., Zetler, G. 1971. Substance P. In *Pharmacology of Naturally Occurring Polypeptides and Lipid-soluble Acids,* ed. J. M. Walker, pp. 29–72. Oxford: Pergamon. 305 pp.

Lenn, N. J. 1976. Synapses in the interpeduncular nucleus: Electron microscopy of normal and habenula lesioned rats. *J. Comp. Neurol.* 166:73–100

Léránth, C. S., Brownstein, M. J., Záborszky, L., Járányi, Z. S., Palkovits, M. 1975. Morphological and biochemical changes in the rat interpeduncular nucleus following the transection of the habenulo-interpenduncular tract. *Brain Res.* 99:124–28

Livett, B. G., Kozouzek, V., Mizobe, F., Dean, D. M. 1979. Substance P inhibits nicotinic activation of chromaffins cells. *Nature* 278:256–57

Ljungdahl, Å., Hökfelt, T., Nilsson, G. 1978a. Distribution of substance P-like immunoreactivity in the central nervous system of the rat. I. Cell bodies and nerve terminals. *Neuroscience* 3:861–943

Ljungdahl, Å., Hökfelt, T., Nilsson, G., Goldstein, M. 1978b. Distribution of substance P-like immunoreactivity in the central nervous system of the rat. II. Light microscopic localization in relation to catecholamine-containing neurons. *Neuroscience* 3:945–76

Lundberg, J. M., Hökfelt, T., Nilsson, G., Terenius, L., Rehfeld, J. F., Elde, R. P., Said, S. 1978. Peptide neurons in the vagus, splanchnic, and sciatic nerves. *Acta Physiol. Scand.* 104:499–501

MacDonald, R. L., Nelson, P. G. 1978. Specific opiate-induced depression of transmitter release from dorsal root ganglion cells in culture. *Science* 199:1449–51

Marier, S. H., Putney, J. W. Jr., Van De-Walle, C. M. 1978. Control of calcium channels by membrane receptors in the rat parotid gland. *J. Physiol.* 279:141–51

Michell, R. H. 1975. Inositol phospholipids and cell surface receptor function. *Biochim. Biophys. Acta* 415:81–147

Michelot, R., Leviel, V., Giorguieff-Chesselet, M. F., Chéramy, A., Glowinski, J. 1979. Effects of the unilateral nigral modulation of substance P transmission on the activity of the two nigro-striatal Dopaminergic pathways. *Life Sci.* 24:715–24

Morley, B. J., Kemp, G. E., Salvaterra, P. 1979. Minireview: α-bungarotoxin binding sites in the CNS. *Life Sci.* 24:859–72

Mroz, E. A., Brownstein, M. J., Leeman, S. E. 1976. Evidence for substance P in the habenulo-interpeduncular tract. *Brain Res.* 113:597–99

Mroz, E. A., Brownstein, M. J., Leeman, S. E. 1977a. Evidence for substance P in the striato-nigral tract. *Brain Res.* 125:305–11

Mroz, E. A., Brownstein, M. J., Leeman, S. E. 1977b. Distribution of immunoassayable substance P in the rat brain: Evidence for the existence of substance P-containing tracts. In *Substance P,* ed. U.S. von Euler, B. Pernow, pp. 147–54. New York: Raven. 344 pp.

Mroz, E. A., Leeman, S. E. 1977. Substance P. In *Vitamins and Hormones,* ed. R. S. Harris, P. L. Munson, E. Diczfalusy, J. Glover, 35:209–81. New York: Academic. 320 pp.

Mudge, A. W., Leeman, S. E., Fischbach, G. D. 1979. Enkephalin inhibits release of substance P from sensory neurons in culture and decreases action potential duration *Proc. Nat. Acad. Sci. USA* 76:526–30

Nakata, Y., Kusaka, Y., Segawa, T., Yajima, H., Kitagawa, K. 1978. Substance P: Regional distribution and specific binding to synaptic membranes in rabbit central nervous system. *Life Sci.* 22:259–68

Neild, T. O. 1978. Slowly-developing depolarization of neurons in the guinea pig inferior mesenteric ganglion following repetitive stimulation of the preganglionic nerves. *Brain Res.* 140:231–39

Nicoll, R. A. 1971. Pharmacological evidence for GABA as the transmitter in granule cell inhibition in the olfactory bulb. *Brain Res.* 35:137–49

Nicoll, R. A. 1976. Promising Peptides. In *Neurotransmitters, Hormones, Receptors: Novel Approaches,* ed. J. A. Ferrendelli, B. S. McEwen, S. H. Snyder, 1:99–122. Neurosci. Symp.

Nicoll, R. A. 1978. The action of thyrotropin releasing hormone, substance P, and related peptides on frog spinal motoneurons. *J. Pharmacol. Exp. Ther.* 207:817–24

Nicoll, R. A., Alger, B. E. 1979. Presynaptic inhibition: Transmitter and ionic mechanisms. *Int. Rev. Neurobiol.* In press

Nieoullon, A., Chéramy, A., Glowinski, J. 1977a. Release of Dopamine in vivo from cat substantia nigra. *Nature* 266:375–77

Nieoullon, A., Chéramy, A., Glowinski, J. 1977b. An adaptation of the push-pull cannula method to study the in vivo release of [³H] Dopamine synthesized from [³H] tyrosine in the cat caudate nucleus: Effects of various physical and pharmacological treatments. *J. Neurochem.* 28:819–28

Nieoullon, A., Chéramy, A., Glowinski, J. 1977c. Interdependence of nigrostriatal Dopaminergic systems on the two sides of the brain. *Science* 198:416–18

Nilsson, G., Brodin, E. 1977. Tissue distribution of substance P-like immunoreactivity in dog, cat, rat, and mouse. In *Substance P,* ed. U.S. von Euler, B. Pernow, pp. 49–54. New York: Raven. 344 pp.

Nilsson, G., Hökfelt, T., Pernow, B. 1974. Distribution of substance P-like immunoreactivity in the rat central nervous system as revealed by immunohistochemistry. *Med. Biol.* 52:424–27

Nilsson, G., Pernow, B., Fischer, G. H., Folkers, K. 1975. Presence of substance P-like immunoreactivity in plasma from man and dog. *Acta Physiol. Scand.* 94:542–44

Ogata, N. 1979. Substance P causes direct depolarization of neurones of guinea pig interpeduncular nucleus in vitro. *Nature* 277:480–81

Olgart, L., Gazelius, B., Brodin, E., Nilsson, G. 1977a. Release of Substance P-like immunoreactivity from the dental pulp. *Acta Physiol. Scand.* 101:510–12

Olgart, L., Hökfelt, T., Nilsson, G., Pernow, B. 1977b. Localization of substance P-like immunoreactivity in nerves in the tooth pulp. *Pain* 4:153–59

Oron, Y., Lowe, M., Selinger, Z. 1975. Incorporation of inorganic p³² phosphate into rat parotid phosphatidylinositol. Induction through activation of α-adrenergic and cholinergic receptors and relation to K⁺ release. *Mol. Pharmacol.* 11:79–86

Osborne, N. N. 1977. Do snail neurones contain more than one neurotransmitter? *Nature* 270:622–23

Otsuka, M. 1978. In *Neurobiology of Peptides,* ed. L. L. Iversen, R. A. Nicoll, W. Vale. *Neurosci. Res. Program Bull.* 16(2). 370. pp.

Otsuka, M., Konishi, S. 1976. Release of substance P-like immunoreactivity from isolated spinal cord of newborn rat. *Nature* 264:83–84

Otsuka, M., Konishi, S., Takahashi, T. 1972. The presence of a motoneuron-depolarizing peptide in bovine dorsal roots of spinal nerves. *Proc. Jpn. Acad.* 48: 342–46

Otsuka, M., Yanagisawa, M. 1979. The action of substance P on motoneurons of the isolated rat spinal cord. *Proc. 7th Congr. Pharmacol.* In press

Patterson, P. H. 1978. Environmental determination of autonomic neurotransmitter function. *Ann. Rev. Neurosci.* 1:1–17

Pelletier, G., Leclerc, R., Dupont, A. 1977. Electron microscope immunohistochemical localization of substance P in the central nervous system of the rat. *J. Histochem. Cytochem.* 25:1373–80

Pernow, B. 1953. Studies on substance P: Purification, occurrence, and biological actions. *Acta Physiol. Scand.* 29:Suppl. 105, pp. 1–90

Pernow, B. 1963. Pharmacology of substance P. *Ann. NY Acad. Sci.* 104:393–402

Pickel, V. M., Joh, T. H., Reis, D. J., Leeman, S. E., Miller, R. J. 1979. Electron microscopic localization of substance P and enkephalin in axon terminals related to dendrites of catecholamineric neurons. *Brain Res.* 160:387–400

Pickel, V. M., Reis, D. J., Leeman, S. E. 1977. Ultrastructural localization of substance P in neurons of rat spinal cord. *Brain Res.* 122:534–40

Precht, W., Yoshida, M. 1971. Blockade of caudate-evoked inhibition of neurons in the substantia nigra by picrotoxin. *Brain Res.* 32:229–33

Preston, R. J., McCrea, R. A., Chang, H., Kitai, S. T. 1978. Antidromic and synaptic activation of HRP identified substantia nigra and retrorubral neurons in the cat: Anatomical and electrophysiological study. *Soc. Neurosci. Abstr.* 4: Abstr. No. 147, p. 49

Putney, J. W. Jr., 1976. Biphasic modulation of potassium permeability in rat parotid gland by carbachol and phenylephrine. *J. Pharmacol. Exp. Ther.* 198:375–84

Putney, J. W. Jr. 1977. Muscarinic, Alpha-adrenergic, and peptide receptors regulate the same calcium influx sites in the parotid gland. *J. Physiol.* 268:139–49

Putney, J. W. Jr. 1978. Role of calcium in the fade of the potassium release response in the rat parotid gland. *J. Physiol.* 281:383–94

Putney, J. W. Jr. 1979. Stimulus-permeability coupling: role of calcium in the receptor regulation of membrane permeability. *Pharmacol. Rev.* In press

Putney, J. W. Jr., Van De Walle, C. M., Leslie, B. A. 1978. Receptor control of calcium influx in parotid acinar cells. *Mol. Pharmacol.* 14:1046–53

Randić, M., Miletić, V. 1977. Effect of sub-

stance P in cat dorsal horn neurones activated by noxious stimuli. *Brain Res.* 128:164–69

Reubi, J. C., Emson, P. C. 1978. Release and distribution of endogenous 5-HT in rat substantia nigra. *Brain Res.* 139:164–68

Reubi, J. C., Emson, P. C., Jessell, T. M., Iversen, L. L. 1978. Effects of GABA, Dopamine and substance P on the release of newly synthesized ³H-5-hydroxytryptamine from rat substantia nigra in vitro. *Naunyn-Schmiedebergs Arch. Pharmacol.* 304:271–75

Reubi, J. C., Iversen, L. L., Jessell, T. M. 1977. Dopamine selectively increases ³H-GABA release from slices of rat substantia nigra in vitro. *Nature* 268: 652–54

Reubi, J. C., Jessell, T. 1978. Distribution of substance P in the pigeon brain. *J. Neurochem.* 31:359–61

Robinson, S. E., Schwartz, J. P., Costa, E. 1979. Substance P in the superior cervical ganglion and the submaxillary gland of the rat. *Brain Res.* In press

Rudich, L., Butcher, F. R. 1976. Effect of substance P and eledoisin on K efflux, amylase release, and cyclic nucleotides in slices of rat parotid gland. *Acta Biochem. Biophys.* 44:704–11

Ryall, R. W., Belcher, G. 1977. Substance P selectively blocks nicotinic receptors in Renshaw cells: A possible synaptic inhibitory mechanism. *Brain Res.* 137: 376–80

Saito, K., Konishi, S., Otsuka, M. 1975. Antagonism between Lioresal and substance P in rat spinal cord. *Brain Res.* 97:177–80

Sastry, B. R. 1978a. Effects of substance P, acetylcholine, and stimulation of habenula on rat interpeduncular neoronal activity. *Brain Res.* 144:404–10

Sastry, B. R. 1978b. Morphine and met-enkephalin effects on sural Aδ afferent terminal excitability. *Eur. J. Pharmacol.* 50:269–73

Schenker, C., Mroz, E. A., Leeman, S. E. 1976. Release of substance P from isolated nerve endings. *Nature* 264:790–92

Schultzberg, M., Dreyfus, C. F., Gershon, M. D., Hökfelt, T., Elde, R. P., Nilsson, G., Said, S., Goldstein, M. 1978a. VIP-, Enkephalin-, substance P-, and somatostatin-like immunoreactivity in neurons intrinsic to the intestine: Immunohistochemical evidence from organotypic tissue cultures. *Brain Res.* 155:239–48

Schultzberg, M., Ebendal, T., Hökfelt, T., Nilsson, G., Pfenninger, K. 1978b. Substance P-like immunoreactivity in cultured spinal ganglia from chick embryos. *J. Neurocytol.* 7:107–17

Selinger, Z., Eimeri, S., Schramm, M. 1974. A calcium ionophore simulating the action of epinephrine on the α-adrenergic receptor. *Proc. Nat. Acad. Sci. USA* 71:128–31

Segawa, T., Nakata, Y., Nakamura, K., Yajima, H., Kitagawa, K. 1976. Substance P in the central nervous system of rabbits: Uptake system differs from putative transmitters. *Jpn J. Pharmacol.* 26:757–60

Segawa, T., Nakata, Y., Yajima, H., Kitagawa, K. 1977. Further observation on the lack of active uptake system for substance P in the central nervous system. *Jpn J. Pharmacol.* 27:573–80

Sheperd, G. M. 1974. *The Synaptic Organization of the Brain.* London & New York: Oxford. 364 pp.

Shoulson, I., Chase, T. N. 1975. Huntington's disease. *Ann. Rev. Med.* 26: 419–26

Singer, E., Sperk, G., Placheta, P., Leeman, S. E. 1979. Reduction of substance P levels in the ventral cervical spinal chord of the rat after intracisternal 5,7-dihydroxytryptamine injections. *Brain Res.* 174:362–65

Skrabanek, P., Cannon, D., Kirrane, J., Legge, D., Powell, D. 1977. Circulating immunoreactive substance P in man. *Ir. J. Med. Sci.* 145:399–408

Skrabanek, P., Powell, D. 1977. Substance P. In *Annual Research Reviews*, Vol. 1, ed. D. F. Horrobin. Montreal: Eden. 181 pp.

Starr, M. S. 1979. Gaba-mediated potentiation of amine release from nigrostriatal dopamine neurones in vitro. *Eur. J. Pharmacol.* 53:215–26

Steinacker, A. 1977. Calcium-dependent presynaptic action of substance P at the frog neuromuscular junction. *Nature* 267:268–70

Steinacker, A., Highstein, S. M. 1976. Pre- and postsynaptic action of substance P at the Mauthner fiber-giant fiber synapse in the Hatchet fish. *Brain Res.* 114:128–133

Sternberger, L. A., Hardy, P. H. Jr., Cuculis, J. J., Meyer, H. G. 1970. The unlabelled antibody enzyme method of immunohistochemistry. Preparation and properties of soluble antigen-antibody complex (horseradish peroxidase-antihorseradish peroxidase) and its use in identification of spirochetes. *J. Histochem. Cytochem.* 18:315–33

Sundler, F., Alumets, J., Brodin, E., Dahlberg, K., Nilsson, G. 1977. Perivascular substance P immunoreactive nerves in tracheobronchial tissue. In *Substance P*, ed. U.S. von Euler, B. Pernow, pp. 271–73. New York: Raven. 344 pp.

Takahashi, T., Konishi, S., Powell, D., Leeman, S. E., Otsuka, M. 1974. Identification of the motoneuron-depolarizing peptide in bovine dorsal root as hypothalamic substance P. *Brain Res.* 73:59–69

Takahashi, T., Otsuka, M. 1975. Regional distribution of substance P in the spinal cord and nerve roots of the cat and the effect of dorsal root section. *Brain Res.* 87:1–11

Theriault, E., Otsuka, M., Jessell, T. 1979. Capsaicin-evoked release of substance P from primary sensory neurons. *Brain Res.* 170: 209–213

Tregear, G. W., Niall, H. D., Potts, J. T. Jr., Leeman, S. E., Chang, M. M. 1971. Synthesis of substance P. Nature New Biol. 232:87–89

von Euler, U. S., 1936. Untersuchungen über Substanz P., die atropinfeste, darmerregende und gefässerweiternde Substanz aus Darm und Hirn, Naunyn-Schmiedebergs. *Arch. Exp. Pathol. Pharmakol* 181:181–97

von Euler, U. S., Gaddum, J. H. 1931. An unidentified depressor substance in certain tissue extracts. *J. Physiol. London* 72:74–87

von Euler, U. S., Pernow, B. 1977. *Substance P.* New York: Raven. 344 pp.

Walker, R. J., Kemp, J. A., Yajima, H., Kitagawa, K., Woodruff, G. N. 1976. The action of substance P on mesencephalic reticular and substantia nigral neurones of the rat. *Experientia* 32:214–15

Wheeler, C. S., Van De Walle, C. M., Putney, J. W. Jr. 1979. Radioligand binding to substance P receptors on parotid acinar cells. *FASEB* Abstr. No. 4293, p. 1039

Wright, D. M., Roberts, M. H. T. 1978. Supersensitivity to a substance P analog following dorsal root section. *Life Sci.* 22:19–24

Yoshida, M., Precht, W. 1971. Monosynaptic inhibition of neurons of the substantia nigra by caudato-nigral fibers. *Brain Res.* 32:225–28

Zieglgänsberger, W., French, E. D., Siggins, G. R., Bloom, F. E. 1979. Opioid peptides may excite hippocampal pyramidal neurons by inhibiting adjacent inhibitory interneurons. *Science* 205: 415–17

Zieglgänsberger, W., Tulloch, I. F. 1979. Effects of substance P on neurones in the dorsal horn of the spinal cord of the cat. *Brain Res.* 166:273–82

Ann. Rev. Neurosci. 1980. 3:269–78

TROPHIC INTERACTIONS ❖11542
IN NEUROGENESIS: A PERSONAL
HISTORICAL ACCOUNT

Viktor Hamburger

Department of Biology, Washington University, St. Louis, Missouri 63130

INTRODUCTORY REMARKS

Instead of introducing the topics in developmental neurobiology reviewed in this volume, I decided to focus on one issue that has intrigued me for more than half a century and provide a historical account by an eyewitness of the analysis of a limited but important issue that unfolded gradually and ramified unexpectedly. The story illustrates a particular style of scientific pursuit in biology and the inexorable historical continuity of the scientific process, the vision of which we often lose.

NEUROLOGICAL FOUNDATIONS

Modern neuroembryology is based on two fundamental principles that were established independently by two great innovators in the 1880s and early 1890s, the German anatomist and embryologist, Wilhelm His (1831–1904), and the Spanish neurologist, S. Ramòn y Cajal (1852–1934). The first principle established the nerve cells, or neurons, as the autonomous elementary constituents of the nervous system. Known later as the neuron doctrine, this principle liberated neurology from the fallacy of the then prevailing network theory, which considered the nervous system as a syncytium. The discovery of *contiguity* (rather than continuity) at axon terminals was made by Cajal in 1888, when he observed that the terminal axon branches of the stellate cells in the cerebellum of birds and mammals contact the perikarya of Purkinje cells in basket fashion, but do not fuse with them.

The second fundamental principle is closely related to the first. His and Cajal, in their independent studies of the embryonic development of the

269

0147-006X/80/0301-0269$01.00

vertebrate nervous system, demonstrated that axons (and dendrites) are outgrowths of embryonic nerve cells or neuroblasts. However, despite the clear demonstration of axonal growth cones by Cajal as early as 1890, decades passed before the competing theories of the origin of nerve fibers from cell chains or plasmodesmas, respectively, were abandoned. The issue was finally settled in 1907 by the classical tissue culture experiment of R. G. Harrison. Under the microscope, he observed the outgrowth of axons from living neuroblasts in isolated pieces of frog embryo spinal cord.

EXPERIMENTAL EMBRYOLOGY

In 1888, the year of the discovery of the synapse, an equally momentous event occurred in the field of embryology. The German anatomist, Wilhelm Roux (1850–1924), a remarkably independent mind, performed one of the first, and certainly one of the most influential, experiments on an embryo: the separation of the first two blastomeres of the frog's egg. This signaled the birth of experimental embryology, or Entwicklungsmechanik, as Roux called it. What occurred here was a conceptual revolution: a novel, causal-analytical approach to problems of embryology. Questions were now directed to the immediate causes or factors which determine the fate of parts of the embryo, and Roux's experiment addressed the primal question of the interrelations of the two daughter cells of the egg. The obvious corollary to this breakthrough was the shift from description to the analytical experiment.

Roux's was essentially a theoretical mind, and the leadership in the new field was taken over by Hans Spemann (1869–1941) in Germany and Ross G. Harrison (1870–1959) in the United States. They matched W. Roux in depth of analytical thought but surpassed him in designing experiments and in the art of microsurgery on the embryo. In the 1920s and 1930s, experimental embryology held as much fascination as molecular biology did a generation later. When I came, as a graduate student, to Spemann's laboratory, his work was approaching its climax in the *organizer* experiment published by Spemann & Hilde Mangold in 1924. One of the key concepts was *embryonic induction:* the initiation of developmental processes in a particular region of the embryo by (presumably chemical) interaction with adjacent contiguous structures; illustrated, for example, by lens induction by the optic vesicle. A major aspect of the organizer experiment is the induction of the neural plate—the anlage of the central nervous system— by the underlying mesoderm mantle.

If the topic of my PhD thesis had been related directly to neural induction, as was the case in most other PhD theses of that time in the Freiburg laboratory, my acquaintance with the nervous system would have ended probably with the closure of the neural tube. In the atmosphere of Spemann's laboratory I caught the spirit of the experimental-analytical approach and the awareness of the crucial importance of embryonic interactions of all kinds (not just embryonic induction) in the process of differentiation. I soon found out that the developing nervous system is an ideal playground for the study of a variety of intricate embryonic relationships.

R. G. HARRISON AND EXPERIMENTAL NEUROEMBRYOLOGY

Spemann suggested the influence of innervation on amphibian limb development, which was then a controversial issue, as a topic for my PhD thesis. Thus I moved into the orbit of R. G. Harrison, who was then the leading neuroembryologist. Since the turn of the century, he had applied the conceptualizations and methodology of experimental embryology to the special problems raised by the distinctive features of neural development and, in particular, by the unique partnership between nervous and non-nervous structures. Almost single-handedly, he established experimental neuroembryology as a field in its own right. He applied, most ingeniously, the extirpation and transplantation methods to such problems as nerve pattern formation and the origins of Schwann cells and lateral line organs. In 1907 he introduced into neuroembryology the limb transplantation experiment, which is an invaluable tool to this day.

Harrison was a friend of Spemann's. I became well acquainted with him during his summer visits to Freiburg. I consulted him freely and his encouragement and interest in my work continued after my transplantation to the United States. To have two mentors of such stature is indeed a rare gift for a budding experimentalist.

NERVELESS LIMBS AND LIMB-DEPRIVED NERVE CENTERS

My doctoral thesis led, by some detours, to a critical test of the problem of the role of innervation in the development of limbs in frog embryos. The nerve ingrowth into limbs was prevented by early extirpation of the lumbar segments of the spinal cord in neurula stages. The nerveless legs developed perfectly normally, except for secondary degeneration of the musculature. (Much later, I produced nerveless wings and legs in chick embryos by the simple method of implanting limb buds into the coelomic cavity. The results were the same.)

For the experimental embryologist, independent development is much less interesting than dependency. Would the reciprocal experiment of depriving the developing nerve centers of their peripheral target fields lead to a more promising road? At this point, my interests intersected with those of the Harrison school. At that time, in the late 1920s, Harrison had turned to other problems unrelated to neurogenesis. But one of his most active students, S. R. Detwiler, then at Columbia University, answered my question in part. Around 1920, he had started a series of experiments that involved extirpating and transplanting limb primordia in early stages of salamander embryos, prior to nerve outgrowth. He observed that (a) dorsal root ganglia deprived of their peripheral target area were hypoplastic and (b) *peripherally overloaded* ganglia that had to supply the normal and a supernumerary limb were hyperplastic. The motor area of the spinal cord seemed to be unaffected.

Through a fateful combination of circumstances, my further preoccupation with the trophic interactions between nerve centers and their targets became intertwined with my move to the United States and, related to this, with my conversion from amphibian to chick embryos. In 1932 I received a traveling fellowship from the Rockefeller Foundation to work for one year in the laboratory of Dr. Frank Lillie, a friend of Spemann's, at the University of Chicago. My objective was to try to apply Spemann's glass needle technique of microsurgery to chick embryos. I was thoroughly familiar with the technique, but my acquaintance with the chick embryo was minimal. My stay in the United States became permanent for political reasons. By personal choice, I adopted the chick embryo for the rest of my life.

At that time, the laboratory of Dr. Lillie was the center of chick embryology; he himself had written the classical book on this topic. However, the only experimental technique then available was the implantation of tissues onto the chorioallantoic membrane. Despite its severe limitations, the method yielded important results in the hands of Lillie's student, B. H. Willier, and his group, and others. Dr. Willier was then in charge of the Embryology Division of the Zoological Institute, while Dr. Lillie served as Dean. Dr. Willier and his associate, Dr. Mary Rawles, introduced me to the secrets of handling chick embryos. Their expertise was unexcelled and their generosity in sharing it was of great help. After a few months, I accomplished limb extirpations and transplantations using the glass needle technique. Many other types of experiments became possible. The chick embryo soon achieved equal rank with the amphibian embryo as a subject of investigations in experimental embryology.

My plan to use wing bud extirpations for the study of the effects of this operation on sensory and motor systems was welcomed by Dr. Lillie. As it happened, 25 years earlier he had suggested the same experiment to one of his students, Miss M. Shorey, with a similar question in mind. She had managed to do the operation successfully with the rather crude technique of electrocautery and, in 1909, she presented evidence for hypoplasia in sensory ganglia and motor columns. But the time was not ripe; her experiment fell into oblivion. Dr. Lillie was interested in seeing it revived and the discrepancy between her findings and those of Detwiler, concerning the motor system, resolved.

My experiments were performed on three-day embryos and the effects studied five to six days later. A substantial hypoplasia of the spinal ganglia and a conspicuous reduction of the brachial motor column were recorded. (In later, more radical extirpations, a total absence of the lateral motor column was observed.) Thus it was firmly established that primary nerve centers, after a few days of self-differentiation, become critically dependent for their survival on their target areas. Detwiler had probably not recognized the striking effect on motor cells because they are difficult to identify in urodeles. For the first time, the advantage of the higher organization of the avian nervous system became obvious.

Three questions of a fundamental nature were raised by these results:

1. What is the nature of the response of the nerve centers?
2. What is the basis of the trophic action of the target tissues?
3. How is the interaction of the two partners mediated?

Some observations I had made suggested an answer to the third question. Semiquantitative data showed that the reduction of the motor column was proportional to the amount of muscle loss and, furthermore, the percentage loss of motor center and ganglia varied independently for each other. From this I concluded that we are not dealing with a global effect of the periphery, i.e. the wing, rather, each neuronal component receives a "message" from its own specific target tissue: motor neurons from muscles and spinal ganglia from sensory fields. This, in turn, suggests that the most likely candidates for the *mediation* of trophic effects from periphery to nerve center are the sensory and motor axons, respectively. Thus, retrograde transmission of a "message," if not that of a trophic agent, was clearly anticipated.

I had no answer to the second question, and my speculations concerning the first turned out to be wrong. But by formulating the questions, I had provided the frame of reference for the next steps of the analysis.

FOCUS ON DORSAL ROOT GANGLIA

Through a strange set of circumstances, the publication of the limb extirpation experiments in 1934 had a decisive influence on the scientific life of one reader and, in the course of events, on the direction of my own research. The paper came to the attention of Dr. Rita Levi-Montalcini in Italy, and it kindled her interest in this facet of neuroembryology. In collaboration with her teacher and friend, Guiseppe Levi, she repeated the experiment during the war under trying circumstances recalled in her review of 1975. My results were confirmed, but there was disagreement on an important theoretical issue. The Italian investigators had come to the conclusion, based on cell counts in spinal ganglia, that the hypoplasia is due to the loss of differentiated neurons. I had proposed a *recruitment-hypothesis,* according to which the size of the neuron populations would be adjusted to the demands of the periphery by recruitment of neurons from a pool of (hypothetical) undifferentiated cells. At my invitation, Dr. Levi-Montalcini came to St. Louis in 1947, to resolve our differences of opinion.

We immediately began an extensive reinvestigation of the spinal ganglia that involved their normal development and their responses to limb extirpation and transplantation. We realized that the key to the full understanding of what was going on was a closely timed series of developmental stages. In our earlier work we had independently observed that the spinal ganglia of the chick embryo consist of two populations: early-differentiating, large neurons in a ventrolateral position (VL); and later-differentiating, smaller, dorsomedial cells (MD).

The results published in 1949 represent substantial progress over previous investigations. The controversial issue was resolved by the observation of large numbers of frankly degenerating cells, clearly identifiable as neurons, in the brachial ganglia of operated embryos. Degenerating neurons were largely confined to the VL population. The degeneration process showed a distinct time pattern with the peak at five and six days of incubation. The MD population showed little cell loss but severe cell atrophy. Thus the earlier position of Levi-Montalcini and Levi was vindicated; my first question (above) was answered unequivocally in terms of secondary degeneration of differentiated cells. The *trophic* factors were thus characterized as *maintenance* factors. (Of course, we realized that both trophic and maintenance were provisional designations which would eventually have to be defined in physiological and molecular terms.) We also observed a reduction in the number of mitoses in cases of limb ablation and an increase after implantation of a supernumerary limb. We concluded that target areas also regulate proliferation and thus population size. However, these data are inconclusive because the proliferating cells could be either prospective neurons or glia. The critical test, using autoradiography, has not been made.

I should like to make a brief digression, turning to the spinal motor system, where the overall situation is in some respects clearer than in spinal ganglia. A reinvestigation of the effects of limb bud ablation in 1958 again showed a massive degeneration process which wiped out the entire motor column between six and nine days of incubation. Again degenerating neurons were clearly observable. However, the proliferation process that produces the motor neurons in the ventral part of the spinal cord is already terminated at six days; in fact, the lateral motor column is numerically complete before the degeneration process begins. Hence it is clear that in this instance, the target area has no control over proliferation, migration, and initial differentiation—including axon outgrowth.

During the 1960s, A. Hughes, J. Kollros, M. Prestige, and their co-workers made extensive investigations of the same problems in sensory and motor systems of amphibians, with essentially similar results.

NORMALLY OCCURRING NEURONAL DEATH

Unexpectedly, degenerating neurons were found in substantial numbers in cervical and thoracic ganglia of normal embryos, whereas the limb-innervating ganglia showed only few degenerating cells. Sporadic death of embryonic cells had been known for a long time in most embryonic tissues, including nervous tissue; but we were dealing here with a pattern that seemed to account, at least in part, for segmental differences in the size of spinal ganglia along the rostro-caudal axis. Furthermore, the massive degeneration occurred during a limited period, with the peak at six and seven days. Clearly, we were dealing here with a novel phenomenon: selective neuronal loss as part of the normal process of neurogenesis. This implied an overproduction of neurons, followed by secondary depletion. In other words, the final population size of a ganglion was determined by a two-step process. A year later, Dr. Levi-Montalcini discovered a similar phenomenon in the embryonic motor system: a massive degeneration of neuroblasts, limited to the cervical level and to a brief time span. It seemed then that we were dealing with a more general phenomenon.

This discovery illuminated the mechanism of trophic interactions. We observed startling resemblances between naturally occurring and experimentally induced neuronal death. First, the VL and MD populations in spinal ganglia were affected in the same way: massive degeneration in the former and mostly atrophy in the latter. Second, the degeneration process occurred in both instances at the same developmental stages, with peaks at five and six days. We argued that

in the experimental situation, the reduction of the peripheral area is definitely responsible for the process of degeneration. It is possible that the same mechanism operates in the case of normal cervical and thoracic VL cells. This would imply that in early stages cervical and thoracic ganglia send out more fibers than the periphery can support. The excess neurons would break down at the stages at which the VL cells are highly susceptible to environmental conditions (1949, p. 495).

Then we realized that the degeneration process occurred approximately at the time at which axons establish contacts with their peripheral target organs (according to Tello and other earlier investigators). We concluded that "adequate connections with the periphery are necessary for the maintenance of sensory neurons" (1949, p. 492). Here, then, we had the first inkling that peripheral connections are implicated in the trophic mechanism.

We did not pursue this matter further and a decade elapsed before interest in this phenomenon of natural death was revived by others. The investigations of Hughes & Prestige on amphibians and of Cowan & coworkers, Landmesser & Pilar, Oppenheim, and others on chick embryos broadened our knowledge of this phemonenon substantially; it is now recognized as of widespread, if not universal, occurrence. Quantitative studies on units with manageable population size have shown that the cell loss is not a trifling matter: it ranges anywhere from 40 to 75%. The earlier observation that the period of degeneration coincides with the time of provisional synapse formation was confirmed for several other systems. It is the basis for the now widely accepted *competition hypothesis,* which postulates a competition of axons at the target areas and degeneration of the unsuccessful neurons. The premise of this hypothesis that originally all neurons, including the losers, send out their axons to the periphery, has been verified in several instances. Perhaps the most convincing evidence for the competition hypothesis comes from the findings of Hollyday & Hamburger that the normal death rate of the lateral motor column can be reduced significantly if the available mass of muscle tissue is increased by the implantation of a supernumerary leg. (The experiment shows the fallacy of equating a numerical surplus with hyperplasia, meaning overgrowth. We have used the term *hypothanasia* for reduction of cell loss.) The competition hypothesis requires much further refinement and more checks.

DISCOVERY OF THE NERVE GROWTH FACTOR (NGF)

Although the question of the responses of nerve centers to the manipulation of their targets was answered, at least in the first approximation, the major problem of the nature of the trophic agents released by the target organs had not been touched. We had only two small clues which helped us further: 1. We were impressed by the hyperplasia of ganglia, which demonstrated their growth potential beyond their normal range. 2. The differential responses of VL and MD cells strengthened our belief in the existence of specific growth promoting agents. We realized that the limbs with their heterogeneous histological structure were of no further use and that the search for specific trophic agents would have to extend to homogeneous tissues that eventually could be obtained in bulk. Pilot experiments substituting muscle, liver, and other tissues for the limb bud gave negative results. We turned then to an experiment which had been done by my former student, E. Bueker, whose PhD thesis had involved coelomic grafts of spinal cord segments. Starting from considerations similar to ours, he hit upon the idea of implanting mouse tumors into the coelom of three-day chick embryos. The sarcomas grew well and were invaded by sensory fibers. But the hyperplasia of the corresponding sensory ganglia was not impressive and he turned to other projects. With his consent we repeated the experiment (apparently with a more potent sample of sarcoma 180) and immediately obtained dramatic results: a neurotization of the tumor by very dense fiber masses whose origin could be traced to both sensory and sympathetic chain ganglia. The latter were hyperplastic up to six times their normal size and the former were more than double in size. However, the motor column was refractory.

Up to that point, the traditional model of trophic target effects was still applicable, but on a larger scale. However, the next discovery changed the picture drastically. Dr. Levi-Montalcini observed that many viscera, such as the meso-, metanephros, and the gonads, were invaded by massive fiber bundles, and that they even invaded blood vessels. Normally, at corresponding stages, these viscera are barely innervated. The search for the origin of these fiber masses led her to examine preganglionic sympathetic complexes. They were found to be enormously enlarged. More importantly, they were located at a considerable distance from the tumor and not directly connected with it. This gave the decisive clue that the basis for the trophic effect was the production of a diffusible nerve growth promoting agent. Conclusive evidence was obtained by Dr. Levi-Montalcini in two experiments: 1. Pieces of sarcoma grown on the chorioallantoic membrane produced the same effects as intraembryonic tumors. Therefore, a tumor agent is transmitted via the circulation. 2. Spinal and sympathetic ganglia isolated from seven-day embryos were grown in vitro at some distance from pieces of sarcoma. A dense halo of nerve fibers grew out within 10 hr, whereas control ganglia produced only a few fibers. The evidence for a systemic route of distribution of a trophic agent by no means affects our previous contention that, in other instances, retrograde transport in axons makes the trophic agent accessible. In fact, as was later demonstrated, NGF can be transmitted to the responsive neurons by either route.

The tissue culture experiment later proved invaluable in the obvious next step: the identification of the active principle. This was undertaken by the biochemist, Dr. Stanley Cohen, who joined our laboratory in 1953. In the following year, he obtained a nucleoprotein fraction that was effective in the in vitro bioassay test, and shortly thereafter the active agent was identified as a pure protein, which was christened *nerve-growth promoting factor,* or NGF. The success story of NGF from here on is common knowledge among neurobiologists. The chapter of Greene & Shooter in this volume is the latest of many reviews and will bring the reader up-to-date.

CONCLUDING REMARKS

In the history of the problem of trophic interactions, a definite turning point occurs approximately in the middle of the century. Until then, the old paradigms taken from experimental embryology, although suitably adapted to the special demands of neurogenesis, prevailed. Concepts and experimental designs carried the stamp of causal-analytical thinking that dated back to Spemann and Harrison. A new era began with the discovery of NGF and the experiments leading up to it. The concept of trophic factors had obtained a specific meaning in molecular-biochemical terms. As a result, new questions can now be answered using new techniques that make the ones we used in the 1930s and 1940s look rather unsophisticated. Undoubtedly, other specific nerve growth factors will be discovered. However, one of the key issues in the field of trophic interactions—the primary physiological response of nerve cells to the message from the target area—is still unresolved.

ACKNOWLEDGMENTS

The research of the author was supported by the Rockefeller Foundation until 1953 and from then on by the National Institutes of Health. Further support was provided by a Center Grant of the Muscular Dystrophy Association to Washington University. I thank Drs. Florence Moog and Dale Purves for helpful suggestions and Ms. Doris Suits for typing the manuscript.

Literature Cited

All references in this essay will be found in the following reviews and articles:

Hamburger, V. 1934. The effects of wing bud extirpation on the development of the central nervous system in chick embryos. *J. Exp. Zool.* 68:449–94

Hamburger, V. 1975. Changing concepts in developmental neurobiology. *Perspect. Biol. Med.* 18:162–78

Hamburger, V. 1977. The developmental history of the motor neuron. The F. O. Schmitt Lecture in Neuroscience 1976. *Neurosci. Res. Program Bull.* 15:1–37 (Supp.)

Hamburger, V., Levi-Montalcini, R. 1949. Proliferation, differentiation, and degeneration in the spinal ganglia of the chick embryo under normal and experimental conditions. *J. Exp. Zool.* 111: 457–501

Levi-Montalcini, R. 1975. NGF: An uncharted route. In *Neurosciences: Paths of Discovery,* ed. F. G. Worden, J. P. Swazey, G. Adelman, pp. 245–65. Cambridge & London: MIT. 622 pp.

Ann. Rev. Neurosci. 1980. 3:279–302

THE GENERATION OF
NEUROMUSCULAR SPECIFICITY

❖11543

Lynn T. Landmesser

Biology Department, Yale University, New Haven, Connecticut 06520

Throughout an individual's life, the activation of a large number of skeletal muscles in the proper sequence and intensity is required for the essential activities of living. Considerable work has been devoted to understanding how the nervous system brings about the proper sequence of muscle activation. Implicit in all of these studies in the acceptance of the fact that specific motoneurons are connected in precise and highly reproducible patterns to certain muscles, sensory neurons, and interneurons.

Virtually all skeletal muscles appear to be innervated by anatomically discrete groupings of motoneurons or motoneuron pools that occur in characteristic locations. The motoneuron pool and muscle thus form a functional unit; the basic problem of myospecificity is to understand how the matching of motoneuron pool and appropriate muscle is brought about during development and, in cases in which selective re-innervation occurs, following regeneration. Further, one would like to understand how the nervous system selectively activates different motoneuron pools. Do motoneuron pool and muscle possess a distinct biochemical identity that allows the axons of the motoneurons to synapse with the appropriate muscle, while the motoneurons themselves receive synapses from appropriate sensory and interneurons? If biochemical individuality does not exist, what alternative mechanisms can explain the specific connectivity and functional activation pattern of muscles? In addressing these questions, this review focuses primarily on vertebrate limb systems from which most of the experimental studies and classical concepts have come. Due to space limitations it has not been possible to include everything of relevance, and the studies that are cited illustrate specific points directly related to myospecificity.

279

0147-006X/80/0301-0279$01.00

DEVELOPMENTAL STUDIES

The Origin of Muscles and Motoneuron Pools

Most studies on muscle development have focused on cellular differentiation of the muscle fiber (for review see Fambrough 1976); few have considered how whole muscles form. Most vertebrate muscles arise as condensations of mesenchyme cells that differentiate into myoblasts and then fuse to form myotubes (Hilfer et al 1973). In some cases, individual condensations may give rise to individual muscles (Adelman 1927). However, the muscles of the tetrapod limb arise secondarily by a process of cleavage from two primary muscle masses, one dorsal and one ventral (Romer 1927, Milaire 1965, Dunlap 1966, Shellswell 1977).

In the chick limb, cleavage occurs following the formation of at least some myotubes, which can be functionally activated by motoneuron stimulation (Landmesser & Morris 1975, Landmesser 1978b). Little is known about the mechanism of cleavage, but it occurs in the absence of nerves (Shellswell 1977) and therefore does not require muscle contraction. If it is simply a physical process that results from differential growth of skeletal elements as proposed by Horder (1978), muscles could lack biochemical identities. Alternatively, muscles could develop distinctive properties based on their position in accord with positional information models of limb development (Wolpert et al 1974, Shellswell & Wolpert 1977). These would include differences in surface membrane molecules that alter adhesivity and result in muscle cleavage. Such differences could also be used by neurons in recognizing specific muscles. Although there is some evidence against a purely physical model (Shellswell & Wolpert 1977), this important process needs further study.

The most detailed knowledge of motoneuron development comes from the limb-moving segments of the amphibian and avian spinal cord. In the bird, lumbosacral motoneurons are generated from the ependymal layer of the basal plate (Wenger 1950) between Stage (St) 17–23, or day 3–4 of a 21 day incubation period (Hollyday & Hamburger 1977). An anterior to posterior gradient in the time of motoneuron origin exists; however, there is considerable overlap within the lumbosacral cord, most motoneurons undergoing their last division within a 20 hr period (Hollyday & Hamburger 1977). At any segmental level, motoneurons that occupy a medial position appear for the most part to be generated before those in a lateral position. The generation and differentiation of the large numbers of motoneurons characteristic of limb-moving segments appear to be intrinsic properties of the cord at that level, and are independent of the limb (Hamburger 1958), sensory input (Wenger 1951), or descending input (Hamburger 1946).

The motoneurons innervating each muscle are situated in coherent, elongate groups in highly characteristic positions, both in the longitudinal and transverse axes of the spinal cord (Landmesser 1978a), as is the case for the frog (Cruce 1974), cat (Romanes 1964), and human (Sharrard 1955). This is experimentally useful because it allows developing motoneurons to be characterized by something other than their ultimate connectivity. One can therefore determine how a given set of motoneurons will behave under various experimental conditions. As is shown later, the lack of detailed motoneuron pool topography makes many classical embryological studies difficult to interpret.

The process of axon initiation is relatively rapid and synchronous and by St 23–24, all eight spinal nerves have reached the base of limb and are beginning to form the crural and ischiadic plexuses characteristic of the hind limb (Fouvet 1973, Lance-Jones & Landmesser 1979, and unpublished observations). Axons invade the limb bud and begin to form functional connections at St 27 (Landmesser & Morris 1975, Landmesser 1978b) prior to muscle cleavage. At this time motoneurons become dependent on their peripheral target (Hamburger 1958) and during a period of naturally occurring cell death (St 29–35), their number is reduced by more than half (Hamburger 1975, Chu-Wang & Oppenheim 1978). There is no direct evidence for sensory connections before St 34 although firm data are lacking about the time at which specific central connections are formed. Finally, shortly before hatching there is a transient period of polyneuronal innervation (Bennett & Pettigrew 1974) similar to that described for mammalian muscles (Brown et al 1976). Because the elimination of polyneuronal innervation occurs after specific connections have formed (Landmesser 1978b), it is presumably not related to the matching of muscles and motoneuron pools and is not discussed further.

A similar sequence of events has been found during limb innervation in *Xenopus* spinal cord, largely by Prestige and his co-workers (Prestige 1967, Prestige 1973, Prestige & Wilson 1974, Lamb 1976), and in avian intraocular (Landmesser & Pilar 1972, 1976) and extraocular muscles (Sohal et al 1978). Although less is known about mammalian development (Romanes 1941, Harris-Flanagan 1969, Nornes & Das 1974), there is no reason to suspect major differences (the reader is referred to Hughes 1968 for a good comparative review of the older literature).

The Matching of Motoneurons and Muscles

Over a period of years numerous studies have attempted to determine how developing motoneurons behave when they encounter an altered environment as the result of embryonic surgery (e.g. Detweiler 1930, Weiss 1937, Hamburger 1939, Narayanan 1964). Such studies are necessary in defining the principal mechanisms used by the nervous system to achieve specific connectivity and are required before a rational analysis of the mechanisms on a molecular level can proceed.

However, most of the earlier studies did not determine precisely which motoneurons innervate a given muscle, either normally or following some experimental manipulation. Because each spinal nerve contains motoneurons for a number of different muscles and because the spinal nerves themselves are joined into a plexus at the base of the limb, virtually nothing can be gleaned from the anatomical nerve branching patterns. Thus the results of many studies based on such data are not interpretable. Fortunately, it has recently become possible to combine precise definition of motor projection patterns using electrophysiological or specific anatomical tracing techniques with some of the previous embryonic manipulations. A major technical advance has been the use of horseradish peroxidase (HRP) to retrogradely label motoneurons (Lamb 1976, Hollyday et al 1977, Landmesser 1978a). It has therefore become possible to determine where defined groups of motoneurons project during different stages of development and following experimental alterations, which leads to a significant increase in our understanding of how motor projections develop.

A number of different hypotheses have been proposed and several of these can now be excluded or at least rendered unlikely by experimental tests. At one extreme is the assumption that the motoneuron initially lacks an identity and is somehow labeled by whatever muscle with which it comes into contact so that appropriate central connections will form. This concept of myotypic specification was proposed by Weiss (1937) to explain how supernumerary axolotl limbs were able to move in synchrony with the host limb (homologous response), although apparently innervated by only part of the normal nerve supply. Homologous response has also been observed in *Xenopus* (Hollyday & Mendell 1976). This hypothesis has the advantage of not requiring specific recognition between motoneuron and muscle, or any special means of guiding growing axons to the appropriate targets. Motoneurons could simply invade the limb in a random or haphazard fashion.

However, a more likely explanation of Weiss' results is that the muscles in the supernumerary limb are actually innervated by appropriate motoneurons. Due to the extensive overlap of motoneuron pools in the amphibian cord (Székely & Czéh 1967, Cruce 1974), each spinal nerve probably contains motoneurons appropriate for most muscles. Further, in lower vertebrates motoneurons have a strong tendency to selectively reinnervate their appropriate muscles, even following surgical deflection of nerves (Mark 1965, Sperry & Arora 1965, Grimm 1971, Cass & Mark 1975). Yet, it is worth pointing out that a precise definition of the motoneurons innervating axolotl supernumerary limb muscles has not been made.

Nonetheless, the high degree of specificity between discrete motoneuron pools and muscles indicates that motoneurons do not innervate the limb in a haphazard or random fashion (Romanes 1964, Cruce 1974, Landmesser 1978a,b). Furthermore, following successful cross-union of motor nerves, there are no detectable alterations in their central connectivity (Mendell & Scott 1975, Mark & Marotte 1972, Scott 1977, Kleinebeckel 1978, but see Cohen 1978). These observations are often taken to refute the idea of myotypic specification, but it is still possible that the muscle influences the type of central connections formed onto motoneurons during initial development. This idea could easily be tested following the anterior-posterior reversal of chick limb buds (B. Ferguson, unpublished observations, Stirling & Summerbell 1979) or the addition of supernumerary limbs (Hollyday et al 1977, Morris 1978). In these situations motoneurons innervate muscles with which they would not normally connect. Characterization of the central connectivity of such motoneurons should answer this question definitively.

Studies of normal development of the chick hind limb have led to the exclusion of another hypothesis; that of diffuse or random outgrowth of motoneurons, followed by the death of those that fail to connect with appropriate muscles (Landmesser & Morris 1975, Landmesser 1978a,b). As early as St 28 (5 days) and prior to the period of motoneuron cell death (Hamburger 1975, Chu-Wang & Oppenheim 1978), muscle nerves can be shown by electrophysiological evidence to contain axons almost entirely from appropriate spinal nerves (Landmesser & Morris 1975). At most, a few percent of the axons may be inappropriate (Landmesser 1978b). In addition, retrograde labeling of motoneurons with HRP injections into specific limb regions did not reveal any projection errors. In other words, motoneurons projected only to regions of the primary muscle mass that would give rise to their adult muscles (Landmesser 1978a,b). Minor errors, small numbers of motoneurons erring only slightly in their termination sites, may have been missed. It is nevertheless clear that the process of cell death does not create a specific pattern out of an initially diffuse one (but see Pettigrew et al 1979).

However, the experiments just described could have missed very transient errors, occurring between St 24 (4 days), when axons first reach the limb base (Fouvet 1973, Lance-Jones & Landmesser 1979, and unpublished observations), and St 27 (5 days), when observations were begun. Early projection errors have been described by Lamb (1976) in *Xenopus* hind limb, using the HRP retrograde labeling technique. More recently, the application of an orthograde labeling technique to the chick lumbosacral cord has allowed axonal trajectories to be followed from the time of initial outgrowth (Lance-Jones & Landmesser 1979). Following injection of HRP into the ventral horn of specific spinal cord segments, large numbers of motoneurons are diffusely stained so that their axons can be traced into the limb. From the earliest time, axons from a given segment grow as far as the base of the limb, with little deviation along the anterior-posterior (a-p) axis, thus maintaining topographical order with respect to axons from other spinal segments. At this point, axons destined for different muscles appear to diverge in discrete fascicles, often crossing over axons destined for other muscles. This process results in plexus formation. At no time were motoneurons found to send axons into inappropriate nerve trunks or muscle nerves (Lance-Jones & Landmesser 1979 and unpublished observations). Thus, initial axonal outgrowth appears to be highly selective.

Further, the position of a motoneuron cell body was found not to be correlated with its function or with the topographical position of its mature muscle, but with its axonal termination site in the embryonic limb bud (Landmesser 1978a). Medially situated motoneurons projected only to the ventral muscle mass; lateral motoneurons to the dorsal muscle mass. There is also a general correspondence within each primary muscle mass between the antero-posterior position of a motoneuron in the cord and its axonal termination site in the limb. Similar results have been found for the amphibian cord (Cruce 1974, Lamb 1976, 1977) and it has been proposed that some quality of the motoneuron, tightly coupled with its position, may be instrumental in achieving the specific connectivity pattern.

The early specificity of connections in the chick suggests first that motoneurons are specified, or possess an identity, prior to their innervation of the limb; secondly, that they may be able to use environmental cues to reach their target muscle with little error (Landmesser 1978b). However, other passive mechanisms that have been proposed to explain the specific innervation pattern cannot be excluded. These do not require that motoneurons actively respond to environmental cues during outgrowth of their axons, or that they recognize specific muscles once they have reached them.

Passive mechanisms based on spatio-temporal factors may fully or partially explain the specific connectivity patterns observed in the arthropod eye (Anderson 1978, Macagno 1978). Jacobson (1970) proposed a timed outgrowth mechanism for the innervation of the vertebrate limb. He postulated that axons from motoneurons in the most anterior cord segments would invade the limb first and synapse in a nonspecific fashion with the most proximal muscles. Later-arriving axons from more posterior segments would, upon finding this territory taken, be constrained to synapse with the next most distal muscles and so on. Such a model is not supported, however, by the detailed connectivity pattern of the chick hind limb because some proximal muscles are innervated by the most caudal segments (Landmesser & Morris 1975, Landmesser 1978a). Further, although there is a slight antero-posterior (a-p) gradient in time of motoneuron origin (Hollyday & Hamburger 1977), axon outgrowth is relatively synchronous, so that all spinal nerves reach the base of the limb at about the same time (Fouvet 1973, Lance-Jones & Landmesser 1979, and unpublished observations). An a-p gradient in timing of limb innervation has also not been observed in *Xenopus* lumbo sacral cord (Lamb 1976). Finally, recent experiments have conclusively excluded such a model (Lance-Jones & Landmesser 1978, 1979): When anterior spinal cord segments were removed from chick embryos at St 15–16, the proximal muscles, which would have received innervation from these segments, remained uninnervated throughout the developmental period. Axons from the remaining motoneurons bypassed these muscles, synapsing only with their normal targets.

Clearly, a simple, timed outgrowth mechanism cannot be responsible for the observed specific innervation pattern. However, Horder (1978) has recently proposed a more refined spatio-temporal hypothesis. This suggests that motoneuron axons retain the same topographical relationships with each other from their exit from the spinal cord. As they grow into the limb, their divergence into nerves is controlled largely by nonspecific mechanical factors associated with limb morphogenesis. Axons are channeled into certain nerves only because they occupy a certain position in space and time, not because they possess any real identity. This model is appealing because it is economical and does not require complicated labeling of motoneurons and muscles. It also explains why a grossly normal nerve pattern can be produced when lumbosacral spinal nerves enter the limb from abnormal positions following either addition of a supernumerary limb (Morris 1978) or a-p limb reversals (Narayanan 1964, Stirling & Summerbell 1979). However, this model does not adequately account for how axons from motoneuron pools that overlap extensively along the a-p axis of the cord are able to reach their appropriate muscles. Due to this overlap, ventral root filaments at each a-p level contain a number of different species of axons (Landmesser 1978a and unpublished observations). Yet by the time the base of the limb is reached, axons have sorted out so that those destined for anterior muscles are situated in the anterior portion of the plexus and those destined for posterior muscles are in the posterior portion (Lance-Jones & Landmesser 1979 and unpublished observations). This necessitates that axons alter their topographical relationships to one another, and it is difficult to see how this would come about by purely mechanical factors.

A direct test of whether axons are passively guided or respond actively to positional cues has recently been performed. Lance-Jones & Landmesser (1978, 1979) rotated portions of the chick lumbosacral spinal cord about the a-p axis at St 15–16. They found, using electrophysiological as well as orthograde and retrograde HRP labeling, that such motoneurons displaced along the a-p axis were able to project to their appropriate muscles. The gross anatomy of the plexus and nerve-branching pattern within the limb were normal. Nonetheless, the tracing of labeled axons revealed that they had altered their course within the plexus and major nerve trunks to reach their original muscles. Clearly, observations of grossly normal plexus and nerve-branching pattern made in many earlier studies are of little value in determining the projection sites of motoneurons.

Earlier studies had revealed intrinsic differences between thoracic, brachial, and lumbosacral levels of the spinal cord (Wenger 1950, Székely 1963, Narayanan & Hamburger 1971). However, this study is the first to experimentally demonstrate that lumbosacral motoneurons are specified with respect to their peripheral targets prior to axon outgrowth and, in fact, prior to their cell birthdates. This study additionally rules out simple mechanical or contact guidance and shows that axons can actively respond to environmental cues by altering their direction of growth to reach appropriate muscles.

A similar early specification of motoneurons in the medio-lateral axis of the cord is also suggested. Following a-p limb reversals (B. Ferguson, personal communication) or the addition of a supernumerary limb (Hollyday et al 1977, Hollyday 1978, Morris 1978), medial motoneurons always project to muscles derived from the ventral muscle mass and lateral motoneurons to muscles derived from the dorsal muscle mass. This occurs even though muscles are now innervated by inappropriate cord segments. The possibility of early specification was tested by rotation of the limb about the dorso-ventral (d-v) axis (Ferguson 1978). Medial and lateral motoneurons were still found to reach their appropriate muscles and did so by altering their pathway between the spinal cord and the limb. Again the nerve branching pattern within the limb itself was not altered.

It seems that whereas the sorting out of axons in the region of the plexus and limb base may be an active process based on their identity, the branching pattern within the limb may be at least a partially passive process directed by morphogenetic events as proposed by Horder (1978). When limbs are rotated about the a-p (Narayanan 1964, Stirling & Summerbell 1979, B. Ferguson, personal communication) or d-v (Ferguson 1978) axis; or when deletions, truncations (Stirling & Summerbell 1977), or duplications (Stirling & Summerbell 1977, Lewis 1978) of limb parts are experimentally produced, the nerve branches are appropriate to the specific region of limb, even though they may now contain inappropriate axons. Stirling & Summerbell (1979) have suggested that axons may be passively channeled into certain nerves, based only on their position in the plexus. It is, however, through an active process that axons achieve this position in the first place (Lance-Jones & Landmesser 1978, 1979, Ferguson 1978, and unpublished observations).

There is only one line of evidence that is difficult to reconcile with early specification of motoneurons followed by an active sorting out of their axons; this derives from experiments with a-p limb reversals (Morris 1978, Stirling & Summerbell 1979, B. Ferguson unpublished observations) or the addition of supernumerary limbs (Hollyday et al 1977, Hollyday 1978, Morris 1978), in which muscles are innervated by segmentally inappropriate motoneurons. Why have motoneurons been unable to project to their appropriate muscles in these cases although they do so following a-p cord reversals?

Several explanations come to mind. One possibility is that axial cues from the limb could respecify motoneurons in the a-p axis (Morris 1978). However, this suggestion is difficult to reconcile with the cord reversal experiments in which respecification does not occur (Lance-Jones & Landmesser 1978, 1979). The cord reversals are done at earlier developmental stages (St 15–16) than either the limb rotations or additions (St 17–20). Thus, if the limb were capable of respecifying motoneurons, it should do so in this case as well.

Alternatively, axons might sort out in the a-p axis on the basis of axial cues, not from the limb, but contained in the region between the spinal cord and limb base. The relationship between motoneurons and such cues would not be altered following an a-p limb rotation, but would be altered following a-p spinal cord rotation. In the limb rotation case, this would result in axons occupying a position in the plexus appropriate for the old but not the new limb orientation. Such axons could then be channeled to inappropriate muscles by passive mechanisms discussed above.

Yet another possibility is that axial cues may still be supplied by the limb, but axons may simply be limited in the distance that they can be displaced along the a-p axis and still reach their appropriate muscles. In fact, in the cord reversal experiments in which motoneurons managed to reach their appropriate muscles, only three to four of the eight lumbosacral segments were rotated (Lance-Jones & Landmesser 1978, 1979). In a-p limb reversals and in most supernumerary limb experiments, the displacement of a motoneuron from its original position relative to the limb is much greater. This limit may be imposed by the inability of an axon to alter its pathway sufficiently within the space and time constraints of the developing system. Alternatively, it could result from the inability of an axon to properly respond to distant or inappropriate environmental cues other than those it would normally encounter. Passive mechanisms would then channel axons to inappropriate muscles.

Considerations of this sort could explain why axons are able to reach their appropriate targets following d-v but not a-p limb rotations. In the former case, medial and lateral motoneurons at each segmental level reach the base of the limb at the midpoint of the dorso-ventral limb axis, as they would normally. They are thus neither in a foreign environment, nor must they greatly alter their path to achieve an appropriate projection. Whether the axial cues are supplied by the limb or by the embryonic axis, axons in both cases sort out within the plexus region at the base of the limb, prior to the formation of muscle nerves and contact with their appropriate muscles.

A somewhat different explanation has been advanced by Hollyday et al (1977) based on their supernumerary limb studies. They suggest a hierarchy of neuronal specificities, such that when a motoneuron is unable to project to its first choice, it makes a second choice based on some broader category, such as flexor vs extensor or dorsal vs ventral muscle mass etc. Whether there is a hierarchy of specificities or a hierarchy of mechanisms, some active and some passive, cannot yet be decided. However, such questions can be answered with present techniques by careful definition of motoneuron pool topography and axon trajectories following various experimental alterations: a-p and d-v limb reversals, the addition of supernumerary limbs, a-p cord reversals of different sizes, and the creation of deletions or duplications of limb parts. Experiments of this design are currently in progress in several laboratories.

Developmental Errors and their Relation to Myospecificity

It seems inevitable that some errors in initial connectivity would occur, and this is supported by several studies (Lamb 1976, McGrath & Bennett 1979, Pettigrew et al 1979). Of greater interest than the occurrence of errors per se, is an understanding of their magnitude (percentage of motoneurons projecting correctly and incorrectly) and their mechanism of removal. If only small numbers of errors are made, error removal would be a way of refining specific connectivity patterns produced by other mechanisms, such as axonal sorting out and cellular recognition. Alternatively, error removal following a period of random axon outgrowth and synapse formation could be the primary mechanism for matching motoneurons with appropriate muscles.

In the most quantitative of the above studies, Lamb (1976) showed that the earliest motor projections to *Xenopus* hind limb were not random; there were, however, systematic errors in the projection pattern to two thigh regions. These inappropriate projections were removed by death of the motoneurons (Lamb 1977) at the onset of the normal cell death period (Prestige 1967). These errors consisted of a relatively small number of motoneurons that were the earliest to send axons into the limb. The majority of later-projecting neurons to these regions, as well as initial projections to more distal limb regions, were correct from the onset. However, the very existence of such errors and their subsequent removal suggests that, in addition to some means for getting axons to the appropriate muscle, there must be an additional mechanism based on specific recognition between axon and muscle.

More recently, errors have been described in both developing axolotl hind limb (McGrath & Bennett 1979) and chick wing (Pettigrew et al 1979). These authors have proposed that a process of competition, similar to that hypothesized to explain selective reinnervation in lower vertebrates (Marotte & Mark 1970a,b), may be of major importance in the establishment of initial connections. Thus, following a period of somewhat random axon outgrowth and synapse formation, competition between correct and incorrect axons would ensue (see the section on reinnervation), resulting in the regression of the wrong connections. It is not possible to determine the percentage of inappropriately projecting motoneurons from these studies. However, the results do appear to be in conflict with the data on the chick hind limb (Landmesser 1978b, Lance-Jones & Landmesser 1979, and unpublished observations), and the reason for this discrepancy is not yet obvious. Clearly competition is not necessary to achieve specific connections. In the chick (Lance-Jones & Landmesser 1978, and unpublished observations), muscles that have been deprived of their correct innervation by partial deletions of the neural tube remain uninnervated even though adjacent muscles are innervated by the remaining cord segments. These cord segments cannot then normally be prevented from synapsing with the muscle in question by competition with the correct innervation. An even more striking example of this has been made recently by Lamb (1979). In this case the removal of projection errors to flexor muscles occurs, even though these muscles are deprived of their correct innervation by early removal of the anterior portion of the spinal cord.

These results suggest that a specific recognition process between axon and muscle cell may take place once axons have reached the muscle. Compatible with this idea is ·the observation that the axons of chick lumbosacral motoneurons only ramify and make synapses within portions of the primary muscle mass generally corresponding to their adult projections (Landmesser 1978b). However, this study was unable to rule out the possibility that some motoneurons could err by slightly projecting over presumptive muscle boundaries to foreign regions. These errors could be removed, not necessarily by an active recognition process, but simply by the physical process of muscle cleavage (Landmesser 1978b). Cleavage planes could interrupt inappropriately projecting axons from their somas, resulting in degeneration. The removal of errors in the amphibian hind limb (Lamb 1976) and chick wing (Pettigrew et al 1979) is temporally correlated with muscle cleavage (Sullivan 1962, Dunlap 1966). Therefore, it will be important to determine if such errors are due to a failure of axonal sorting at the limb base, or to axons sprouting over premuscle boundaries. If the latter is true, careful analysis of the boundary regions will be necessary to determine the mechanism of error removal.

Finally, Harris & Dennis (1977) have described possible connectivity errors in embryonic rat intercostal muscle. These errors result because axons from one cord segment cross over to innervate intercostal muscles usually innervated by neighboring cord segments. The observed reduction in such projections in late embryonic development (Harris & Dennis 1977) has recently been found to coincide with the normal reduction of polyneuronal innervation (M. J. Dennis, unpublished observations; for a discussion of the phenomenon of transient polyneuronal innervation see Jansen et al 1978). Since both are reduced to a similar extent, these projections may not represent errors in the strict sense of the word (M. J. Dennis, personal communication). However, errors may occur between the projections to intercostal and extracostal muscles, a possibility under current investigation (M. J. Dennis, personal communication).

In summary, all existing evidence is at least compatible with an early specification of motoneurons, based on their position in the neural tube. Axons reach appropriate muscles, first by an active process of sorting out at the base of the limb, which involves recognition of environmental cues and possibly interaction between axons as well. The actual channeling of axons into muscle nerves may be a passive process. A subsequent, specific recognition step between axon and muscle is probable but not yet proved by available data.

The Formation of Appropriate Central Connections

It is unfortunate that virtually nothing is known about the mechanisms governing the development of specific central connections. It is through these connections that motoneuron pools can be selectively activated, resulting in coordinated behavior (Grillner 1975, 1976).

Within a single hind limb of the chick embryo, Bekoff (1976) was able to show that the central connections responsible for alternation of flexor and extensor muscles during spontaneous activity develop very early, by St 30 (7 days), shortly after motoneurons have established peripheral connections. It was suggested above that some aspect of the motoneuron, tightly correlated with its position, is probably important in the development of its peripheral connections. However, because the functional activation of motoneurons (Engberg & Lundberg 1969, Edgerton et al 1976, Rasmussen et al 1978, Landmesser 1978a), and therefore their aggregate central connectivity does not follow this relationship (Romanes 1964, Landmesser 1978a), some other mechanism(s), not based simply on the position of the motoneurons in the spinal cord, must ensure proper central connectivity (Landmesser 1978a). At this time it is not possible to exclude either functional activity or some influence from the periphery, similar to myotypic specification, as playing a role.

EVIDENCE DERIVED FROM REINNERVATION EXPERIMENTS

Whole Muscle Matching

If muscles possess a biochemical identity at some stage during ontogeny, then this identity could be retained in the mature animal and be detected by properly designed reinnervation experiments. These would be technically simpler to perform and easier to interpret than studies of the initial formation of neuromuscular connections. However, the design of many of these experiments makes it difficult to relate the findings to the nature of myospecificity and especially to the mechanisms by which specific connections are initially achieved.

An overwhelming amount of evidence shows that in higher vertebrates, especially mammals, specific reinnervation does not occur following nerve transection and various experimental manipulations (for reviews see Sperry 1945, Fambrough 1976, Purves 1976). These experiments are generally of two types. In the first, exemplified by that of Bernstein & Guth (1961), a nerve trunk is transected proximal to the divergence of two muscle nerves and, following regeneration, it is determined whether axons have specifically reconnected with their original muscle. This experimental design requires that one be capable of distinguishing between the two species of motoneurons; in this case, use was made of the fact that spinal nerves L_4 and L_5 normally contribute different proportions to the total tension generated by each muscle. Following reinnervation, L_4 and L_5 contributed equally to the tension of both muscles, and a reasonable interpretation was that selective reinnervation did not occur. However, other explanations for these results are possible, and it was never clearly shown that axons that had previously connected with one muscle were now innervating the opposite muscle.

Nonetheless, the poor recovery of function following transection of mixed nerves (Sperry 1945), together with experiments similar to that just described, suggests that in higher vertebrates, including frogs (Hoh 1971), regenerating motoneurons are not capable of selecting the correct muscle nerve. This is not surprising when one considers that the decision must be made by the growing axon some distance (often centimeters) from the muscle and that complete degeneration of the distal axonal stump occurs precluding specific reconnection, as occurs in some invertebrates. Furthermore, possibly because they elongate better on more adhesive substrates (Letourneau 1975), axons have a propensity to follow mechanical guides nonspecifically (Weiss & Hoag 1946). Therefore, unless one imagines the unlikely possibility that glial and supporting cells in muscle nerves possessed a specific biochemical identity, there is no reason to expect regenerating motoneurons to select the proper muscle nerve and consequently find their appropriate muscle under these experimental conditions.

Additional observations following surgical transposition of muscles or nerves show that vertebrate muscles are readily reinnervated by a variety of foreign axons (see Purves 1976 for a review). Myospecificity, if it exists, is not absolute, and one comes to the conclusion that any cholinergic axon that contacts a muscle is capable of forming at least some synapses. Yet in all of these cases, the individual neuron is not given a choice between the proper and foreign target, but rather between synapsing with the wrong target or with nothing at all. A strong tendency for a neuron to form synapses could possibly override any constraints imposed by specificity.

This same problem recurs in experiments in which both a foreign and correct nerve are allowed to simultaneously compete for innervation of the same muscle. Recent electrophysiological evidence has convincingly demonstrated that the foreign nerve is capable of forming synapses under such conditions (Frank et al 1975, Frank & Jansen 1976, Scott 1977) and that foreign and appropriate synapses can persist for long periods of time on the same muscle fiber. In this situation, the neuron is once again given no choice and the muscle fiber itself does not appear to possess any mechanism to reject the foreign synapses.

In apparent contrast to these results are a number of observations, beginning with those of Sperry (Sperry & Arora 1965) and of Mark (1965), that there is a strong tendency for fish and urodele muscle (Grimm 1971, Cass & Mark 1975) to be specifically reinnervated even following attempts at surgical cross-innervation. Mark extended these observations with experiments on goldfish extraocular muscle. Based mostly on behavioral observations, he concluded that following reinnervation by the correct nerve (NIV), the foreign nerve (NIII) was no longer capable of causing activation of the superior oblique muscle with which it had previously established functional connections (Marotte & Mark 1970a,b). Mark hypothesized that due to competitive interactions, the foreign nerve terminals had become *repressed*, although they remained morphologically intact. The latter, very novel, suggestion was based on the rapidity with which the foreign repressed nerve reinnervated the muscle following a second transection of the appropriate nerve, as well as by failure to observe degenerating synapses during the period of repression. Competition with ultimate repression of foreign inputs was proposed as a general mechanism to explain selective reinnervation of

multiply innervated lower vertebrate muscles, in contrast to focally innervated mammalian muscles in which selectivity is not found.

However, in an extension of the investigation of competition in the fish
extraocular system, Scott (1977), utilizing electrophysiological techniques,
found that appropriate and foreign synapses could persist on single muscle
fibers indefinitely. Thus displacement or regression of foreign innervation
was not an inevitable process, even on the multiply innervated muscles of
lower vertebrates. (See also Frank & Jansen 1976).

Nonetheless, additional observations made on *Xenopus* extraocular muscle (Fangbonner & Vanable 1974) and urodele limb muscle (Cass & Mark
1975, Bennett & Raftos 1977, Genat & Mark 1977) indicate some form of
competition that ultimately favors the appropriate nerve. The question to
be answered: What is the mechanism of this competition? Does it involve
a specific recognition process that distinguishes between appropriate and
foreign nerves? An explanation, consistent with this last group of experiments, suggests that it need not. For in all of these cases (see also Guth &
Bernstein 1961), the foreign target was innervated by sprouts from axons
that were already in contact with their normal peripheral target. Such
sprouts may be inherently less capable of maintaining synapses in the face
of competition, because their parent cell body has overexpanded its peripheral territory. Similar views have been advanced by others (Jansen et al
1978, Purves & Lichtman 1978). In fact, Bennett & Raftos (1977) observed
that the foreign synapses on salamander limb muscle took longer to mature
with respect to size and quantal content than did synapses from the original
nerve, and attributed this to mismatching. An equally plausible explanation
is that these synapses were being formed by motoneurons that had already
extended their peripheral field beyond that which they could efficiently
maintain. That motoneurons have limits to the size of their peripheral field
has been shown by partial denervation experiments (Thompson & Jansen
1977). Additional observations have recently been made that are consistent
with the idea that motoneurons with an enlarged peripheral field are at a
disadvantage in competitive situations (Bixby & Van Essen 1978, Brown &
Ironton 1978).

However, using a different experimental design, Dennis & Yip (1978)
have shown regression of foreign urodele limb innervation following regeneration of the correct nerve. In this case the foreign synapses were formed
by axons that had been surgically transposed to the foreign muscle following excision of their own muscle. After reinnervation by the correct nerve,
the foreign nerve lost its ability to evoke synaptic potentials in two thirds
of the muscle fibers and the size of such potentials was reduced in the
remaining fibers. This reduction in efficacy was due to a reduction in quantal
content and was therefore presynaptic in nature. Because 94% of the nerve

terminals could be labeled above background levels with horseradish perox-idase following stimulation of the original nerve, and because the foreign innervation did not rapidly return following a second section of the original nerve, the authors concluded that a physical retraction of the nerve endings had probably occurred. There is in fact no convincing evidence for function-ally repressed but morphologically normal nerve terminals as originally proposed by Mark. In fact, because the massive loss of synapses during developmental regression of polyneuronal innervation occurs without any evidence of degeneration at the ultrastructural level (Korneliussen & Jansen 1976), lack of degeneration as originally observed by Mark cannot be used in support of this hypothesis (see also Scott 1977).

Whereas the results of Dennis & Yip (1978) are not easily explained by the overextension hypothesis, they did note that the excised foreign muscle regenerated in some cases. This was also found by Scott in the goldfish extraocular system (1977) and may partly explain the discrepancy between her experiments and those of Mark. In future experiments of this sort it will be important to determine whether the foreign nerve has managed to rein-nervate its original or any additional target. The fact that foreign synapses do not spontaneously regress, but do so only when the original nerve regrows (Dennis & Yip 1978), may merely indicate that axons are able to sustain an expanded periphery only in the absence of competition.

Another explanation, not based on specific recognition, exists for compe-tition. This proposes that the synapses formed by one nerve (not necessarily the foreign) are inherently less stable because they are smaller or less mature. Alternatively, one nerve may not have sufficient time to adequately establish itself so that it can compete effectively with or even prevent the opposite nerve from forming synapses. Slack put forth this explanation to reconcile the conflicting observations on fish extraocular muscle. He was able to prevent reinnervation of urodele limb muscle by the correct nerve by delaying its return and allowing the foreign nerve more time to establish itself (Slack 1978). This cannot account for Scott's findings, however, for in some experiments in which foreign and correct nerves were simulta-neously sutured to the muscle, the foreign nerve established synapses. Fur-ther, it prevented the correct nerve from innervating as much of the muscle as it would have following simple reinnervation (Scott 1977). However, this does point to another variable that must be controlled.

Similar competitive interactions have even been observed between foreign nerves (Grinnell et al 1979). When two foreign nerves are allowed to inner-vate a surgically transposed frog muscle, either simultaneously or in a staggered fashion, each is able to take over a portion of the muscle and prevent or cause regression of synapses by the opposite nerve. This study suggests that competition need not involve a recognition of foreignness and is in fact compatible with the overextension hypothesis. In this case the first foreign nerve to establish synapses would be at a quantitative competitive disadvantage when the second nerve began to innervate the muscle (see also Bixby & Van Essen 1978).

In summary, there is good evidence for competition between neuronal inputs (Marotte & Mark 1970a,b, Cass & Mark 1975, Bennett & Raftos 1977, Genat & Mark 1977, Scott 1977, Grinnell et al 1979), but only one study (Dennis & Yip 1978) suggests that competition involves the selection of the correct input by a mechanism that distinguishes synapses on the basis of their "appropriate" or "foreign" nature. In view of the importance of this mechanism, both to the development of specific neuromuscular connections and to the concept of myospecificity, additional experiments that adequately control for a number of variables are desirable.

First, it is necessary to be certain that a given nerve supply is actually foreign, especially when using limb muscles. Because motoneuron pools usually extend over several segments and vary somewhat in position in different individuals, attaching undue significance to spinal nerve source is unwise. If a spinal nerve makes even a minor contribution to a given muscle, possibly not detected if sampling is insufficient, these few axons could sprout and take over the muscle following transection of the major nerve input.

A more important problem arises when we try to precisely define foreign. For example, in the salamander limb, axons from nerve 16, which normally innervate only a small region of the iliotibialis muscle, sprout to occupy the whole muscle following transection of nerve 17 (Slack 1978). Should these nerve 16 axons be considered foreign because they normally are restricted to a different part of the muscle? Differences in segmental innervation can occur within a single muscle (Burke et al 1977, Genat & Mark 1977, Lance-Jones and Landmesser, unpublished observations) and this may result from the same mechanism that produces the specific innervation of a whole muscle. Alternatively, there may be different levels of specificity involving different mechanisms; some may be based on specific recognition, others on nonspecific spatio-temporal factors.

Second, in competitive innervation experiments, it will be necessary to rule out alternative mechanisms not requiring specific recognition. In considering the validity of the overextension hypothesis, the number and size of motor units in each muscle, both prior to and following any experimental manipulation must be known. It will be especially important to insure that the foreign motoneurons have not sprouted to reinnervate their original targets. Therefore, the point of reference needs to be a single motoneuron and its total peripheral innervation field, rather than an axon or nerve ending in isolation.

Finally, the difficulty in relating many muscle reinnervation studies to development is that in the former case the neuron is rarely given a choice of targets. Most developmental models are based on an axon optimizing its position, either with respect to other axons or to some environmental markers (Prestige & Willshaw 1975, Willshaw & Von der Marlburg 1976). If in vivo axonal outgrowth is similar to that in vitro, during development a neuron would presumably send out a number of collaterals, some of which would be stabilized by adhesive (Letourneau 1975), chemotrophic (Campenot 1977), chemotactic (Letourneau 1978), or other factors. Given an upper limit to the total arborization that a neuron can maintain, less optimally situated collaterals would be retracted. Decisions could be made continuously or at a number of points prior to actual contact with the muscle membrane. In reinnervation studies, the environment encountered by axons is so radically different from that during their initial outgrowth during embryogenesis that fruitful comparisons are difficult. Once an axon reaches a target, it is unlikely that it (or a collateral from the same neuron) will be able to simultaneously interact with both more and less appropriate targets.

Considerations of this sort may reconcile the selective reinnervation seen in some lower vertebrates with the lack of selectivity seen in mammals. In fish and urodeles, the extensive sprouting following nerve transection (Mark 1965, Sperry & Arora 1965, Grimm 1971, see also Hoh 1971) would ensure that at least some collaterals of a motoneuron would reach the appropriate target. As these contacts become stabilized and expanded, less optimally situated collaterals could regress. This latter process could also be intensified by competition. In cases in which sprouting is less extensive or successful cross-innervation has been brought about, competition would result only from the various nonspecific factors already discussed. Indeed, in autonomic ganglia where selective reinnervation occurs (Guth & Bernstein 1961, Landmesser & Pilar 1970, Nja & Purves 1977), species of ganglion cells are intermingled allowing regenerating preganglionic fibers of a choice between correct and incorrect targets.

In summary, in cases where selective reinnervation of muscles has been carefully documented, the mechanism is not yet fully understood. On the other hand, the lack of selectivity shown in many other experiments tells us little about myospecificity or how initial connections can develop.

Fiber Type Matching

In addition to matching between motoneuron pool and whole muscle, in the amphibian, fast and slow motoneurons selectively reinnervate fast-twitch and slow graded muscle fibers within the same muscle (Hoh 1971). Such selectivity can ultimately occur, even when slow graded muscle fibers are first innervated by fast motoneurons (Schmidt & Stefani 1976). Thus, fiber type specificity is also not absolute, but given the appropriate choices, fast and slow motoneurons are capable of recognizing and of showing a preference for their appropriate targets. Fiber type specificity has also been observed in the bird (Feng et al 1965) but not apparently in mammals (Romanul & Van der Meulen 1967, but see Hoh 1975). An elucidation of the mechanism of fiber type matching will be of considerable interest.

In summary, observations derived from both developmental and reinnervation studies indicate that the highly specific neuromuscular connections found in vertebrates cannot be entirely accounted for by passive processes. Rather, these studies suggest that motoneurons and muscles possess identities that should eventually be definable at the biochemical level. An attempt to begin to understand the molecular nature of these identities as well as the cellular mechanisms involved in matching between motoneurons and muscles would seem fruitful. Such studies have recently been undertaken for the chick retinotectal system (Barbera 1975, Marchase 1977, see also Denberg 1975, and Denberg et al 1977 for the cockroach neuromuscular system). Although it is not yet certain that individual muscles possess distinct biochemical labels, larger groupings such as dorsal and ventral muscle masses or medial and lateral motoneurons probably do. It is possible that such labels are continuously graded (for instance along the a-p axis of the cord or primary muscle mass), as has been proposed for the chick retinotectal system. Alternatively, motoneuron pools and individual muscles could be labeled in a discontinuous fashion. Additional studies, both in vivo and in vitro, should allow one to decide between these possibilities. A biochemical investigation of the molecular basis of specificity could profitably be carried out, then, in vitro or on isolated cells provided that the motoneurons and muscle fibers used are carefully defined populations based on the in vivo developmental studies.

However, other mechanisms of possibly equal importance in the development of specific neuromuscular connections, such as axonal guidance, will probably not be understandable on the basis of the biochemical properties of isolated motoneurons and muscles. They need to be more precisely defined first at the level of the in vivo system. Knowledge derived from these studies can then be combined with an understanding of the cellular mechanisms responsible for directed axonal growth in vitro (Letourneau 1975, 1978, Campenot 1977) to allow for the reasonable design of future experiments.

Finally, the role, if any, that activity plays will need to be determined. Changeux & Danchin (1976) have proposed the selective stabilization of synapses based on functional activity as a mechanism to generate specific connections. Neither the very early directed axonal outgrowth seen in the developing limb nor, perhaps, the establishment of specific peripheral motor connections seem explicable on this basis. Recent preliminary findings (R. W. Oppenheim, personal communication) suggest that when neuromuscular activity is blocked throughout the time that connections normally develop in the chick limb, motoneuron death is prevented (Pittman & Oppenheim 1978), but the motoneuron pools appear similar if not identical to those in normal embryos. However, activity may still be involved in the generation of more subtle aspects of peripheral connectivity, as well as playing a major role in the formation of central connections.

ACKNOWLEDGMENTS

I would like to thank my colleagues, G. Pilar, C. Lance-Jones, M. Honig, and B. Ferguson for helpful discussion and for critical reading of the manuscript, and a number of authors who sent me their recent work, including unpublished material. I would also like to thank Mrs. F. Hunihan for typing the manuscript and NINCDS for support.

Literature Cited

Adelmann, H. B. 1927. The development of the eye muscles of the chick. *J. Morphol.* 44:29–87

Anderson, H. 1978. Postembryonic development of the visual system of the locust, *Schistocerca gregaria. J. Embryol. Exp. Morph.* 46:147–70

Barbera, A. J. 1975. Adhesive recognition between developing retinal cells and the optic tecta of the chick embryo. *Dev. Biol.* 46:167–91

Bekoff, A. 1976. Ontogeny of leg motor output in the chick embryo: A neural analysis. *Brain Res.* 106:271–91

Bennett, M. R., Pettigrew, A. G. 1974. The formation of synapses in striated muscle during development. *J. Physiol. London* 241:515–45

Bennett, M. R., Raftos, J. 1977. The formation and regression of synapses during the re-innervation of axototl striated muscles. *J. Physiol.* 265:261–95

Bernstein, J. J., Guth, L. 1961. Non-selectivity in establishment of neuromuscular connections following nerve regeneration in the rat. *Exp. Neurol.* 4:262–75

Bixby, J. L., Van Essen, D. C. 1978. Suppression of original nerve inputs to a mammalian skeletal muscle by a foreign motor nerve. *Soc. Neurosci.* 4:367 (Abstr.)

Brown, M. C., Ironton, R. 1978. Sprouting and regression of neuromuscular synapses in partially denervated muscles. *J. Physiol.* 278:325–48

Brown, M. C., Jansen, J. K. S., Van Essen, D. 1976. Polyneuronal innervation of skeletal muscle in new-born rats and its elimination during maturation. *J. Physiol.* 261:387–422

Burke, R. S., Strick, P. L., Kanda, I. K., Kim, C. C., Walmsley, B. 1977. Anatomy of medial gastrocnemius and soleus motor nuclei in cat spinal cord. *J. Neurophysiol.* 40:667–80

Campenot, R. B. 1977. Local control of neurite development by nerve growth factor. *Proc. Natl. Acad. Sci. USA* 74:4516–19

Cass, D. T., Mark, R. F. 1975. Re-innervation of axolotl limbs. I. Motor nerves. *Proc. R. Soc. London Ser. B.* 190:45–58

Changeux, J.-P., Danchin A. 1976. Selective stabilization of developing synapses as a mechanism for the specification of neuronal networks. *Nature* 264:705–12

Chu-Wang, I. W., Oppenheim, R. W. 1978. Cell death of motoneurons in the chick embryo spinal cord. II. A quantitative and qualitative analysis of degeneration in the ventral root including evidence for axon outgrowth and limb innervation prior to cell death. *J. Comp. Neurol.* 177:59–86

Cohen, A. 1978. Functional recovery following cross-reinnervation of antagonistic forelimb muscles in rats. *Acta Physiol. Scand.* 103:331–33

Cruce, W. L. R. 1974. The anatomical organization of hindlimb motoneurons in the spinal cord of the frog. *J. Comp. Neurol.* 153:59–76

Denburg, J. L. 1975. Possible biochemical explanation for specific reformation of synapses between muscle and regenerating motoneurons in cockroach. *Nature* 258:535–37

Denburg, J. L., Seecof, R. L., Horridge, G. A. 1977. The path and rate of growth of regenerating motor neurons in the cockroach. *Brain Res.* 125:213–26

Dennis, M. J., Yip, J. W. 1978. Formation and elimination of foreign synapses on adult salamander muscle. *J. Physiol.* 274:299–310

Detweiler, S. R. 1930. Some observations upon the growth, innervation, and function of heteroplastic limbs. *J. Exp. Zool.* 57:183–203

Dunlap, D. G. 1966. The development of the musculature in the hindlimb in the frog, *Rana pipiens. J. Morphol.* 119:241–58

Edgerton, V. R., Grillner, S., Sjostrom, A., Zangger, P. 1976. Central generation of locomotion in vertebrates. *Adv. Behav. Biol.* 18:439–64

Engberg, I., Lundberg, A. 1969. An electromyographic analysis of muscular activity in the hindlimb of the cat during unrestrained locomotion. *Acta Physiol. Scand.* 75:614–30

Fambrough, D. M. 1976. Specificity of nerve-muscle interactions. In *Neuronal Recognition,* ed. S. Barondes, F. E. Bloom, pp. 25–67. New York: Plenum. 367 pp.

Fangbonner, R. F., Vanable, J. W. 1974. Formation and regression of inappropriate nerve sprouts during trochlear nerve regeneration in *Xenopus laevis. J. Comp. Neurol.* 157:391–406

Feng, T. P., Wu, W. Y., Yang, F. Y. 1965. Selective reinnervation of a "slow" or "fast" muscle by its original motor supply during regeneration of a mixed nerve. *Scientia Sinica* 14:1717–20

Ferguson, B. 1978. Effect of dorsal-ventral limb rotations on the development of motor connections. *Soc. Neurosci.* 4:111 (Abstr.)

Fouvet, P. B. 1973. Innervation et morphogenèse de la patte chez l'embryon de poulet. I. Mise en place de l'innervation normale. *Arch. Anat. Microsc. Morphol. Exp.* 62:269–80

Frank, E., Jansen, J. K. S. 1976. Interaction between foreign and original nerves innervating gill muscles in fish. *J. Neurophysiol.* 39:84–90

Frank, E., Jansen, J. K. S., Lomo, T., Westgaard, R. H. 1975. The interaction between foreign and original motor nerves innervating the soleus muscle of rats. *J. Physiol.* 247:725–43

Genat, B. R., Mark, R. F. 1977. Electrophysiological experiments on the mechanism and accuracy of neuromuscular specificity in the axolotl. *Phil. Trans. R. Soc. London Ser. B* 278:335–47

Grillner, S. 1975. Locomotion in vertebrates-central mechanisms and reflex interactions. *Physiol. Rev.* 55:247–304

Grillner, S. 1976. Some aspects on the descending control of the spinal circuits generating locomotor movements. *Adv. Behav. Biol.* 18:351–75

Grimm, L. 1971. An evaluation of myotypic respecification in axolotls. *J. Exp. Zool.* 178:479–96

Grinnell, A. D., Letinsky, M. S., Rheuben, M. B. 1979. Competitive interaction between foreign nerves innervating frog skeletal muscle. *J. Physiol.* 289:241–62

Guth, L., Bernstein, J. J. 1961. Selectivity in the re-establishment of synapses in the superior cervical ganglion of the cat. *Exp. Neurol.* 4:59–69

Hamburger, V. 1939. The development and innervation of transplanted limb primordia of chick embryos. *J. Exp. Zool.* 80:347–89

Hamburger, V. 1946. Isolation of the brachial segments of the spinal cord of the chick embryo by means of tantalum foil blocks. *J. Exp. Zool.* 103:113–42

Hamburger, V. 1958. Regression versus peripheral control of differentiation in motor hypoplasia. *Am. J. Anat.* 102:365–410

Hamburger, V. 1975. Cell death in the development of the lateral motor column of the chick embryo. *J. Comp. Neurol.* 160:535–46

Harris, A. J., Dennis, M. J. 1977. Deletion of "mistakes" in nerve muscle connectivity during development of rat embryos. *Soc. Neurosci.* 3:107 (Abstr.)

Harris-Flanagan, A. 1969. Differentiation and degeneration in the motor horn of the foetal mouse. *J. Morph.* 129:281–306

Hilfer, S. R., Searls, R. L., Fonte, V. G. 1973. An ultrastructural study of early myogenesis in the chick wing bud. *Dev. Biol.* 30:374–91

Hoh, J. F. Y. 1971. Selective reinnervation of fast-twitch and slow-graded muscle fibers in the toad. *Exp. Neurol.* 30:263–76

Hoh, J. F. Y. 1975. Selective and non-selective reinnervation of fast-twitch and slow twitch rat skeletal muscle. *J. Physiol.* 251:791–801

Hollyday, M. 1978. Target selectivity of motor pools in chick embryos. *Soc. Neurosci.* 4:115 (Abstr.)

Hollyday, M., Hamburger, V. 1977. An autoradiographic study of the formation of the lateral motor column in the chick embryo. *Brain Res.* 132:197–208

Hollyday, M., Hamburger, V., Farris, J. 1977. Localization of motor neuron pools supplying identified muscles in normal and supernumerary legs of chick embryo. *Proc. Natl. Acad. Sci. USA* 74:3582–86

Hollyday, M., Mendell, L. 1976. Analysis of moving supernumerary limbs of *Xenopus laevis. Exp. Neurol.* 51:316–29

Horder, T. J. 1978. Functional adaptability and morphogenetic opportunism, the only rules for limb development? *Zoon Suppl.* 6:181–92

Hughes, A. F. W. 1968. *Aspects of Neural Ontogeny.* New York: Academic. 249 pp.

Jacobson, M. 1970. *Developmental Neurobiology.* p. 272. New York: Holt, Rinehart, Winston. 465 pp.

Jansen, J. K. S., Thompson, W., Kuffler, D. P. 1978. The formation and maintenance of synaptic connections as illustrated by studies of the neuromuscular junction. *Prog. Brain Res.* 48:3–18

Kleinebeckel, D. 1978. Überprüfung des Modulations prinzips an regenerierenden Hinterbeinnerven bei Larven des Krallenfrosches *Xenopus laevis. Wilhelm Roux Arch. Entwicklungsmech. Org.* 185:1–17

Korneliussen, H., Jansen, J. K. S. 1976. Morphological aspects of the elimination of polyneuronal innervation of skeletal muscle fibers in newborn rats. *J. Neurocytol.* 5:591–604

Lamb, A. H. 1976. The projection patterns to the ventral horn to the hind limb during development. *Dev. Biol.* 54:82–99

Lamb, A. H. 1977. Neuronal death in the development of the somatotopic projections of the ventral horn in *Xenopus. Brain Res.* 134:145–50

Lamb, A. H. 1979. Evidence that some developing limb motoneurones die for reasons other than peripheral competition. *Dev. Biol.* 71:8–21

Lance-Jones, C., Landmesser, L. 1978. Effect of spinal cord deletions and reversals on motoneuron projection patterns in the embryonic chick hindlimb. *Soc. Neurosci.* 4:118 (Abstr.)

Lance-Jones, C., Landmesser, L. 1979. Pathway selection by embryonic chick lumbosacral motoneurons. *Soc. Neurosci.* 5: In press (Abstr.)

Landmesser, L. 1978a. The distribution of motoneurons supplying chick hind limb muscles. *J. Physiol.* 284:371–89

Landmesser, L. 1978b. The development of motor projection patterns in the chick hind limb. *J. Physiol.* 284:391–414

Landmesser, L., Morris, D. G. 1975. The development of functional innervation in the hind limb of the chick embryo. *J. Physiol.* 249:301–26

Landmesser, L., Pilar, G. 1970. Selective reinnervation of two cell populations in the adult pigeon ciliary ganglion. *J. Physiol.* 211:203–16

Landmesser, L., Pilar, G. 1972. The onset and development of transmission in the chick ciliary ganglion. *J. Physiol.* 222:691–713

Landmesser, L., Pilar, G. 1976. Fate of ganglionic synapses and ganglion cell axons during normal and induced cell death. *J. Cell Biol.* 68:357–74

Letourneau, P. C. 1975. Cell-to-substratum adhesion and guidance of axonal elongation. *Dev. Biol.* 44:92–101

Letourneau, P. C. 1978. Chemotactic response to nerve fiber elongation to nerve growth factor. *Dev. Biol.* 66:183–196

Lewis, J. 1978. Pathways of axons in the developing chick wing: Evidence against chemo-specific guidance. *Zoon Suppl.* 6:175–79

Macagno, E. R. 1978. Mechanism for the formation of synaptic projections in the arthropod visual system. *Nature* 275: 318–20

Marchase, R. B. 1977. Biochemical investigations of retinotectal adhesive specificity. *J. Cell Biol.* 75:237–57

Mark, R. F. 1965. Fin movements after regeneration of neuromuscular connections: An investigation of myotypic specificity. *Exp. Neurol.* 12:292–302

Mark, R. F., Marotte, L. R. 1972. The mechanism of selective reinnervation of fish eye muscles. III. Functional electrophysiological and anatomical analysis of recovery from section of the IIIrd and IVth nerves. *Brain Res.* 46:131–48

Marotte, L. R., Mark, R. F. 1970a. The mechanism of selective reinnervation of fish eye muscle. I. Evidence from muscle function during recovery. *Brain Res.* 19:41–51

Marotte, L. R., Mark, R. F. 1970b. The mechanism of selective reinnervation of fish eye muscle. II. Evidence from electron microscopy of nerve endings. *Brain Res.* 19:53–62

McGrath, P. A., Bennett, M. R. 1979. The development of synaptic connections between different segmental motoneurones and striated muscles in the axolotl limb. *Dev. Biol.* 69:133–45

Mendell, L. M., Scott, J. G. 1975. The effect of peripheral nerve cross-union on connections of single 1a fibers to motoneurons. *Exp. Brain Res.* 22:221–34

Milaire, J. 1965. Aspects of limb morphogenesis in mammals. In *Organogenesis,* ed. R. L. DeHaan, H. Usprung, pp. 283–300. New York: Holt, Rinehart, Winston. 804 pp.

Morris, D. G. 1978. Development of functional motor innervation in supernumerary hind limbs of the chick embryo. *J. Neurophysiol.* 41:1450–65

Narayanan, C. H. 1964. An experimental analysis of peripheral nerve pattern development in the chick. *J. Exp. Zool.* 156:49–60

Narayanan, C. H., Hamburger, V. 1971. Motility in chick embryos with substitution of lumbosacral by brachial and brachial by lumbosacral cord segments. *J. Exp. Zool.* 178:415–32

Nja, A., Purves, D. 1977. Re-innervation of guinea-pig superior cervical ganglion cells by preganglionic fibers arising from different levels of the spinal cord. *J. Physiol. London* 273:633–51

Nornes, H., Das, G. 1974. Temporal pattern of neurogenesis in spinal cord of rat. I. An autoradiographic study-time and sites of origin and migration and settling patterns of neuroblasts. *Brain Res.* 73:121–38

Pettigrew, A. G., Lindeman, R., Bennett, M. R. 1979. Development of the segmental innervation of the chick forelimb. *J. Embryol. Exp. Morph.* 49:115–137

Pittman, R. H., Oppenheim, R. W. 1978. Neuromuscular blockade increases motoneuron survival during normal cell death in the chick embryo. *Nature* 271:364–66

Prestige, M. C. 1967. The control of cell number in the lumbar ventral horns during the development of *Xenopus laevis* tadpoles. *J. Embryol. Exp. Morph.* 18:359–87

Prestige, M. C. 1973. Gradients in the time of origin of tadpole motoneurones. *Brain Res.* 59:400–4

Prestige, M. C., Willshaw, D. J. 1975. On a role for competition in the formation of patterned neuronal connexions. *Proc. R. Soc. London Ser. B* 190:77–98

Prestige, M. C., Wilson, M. A. 1974. A quantitative study of the growth and development of the ventral root in normal and experimental conditions. *J. Embryol. Exp. Morph.* 32:819–33

Purves, D. 1976. Long term regulation in the vertebrate peripheral nervous system. *Int. Rev. Physiol.* 10:125–77

Purves, D., Lichtman, J. W. 1978. Formation and maintenance of synaptic connections in autonomic ganglia. *Physiol. Rev.* 58:821–62

Rasmussen, S., Chan, A. K., Goslow, G. E. 1978. The cat step cycle: Electromyographic pattern for hindlimb muscles during posture and unrestrained locomotion. *J. Morph.* 155:253–70

Romanes, G. J. 1941. The development and significance of the cell columns in the ventral horn of the cervical and upper thoracic spinal cord of the rabbit. *J. Anat. London* 76:112–30

Romanes, G. J. 1964. The motor pools of the spinal cord. *Prog. Brain Res.* 11:93–119

Romanul, C. A., Van der Meulen, J. P. 1967. Slow and fast muscles after cross-innervation: Enzymatic and physiological changes. *Arch. Neurol.* 17:387–402

Romer, A. S. 1927. The development of the thigh musculature of the chick. *J. Morph. Physiol.* 43:347–85

Schmidt, H., Stefani, E. 1976. Re-innervation of twitch and slow muscle fibres of the frog after crushing the motor nerves. *J. Physiol. London* 258:99–123

Scott, S. A. 1977. Maintained function of foreign and appropriate junctions on reinnervated goldfish extraocular muscles. *J. Physiol.* 268:87–109

Sharrard, W. J. W. 1955. The distribution of the permanent paralysis in the lower limb in poliomyelitis. *J. Bone Jt. Surg. B. Vol.* 37:540–558

Shellswell, G. B. 1977. The formation of discrete muscles from the chick wing dorsal and ventral muscle masses in the absence of nerves. *J. Embryol. Exp. Morph.* 41:269–77

Shellswell, G. B., Wolpert, L. 1977. The pattern of muscle and tendon development in the chick wing. In *Vertebrate Limb and Somite Morphogenesis,* ed. D. A. Ede, J. R. Hinchcliffe, M. Balls, pp. 71–86. Cambridge: Cambridge Univ. Press. 498 pp.

Slack, J. R. 1978. Interaction between foreign and regenerating axons in axolotl muscle. *Brain Res.* 146:172–76

Sohal, G. S., Weidman, T. A., Stoney, S. D. 1978. Development of the trochlear nerve: Effects of early removal of periphery. *Exp. Neurol.* 59:331–41

Sperry, R. W. 1945. The problem of central nervous reorganization after nerve regeneration and muscle transposition. *Q. Rev. Biol.* 20:311–69

Sperry, R. W., Arora, H. L. 1965. Selectivity in regeneration of the oculomotor nerve in the cichlid fish, *Astronotus ocellatus.* *J. Embryol. Exp. Morph.* 14:307–17

Stirling, R. V., Summerbell, D. 1977. The development of functional innervation in the chick wing-bud following truncations and deletions of the proximal-distal axis. *J. Embryol. Exp. Morph.* 41:189–207

Stirling, R. V., Summerbell, D. 1979. The segmentation of axons from the segmental nerve roots to the chick wing. *Nature* 278:640–42

Sullivan, G. E. 1962. Anatomy and embryology of the wing musculature of the domestic fowl (*Gallus*). *Aust. J. Zool.* 10:458–518

Székely, G. 1963. Functional specificity of spinal cord segments in the control of limb movements. *J. Embryol. Exp. Morph.* 11:431–44

Székely, G., Czéh, G. 1967. Localization of motoneurones in the limb moving spinal cord segments of *Amblystoma. Acta Physiol. Acad. Sci. Hung.* 32:3–18

Thompson, W., Jansen, J. K. S. 1977. The extent of sprouting of remaining motor units in partially denervated immature and adult rat soleus muscles. *Neuroscience* 2:523–35

Weiss, P. 1937. Further experimental investigations on the phenomenon of homologous response in transplated amphibian limbs. II. Nerve regeneration and the innervation of transplanted limbs. *J. Comp. Neurol.* 66:481–535

Weiss, P., Hoag, A. 1946. Competitive reinnervation of rat muscles by their own and foreign nerves. *J. Neurophysiol.* 9:413–18

Wenger, B. S. 1951. Determination of structural patterns in the spinal cord of the chick embryo studied by transplantations between brachial and adjacent levels. *J. Exp. Zool.* 116:123–63

Wenger, E. L. 1950. An experimental analysis of relations between parts of the brachial spinal cord of the embryonic chick. *J. Exp. Zool.* 114:51–86

Willshaw, D. J., Von der Marlburg, C. 1976. How patterned neural connections can be set up by self-organization. *Proc. R. Soc. London Ser. B* 194:431–45

Wolpert, L., Lewis, J., Summerbell, D. 1974. Mophogenesis of the vertebrate limb. In *Cell Patterning,* ed. R. Porter, J. Rivers, pp. 95–130. Amsterdam: Elsevier. 356 pp.

Ann. Rev. Neurosci. 1980. 3:303–18
Copyright © 1980 by Annual Reviews Inc. All rights reserved

CELLULAR RECOGNITION ❖11544
DURING NEURAL DEVELOPMENT

David I. Gottlieb and Luis Glaser[1]

Departments of Biological Chemistry and Neurobiology, Division of Biology and Biomedical Sciences, Washington University School of Medicine, St. Louis, Missouri 63110

It now appears that the complicated nerve fiber circuits of the brain grow, assemble and organize themselves through the use of intricate chemical codes under genetic control. Early in development, the nerve cells, numbering in the billions, acquire and retain thereafter, individual identification tags, chemical in nature by which they can be recognized and distinguished from one another.

Roger W. Sperry, 1965

Roger Sperry's chemoaffinity hypothesis, elegantly summarized in the preceding quotation (Sperry 1965), is one of the central ideas of contemporary neurobiology. When Sperry first proposed the chemoaffinity hypothesis, the experimental evidence for it consisted entirely of behavioral and anatomical studies. The proposed identification tags were necessary to explain the way axons grow during the regeneration of altered nervous pathways; the hypothesis gained widespread support because it was extremely difficult to imagine other reasonable explanations for the anatomical and behavioral results.

Experimental support for the chemoaffinity hypothesis could theoretically be of two distinct kinds. In the first, the nervous system is altered and the assembly of nerve circuits is observed and compared to that in normal animals. If chemical identification tags offer a parsimonious explanation for the observed patterns of growth and connections, then the theory gains support. More direct support must come from experiments of a second kind, which study actual molecular content, in order to test the proposition

[1]Work in the authors' laboratory has been supported by Grants GM18405 and NSF PCM 77-15972 to L.G. and NS 12867 to D.I.G.

303

that certain macromolecules mediate the formation of specific connections. The detailed structure and localization of the molecules must be given and perturbation of the molecules shown to lead to perturbations in the formation of pathways.

Research on neuronal specificity is now at an exciting juncture. A large body of experiments of the first kind described above have greatly strengthened and extended the experimental support for chemoaffinity as the basis for the normal assembly of nerve circuits. Simultaneously, a number of research groups have made impressive gains in studying the molecular basis of in vitro models for cell recognition in the nervous system. In light of these recent developments, we address ourselves to three tasks in our review: 1. We cite a number of experiments on neurospecificity done since Sperry's formulation of the chemoaffinity hypothesis, which add strong and novel support to that view. 2. We review recent work dealing with the molecular basis of embryonal cell adhesion and recognition in developing nervous tissues. 3. We enumerate some of the technical obstacles which must be overcome if molecular studies of synaptic recognition are to proceed and review hopeful developments in this area.

FURTHER EVIDENCE IN FAVOR OF CHEMOAFFINITY

Sperry's pioneering studies spawned a large field of research which has been treated in a number of comprehensive reviews (Gaze 1970, Hunt & Jacobson 1974, Jacobson 1978, Cowan 1978). Of the many experiments we will cite a few that provide particularly strong support for the view first enunciated by Sperry. Attempts to characterize synaptic recognition at a molecular level are likely to be frustrated in the short-term by technical constraints, but by marshalling the strongest evidence for chemoaffinity, we hope to underscore the incentive for studying various model systems of cell-cell recognition.

Sperry's original data suffered from two major limitations. The first was that all of his critical experiments were performed on animals undergoing regeneration of nervous pathways (Sperry 1943, 1944, 1945, Attardi & Sperry, 1960, 1961). Although the extension of these results to original embryonic outgrowth seemed natural, direct evidence that the foundation of embryonic pathways and their regeneration follow the same principles was lacking. Second, in many cases the patterns of pathway regeneration was inferred from behavioral results rather than from direct anatomical analysis. In some cases anatomy was done, but only to the level of following major fiber bundles with stains for normal material (Attardi & Sperry 1961). The studies of Jacobson and his colleagues on the developing frog visual system (reviewed in Hunt & Jacobson 1974) have shown that primary

CELLULAR RECOGNITION   305

embryonic outgrowths do indeed show the same principles first discovered for regenerating pathways. Studies by Cowan and his colleagues on the visual system of the chick have made a number of important contributions to this issue. In one of these studies (Crossland et al 1974) parts of the early developing retina were surgically removed. The remaining intact part of the retina projected to the tectum. Several crucial observations were made:

1. The remaining parts of the retina projected to the appropriate parts of the tectum.
2. Fibers growing from partial retinas crossed over large portions of the uninnervated tectum, ignoring countless opportunities to form synapses with sites normally innervated by excised portions of the retina.
3. The actual terminals of the axons from partial retinae were studied and shown to be indistinguishable from normal retinal terminals. These facts find their most plausible explanation in the chemoaffinity hypothesis.

In its simplest form, the chemoaffinity hypothesis also predicts that cells of the tectum should also be specified. To determine if this is so, experiments have been performed in which portions of the adult tectum in frogs and fishes have been excised and repositioned. In many cases, there was a clearcut reorientation of the retinal projection as would be predicted from chemoaffinity (Yoon 1975, Jacobson & Levine 1975a,b). Finally, Chung & Cooke (1975) succeeded in rotating the optic tectum of the developing frog before the stage at which innervation by the optic nerve occurs. In cases in which a portion of the diencephalon was included with the rotated piece of tectum, inverted retinal maps were formed. Thus some positional information must reside in the central nervous system.

Recent studies show that the innervation of the embryonic limb seems to proceed by similar mechanisms as those discussed for the visual system. When motoneuron axons first innervate the developing limb, they immediately go to the appropriate developing muscle mass rather than search for it by trial and error (Landmesser & Morris 1975).

It has long been clear that each class of axon in the central nervous system chooses to synapse with only a very limited repertoire of cell types. Until recently this ability could be explained either by chemoaffinity or by elaborate timing mechanisms of axon and dendrite outgrowth. However, thalamocortical afferents in the reeler mouse find their appropriate target cells in spite of the gross rearrangement of cortical cellular laminae (Caviness & Rakic 1978). The thalamocortical afferents follow curved trajectories to their target cells; the suggestion that the axons are following chemical cues is strong, and it is difficult to see how a timing mechanism could generate this pattern. Similar conclusions are drawn from the study of regenerating nerve pathways in invertebrates. In the case of the leech, certain central axons regenerate and reinnervate their original targets, while ignoring countless inappropriate cells (Baylor & Nicholls 1971, Jansen & Nicholls 1972). In the house cricket, cercal afferents whose development has been experimentally delayed establish connections with appropriate central neurons (Edwards & Palka, 1971, Palka & Edwards 1974). It is difficult to reconcile these observations with a mechanism that assigns a central role to timing of axon outgrowth in the specification of nerve circuits.

Although we have emphasized axonal connections, there are many other patterns of cell-cell association in the nervous system that suggest specific affinities between complementary cell surfaces. The dramatic, characteristic lamination of neurons in many parts of the central nervous system and the specific association between glia and neurons come readily to mind. The fact that some of these patterns can be reconstructed from dissociated cells in vitro is most easily explained by the chemoaffinity theory (Garber & Moscona 1972a,b, Delong & Sidman 1970).

STUDIES OF CELL ADHESION

As discussed in the previous section, cell recognition in the nervous system is presumed to be an important component for the three-dimensional arrangement of cells in any given region of the nervous system as well as for synaptogenesis. These two situations probably represent different degrees of complexity in that in the former, one need only postulate a very limited number of cell surface components responsible for cell recognition, whereas in the latter case a large number of cell surface components or combination of components have to be invoked in order to account for specific patterns of synapse formation. As detailed above, the specificity for these interactions is determined at least initially by complementary surface molecules present on the interacting cells, a model first presented by Sperry (1963) and elaborated into specific chemical models by other investigators (e.g. Barondes 1970, Barbera et al 1973, Barbera 1975). This hypothesis is by no means proven, but would seem most reasonable in light of our knowledge of cell surfaces. It appears less likely that cellular organization is totally the consequence of gradients of soluble molecules, although these may be important in specific instances (Crick 1970, Wolpert 1971).

Recently it has been suggested that reinnervation of muscle at the original synaptic site may be determined by the basal lamina rather than by the surface of the muscle cell (Sanes et al 1978). While regeneration of a neuromuscular junction may be a special situation not directly applicable to normal neuronal development, the possible contribution of the extracellular matrix to the organization of cells is an important topic for histotypic differentiation, which is outside the scope of this review.

The Sperry model fundamentally assumes not only that the presence of complementary molecules on the surface of interacting cells governs cellular adhesion, but also that during development, equilibrium is reached such that within a cell population, those cells will bind to each other that have either the largest number of complementary ligands, or the highest affinity, or both. Implied in such a model is functional reversibility of adhesion. The high specificity required by such a model implies that at least one of the specific cell surface ligands involved in cell adhesion is a protein.

The study of the molecules involved in cellular recognition, either in the nervous system or elsewhere, is complicated by the fact that standard biochemical methods apply, to a large extent, only to large numbers of cells, or components extracted from such cells. Currently this means that within the nervous system, cellular recognition can only be studied in very heterogenous collections of cells. This contrasts strikingly with developmental studies that suggest precise specification of cell interactions and allow distinctions to be made between similar cells in adjacent regions, e.g. the neuroretina as discussed in the previous section.

An ideal system for the study of cell recognition would require the availability of large populations of homogenous cells, appropriately differentiated, wherein specific cell recognition can be recognized by the presence of either physiological or morphological alterations in the interacting cells. With very few exceptions this ideal situation is at present unattainable. Therefore a large amount of work in this area has been devoted to the study of two types of model systems utilizing (a) heterogenous cell populations of known anatomical origin and (b) homogenous populations of cells in culture, whose anatomical origin is not always known. An additional complication arises in these studies from the fact that alterations of the cell surface brought about by the procedures involved in cell dissociation etc cannot always be adequately controlled.

Any given cell may interact with a variety of other cells; different cell surface ligands are likely to be involved in these interactions. For example, a given cell specifically interacts with adjacent cells to form a particular region or layer in the nervous system; this interaction presumably involves ligands different from the interaction of this same cell with axons derived from other regions of the nervous system.

It is convenient to discuss the model systems that have been studied based on the methodology used in the investigation:

Studies of Cell Recognition Using Intact Cells

These methods all seek to discover whether a particular cell can adhere to another cell of either homologous or heterologous type. The measurement that is usually but not always made (e.g. see Beug & Gerisch 1972) is one of rate of adhesion, which is assumed to reflect the number of sites and the types of adhesive sites present on the two cell surfaces. A number of methods have been devised for such studies, most of which immobilize a large number of cells of a given type and ask whether cells of other types can adhere to these immobilized cells (Roth 1968, Roth et al 1971, Grady & McGuire 1976a,b, McGuire & Burdick 1976, Walther et al 1973, Gottlieb & Glaser 1975). These methods have been reviewed recently and are not discussed in detail (Frazier & Glaser 1979). The adhesion of one cell to another under these conditions appears to be a complex phenomenon which can be seen as a reversible step followed by functionally irreversible events (Umbreit & Roseman 1975). Some of the existing data suggest that the first reversible step represents the binding of a small number of ligands to each other, whereas the irreversible step may represent the binding of a large number of ligands from one cell to the other (Moya et al 1979). While the minimum number of molecules required to obtain specific cell adhesion is one or two different molecules, the complexity of the process suggests that a larger number of molecules will be involved.

One of the favorite systems for the study of cell adhesion has been the embryonal nervous system, and in particular the retino-tectal system. This is due in part to the Sperry hypothesis for the establishment of retino-tectal connectivity, but also because pioneering studies by Moscona and co-workers showed that embryonic chick neural retina can readily be dissociated into single cells which under appropriate conditions will reaggregate and undergo histiotypic differentiation (Moscona 1965, Fujisawa 1973, Moscona & Hausman 1977). Such aggregates will also show spatial segregation, not only between cells from different organs, e.g. liver and retina, but also between cells from different regions of the nervous system (Garber & Moscona 1972a,b). This classical demonstration of cell specificity does not readily provide a tool for the identification of the molecules involved in this sorting out process because of the complexity of the systems involved and the relatively long time period required for segregation.

The available evidence suggests that neuronal cells can be easily distinguished from non-neuronal cells in adhesion assays, e.g. liver or heart, but the distinction between different regions of the embryonal nervous system is less apparent. A discussion of these differences should start with a discussion of the studies of adhesive specificity in the neuroretina of the chick embryo.

Neural retina cells, obtained by gentle dissociation with trypsin, will rapidly adhere to each other to form large aggregates. Because of the general interest in retino-tectal connectivity, Roth and co-workers examined the adhesion of such cells to tectal halves (Barbera et al 1973, Barbera 1975, Marchase et al 1975, Marchase 1977). They found that retinal cells prepared from the ventral half of the neural retina adhered preferentially (faster) to dorsal tectal halves as compared to ventral tectal halves; the converse was true for cells obtained from the dorsal half of the neural retina. It was therefore tempting to conclude that this asymmetry is related to retino-tectal connectivity in that axons from the dorsal half of the neuroretina connect to the ventral half of the tectum, and conversely axons from the ventral half of the neuroretina connect to the dorsal half of the tectum. [Similar results have been obtained using monolayers of tectal cells rather than tectal halves (Gottlieb & Arington 1979).] Implicit in this interpretation is the notion that at least at early stages of development all neural retina cells carry the same cell surface markers.

In related studies, Gottlieb et al (1976) found an adhesive gradient within the neural retina such that cells from the extreme dorsal area of the neural retina adhere preferentially (faster) to a monolayer of cells from the extreme ventral area of the neural retina, and then adhere progressively less well (slower) to cells from increasingly dorsal areas of the retina. The converse is true of cells from the extreme ventral area of the neural retina. Neuraminic acid residues on glycoproteins may be involved in this preferential adhesion (Cafferata et al 1979).

A more extensive series of studies of adhesive preference of dissociated cells obtained from the nervous system has led to the following conclusion (Gottlieb & Arington 1979): Homotypic cell adhesion is not necessarily the preferred mode of cell adhesion, and within any defined anatomical region there occur significant differences in cellular adhesion rate (which may represent qualitative or quantitative changes in the adhesive component). These differences occur along one of the major axes of the embryo, and reflect the presence of an adhesive gradient which may have considerable developmental significance in that it may provide embryonal cells with information regarding the orientation of the embryo. Within any given area of the nervous system, adjacent cells do not necessarily show the highest adhesive preference to each other. Finally, adhesive preference is shown by cells which normally are not synaptically related.

The chemical basis for the adhesive specificity shown in these tests is not understood, and is unlikely to be understood entirely unless some attempt can be made to isolate the cell surface components responsible for cell adhesion. A preliminary attempt has been made to identify such molecules on the basis of their susceptibility to hydrolytic enzymes (Marchase 1977). These observations have suggested that adhesion between retinal cells and tectal halves is mediated in part by a carbohydrate-binding protein specific for N–acetyl–D–galactosamine (GalNAc) residues and a GalNAc-containing molecule, possibly the ganglioside GM-2. These two molecules are assumed to be distributed assymetrically in the retina and tectum with highest concentration of the binding protein on the ventral half of each structure. It should be noted that the adhesive characteristic of these cells cannot be explained entirely by these ligands, because treatment with appropriate hydrolytic enzymes prevents selective cell adhesion but allows adhesion to take place at the same rate for dorsal and ventral cells.

The data obtained by these methods are clearly of interest in that they define differences in cell surface adhesive components and reveal the presence of a gradient or gradients of those components that are likely to be of importance for cell orientation during development. The methods also have serious limitations in that mixtures of cell types are used, and there is some uncertainty as to whether the adhesive components revealed by these in vitro assays are biologically functional. The amount of chemical information obtained in this type of study is limited, although, as discussed below, specific antibodies may prove to be extremely useful in identifying adhesive molecules.

Studies Using Plasma Membranes

Alternative approaches to the study of cell recognition require the development of systems in which the binding of a cell surface component to cells is measured; such systems would allow purification and identification of these surface components.

In several systems it has been possible to demonstrate that plasma membrane fractions retain the ability to bind to cells and do so with specificity, i.e. they bind very much faster to one cell type than to another (Merrell & Glaser 1973, Gottlieb et al 1974, Obrink et al 1977a,b, Santala et al 1977). In addition to binding assays, membranes have also been tested as agents which prevent cell aggregation,[2] presumably by competing for cell adhesive sites (Merrell & Glaser 1973, Gottlieb et al 1974), or in the case of liver, promote aggregation by crosslinking cells (Obrink et al 1977a). In the nervous system these membranes have indicated the presence of temporal changes in cell surface properties of retina and tectum and have demonstrated an interesting unidirectional reactivity between retina and tectum in that retinal membranes inhibit aggregation of both retinal and tectal cells, but tectal membranes appear to only inhibit tectal cell aggregation (Gottlieb et al 1974).

Until recently only modest success was obtained in isolating the membrane components responsible for specific adhesion in these systems (Merrell et al 1975a). However, recently a small molecular weight protein that is at least partly responsible for the inhibition by retinal cell membranes of cell-to-cell adhesion has been isolated as a pure component and partially characterized (Jakoi & Marchase 1979).

---

[2] The limitation to these inhibition assays have been discussed in detail previously (Merrell et al 1976).

Specific adhesion can also be demonstrated with permanent cell lines derived from the nervous system and with plasma membranes derived from these cells (Santala et al 1977). The origin within the nervous system of these established cell lines is not known and therefore the relevance of these observations to adhesive components in the nervous system is not always apparent. Work with such cell lines has, however, led to several conclusions which are of general applicability. These conclusions could have been obtained only with these cell lines because they are homogenous cell populations.

Any given cell contains a number of different molecules on its surface that can participate in specific cell adhesion. Molecules may be present on a cell that allow it to adhere to heterologous cells but that are not involved in homologous cell adhesion (Santala et al 1977, Stallcup 1977). Cell adhesion can be influenced by trophic factors (Merrell et al 1975b) and the nature of the adhesive factors present on the cell surface can be altered by culture conditions (Santala & Glaser 1977)—most notably cell density and cell-to-cell contact.[3] When extrapolated to the nervous system these observations suggest that cells' adhesive properties will alter during development and may alter as a consequence of contact with either heterologous or homologous cells. This last point is of particular interest in that normal development must by necessity involve both adhesion between cells and the breaking of such connections. "De-adhesion" may therefore arise as a consequence of changes in cell surface compositions induced by trophic factors or contacts with other cells (see also Changeux & Danchin 1976).

Very little information is available regarding the reversal of adhesion. Observations in this regard have been made in *Chlamydomonas* (Snell & Roseman 1977) but this system may not be an appropriate model for the nervous system. In culture what appears to be "synapse" formation between retina cells and muscle has been noted, but this is reversed with time. The survival of appropriate cells, or their ability to synthesize transmitter, is not well enough understood under these culture conditions to be certain that this really represents a reversal of adhesion between cells for which synapse formation is inappropriate (Ruffolo et al 1978). The suggestion that under the same culture conditions stable synapses are formed between cells which normally synapse in vivo has not yet been reported in full.

[3]Nerve growth factor has been shown to alter the adhesion of PC-12 pheochromocytomic cells not only to each other but also to substratum under culture conditions (Schubert & Whitlock 1977). The importance of substratum for tissue culture cells in general has recently been reviewed and is outside the scope of this article (Grinnell 1978, Grinnel & Hays 1978). It is a subject of particular interest in the study of the nervous system, because successful culture of cells (as well as some cell separation methods) depend on the presence of a suitable surface on which cells can grow.

Aggregation Promoting Factors

Factors that promote cell aggregation can generally be considered in two categories. The first are lectins or antibodies, which are multivalent ligands that can aggregate cells. A discussion of such ligands is outside the scope of this review, except to point out that multivalent carbohydrate binding proteins have been found in vertebrate cells (see Frazier & Glaser 1979), including cells in the nervous system; although in principle these could be involved in cell-cell adhesion, evidence in this regard (except for liver) (Obrink et al 1977a, Schnaar et al 1978) is lacking.

Proteins that specifically promote the formation of large aggregates of neuronal cells and that show regional specificity within the nervous system have been described through extensive work in the laboratory of A. A. Moscona (Hausman & Moscona 1975, 1976, Moscona & Hausman 1977). One of these proteins has been obtained from neural retina in pure form; it stimulates the formation of large retinal cell aggregates, after 24 hr incubation, while the same cells incubated in the absence of the protein show only the formation of small aggregates. The precise function of this protein is not known, although it has been shown to be located on the surface of the retinal cells (Hausman & Moscona 1979) and can be isolated from a surface membrane fraction. Whether this and other factors represent adhesive molecules or trophic agents, they provide a striking molecular correlate of anatomical differences between major areas of the nervous system. Taurine, a small molecular weight amino acid, functions as an aggregation-promoting factor for liver (Sankaran et al 1977).

Use of Antibodies to Study Cell Recognition

Antibodies have generally been prepared against determinants in different regions of the nervous system with the expectation of generating antibodies specific for a unique cell type. In a few examples antibodies have been obtained that interfere with cell adhesion (Rutishauser et al 1976, Brackenbury et al 1977, Thiery et al 1977, McClay et al 1977, Rutishauser et al 1978a,b). The most extensively studied of these is an antibody against a cell surface molecule designated as CAM (cell adhesion molecule). CAM is widely distributed in the nervous system and anti-CAM Fab' fragments prevent cell adhesion in a variety of assays. Anti-CAM Fab' also prevents the formation of neurite bundles in cultures of dorsal root ganglia. CAM appears to be particularly enriched in neurite membranes.

This antibody defines a unique molecule (mol wt 140,000) present in the surface of neuronal cells and is a component of the adhesive mechanism. Since this molecule does not show regional specificity, it most likely represents a common component of the cell surface adhesive mechanism.

A related approach is the use of lectins, or antibodies to yeast or bacterial carbohydrates, to block cellular interactions. Interesting data in these systems have been obtained in studies of cerebellar development in normal and mutant mice. Whether such antibodies are specific enough to identify unique cell surface ligands remains to be established, as it seems likely that such reagents will react with a variety of cell surface molecules, although only a few of these will be responsible for the observed developmental change (Trinkner 1978).

Biological Consequences of Cell Adhesion

A totally different approach would measure adhesive events that result in functional consequences to cells. This approach has the unique advantage that the biological meaning of the adhesive molecules is never in doubt and the biological assay provides an amplification of the binding event. This approach has only been used in a limited number of systems, and we will discuss only one of these in detail.

In 1975, Wood & Bunge (see also Wood 1976) showed that rat dorsal root ganglia could be freed of fibroblasts and that such ganglia showed extensive developments of neurites and proliferation of Schwann cells. Removal of the ganglia from this culture left a bed of Schwann cells that ceased to divide but remained viable and could be induced to resume division if a new ganglion (free of Schwann cells) was added to the culture. The mitogenic effect of neurons on Schwann cells appears to require contact between the Schwann cells and neurites.

It has been possible to prepare a neurite membrane fraction which is mitogenic to Schwann cells (Salzer et al 1977, Bunge et al 1979). The data obtained so far indicate that the mitogenic signal resides in the surface of the neurites (it is inactivated by mild trypsin treatment of intact cells and neurites) and is absent from the neuritic cytosol. The neurite membrane component appears to be highly specific since it is absent from a variety of other plasma membrane fractions obtained from cultured neuronal and non-neuronal cell lines. Mitogenic activity for Schwann cells prepared by different procedures have been noted for cholera toxin as well as components present in pituitary extracts and other cells (Raff et al 1978a, Hanson & Partlow 1978). The precise relation of such cells to those isolated according to the protocol of Wood (1976) is not yet clear. It should also be understood that the presence of mitogenic factors for Schwann cells in pituitary extracts is in no way incompatible with the presence of a specific membrane bound mitogenic factor on the surface of neurites.

Although this system suffers from the disadvantage that the relevant cells can only be obtained in small quantities, and are thus difficult objects for biochemical investigation, the system is ideal from other points of view— it uses defined cells and it measures a physiological consequence of cell adhesion. The fact that the biological effect can be reproduced at the membrane level strongly suggests that further purification and characterization of the membrane components should be feasible. A similar approach in a non-neuronal system, which has also made considerable progress, is the study of contact inhibition of growth of Swiss 3T3 cells (Whittenberger & Glaser 1977, Whittenberger et al 1978, Whittenberger et al 1979, Bunge et al 1979) wherein it has been possible to extract the membrane components responsible for growth control, but these have not yet been obtained in purified form.

FUTURE APPROACHES TO CELLULAR RECOGNITION

In the preceding sections, we have reviewed some of the successful attempts to probe the molecular basis of cell-cell recognition in the early developing nervous system. These studies have been successful insofar as they have given evidence for the existence of cell surface molecules involved in cell surface recognition. Nevertheless, the studies have important limitations. In many cases, heterogeneous cellular systems are being used and it is unclear on which particular cells the recognition molecules in question actually reside. In other cases, activities have been identified on whole cells or derivatives from them, but have not been purified to homogeneity. Two important technical problems tend to limit further progress in this area. The first is the difficulty of obtaining purified populations of cells from the developing nervous system. Recent advances in the culture of both normal and transformed cells from the nervous system make it appear that this limitation will gradually be reduced. But even if we assume the availability of reasonable quantities of pure neuronal cells, there is a further problem that limits progress in this area, and that is the extreme complexity of the repertoire of cell surface proteins on any vertebrate cell. For this reason, progress in the study of cell recognition will be intimately related to progress in being able to characterize cell surfaces at the molecular level. Fortunately, there have been several recent advances in this regard that are pertinent to the present discussion. The ability to discriminate proteins on two-dimensional polyacrylamide gel electrophoresis systems has enabled investigators to definitively separate complex mixtures of proteins, such as we would expect to find in growing axons and dendrites. This technique has been applied to developing systems in general (e.g. Levinson et al 1978), and to neurons in particular (Stone et al 1978), with great success. The problem of mapping cell surface macromolecules in particular has been addressed by separating proteins by SDS-PAGE electrophoresis and staining glycoproteins with radioactive lectin preparations. Although first applied to non-neuronal cells (Burridge 1976), this method has now been extended to the developing nervous system (Mintz & Glaser 1978). It has been clearly shown that in the chick retina, five glycoproteins are developmentally regulated. Three glycoproteins can be shown to be present in later stages of development although they are absent in earlier stages, and in the case of two other proteins, the converse is true. These are present during early development and disappear somewhat later. Since these studies are performed with very small masses of tissue, they can be extended to a number of developmental situations in nervous tissues.

Immunological methods also could potentially be of great value in the study of cell-cell recognition in the nervous system. These methods have already provided probes that clearly distinguish between different cell types in the nervous system. For example, Brockes et al (1977) have shown that a particular antigen called RAN 1 is present on Schwann cells, but not on fibroblasts; conversely, the antigen, theta, is present on fibroblasts and neurons but not on Schwann cells. Using immunological techniques, Raff and his colleagues (Raff et al 1978b) have also shown that the glycolipid galactocerebrocide is a rather specific marker for developing oligodendrocytes. The potential of immunological techniques has been greatly heightened by the introduction of the hybridoma method for obtaining antibodies. The unique advantage of this methodology is that purified antibodies can be obtained from antigens which have themselves not been purified. This method has already proven to be very useful in obtaining antibodies against specific subpopulations of lymphocytes (Williams et al 1977), and its extension into the nervous system should provide very refined maps of the cell surface of various types of central nervous system cells.

Literature Cited

Attardi, G., Sperry, R. W. 1960. Central routes taken by regenerating optic fibers. *Physiologist* 3:12

Attardi, G., Sperry, R. W. 1961. Preferential selection of central pathways by regenerating optic fibers. *Exp. Neurol.* 4: 262–75

Attardi, G., Sperry, R. W. 1963. Preferential selection of central pathways by regenerating optic fibers. *Exp. Neurol.* 7: 46–64

Barbera, A. J. 1975. Adhesive recognition between developing retinal cells and the optic tecta of the chick embryo. *Dev. Biol.* 46:167–91

Barbera, A. J., Marchase, R. B., Roth, S. 1973. Adhesive recognition and retinotectal specificity. *Proc. Natl. Acad. Sci. USA* 70:2482–86

Barondes, S. H. 1970. Brain glycomacromolecules and interneuronal recognition. In *Neurosciences Second Study Program,* ed. F. O. Schmitt, pp. 747–60. New York: Rockefeller

Baylor, D. A., Nicholls, J. G. 1971. Patterns of regeneration between individual nerve cells in the central nervous system of the leech. *Nature* 232:268–70

Beug, H., Gerisch, G. 1972. A micromethod for routine measurement of cell agglutination and dissociation. *J. Immunol. Methods* 2:49–57

Brackenbury, R., Thiery, J. P., Rutishauser, U., Edelman, G. M. 1977. Adhesion among neural cells of the chick embryo. I. An immunological assay for molecules involved in cell-cell binding. *J. Biol. Chem.* 252:6835–40

Brockes, J. P., Fields, K. L., Raff, M. C. 1977. A surface antigenic marker for rat Schwann cells. *Nature* 266:364–66

Bunge, R., Glaser, L., Lieberman, M., Raben, D., Salzer, J., Whittenberger, B., Woolsey, T. 1979. Growth control by cell to cell contact. *J. Supramol. Struct.* In press

Burridge, K. 1976. Changes in cellular glycoproteins after transformation: Identification of specific glycoproteins and antigens in sodium dodecyl sulfate gels. *Proc. Natl. Acad. Sci. USA* 73:4457–61

Cafferata, R., Panosian, J., Bordley, G. 1979. Developmental and biochemical studies of adhesive specificity among embryonic retinal cells. *Dev. Biol.* 69:108–17

Caviness, V. S. Jr., Rakic, P. 1978. Mechanisms of cortical development: A view from mutations in mice. *Ann. Rev. Neurosci.* 1:297–326

Changeux, J. P., Danchin, A. 1976. Selective stabilization of developing synapses as a mechanism for the specification of neuronal networks. *Nature* 264:705–12

Chung, S. H., Cooke, J. 1975. Polarity of structure and of ordered nerve connections in the developing amphibian brain. *Nature* 258:126–32

Cowan, W. M. 1978. Aspects of neural development. In *International Review of Physiology Neurophysiology III,* 17: 149–89 ed. R. Porter, Baltimore: Univ. Park Press

Crick, F. 1970. Diffusion in embryogenesis. *Nature* 225:420–22

Crossland, W. J., Cowan, W. M., Rogers, L. A., Kelly, J. P. 1974. The specification of the retino-tectal projection in the chick. *J. Comp. Neurol.* 155:127–64

DeLong, G. R., Sidman, R. L. 1970. Alignment defect of reaggregating cells in cultures of developing brain of reeler mutant mice. *Dev. Biol.* 22:584–600

Edwards, J. S., Palka, J. 1971. Neural regeneration: Delayed formation of central contacts by insect sensory cells. *Science* 172:591–94

Frank, E., Jansen, J. K. S., Rinvik, E. 1975. A multisomatic axon in the central nervous system of the leech. *J. Comp. Neurol.* 159:1–14

Frazier, W., Glaser, L. 1979. Surface components and cell recognition. *Ann. Rev. Biochem.* 48:491–523

Fujisawa, H. 1973. The process of reconstruction of histological architecture from dissociated retina cells. *Wilhelm Roux Arch. Entwicklungsmech. Org.* 171:312–30

Garber, B. B., Moscona, A. A. 1972a. Reconstruction of brain tissue from cell suspensions. I. Aggregation patterns of cells dissociated from different regions of the developing brain. *Dev. Biol.* 27:217–34

Garber, B. B., Moscona, A. A. 1972b. Reconstruction of brain tissue from cell suspensions. II. Specific enhancement of aggregation of embryonic cerebral cells by medium from homologous cell cultures. *Dev. Biol.* 27:235–43

Gaze, R. M. 1970. *The Formation of Nerve Connections,* pp. 1–128. London & New York: Academic

Gottlieb, D. I., Arington C. 1979. Patterns of adhesive specificity in the developing central nervous system of the chick. *Dev. Biol.* 71:260–73

Gottlieb, D. I., Glaser, L. 1975. A novel assay of neuronal cell adhesion. *Biochem. Biophys. Res. Commun.* 63:815–21

Gottlieb, D. I., Merrell, R., Glaser, L. 1974. Temporal changes in embryonal cell surface recognition. *Proc. Natl. Acad. Sci. USA* 71:1800–2

Gottlieb, D. I., Rock, K., Glaser, L. 1976. A gradient of adhesive specificity in developing avian retina. *Proc. Natl. Acad. Sci. USA* 73:410–14

Grady, S. R., McGuire, E. J. 1976a. Intercellular adhesive selectivity. III. Species selectivity of embryonic liver intercellular adhesion. *J. Cell Biol.* 71:96–106

Grady, S. R., McGuire, E. J. 1976b. Tissue selectivity of the initial phases of cell adhesion. *J. Cell Biol.* 70:346a

Grinnell, F. 1978. Cellular adhesiveness and extracellular substrata. *Int. Rev. of Cytol.* 53:65–144

Grinnell, F., Hays, D. G. 1978. Cell adhesion and spreading factor. *Exp. Cell Res.* 115:221–29

Hanson, G. R., Partlow, L. M. 1978. Stimulation of non-neuronal cell proliferation in vitro by mitogenic factors present in highly purified sympathetic neurons. *Brain Res.* 159:195–210

Hausman, R. E., Moscona, A. A. 1975. Purification and characterization of a retina specific aggregation factor. *Proc. Natl. Acad. Sci. USA* 72:916–20

Hausman, R. E., Moscona, A. A. 1976. Isolation of retina specific aggregating factor from membranes of embryonic neural retina. *Proc. Natl. Acad. Sci. USA* 73:3594–96

Hausman, R. E., Moscona, A. A. 1979. Immunological detection of retina cognin on the surface of embryonic cells. *Exp. Cell Res.* 119:191–207

Hunt, R. K., Jacobson, M. 1974. Neuronal specificity revisited. *Curr. Top. Dev. Biol.* 8:203–59

Jacobson, M. 1978. *Developmental Neurobiology,* pp. 345–433. New York: Plenum.

Jacobson, M., Levine, R. L. 1975a. Plasticity in the adult frog brain: Filling the visual scotoma after excision or translocation of parts of the optic tectum. *Brain Res.* 88:339–45

Jacobson, M., Levine, R. L. 1975b. Stability of implanted duplicate tectal positional markers serving as targets for optic axons in adult frogs. *Brain Res.* 92:468–71

Jakoi, E. R., Marchase, R. B. 1979. Ligatin

from embryonic chick neuroretina. *J. Cell Biol.* 80:642–50

Jansen, J. K. S., Nicholls, J. G. 1972. Regeneration and changes in synaptic connections between individual nerve cells in the central nervous system of the leech. *Proc. Natl. Acad. Sci. USA* 69:636–39

Landmesser, L., Morris, D. G. 1975. The development of functional innervation in the hind limb of the chick embryo. *J. Physiol. London* 249:301–26

Levinson, J., Goodfellow, P., Vadeboncoeur, M., McDevitt, H. 1978. Identification of stage-specific polypeptides synthesized during murine preimplantation development. *Proc. Natl. Acad. Sci. USA* 75:3332–36

Marchase, R. B. 1977. Biochemical studies of retino-tectal specificity. *J. Cell Biol.* 75:237–57

Marchase, R. B., Barbera, A. J., Roth, S. 1975. A molecular approach to retina-tectal specificity. *Ciba Symp.* 29:315–27

McClay, D. R., Gooding, L. R., Fransen, M. E. 1977. A requirement for trypsin sensitive cell surface components for cell-cell interaction of embryonic neural retina cells. *J. Cell Biol.* 75:56–66

McGuire, E. J., Burdick, C. L. 1976. Intercellular adhesive selectivity. I. An improved assay for the measurement of embryonic chick intercellular adhesion (liver and other tissues). *J. Cell Biol.* 68:80–89

Merrell, R., Glaser, L. 1973. Specific recognition of plasma membranes by embryonic cells. *Proc. Natl. Acad. Sci. USA* 70:2794–98

Merrell, R., Gottlieb, D. I., Glaser, L. 1975a. Embryonal cell surface recognition, extraction of an active plasma membrane component. *J. Biol. Chem.* 250:5655–59

Merrell, R., Gottlieb, D. I., Glaser, L. 1976. Membranes as a tool for the study of cell surface recognition. In *Neuronal Recognition*, ed. S. H. Barondes, pp. 249–73. New York: Plenum

Merrell, R., Pulliam, M. W., Randono, L., Boyd, L. F., Bradshaw, R. A., Glaser, L. 1975b. Temporal changes in tectal cell surface specificity induced by nerve growth factor. *Proc. Natl. Acad. Sci. USA* 72:4270–74

Mintz, G., Glaser, L. 1978. Specific glycoprotein changes during development of the chick neural retina. *J. Cell Biol.* 79:132–37

Moscona, A. A. 1965. Recombination of dissociated cells and the development of cell aggregates. In *Cells and Tissues in Culture*, ed. E. N. Willmar, pp. 489–529. New York: Academic

Moscona, A. A., Hausman, R. E. 1977. Biological and biochemical studies on embryonic cell recognition. In *Cell and Tissue Interactions*, ed. J. W. Lash, M. M. Burger, pp. 173–86. New York: Raven

Moya, F., Silbert, D. F., Glaser, L. 1979. The relation of temperature and lipid composition to cell adhesion. *Biochim. Biophys. Acta* 550:485–99

Obrink, B., Kuhlenschmidt, M. S., Roseman, S. 1977a. Adhesive specificity of juvenile rat and chicken liver cells and membranes. *Proc. Natl. Acad. Sci. USA* 74:1077–81

Obrink, B., Warmergard, B., Pertoft, H. 1977b. Specific binding of rat liver plasma membranes by rat liver cells. *Biochem. Biophys. Res. Commun.* 77:665–70

Palka, J., Edwards, J. S. 1974. The cerci and abdominal giant fibers of the house cricket. *Acheta domesticus.* II. Regeneration and effects of chronic deprivation. *Proc. R. Soc. London Ser. B* 185:105–21

Raff, M. C., Abney, E., Brockes, J. P., Smith, A. H. 1978a. Schwann cell growth factors. *Cell* 15:813–22

Raff, M. C., Mirsky, R., Fields, K. L., Lisak, R. P., Dorfman, S. H., Silberberg, D. H., Gregson, N. A., Leibowitz, S., Kennedy, M. C. 1978b. Galactocerebroside is a specific cell surface antigenic marker for oligodendrocytes in culture. *Nature* 274:813–16

Roth, S. 1968. Studies on intercellular adhesive specificity. *Dev. Biol.* 18:602–31

Roth, S., McGuire, E. J., Roseman, S. 1971. An assay for intercellular adhesive specificity. *J. Cell Biol.* 51:525–35

Ruffolo, R. R. Jr., Eisenbarth, S. S., Thompson, J. M., Nirenberg, M. 1978. Synapse turnover: A mechanism for acquiring synaptic specificity. *Proc. Natl. Acad. Sci. USA* 75:2281–85

Rutishauser, U., Thiery, J. P., Brackenbury, R., Sela, B. H., Edelman, G. M. 1976. Mechanism of adhesion among cells from neural tissue of the chick embryo. *Proc. Natl. Acad. Sci. USA* 73:577–81

Rutishauser, U., Thiery, J. P., Brackenbury, R., Edelman, G. M. 1978a. Adhesion among neural cells of the chick embryo. III. Relationship of the surface molecule CAM to cell adhesion and the development of histotypic patterns. *J. Cell Biol.* 79:371–81

Rutishauser, U., Gall, W. E., Edelman, G. M. 1978b. Adhesion among neural cells of the chick embryo. IV. Role of the cell surface molecule CAM in the formation of neurite bundles in cultures of spinal ganglia. *J. Cell Biol.* 79:382–93

Salzer, J., Glaser, L., Bunge, R. P. 1977. Stimulation of Schwann cell proliferation by a neurite membrane fraction. *J. Cell Biol.* 75:75

Sanes, J. R., Marshall, L. M., McMahan, U. J. 1978. Reinervation of muscle fiber basal lamina after removal of myofibers. *J. Cell Biol.* 78:176–98

Sankaran, L., Proffitt, R. T., Petersen, J. R., Pogell, B. M. 1977. Specific factors influencing histotypic aggregation of chick embryo hepatocytes. *Proc. Natl. Acad. Sci. USA* 74:4486–90

Santala, R., Glaser, L. 1977. The effect of cell density on the expression of cell adhesion properties in a cloned rat astrocytoma (C-6). *Biochem. Biophys. Res. Comm.* 79:285–91

Santala, R., Gottlieb, D. I., Littman, D., Glaser, L. 1977. Selective cell adhesion of neuronal cell lines. *J. Biol. Chem.* 252:7625–34

Schnaar, R. L., Weigel, P. H., Kuhlenschmidt, M. S., Lee, Y. C., Roseman, S. 1978. Adhesion of chicken hepatocytes to polyacrylamide gels derivatized with N-acetylglucosamine. *J. Biol. Chem.* 253:7940–51

Schubert, D., Whitlock, C. 1977. Alteration of cellular adhesion by nerve growth factor. *Proc. Natl. Acad. Sci. USA* 74:4055–58

Snell, W. J., Roseman, S. 1977. A quantitative assay for the adhesion and deadhesion of *Chlamydomonas reinhardii.* *Fed. Proc.* 36:811

Sperry, R. W. 1943. Visuomotor coordination in the newt (*Triturus viridescens*) after regeneration of the optic nerve. *J. Comp. Neurol.* 79:33–55

Sperry, R. W. 1944. Optic nerve regeneration with return of vision in anurans. *J. Neurophysiol.* 7:57–69

Sperry, R. W. 1945. Restoration of vision after crossing of optic nerves and after contralateral transplantation of eye. *J. Neurophysiol.* 8:15–28

Sperry, R. W. 1963. Chemoaffinity in the orderly growth of nerve fiber patterns and connections. *Proc. Natl. Acad. Sci. USA* 50:703–10

Sperry, R. W. 1965. In *Organogenesis,* ed. R. L. DeHaan, H. Ursprung, p. 170. New York: Holt, Rinehart, Winston

Stallcup, W. B. 1977. Specificity of adhesion between cloned neural cell lines. *Brain Res.* 126:475–84

Stone, G. C., Wilson, P. L., Hall, M. E. 1978. Two-dimensional gel electrophoresis of proteins in rapid axoplasmic transport. *Brain Res.* 144:287–302

Thiery, J. P., Brackenbury, R., Rutishauser, U., Edelman, G. M. 1977. Adhesion among neural cells of the chick embryo. II. Purification and characterization of a cell adhesion molecule from neural retina. *J. Biol. Chem.* 252:6841–45

Trinkner, E. 1978. Postnatal cerebellar cells of staggerer express immature components on their surface. *Nature* 277:566–67

Umbreit, J., Roseman, S. 1975. A requirement for reversible binding between aggregating embryonic cells before stable adhesion. *J. Biol. Chem.* 250:9360–68

Van Essen, D., Jansen, J. K. S. 1976. Repair of specific neuronal pathways in the leech. *Cold Spring Harbor Symp. Quant. Biol.* 40:495–502

Walther, B. T., Ohman, R., Roseman, S. 1973. A quantitative assay for intercellular cell adhesion. *Proc. Natl. Acad. Sci. USA* 70:1569–73

Whittenberger, B., Glaser, L. 1977. Inhibition of DNA synthesis in cultures of 3T3 cells by isolated surface membranes. *Proc. Natl. Acad. Sci. USA* 74:2251–55

Whittenberger, B., Raben, D., Lieberman, M. A., Glaser, L. 1978. Inhibition of growth of 3T3 cells by extract of surface membranes. *Proc. Natl. Acad. Sci. USA* 75:5457–61

Whittenberger, B., Raben, D., Glaser, L. 1979. Regulation of the cell cycle of 3T3 cells in culture by a surface membrane enriched cell fraction. *J. Supramol. Struct.* 10: In press

Williams, A. F., Galfre, G., Milstein, C. 1977. Analysis of cell surfaces by xenogeneic myeloma-hybrid antibodies: Differentiation antigens of rat lymphocytes. *Cell* 12:663–73

Wolpert, L. 1971. Positional information and pattern formation. *Curr. Top. Dev. Biol.* 3:183–225

Wood, P. M. 1976. Separation of functional Schwann cells and neurons from normal peripheral nerve tissue. *Brain Res.* 115:361–75

Wood, P. M., Bunge, R. P. 1975. Evidence that sensory axons are mitogenic for Schwann cells. *Nature* 256:662–64

Yoɯn, M. G. 1975. Topographic polarity of the optic tectum studied by reimplantation of the tectal tissue in adult goldfish. *Cold Spring Harbor Symp. Quant. Biol.* 40:503–19

Ann. Rev. Neurosci. 1980. 3:319–52
Copyright © 1980 by Annual Reviews Inc. All rights reserved

RETINOTECTAL SPECIFICITY: ❖11545
Models and Experiments in Search of a Mapping Function

Scott E. Fraser and R. Kevin Hunt

Jenkins Biophysical Laboratories, The Johns Hopkins University,
Baltimore, Maryland 21218

INTRODUCTION

Embryonic development of nerve patterns has fascinated and frustrated neurobiologists for decades. The orderly connections within the central nervous system and between the central nervous system and peripheral end organs are thought to be formed by a number of different means ranging from selective cell death, through timing and mechanical guidance, to chemospecificity. A system that has gained great popularity for the study of topographic connections between sets of neurons is the retinotectal system of lower vertebrates (amphibia and fish). Although this system has hardly been cooperative in yielding conclusive evidence (Horder & Martin 1978, Hunt & Jacobson 1974), its great popularity has led to a mass of experimental data, which, by sheer bulk and ever more refined experimental design, has begun to crystallize into strong support and strong challenge to various proposed ideas for the assembly of connections between the eye and brain.

The development and continued refinement of good models is an indispensable task (Gaze 1978). Models for the assembly of retinotectal connections must provide a *mapping function* that expresses the correspondence between (map-making) information in the eye (and the tectum) and the pattern of connectivity observed. At this stage in our understanding, it is probably unreasonable to expect models to approach full explanatory power or comprehensive description. Yet even incomplete models, sketched in broad outlines and lacking in fine detail, provide an important focus to research: shaping amorphous issues into testable hypotheses, translating experimental questions into research designs, and guiding the selection of

319

0147-006X/80/0301-0319$01.00

controls in various experiments. One need only reflect upon the will-o'-the-wisp quality of the mechanistic questions behind fifteen years of retinotectal "mismatch" experiments to appreciate, as the late Mac Edds put it, that "Neuronal specificity is a here-not-there, this-not-that kind of concept" (Edds et al 1979). The recent involvement of several experimental laboratories in the refinement of key theoretical issues has already produced encouraging refinements in experimental design (Hunt & Jacobson 1974, Marchase et al 1975, Prestige & Willshaw 1975, Hope et al 1976, Meyer 1978, Mac-Donald 1977, Hunt 1978, Horder & Martin 1978, Schmidt & Easter 1978, Fraser 1979, Bodenstein & Hunt 1979).

The need for good models of map assembly mechanisms transcends the problem of map assembly itself. Because the pairings of fibers and cells that define synaptic specificity are topographic and patterned in space, the retinotectal system has become a model system for nerve patterns as well as synaptic specificity, for the generation of position-dependent cellular properties (here used for mapping) as well as the way such properties mediate connectivity (reviewed in Jacobson 1978, Hunt 1978, Gaze 1979). To the extent that characterizing the differentiation of such properties in the retina depends upon experiments in which the "test retina" is allowed to map into the brain, the mapping function is an intervening variable. One can make overly stringent assumptions about the correspondence of property and connection site, only to see one's inferences dissolve as the mapping function proves more complex than anticipated. Alternatively, one can define a very generous set of possible mapping functions and draw only the most limited inferences about properties from the most unambiguous features of the map—whether it is frontwards or backwards, whether it is duplicated or not—and by controlling for potentially intrusive variables, such as fiber competition, by varying the context of map assembly for individual test retinae (Hunt & Jacobson 1974, Hunt 1977a, Schmidt 1978). Neither approach is endearing: the first holds potential disaster, and the second is, at best, an exercise in the agony of self-restraint. Surely the maps contain much more information about the properties that engendered them than we now allow ourselves to infer, but only with a more refined mapping function can we extend the set of allowed inferences and design the appropriate contexts in which to test them.

In the remainder of the introduction, we briefly survey some of the key features of the system, the issues and assays that have received emphasis from those who study it, and some of the basic categories of experiment that have been performed on it. We then turn to a critique of classic and contemporary theoretical work, try to assemble it into a synthetic model, and, in the final section, assess how well synthetic models can encompass recent experimental data on the mechanisms of map assembly.

Background, Methods, and Normal Development

The retinotectal system of *Xenopus* has gained its popularity for many reasons. The animals themselves are durable and easily raised, and visual stimulus and response are easily managed in physiological experiments. The optic fibers of frogs and fish retain the ability to regenerate orderly tectal connections late into adult life; the eye and brain are also accessible throughout the entire development of the animal. Finally, retinotectal connections show a topographic order, and an important data base was established from perturbations and distortions of pattern long before it was possible to study the synaptic behavior of individual cells. This advantage was not untainted. The early multi-unit studies of retinotectal patterns were plagued by ambiguities about the true source(s) of recorded signals (Hunt & Jacobson 1974). The application of single-unit methods and corroborative anatomical techniques has removed this ambiguity and ascribed the "routine" signals to the terminal arbors of optic nerve fibers (George & Marks 1974, Potter 1972, Chung et al 1973, 1975, Hunt & Jacobson 1974, Hunt 1976, Schmidt 1978, Fraser & Hunt 1978). Apart from making possible a whole new class of experiments (Chung et al 1973, Hunt 1979a), the clean separation of presynaptic and postsynaptic (Schmidt 1978, Freeman 1977) responses has enabled researchers to reap the full benefits of the system's topographic order: distortions in the multicellular pattern provide rapid and convenient inferences to be tested in terms of the behavior of single cells.

Multi-unit electrophysiologic methods reveal the topographic order of the visual projection in the clawed frog, *Xenopus* (Figure 1); the pattern differs little from that seen in Ranids and goldfish. Through the optics of the eye, the retina receives an inverted, topographic image of the visual world; and the ganglion cells of the retina relay that image—topography intact—to the contralateral optic tectum. The pattern of axonal connections reconstructs, across the tectal surface, the spatial pattern of the corresponding ganglion cell bodies in the retinal sheet. The retinotectal map is oriented such that more nasal ganglion cells (subtending more temporal visual field) innervate more caudal regions of the tectum, and more ventral ganglion cells (subtending more superior visual field) innervate more medial regions of the tectum. Single-unit methods (Figure 1) and field-potential methods (Chung et al 1974a,b, Freeman 1977) confirm the topographic order, and reveal (in frogs) a third dimension of ordering to the connectivity: the three classes of retinal ganglion cell (sustained, event, and dimming) each connect at a different depth beneath the tectal surface. There is evidence that an independent informational unit ("specifier") mediates this depth segregation (Chung et al 1973, Hunt 1976, 1979a), and we shall not consider it

further. Intracellular recording methods (Freeman 1977) offer much promise for the future, though the data we review have rarely used them. Analysis of postsynaptic signals—that is, responses of tectal cells to visual stimuli
—have thus far featured multi-unit and single-unit extracellular recordings
from intertectal "visual relay" cells (Figure 1), or from cells judged to be

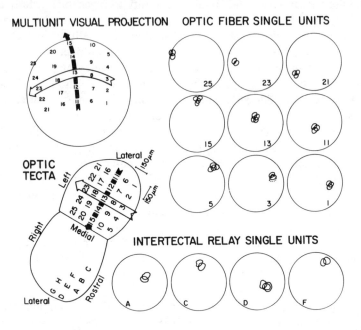

Figure 1   To show the normal retinal projections to both the contralateral and ipsilateral
tectum. The large circle represents 200 degrees of visual field of the right eye. The numbers
represent the center of the area in visual space (receptive field) that elicits neural activity at
the numbered electrode positions on the outline of the optic tectum. This correspondence
between visual space and the multiunit activity recorded in the tectum is what is typically
referred to as the retinotectal map. In actuality it represents the visuotectal map. In the absence
of optical defects, the two maps are synonymous. This analysis can be carried one step further
by determining the area in visual field that excites a single unit in the optic tectum. A number
of these single units are displayed in the small circles in the top right corner of the figure. Note
that the contours (each representing the receptive field for a single, isolated unit) overlap one
another to produce the multiunit projection shown to the left. In addition to this normal direct
projection from eye to contralateral tectum, the intertectal relay supplies visual information
to the ipsilateral tectum. A few of the units recorded in the ipsilateral tectum are displayed.
These responses are mediated first by the connections to the contralateral tectum and then
relayed by an as yet undetermined pathway to the ipsilateral tectum. In a normal animal these
are organized such that for any one spot in the binocular visual field, only one region on each
of the tecta is excited. These intertectal relay units provide a rapid confirmation that the
contralateral optic tectum is functionally innervated by the eye since only functional connections to the contralateral tectum can drive the intertectal relay cells.

tectal by virtue of two cleanly separable ocular inputs (Skarf 1973, Skarf & Jacobson 1974, Hunt & Jacobson 1974, Hunt 1979a, Schmidt 1978).

Behavioral tests, applied cautiously, have proved a useful adjunct to the recording techniques. They allow a direct confirmation that the pattern of connections observed is functional since only functional connections can mold the behavior of the animal. These behavioral tests yield correlative information about the pattern of retinotectal connections only if the behavior is mediated by the optic tectum. Optokinetic responses, for example, are not an appropriate adjunct to retinotectal electrophysiology. The elegant color-vision discrimination tests of Scott (1977), by contrast, have provided a definitive confirmation of the functional significance of compressed optic fiber patterns.

Anatomical methods, despite a pivotal role in the early studies on retinotectal plasticity (Attardi & Sperry 1963, Sperry 1963), have only recently come into widespread use in anuran and goldfish systems. Tritiated (^{3}H-) proline autoradiography, horseradish-peroxidase histochemistry, and three-dimensional reconstruction techniques have begun to broaden the information derived from electrophysiological experiments. The uneven expansion of half retinae over intact tecta in goldfish (Meyer 1975), the formation of "ocular dominance" pseudocolumns under some conditions of coinnervation of a single tectum by two eyes (Levine & Jacobson 1975, Constantine-Patton & Law 1978, Meyer 1979b), and the elimination of mapping errors during development (Longley 1978) are all phenomena that would have proved difficult to detect by recording methods alone. Fiber tracing methods have also helped to refocus attention on the old problem of fasciculation (Ramon y Cajal 1928), its modern analogs of "pathfinding" in the optic nerve and tract (Cook & Horder 1977, Chung & Cooke 1975, Gaze & Grant 1978, Horder & Martin 1978); and to extend retinotectal studies to developing chicks and hamsters, which are less suitable for electrophysiological recording (Crossland et al 1974, Goldberg 1976, Schneider & Jhaveri 1974, Frost & Schneider 1976).

Finally, ^{3}H-thymidine autoradiography and marker-mutant techniques have helped to define the normal descriptive development of the retina and tectum in *Xenopus*. The eye forms a tiny, working retina within four days of fertilization (Jacobson 1968b, Straznicky & Gaze 1971, Bergey et al 1973, Hunt 1975b, Rubin & Grant 1979, Gaze et al 1974). Thereafter, a ring of stem cells resides at the ciliary margin of the retina, and by some process of asymmetric division, adds *annuli* of new neurons and pigment cells to the periphery of the retina. Thus, annuli added early in development surround the optic disc in the adult eye; annuli added slightly later in development occupy more "frontal" positions, further from the disc, on the post-limbic face of the eye; annuli added still later in development are more frontal still, further from the disc and closer to the margin of the pupil, on the pre-limbic face of the adult eye (Straznicky & Gaze 1971, Hollyfield 1971, Jacobson 1976, Beech 1977). During the first two-thirds of larval development annuli are "symmetrical," with respect to cell accretion rate, at various angular positions around the eye; in the late larval and juvenile stages, however, annuli are graded from dorsal to ventral, with as many as ten-fold more cells being added ("per annulus") to ventral regions of the eye (Jacobson 1976, Beech 1977, Straznicky & Tay 1977). Marker mutant mosaics, produced by inoculating a small "arc" on the early eye bud with genetically marked cells, show *polyclones* that are largely contiguous and radially disposed in the pigment retinal epithelium and neural retina (Hunt & Ide 1977, Ide & Hunt 1978, Conway et al 1979). Although some interesting shifts in growth rates (e.g. ventral vs dorsal) and in angular positions (e.g. pushing nasal toward dorsal) occur during the late larval growth of polyclones, their general appearance is thus one of a "pie-slice" extending radially outward from the disc to the ciliary margin of the adult eye (Figure 2).

Shape changes have complicated the analysis of tectal growth, although preliminary data suggest a geometry of cell accretion vastly different from that in the eye. Wedges of cells appear to be added to the caudomedial edge of the tectum (Figure 2; Straznicky & Gaze 1972), and the tectum largely ceases growth at metamorphosis in *Xenopus* as well as *Rana pipiens* and *Hyla regilla* (see also Currie & Cowan 1974, Meyer & Sperry 1973).

Concurrent with growth in the retina and tectum, retinotectal map assembly is taking place. Optic fibers leave the eye bud at late optic cup stages, reaching the brain on the third day of embryonic life (Gaze 1970). Optic-evoked potentials are detectable at early larval stage 44 and by stage 47 (one week after fertilization) the rudimentary visual projection shows crude topographic order. As additional annuli of neurons are added to the growing retina, the ganglion cells in the annuli send their axons across the retina, forming and joining fascicles and exiting through the optic disc. The map continues to mature during larval stages; and during metamorphosis, the visual projections show a shift from nonlinear metrics with confinement to the rostrolateral two-thirds of the tectum (Figure 3b) to the linear metrics and full tectal coverage characteristic of the juvenile and adult map. At least some synapses between retinal fibers and tectal cells are present at all stages from the early larval period onward, as judged from laminar field potentials (Chung et al 1974a,b), Golgi silhouettes (Lazar 1973, Chung et al 1975), and electron microscopic degeneration studies (Longley 1978), although estimates of the relative numbers of connected and unconnected optic fibers at various phases of larval maturation remains to be made.

(a)

(b)

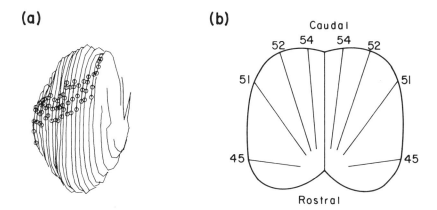

Figure 2   To show the growth patterns of the retina and the tectum of *Xenopus*. (*a*) The eye grows by adding rings of cells to the periphery of the retina. If a small piece of pigmented embryonic retina is implanted into the developing eye of an albino host before stage 32, the adult displays an eye with a stripe of pigment running from the optic nerve head to the ciliary margin. A computer-generated three-dimensional reconstruction of such an eye appears above. The lines in the reconstruction represent every sixth section of a paraffin embedded and sectioned eye. Due to the annular addition of cells to the retina, these lines closely approximate cells of similar birthdates. The small circles mark regions where the pigmented clone was evident. In combination with tritiated thymidine studies, this demonstrates that as the eye grows by adding cells to the ciliary margin, the cell lineage is radial along the eye. (Reconstruction courtesy of K. Conway and K. Feiock.) The growth patterns of the tectum have not been probed by such clonal analyses. The available data comes from tritiated thymidine studies of *Xenopus* tecta, and are diagrammatically represented in (*b*). Larvae were injected with the isotope, which was incorporated in noticeable quantities only by cells undergoing their final few mitoses. The lines represent the position of the labeled cells assayed by autoradiography of animals after metamorphosis. The numbers near the lines indicate the stage at which the animal was injected. These bands of label would have all appeared at the medial margin of the tectum if the animal had been sacrificed immediately. These results seem to indicate that cells are added to the tectum only at its medio-caudal border (after Straznicky & Gaze 1972).

Key Issues and Classes of Experiments

The broadening of methodologies and more refined descriptions of the normal development of the system have led to a renewed assault on several key issues: (*a*) evidence for the presence and nature of locus-specific "discriminator" properties on optic fibers; (*b*) whether, and how much, information exists on the surface of the tectum for guiding optic fibers to the correct terminal sites; (*c*) the extent to which position-dependent (e.g. sorting) and position-independent (e.g. generalized attractive or repulsive) interactions among optic fibers themselves help to order their pattern of tectal connections; and (*d*) the extent to which passive and supervening

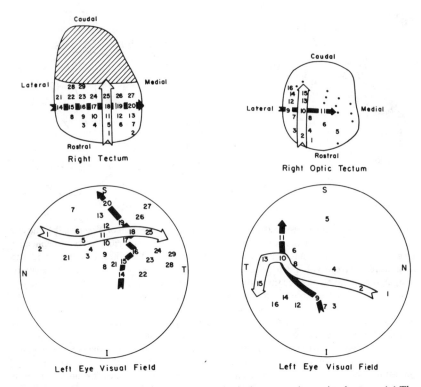

Figure 3   To demonstrate examples of plasticity in the lower vertebrate visual system. (*a*) The projection pattern observed following excision of half of the optic tectum in goldfish demonstrates compression of the entire visual field onto the remaining tectal tissue. Such a compressed projection pattern was the first evidence for plasticity in the lower vertebrate visual system (after Sharma & Gaze 1971). Note that the relative ordering of the retinotectal projection is maintained, only the spacing (metrics) of the projection are altered in the direction of the compression. (*b*) In normal development, *Xenopus* tadpoles demonstrate a compressed visuotectal projection, limited to the lateral and rostral optic tectum. This compressed representation with nonlinear metrics, changes into the normal pattern shown in Figure 1, with the associated equal metrics, at, or near, metamorphosis.

processes, such as mechanical guidance and fasciculation or the spatio-temporal sequence of fiber arrivals in the tectum can explain ordering of the retinotectal map.

Some of these issues, for example passive ordering mechanisms, and the extent to which "active" discriminator properties must be invoked when passive ordering is inadequate, have been pursued in a wide range of experimental settings: *descriptive* studies of cell death and mapping errors, of fasciculation and retinotopy in the optic nerve and tract; and *perturbation* studies that disrupt the spatio-temporal patterns of fiber growth. Such studies have produced the perplexing observation that passive ordering processes clearly exist and must certainly exert a "helper effect" in the normal development of the map (Horder & Martin 1978). However, they are insufficient as the sole, or even predominant, mechanism for producing the final map: cell death is limited, retinotopy in the developing anuran optic nerve is imprecise, fibers jump from one fascicle to another, and many mapping errors occur (Straznicky & Gaze 1971, Hunt 1976, Longley 1978, Rubin & Grant 1979, Horder & Martin 1978). They are also largely unnecessary for the establishment of a precise retinotopic map: optic fibers can assemble normal maps despite perturbations that disrupt their normal pathway to the tectum, their internal topographic order, their selection of one or the other tract brachium, and the normal timetable of arrival of fibers into the tectum (Gaze 1960, Hibbard 1967, Feldman et al 1971, Sharma 1972, Hunt & Jacobson 1972, Horder 1974, Hunt 1975a,b, Udin 1978, Gaze & Grant 1978). Finally, many of the so-called "passive" ordering processes (Horder & Martin, 1978) actually feature active discriminations: errors are selectively eliminated, retinotopy in the regenerating nerve is actively reestablished from an initially scrambled pattern, and fibers that reach the tectum in the wrong group actively track down their correct tectal position (Sperry 1963, Longley 1978, Gaze & Grant 1978, Horder & Martin 1978).

Other issues, however, have been associated with a particular experimental paradigm, which, in various guises and permutations, has come to occupy a large share of the strategic terrain. The most popular type of experiment has been the retinotectal "size-disparity" experiment. Originally developed by Attardi & Sperry (1963), the basic paradigm has been to delete a portion of either the tectum or the retina, and to examine (following optic fiber regeneration) how a partial set of fibers innervates a whole tectum or how a full set of fibers innervates a partial tectum. When normal "pairings" are restored, as occurred in the early studies in which fibers from a partial retina bypassed inappropriate regions of a whole tectum and regenerated only to their usual subregion of tectum (Attardi & Sperry 1963), the results are unambiguous and easily explained by simple models based on patterned fiber-target affinities. The same was true for the reciprocal experiments in which only the appropriate subset of a whole retina's fibers regenerate to a surgically prepared half tectum (Gaze 1970, Straznicky 1973, Meyer & Sperry 1973). In later experiments, a more ambiguous and more intriguing result occurred: a whole retina could compress its map (Figure 3a), maintaining the internal topology of the pattern but altering the metrics, onto the surviving tectal remnant (Gaze & Sharma 1970, Yoon 1972, Cook & Horder 1977). Similarly, a half retina could eventually expand its projection over much or all of the tectal surface (reviewed in Edds et al 1979). These "plasticity" results could reflect a reassignment of discriminator properties ("embryonic regulation"), tissue or cell replacement (tissue regeneration), an unmasking of hidden cytoaffinity preferences (midline fibers always yearned for the edge of the tectum) normally suppressed by competition (peripheral retinal fibers dominate the tectal edge), or the operation of a heretofore unexpected and locus-independent parameter (fiber-fiber repulsion). Subsequent research has shown that the dichotomy of results are two stages in map regeneration in which half-to-half mapping gives way to expansion or compression, and that the two-stage sequence repeats itself on a second regeneration (Cook & Horder 1977, Edds et al 1979). It now appears, to a first approximation, that any mismatch, accommodating all surviving elements, appears possible. Moreover, by controlled competition, it has been possible to show that competitive interactions do occur among optic fibers (see Figure 6, p. 335) and that at least some (and perhaps all) instances of mismatch plasticity do not depend on replacement of cells or their discriminator properties (see Edds et al 1979). That such mismatch plasticity has now been shown in preparations that do not depend upon partial ablation further strengthens the inference that map plasticity depends on the interactions among the fibers, not upon a change in their specificities (see Figure 8, p. 343).

In a second popular paradigm, a piece of tectal tissue is excised, and then either rotated or exchanged with another piece of tectal tissue. If information exists on the tectal surface, the optic nerve fibers should follow the tissue and show small displacements or rotations in the grafted regions. This is also something of a one-way experiment: if the fibers ignore the tectal rearrangement, mapping (normally) to a 180°-rotated region as they do to a nonrotated control, or mapping to translocated fragments as if they were sham-reimplant controls, the result is also compatible with tissue regeneration, reassignment of target properties in the graft (repolarization), absence of target properties in the tectum, or fiber ordering by a process other than fiber-tectum affinities. Even the less ambiguous alternative, in which fibers track down their targets in the graft, is tainted since most tectal rearrangements are done on adult fish or juvenile frogs. The graft contains degenerating optic terminals from the original map; what passes for map assembly by fiber-tectum affinity could really be map assembly by fiber-fiber sorting —here extended to include sorting by degenerating fibers and terminals as well as by live ones (Hunt 1976, Sharma & Romeskie 1977). Both classes of result have been obtained, in individual animals in the same laboratory, in virtually every experiment involving tectal graft rotation or rostrocaudal translocation (*Xenopus:* Levine & Jacobson 1974, Hunt 1976, 1979b, Rho 1978; Goldfish: Sharma & Gaze 1971, Yoon 1975, Sharma 1975; *Rana:* Jacobson & Levine 1975a,b, Hope et al 1976). In the majority of cases, fibers track down their appropriate targets; and in the minority, the fibers ignore the graft. That the targets in the majority result are true tectal properties has only recently been confirmed, in rotation experiments on chronically deafferented tecta in goldfish (Sharma & Romeskie 1977), in rotation experiments on "virgin" *Xenopus* tecta (Straznicky 1979, in press), and in both rotation and rostrocaudal-exchange experiments on *Xenopus* tecta previously innervated by a scrambled retinotectal map (Figure 4; Hunt 1976, 1979b). Recently, Rho (1978) has produced the minority result by fiber-fiber

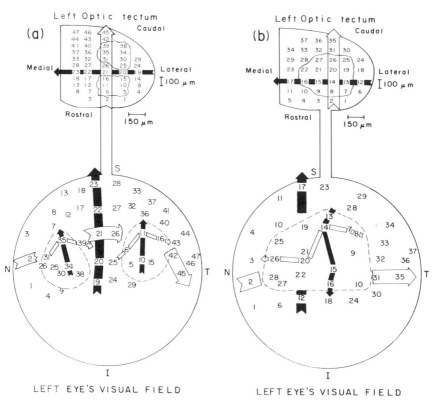

Figure 4 To show tectal graft experiments performed on animals with previously scrambled retinotectal projections. A piece of tectum was excised and then exchanged with another piece of tectal tissue (*a*) or rotated and replaced (*b*). The projection pattern seen after the optic nerve of a normal eye had been deflected to the tectum displays a pattern indicative of information on the tectal surface that the optic nerve fibers use to find their correct tectal locus. That is, the optic nerve fibers followed the moved tectal tissue. Scrambled maps were produced by serially rotating *Xenopus* eyes at two hour intervals for a total of eight to twelve hours. The adult retinotectal projection displays a randomized pattern. Following the tectal graft operation, the fibers of the normal eye were deflected to the grafted tectum. The use of scrambled map animals negates the possibility that the graft results are due to fibers reconnecting to their previously ordered terminals.

ordering in the clear absence of tissue replacement or reassignment of target properties. Following a second translocation of a graft, which had been ignored by regenerating optic fibers after one translocation, fibers regenerated once again, this time mapping in accordance with the original regional properties of the graft tissue. More recent experiments, involving mediolateral translocation of tectal grafts has provided independent support for such fiber-fiber ordering in *Xenopus* (Rho & Hunt 1979). Unfortunately, there does not appear to be a simple two-stage resolution to the bimodal pattern of results: both appear stable for months, although the relative incidence of the minority result rises to 100% when rotations are performed at earlier larval stages (Chung & Cooke 1975, Jacobson 1978).

A final experimental paradigm has featured surgical manipulation and rearrangement of embryonic and early larval eyes. A variety of interactive processes can occur, which produce a modification in the predicted mapping behavior of the eye or parts of the eye (Jacobson 1968a, Hunt & Jacobson 1972, Hunt 1975a, Feldman & Gaze 1975, Berman & Hunt 1975, Hunt & Berman 1975, MacDonald 1975, Hunt & Frank 1975, Cooke 1977, Gaze 1979, Edds et al 1979, Ide et al 1979, Ling et al 1979). Yet a variety of mosaic mappings from local regions of certain kinds of surgically recombinant eyes provide conclusive evidence for local discriminator properties marking regions of the ganglion cell population. A paradigm case is provided by the mosaic "pie-slice": following inoculation of the temporal pole of the eye bud with a small group of nasal or ventral cells, the resultant polyclone (Figure 2a), as it grows out during larval growth of the retina, lays down the appropriate nasal (or ventral) mapping properties and maps autonomously to the tectum. (Figure 5; Hunt & Ide 1977, Ide & Hunt 1978).

Originally, each of these classic paradigms was developed in pursuit of one particular key issue: size-disparities to assess fiber competition and the elasticity of the mapping function itself, tectal grafts to assess target properties in the tectum, etc. Yet as the experiments have become more complex, the results have become more kaleidoscopic. Schmidt (1978), for example, performed a novel mismatch experiment in which half eyes and whole eyes were allowed to follow one another into a single goldfish tectum. His results provide conclusive evidence for the presence and stability of discriminator properties in the (half) retina, and for some form of fiber interaction/competition as the driving force in map expansions. Yet his data also demonstrate a form of target properties in the tectum, and either that they can be modulated by chronic innervation from an orderly but expanded partial map, or that the function can be over-written by a dominant new form of "target" based on debris from degenerating optic terminals. Likewise, Chung & Cooke (1975) performed early tectal rotations and failed to find

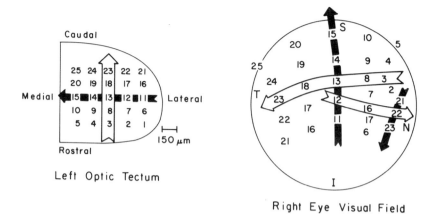

Figure 5 To show a pie-slice projection pattern produced by introducing a nonhomologous piece of retina into an embryonic eye. The basic operation is identical to that performed to obtain the eye shown in Figure 2*b*. The region of retina corresponding with the grafted tissue maps autonomously to the tectum, clearly demonstrating that the position of origin, not the final position of the tissue, determines the retinotectal projection pattern and giving credence to the idea of labels on the retina as well as the tectum.

evidence for stable target properties in the embryonic tectum, leaving open the question of whether such properties are absent, present but rearrangeable, or "yet to be deployed" at early larval stages. Their data also bear upon the issue of passive ordering, indicating that neither the rostrocaudal asymmetry of the growing tectum, nor the direction of the invasion by optic fibers, play a determinative role in map ordering and polarity.

Development and Regeneration

One problem with many studies on the retinotectal system is that they deal with regeneration in an attempt to understand the development of nerve connections. This concern can be partially put to rest by the fact that in most fish and amphibians the retinotectal system continues to develop late into adult life; cells continue to be added to the periphery of the retina and to send axons into the tectum throughout the life of the animal. Thus, the system must still retain the qualities that guide fibers to synapse in normal development late into adult life. Thus, most regeneration experiments are in fact mixed experiments, in which the resultant map includes many regrown fibers and a substantial population of fibers that have innervated the brain for the first time. If it is unwise to ignore the real differences between regeneration and development (e.g. presence of optic fiber debris on the tectum), it seems equally dangerous to throw up one's hands and reject regeneration studies altogether, or worse yet, generate developmental models that only implausibly accommodate regeneration data. Some commentators have extrapolated from the cautiously interpreted findings of Schmidt (1978), for example, to conclude that positional information on the surface of the adult tectum was induced solely by the tectum's developmental association with a normal map (von der Marlsberg & Willshaw 1977, Horder & Martin 1978). While such a consideration is useful from a theoretical standpoint, it leads us into a dilemma, since it implies that the tectum develops the ability to guide optic nerve fibers to their correct locus only after it is no longer required to do so. In short, no regeneration experiment should ignore the antecedent developmental history; but at the same time, legitimate inferences from regeneration studies must inform, and be compatible with, any viable hypothesis about normal developmental mechanisms.

NEURONAL SPECIFICITY AND THE CHEMOAFFINITY MODEL

Experimentation on the retinotectal system (Sperry 1945), cutaneous innervation (Sperry & Miner 1949), and oculomotor innervation (Sperry 1947) prompted Sperry to propose a general theory for the formation of neuronal connections (see Sperry 1950, 1963, 1965, Meyer & Sperry 1973). In all of these systems, the relevant neurons behaved as though they differed from one another qualitatively and displayed an affinity for their correct neuronal "partners." The theory of neuronal specificity was proposed to counter the concurrent models for modulation (Weiss 1942) or resonance (Herring 1913, Head 1920, Weiss 1928) as the responsible factors for orderly neuronal connections. The resonance theory proposed that the central nervous system sends a patterned impulse down most, if not all, fibers of the peripheral nervous system and that this signal serves to activate only those end organs that are "tuned" to receive it. The modulation theory proposed that the peripheral nerves grow and connect indiscriminately to end organs, the end organs then somehow modify the peripheral nerves, which, in turn, make the appropriate central connections. For example, homologous movements of grafted limbs seemed, in early cine analyses, to argue for such a model. The resonance and modulation theories were but two clearly articulated versions of the general view that some aspect of neuronal function —coincidence of firing rates, similarity in the nature of impulse traffic, behavioral utility to the animal—selected the useful permanent nerve connections out of an initially equipotential neural net. In fact, this view had itself displaced the earlier belief that neurons differed from one another chemically and used this chemical difference to "wire" correctly (Langley 1898, Ramon y Cajal 1928, Tello 1915, Harrison 1935). Sperry drew upon those very early notions, formulated in the debate surrounding the neuron doctrine, and also upon the embryological findings of Holtfreter (1939) on tissue affinity in early embryogenesis. But it was against the background of the functionalist view that he proposed his general theory of neuronal specificity.

In his early writings, Sperry sketched out special theories for assembly of nerve connections in a number of systems, including the retinotectal system. The special theory for retinotectal map assembly was elaborated in the early 1960s into what we shall refer to as the *chemoaffinity model* (Sperry 1963). Sperry and his co-workers had found that only half the tectum was innervated following removal of half of the retina and regeneration of the optic fibers from the surviving half. By the careful analysis of fibers that bypassed inappropriate tectal sites or were initially deranged but actively reordered themselves and grew toward their normal targets, Sperry (1963) concluded that retinal specificities are expressed at all stages of fiber growth. The more complete special theory of 1963 thus proposed that fibers are ordered with respect to cues in the nerve, tract, and tectum before they even reach the terminal site for which they have a special (and final) chemoaffinity. It later turned out, of course, that in time, half retinae may expand over whole tecta, that whole retinae may compress onto half tecta, and that almost any combination of retinal or tectal fragments could form patterned connections that (eventually) accommodate all the surviving elements.

These examples of plasticity argued strongly against a narrow interpretation of the chemoaffinity model. The interpretive morass that followed is now merely of historical interest. It is pointless to attempt to legitimize past confusion by juxtaposing quotations from various authors, and equally pointless to attempt to systematize the sources of confusion yet one more time. Rather, we prefer to summarize the positive legacy of this period in four comments:

1. Size disparity experiments have nothing to say about the general theory of neuronal specificity, which does not depend upon the particulars of the special theory for the retinotectal system (the chemoaffinity model) or the details of the retinotectal mapping function, whatever it turns out to be. In fact, the general theory accommodates many clearly articulated mechanisms for the acquisition and expression of synaptic specificities in neurons, including end-organ specification, induction through synaptic contact, and interactions with substrata and other fibers as well as potential synaptic partners (Sperry 1950, 1965).

2. Size disparity experiments, in their simple form, are most easily interpreted by the chemoaffinity model only when the "right" result occurs; namely when previous pairings are reestablished, without expansion or compression or accommodation of "unmatched" elements. That this result

provides strong support for features of the chemoaffinity model is in no way eliminated by the fact that expansions and compressions can also occur. Any model must still explain why the elements initially reestablish their normal pairings.

3. The key issues and experimental criteria for evaluating the alternative result—expansions, compressions, and mismatch, accommodating whatever elements are present—have been clearly articulated by Hunt & Jacobson (1974) and have been met by the carefully controlled experiments of the past five years. By allowing a limited subpopulation of retinal fibers to expand over the tectum, and then displacing them from inappropriate tectal regions by delayed innervation by the (complementary) subpopulation, which were initially absent (Figure 6), it has been established that expansions at least, do not reflect "regulation" of retinal specificities (Sharma & Tung 1979, Hunt 1977b, Schmidt 1978, Meyer 1978). This conclusion has been independently confirmed by the finding that expansions may be uneven (Meyer 1975), that they may occur in settings that do not surgically disrupt the retina (Meyer 1978, Fraser & Hunt 1978, Fraser 1979), and that the same retinal subpopulation may assemble an expanded map to one tectum and its usual part of a normal projection to the other in the same animal (Fraser & Hunt 1978, Fraser 1979). Since true elasticity in the mapping function can now be inferred, the chemoaffinity model would appear to require more explicit elaboration or some refined or additional components.

4. Attempts to deal with retinotectal plasticity led to a number of intriguing ideas and even some alternative models that could be tested in simulation.

In some cases, authors have suggested a new force or component interaction in the map assembly process. For example, a repulsive force between optic fiber terminals (Frost & Schneider 1976, Levine & Jacobson 1975), competition for terminal sites (Meyer 1975, 1978), and arbor size limitations (Frost & Schneider 1976) have all been discussed. Although no attempt has been made to judge their relative or quantitative importance, or to integrate them with other parameters into a testable synthetic model, they represent simple and powerful ideas, clearly articulated.

Other authors have developed more rigorous models that examine the question, "What can be achieved with minimal tectal information?" For example, Prestige & Willshaw (1975) performed theoretical analyses of two different types of chemoaffinity models. Their calculations indicate that a repulsive interaction, or some other modulation of arbor size, must be present in order to fit the data on retinotectal plasticity. In addition, as the only other parameter in a chemoaffinity model such a repulsive force must adapt its strength in peculiar ways to achieve both expansions and compres-

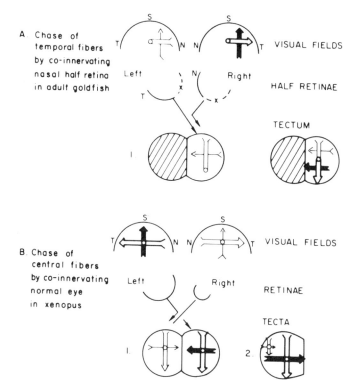

Figure 6   To schematically represent "chases" of the optic fibers of partial retina by the optic fibers of another partial retina or whole retina. (*a*) The optic nerve fibers from two separate half retinae are found to expand over the entire dorsal surface of their contralateral tecta. Following removal of one tectum, both half retinae are forced to innervate the same tectum and "chase" one another back to their proper regions of the tectum. (*b*) A tiny eye, produced by treatment of an embryonic eye with FUdR, consisting of less than 10% of the normal adult eye, spreads to innervate the entire tectum if alone, but is "chased" back to only a small region of the tectum if the tectum is co-innervated by a normal, full-sized, eye. These chases indicate that some fiber-fiber interaction plays a role in the formation of the retinotectal projection.

sions, and is unable to generate the minority results in tectal graft experiments. This work has served an important role in the field, revealing the power of detailed simulations and drawing attention to the need for a repulsive or competitive interaction.

In the arrow model of Hope et al (1976), the only markers on the tectum were polarity markers. Fibers utilized this polarity, along with their own locus-specific properties, to sort amongst themselves into an ordered retinotectal projection. The model fits the mismatch literature very well, demonstrating both expansions and compressions. While it fits tectal rotation data, it fails to explain tectal translocation experiments. This type of minimum information model drew attention to the possibility of a major role for fiber interactions in the selection of targets on the tectum.

Von der Marlsberg & Willshaw (1977) have proposed a model in which the tectum is completely unspecified before fiber ingrowth. The fibers label both themselves and the tectum by possessing a proper blend of molecules. This blend is established by fiber interactions and some initial order or polarity among the optic nerve fibers as they invade the tectum. The model has the problem that the final pattern is dependent on the initial order or polarity of the fibers. In addition, tectal labels are generated in this model only after they are no longer required: the fibers sort themselves and then label the tectum. While this results in a model consistent with most of the tectal graft experiments, one must ask what purpose such tectal markers could serve. The model does, however, have the advantage that it allows the fibers to respecify the tectum (see Schmidt 1978), but these experimental results may be due to guidance from degenerating fiber terminals (Sharma & Romeskie 1977).

Horder & Martin (1978) have proposed that "morphogenetics," in which contact guidance plays a major role, patterns the retinotectal projection. This model accommodates some data on retinotectal mismatches and has refocused attention upon fiber guidance, which surely plays a helper role in the formation of the retinotectal projection.

In short, the theoretical landscape is rich with elements that offer the key to the final image: (a) A general theory of neuronal specificity whose foundation appears rock solid; (b) a chemoaffinity model, that is elegant but not sufficiently detailed to predict a number of experimental results, but the basic components of which are indispensable to explain certain experimental findings (e.g. tectal grafts, place-specific size disparity); (c) a few alternative models, each intriguing but inadequate to explain much of the same data; and (d) a number of simple and powerful ideas, such as systems matching, context-dependent mapping, and fiber-fiber repulsions, which suffer either from inadequate detail or from the fact that no effort has been made to integrate them with quantitative parametric assignments into a single multi-component model.

No one model can predict both the majority and minority results of an experiment on the system. For example, the chemospecificity model predicts the majority result of tectal graft experiments, namely, that the fibers should follow the transposed tissue, even if this produces a discontinuous retinotectal map. The arrow model (Hope et al 1976) predicts only the minority result in which fibers appear to ignore the translocation of the graft. One could be led to the conclusion that both factors operate to form the retinotectal projection. Yet the arrow model fails to predict the minority result after other tectal rearrangements, where fibers ignore the polarity of a tectal graft (Jacobson & Levine 1975a,b). This one-to-one matching of results with mechanisms has many problems beyond simply lacking some of the aesthetic charm of a single model. We must impose a new mechanism for every class of experimental result. While some redundancy in information is desirable for fail-safe reasons, this stacking of mechanism on top of mechanism can become more wasteful than the benefits of reliability they add. A single model that demonstrates multiple "metastable" states would be particularly attractive. If each of the experimental results corresponded to one of these states in which the system may become "trapped," then we are most likely approaching a model closer to the mechanisms utilized in the embryo.

A SYNTHETIC MODEL

We now consider a synthetic model with a single set of rules that remain unchanged for any experiment performed on the system and a set of parameters that can be tuned slightly, depending on the setting (S. E. Fraser, in preparation). The model fits entirely and comfortably within the domain of Sperry's general theory and is an extension of his special chemoaffinity model for retinotectal mapping. As with previous molecular concretizations (Barondes 1970, Barbera 1975, Marchase 1977), it utilizes cell surface chemistry for locus-specific affinity but sets down somewhat different energetic constraints upon what fibers can do to achieve a retina-to-tectum match. The model incorporates two locus-independent forces, a general fiber-fiber repulsive force and a stronger general fiber-tectum adhesion, to the locus-specific "chemoaffinity" interaction. It offers an embarrassingly simple reworking of the chemical matching to allow the same locus-specific markers to mediate two distinct ordering functions. These additions give the model the systems matching type of behavior needed for plasticity results in conjunction with the best-fit behavior indicated by tectal graft experiments. In addition, the model predicts both the majority and minority results for experiments that give mixed results.

The model consists of six invariant rules:

Rule I A strong general adhesive force exists between the cells of the retina and the tectum; it is locus-independent, that is, any pairwise combination of retinal cells and tectal cells (regardless of their address in eye or tectum) generate the same amount of this "C" force.

Rule II A general fiber-fiber repulsion between optic fiber-tips is initiated when synaptic contact is made between optic nerve fibers and cells of the optic tectum; this "R" force is also locus-independent.

Rule III The retinal and tectal cell sheets are each labeled with two quantitative adhesive gradients, which are matched for the retina and tectum. Gradient values are locus specific. One adhesive gradient is displayed anteroposteriorly on both cell sheets (the "AP" gradient); the other dorso-ventrally on the retina and lateromedially on the tectum (the "DV" gradient).

Rule IV The molecules that make up these adhesive gradients interact in a homophilic manner (like molecules attract like molecules) and are organized such that there exists a best-fit site for each fiber on the surface of the tectum.

Rule V The system tends to a configuration of minimum adhesive free energy, by maximizing its energetically favorable adhesive contacts (due to "C," "AP," and "DV") while minimizing the energetically unfavorable "R" force.

Rule VI Each fiber can only overcome a relatively small energy barrier in searching for the lowest free energy configuration.

The parameters of the model place constraints on the rules given above, and unlike the rules, may change from species to species or be tuned slightly, in plausible adaptation to certain experimental perturbations. The parameters of the model assign relative strengths to the various adhesive and repulsive forces. In general, it appears that:

(a) The general eye-tectum adhesive force "C" is the strongest.
(b) The repulsive force "R" is the next strongest.
(c) The graded labels "AP" and "DV" produce the weakest force.
(d) "AP" and "DV" may be of different strengths. In Xenopus, "DV" appears to be stronger than "AP".

The choice of homophilic interactions for the adhesive gradients, and the decision to use linear quantitative gradients and to display them along AP and DV axes in the retina and textum, were made for reasons of economy and general fit with the data: one-axis scrambled maps in *Xenopus* (Hunt 1976, 1979a) suggest the map may in fact be coded in Cartesian coordinates; a homophilic gradient for fiber-tectum affinity will automatically produce a specific fiber-fiber affinity. (See Evidence for Components of the Model below.) In addition, the combination of (a) linear quantitative gradients with Cartesian display, and (b) homophilic interactions enabled the model, with reasonable parametric assignments for "C" and "R" forces, to explicitly generate the observed experimental patterns. While the specifics of the model may change if other coordinates are used, many of the more general predictions of the model are independent of the coordinate system utilized. The proposed interactions all have plausible chemical mechanisms. For example, the repulsive force could reflect a diffusible substance that raises the energy of neighboring fibers perhaps by "robbing" them of "C" contacts. This would effectively move the minimum energy position for each fiber away from its neighbors (see Diamond et al 1976). Alternatively, the

"R" force could be little more than what we commonly call contact inhibition. Axons in culture are commonly seen to branch away from one another following contact. A similar phenomenon of branching following contact (coupled with arbor size limitations) occurring in the tectal neuropil could produce a density dependent repulsion since the likelihood of contact increases with increased numbers.

The homophilic interactions predicted for "AP" and "DV" could be easily produced by Van der Waals forces between identical polysaccharide moieties on cell-surface glycoproteins or glycolipids (Jehle 1963, Edwards 1978, Huang 1978), or any other surface macromolecules (Jehle 1963). While these forces are weak, many of them over the cell surface may sum to produce a substantial net force. They match the available data (Marchase et al 1975) in that these forces are favored at high temperatures as is the retina to tectum adhesion assayed in vitro between tectal fragments and dissociated retinal cells. Such bonds also have the advantage of being somewhat easily broken and replaced, an important advantage in a system that displays plasticity and perhaps "sliding" connections (Chung et al 1974a,b, Jacobson 1976, Scott & Lazar 1976; S. E. Fraser, in preparation).

For example, the "AP" gradient across the retina could utilize the two molecules, "A" and "P," deployed such that the ratio "A"/"P" varies with position but the sum (i.e. density) "A"+"P" remains constant. As given above, "A" will attract other "A" molecules and "P" will attract other "P" molecules. A similar gradient of "D" and "V" exists along the dorsal-ventral axis of the retina. Matching gradients are deployed over the tectum, running in opposite directions to those in the retina (e.g. rostral tectum is enriched in "P" not "A"). Matching in this case means an exact duplicate. This simple arrangement produces a "best-fit" site for each of the fibers at a particular location on the tectum (see Steinberg 1970). Since most cell-surface molecules are able to diffuse in the plane of the cell membrane we must add the constraint that the ratios "A"/"P" and "D"/"V" remain the same over the cell surface even if the densities vary. If "A" and "P" were able to diffuse independently, local differences in the ratio "A"/"P" could result and much of the positional information lost. This constraint is easily met by linking both of the markers to the same molecular backbone. The similarity of this arrangement to a cell surface glycoprotein (or glycolipid) is striking.

Evidence for Components of the Model

We now consider evidence for each component of the model. The evidence is of two types. Some of the components have received rather direct experimental support, while others are merely an economical distillation of a range of experimentation.

The most decisive evidence for target properties (labels) on the surface of the tectum that are capable of guiding optic nerve fibers comes from tectal graft experiments reviewed above. The majority result of these experiments is that the optic fibers "track down" the grafted tissue and synapse in a manner appropriate for the tectal graft, independent of its position with respect to the whole of the tectum (Figure 4). This phenomenon is observed even if the tectum had only entertained a random projection before the grafting experiment (Hunt 1976, 1979b), and the resultant projection patterns are identical to those observed in normal animals following a similar tectal graft. It conclusively demonstrates that the fibers can read some type of positional cue on the surface of the tectum, and that this information is not merely induced by the previous presence of an ordered retinotectal projection. Beyond the tectal graft experiments, a range of other data including the establishment of normally oriented maps when fibers invade the tectum via abnormal routes (Hibbard 1967, Chung & Cooke 1975) requires that at least boundary or polarity markers exist on the tectal surface (Hunt, 1976).

Evidence for labels on the retinal axons derives from a number of experiments, including the many observations that show passive ordering mechanisms to be helpful but insufficient for the task of ordering the retinotectal map (see above). Although initially modifiable by embryonic tissue interactions (reviewed in Jacobson & Hunt 1973, Hunt & Frank 1975, Cooke 1977), a prepattern of retinal mapping properties is present in the eye from the earliest optic vesicle stages (Hunt & Jacobson 1973, Gaze 1979). More directly analogous to the tectal graft experiment is the autonomous mapping of polyclones after translocation of a small nasal or ventral wedge of embryonic eye (Figures 2 and 5; Hunt & Ide 1977). If no labels were present on the optic nerve fibers, such polyclones should be recruited into a single, normal pattern over the surface of the tectum. Instead the projection pattern is clearly mosaic in nature, indicating that the fibers must somehow differ from one another depending on their position of origin. In addition, the majority result from tectal graft experiments (Figure 4) demands that the optic fibers possess the capacity to discriminate and locate tectal targets and associate with those targets in a manner determined by their position of origin. Finally, it is instructive to note that discriminator properties were confirmed conclusively in the studies showing limited modifiability of tectal properties (Schmidt 1978), and were assumed necessarily in all of the "minimum tectal information" models examined to date (Hope et al 1976, Prestige & Willshaw 1975, Hunt 1976, Von der Marlsberg & Willshaw 1977).

These two findings argue strongly for the presence of labels on the retina and tectum. The homophilic nature of these labels is chosen because of the recent work indicating that retinotopic ordering of nerve fibers is present in the optic nerve of all amphibians and fish so far investigated even if the optic nerve had been severed and hence partially scrambled (Sperry 1963, Horder 1974, Horder & Martin 1978). That is, fibers that originate from neighboring retinal ganglion cells and synapse on neighboring tectal cells tend to remain neighbors in the optic nerve. A homophilic gradient will produce a retinotopic nerve, whereas a heterophilic (Marchase et al 1975) gradient produces different ordering in the nerve. Figure 7 outlines the logic of these statements. Second, the homophilic nature of the gradients is consistent with findings in goldfish (Schmidt 1978, Sharma 1979) that fibers may be partially guided in the tectum by the debris of their degenerating terminals. In contrast, debris guidance in response to a heterophilic labeling would produce a reversed map, if it had any effect at all (see Figure 7).

In addition to the above arguments, some recent experimental evidence indicates that fiber affinities may play an important role in the ordering of the retinotectal map. Meyer (1979b, in press) has found that the pattern of an expanded projection in goldfish maximizes neighbor interactions with already present optic nerve fibers at the expense of incorrect fiber tectum ordering. This demonstrates that, in at least some settings, fiber-fiber affinities may play as major a role as fiber-tectum affinities. The only type of labeling consistent with these near neighbor affinities is homophilic (see Figure 7).

The repulsive force "R" is argued for by retinotectal mismatch experiments. As reviewed above, a subset of the normal complement of optic nerve fibers can expand to cover the entire tectum. A force stronger than the affinity between optic nerve fibers and their correct tectal locus could overpower the tectal labels in the mismatch experiments. This would result in an expanded projection in the absence of any regulation.

Two recently devised preparations are particularly useful in considering the driving role of the "R" force in retinotectal expansions. In the late LEO ("left-eye-out") preparation, we have induced a subpopulation of the optic nerve fibers from one eye to make a crossing error at the chiasm and innervate both of optic tecta by removing the left eye of *Xenopus* larvae at late stages of development (Fraser & Hunt 1978, Fraser 1979). Because retinal cells are only added at the periphery of the retina, the only fibers involved in the crossing error are of peripheral origin. The crossing error results in the same subregion of retina (*a*) involved in a portion of the normal contralateral projection and (*b*) innervating a part of the ipsilateral tectum that is otherwise denervated. The projection pattern assayed four months after left eye removal indicates that the region of retina mediates a part of the normal projection to the contralateral tectum and an expanded projection that covers the entire dorsal surface of the ipsilateral tectum.

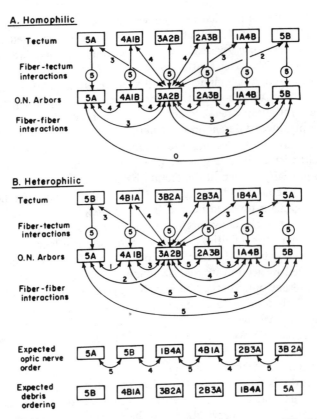

Figure 7 To demonstrate differences between homophilic (*a*) and heterophilic (*b*) labeling of the optic nerve. The small boxes represent cells of the tectum and optic nerve arbors; the numbers inside these boxes represent the chemical labeling of the cells. Arrows and the associated numbers indicate the number of energetically favorable contacts between the cells interconnected by the arrows. For homophilic labeling this is the number of A to A contacts plus B to B contacts. For heterophilic labeling, this is the number of A to B contacts. Note that both types of labeling produce a best fit with the correct cell in the tectum for each fiber. Homophilic labeling produces high affinity neighbor interactions with only nearest neighboring cells, whereas heterophilic labeling does not. The order expected in the optic nerve and the expected debris guidance from heterophilic labeling are shown.

Thus, the same region of retina can map in two different patterns solely on the basis of the presence of fibers from other regions of the retina. In goldfish, similar results have been obtained (Meyer 1978). Following surgical deflection of a bundle of optic nerve fibers to a denervated tectum, the fibers expand to cover the entire surface of the optic tectum (see Figure 8b). The remainder of the retina is involved in a normal projection to the contralateral tectum making regulation an unlikely candidate for the observed results.

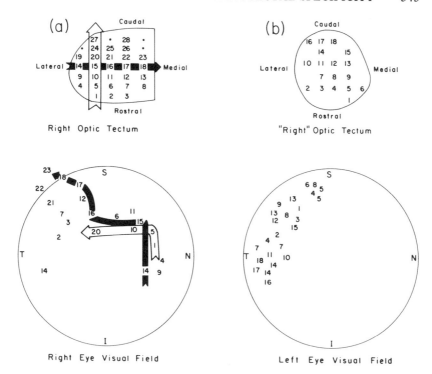

Figure 8 To demonstrate experimental results that show expansion of a portion of the normal retina's optic fibers can expand to cover the entire tectum. (*a*) In *Xenopus* a portion of the peripheral retina can be induced to directly innervate both the normal contralateral tectum and the denervated ipsilateral tectum by removal of the left eye late in larval life. The same subregion of retina projects to the proper region of the contralateral tectum, but expands on the ipsilateral tectum to cover most if not all of the dorsal surface of the tectum. (*b*) In goldfish, surgical deflection of a small bundle of optic nerve fibers to the ipsilateral (denervated) tectum leads to an expansion of these fibers over the surface of the tectum (after Meyer 1978). In both of these preparations, the rest of the retina is involved in a normal projection, reducing the possibility that the retina had undergone some sort of regulation, and instead argues for fiber interaction as the driving force behind expansions.

These two experimental preparations confirm that fibers can expand over the surface of the tectum in the absence of any regulation. Thus, to reconcile the presence of tectal and retinal labels with the phenomenon of expansion we must propose the existence of a repulsive interaction. In addition, both of these preparations have characteristic pattern alterations in the observed retinotectal projection. These patterns provide further tests for any model and more evidence for use in assigning the parameters of the synthetic model considered here.

The evidence for the "C" force is similar to that for a repulsive interaction. There must be an interaction more energetically favorable than the repulsive force to serve as the driving force for compressions. In addition, the amazing ability of the optic nerve fibers to find the optic tectum argues for some type of interaction that makes fiber-tectum interactions highly favorable. Experimental evidence for compression of the retinotectal projection following removal of a part of the optic tectum is depicted in Figure 3a. Note that the entire population of optic nerve fibers appears to be compressed onto the remaining tectal fragment. Such compressed projections have been shown to be functional by both electrophysiology (Yoon 1975) and behavioral testing (Scott 1977). Finally, introduction of the "C" force eliminates the need to tune the "R" force (Prestige & Willshaw 1975), by some ad hoc mechanism to achieve both expansions and compressions.

Applications and Implications of the Synthetic Model

We can now begin a test of this synthetic model. Using physical reasoning we can easily check some of the major predictions of the model. "C" is the largest force and would be expected to dominate. An example of this is a mismatch experiment in which a whole eye projects to a half tectum. The pattern observed immediately after regeneration of the nerve will be that of a half retina connected to the half tectum. This pattern is not stable, however, because the fibers from half of the eye will have no tectal contacts and hence, no "C" interactions. The fibers can gain "C" contacts by invading the tectal half. Both the "R" and "AP" (and "DV") forces will oppose this invasion but since "C" is much larger, the invasion still occurs. Once the fibers have invaded the tectum, "C" will no longer play a role and "R" would be expected to dominate. This would lead to the fibers becoming spaced more or less evenly across the tectum.

In a mismatch experiment between a half retina and a whole tectum, "C" would have no effect since no optic nerve fibers will be deprived of contact with the tectum. Instead, "R" will be the dominant force. Since "R" force starts only after synaptic contact with the tectum, the initial pattern should be the half retina connecting to only the appropriate half tectum. After this map is formed the "R" force would lead to an expansion of the half retina over the entire surface of the tectum. Again, "AP" and "DV" will oppose this expansion but "R" is a much larger force and will dominate. The equilibrium projection pattern should then be a normally ordered, but expanded, projection over the entire tectum. Note that the model predicts both patterns observed at different times after the regeneration of the optic nerve.

We can go beyond these simple thought experiments to exact solutions by means of digital computer simulations (Figure 9). The patterns shown are the final state, lowest energy (majority result) configurations. The simulations were performed by reiterative calculations of the energy of each optic nerve fiber in its present and neighboring positions. If the adhesive free

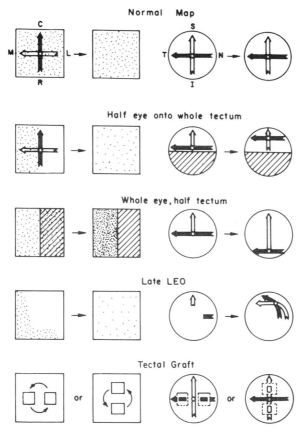

Figure 9 To show some of the computer simulations of the synthetic model. The squares to the left represent the tectum; the stippling represents the density of optic innervation. The exact pattern of the computer simulations have been translated into the visual field maps shown to the right. Notice that the same values for all of the parameters produces a stable normal map, plasticity results, and tectal graft results. In addition, note that the model produces the characteristic pattern of the late LEO preparation (compare with Figure 7*a*).

energy was lower at any of the neighboring positions, the fiber was moved to that position. This amounts to each fiber moving to its local minimum adhesive free energy. The arbors moved within a few iterations to the equilibrium positions. In particular note that the model is consistent with the classic phenomena of retinotectal research: expansions, compressions, and tectal graft data.

One of the major implications of making "AP" and "DV" the minor forces generating the retinotectal projection is that the system can then produce multiple metastable equilibria. The fibers can only overcome a relatively small energy barrier (Rule VI). This means that the fibers could be trapped in a region with a local minimum, much higher than the lowest energy state, but separated from lower energy states by an energy barrier greater than the fibers can cross. A compelling example of this comes from tectal graft experiments. As mentioned above, the majority result of transposing small pieces of tectal tissue is that the fibers follow the tissue and make synapses appropriate for the transposed tissue (Hunt 1979a,b; Hope et al 1976, Rho 1978). In the minority result, the fibers appear not to follow the graft and not observe the polarity of the tectal transplant (Jacobson & Levine 1975a,b; Hunt 1976). More recently, rearrangements that alter both position and orientation of tectal tissue have produced another minority result: the graft fails to attract the fibers that were previously connected with it, yet the new group of (inappropriate) fibers that regenerate to the graft observe the polarity of the grafted piece of tectal tissue (Rho 1978, Martin 1978). These minority results correspond with metastable configurations predicted directly from the model. They are separated by rather large energy barriers from the lower energy states. In the case of the second minority result, fibers can be trapped by the mass of correctly positioned neighboring fibers and thereby kept from finding more appropriate tectal tissue to innervate. If trapped in this way, their minimum energy configuration would be an orientation correct for the graft's polarity. The first minority result corresponds to a higher energy state in which the fibers order with respect to their neighbors outside the graft. To move into a position in accord with the polarity of the graft, the fibers would first have to sacrifice their neighbor contacts in exchange for tectal contacts that are not readily available to them. Thus, they would have to move to a higher energy configuration before they could drop into the lower energy configuration. It should be mentioned that the relative energies for these two minority configurations depends on the size of the tectal graft involved. For extremely small grafts, the amount of neighbor contacts increases with relation to the amount of tectal contact, thereby making the first minority result of lower energy. This correspondence between metastable equilibria and minority results is an important advantage for the model since it allows mixed results to be generated by a single set of rules.

Other classes of results in which fiber-fiber affinities dominate over fiber-tectum affinities are also predicted by the model. As discussed above, Meyer (1979a, in press) has discovered that fiber affinities can produce a reversed projection over half of a tectum. In addition, Martin (Horder & Martin 1978) has found that a reversed projection pattern may be induced in a retinotectal mismatch experiment. He removes one half of the tectum and the nonhomologous half of the retina, leaving a small band of tectal tissue still innervated at the cut edge. The pattern assayed after allowing time for regeneration of the severed optic nerve fibers is not the polarity-appropriate but shifted pattern, typically observed where no band of innervated tectum is spared (Yoon 1972), but instead a backwards projection, as though the pattern was flipped over the innervated band. This result demonstrates the fiber-fiber affinities predicted by the model. The band of innervated tectum plays two important roles. First, since the fibers in this band possess well elaborated arbors and the affinity is mediated by a cell surface molecule, the fibers left in this innervated band will adhere strongly to the tectum. This means that it will be extremely difficult for the incoming, regenerating arbors to displace the already established arbors. Thus, the minimum energy position for the incoming fibers would be the tectal tissue next to the innervated band. Second, the increased surface area of the arbors in the innervated band will increase the amount of fiber-fiber interaction possible with the incoming fibers and should bias the pattern slightly towards that dictated by the fiber-fiber affinities. Since the incoming fibers cannot induce the fibers in the band to move to new synaptic sites, the only means by which the incomers can maximize their fiber-fiber affinities is to project backwards over the remaining half-tectum.

The importance of fiber-fiber affinities is further highlighted by the fascicule transplant experiments of Meyer (1978). Following the surgical translocation of 10% of the normal contralateral projection to a denervated tectum, the expanded projection is found to be much less ordered than normal (see Figure 8). The spreading of the optic nerve terminals would be expected to greatly reduce the amount of neighbor interactions in the expanded projection. The extremely small force produced by fiber-tectum affinities alone could then be partially lost in thermal noise. In the normal projection pattern sufficient fiber-fiber affinities exist to at least double the depth of the "energy well" experienced by any one optic nerve arbor. It is thus predicted that the decreased interaction would yield the decreased order actually observed. In addition, these fiber-fiber interactions, participating with the other ranked forces in the synthetic model, produce not only the slight disorganization but also the systematic topographic distortion seen in the late LEO preparation (Figures 8 and 9).

There have been a number of recent findings that may ultimately lead us to further refine the parameters of the model. The "ocular dominance pseudocolumns" found in lower vertebrates (Levine & Jacobson 1975, Constantine-Patton & Law 1978, Meyer, 1978, 1979b, in press) are not directly predicted by the model in its present form. It is possible that the increase in optic nerve fiber density, due to doubly innervating a single tectum, increases the amount of fiber-fiber interaction and causes fiber sorting, not normally seen, to become apparent. A slight preference for fibers from the same eye would thereby produce the local domains seen in these patterns but a mechanism to generate the orderly patterns seen in some preparations (Constantine-Patton & Law 1978) is not readily apparent. In addition, Freeman's (1977) finding, that applications of α-bungarotoxin to a small area of tectum displaces the optic nerve fibers from the area, is not readily explainable by this model. Unless the toxin provides some nonspecific surface effect, the experiments indicate that the functionality of the synapse plays some role in the fiber's target selection. These findings offer a challenge to further refinements of the model.

CONCLUSION

We have attempted to review some of the key experimental and theoretical issues presented by the visual system of lower vertebrates. The demonstration of plasticity in the absence of regulation, and the presence of tectal labels independent of a previous retinotectal projection, are incompatible with some of the available models. A synthetic model is presented and found to be consistent with both findings. The model predicts the two-stage expansions and compressions in retinotectal mismatch experiments that have been problematic for other models. In addition, this new model predicts the mixed results commonly encountered in experiments on the retinotectal system. Since a single model can fit most of the available data, we can calculate a mapping function. This offers the promise of going beyond simple explanations of the observed retinotectal projection patterns and should allow investigations into the far more intriguing questions concerning the nature of the information that serves to mold these patterns.

ACKNOWLEDGMENTS

We wish to thank Drs. R. A. Cone, M. Steinberg, and R. Meyer for helpful discussions, L. Hentges for expert assistance in assembling the manuscript, K. Conway, M. Duda, and K. Feiock for supplying the chimeric eye reconstruction. Work in the authors' lab supported by grants from NIH (NS12606), NSF (PCM77–26987) to R. K. H., and NIH (GM07231–04) to S. E. F. R. K. H. is a Sloan Foundation fellow. We thank JHU Dean's fund for supplying funds for computer simulations.

Literature Cited

Attardi, D. G., Sperry, R. W. 1963. Preferential selection of central pathways by regenerating optic fibers. *Exp. Neurol.* 7:46–64

Barbera, A. J. 1975. Adhesive recognition between developing retinal cells and the optic tecta of the chick embryo. *Dev. Biol.* 49:338–46

Barondes, S. H. 1970. Brain glycomacromolecules and interneuronal recognition. In *The Neurosciences: Second Study Program,* ed. F. O. Schmitt, pp. 747–60. New York: Rockefeller Univ. Press

Beech, D. H. 1977. PhD thesis. Univ. Miami, Miami

Bergey, G. K., Hunt, R. K., Holtzer, H. 1973. Selective effects of bromodeoxyuridine on developing *Xenopus laevis* retina. *Anat. Rec.* 175:271

Berman, N., Hunt, R. K. 1975. Visual projections to the optic tecta in *Xenopus* after partial extirpation of the embryonic eye. *J. Comp. Neurol.* 162:23–42

Bodenstein, L., Hunt, R. K. 1979. A model for cellular position determination in growing *Xenopus* retina: Components of the model and their operation. *Proc. Natl. Acad. Sci. USA.* In press

Chung, S. H., Cooke, J. 1975. Polarity of structure and of ordered nerve connections in the developing amphibian brain. *Nature* 258:126–32

Chung, S. H., Gaze, R. M.. Stirling, R. V. 1973. Abnormal visual function in *Xenopus. Nature New Biol.* 246:186–89

Chung, S. H., Bliss, T. V. P., Keating, M. J. 1974a. The synaptic organization of optic afferents in the amphibian tectum. *Proc. R. Soc. London Ser. B* 187:421–47

Chung, S. H., Keating, M. J., Bliss, T. V. P. 1974b. Functional synaptic relations during the development of the retinotectal projection in amphibians. *Proc. R. Soc. London Ser. B* 187:449–59

Chung, S. H., Stirling, R. V., Gaze, R. M. 1975. The structural and functional development of the retina in larval *Xenopus. J. Embryol. Exp. Morphol.* 33:915–40

Constantine-Patton, M., Law, M. I. 1978. Eye-specific termination bands in tecta of three-eyed frogs. *Science* 202:639–41

Conway, K., Feiock, K., Hunt, R. K. 1979. Marker mutant mosaics in *Xenopus* retina. *Curr. Top. Dev. Biol.* Vol. 16. In press

Cook, J. E., Horder, T. J. 1977. Multiple factors determining the retinotopic order in the growth of the optic nerve fibers into the optic tectum. *Philos. Trans. R. Soc. London Ser. B* 278:261–76

Cooke, J. 1977. Organizing principles for anatomical patterns and for selective nerve connections in the developing amphibian brain. In *Cell Interactions in Differentiation,* ed. M. Karkinen-Jaaskelainen, L. Saxen, L. Weiss, pp. 111–24. London: Academic

Crossland, W. J., Cowan, W. M., Rogers, L. A., Kelly, J. P. 1974. The specification of the retinotectal projection in the chick. *J. Comp. Neurol.* 155:127–64

Currie, J., Cowan, W. M. 1974. Some observations on the early development of the optic tectum in the frog (*Rana pipiens*) with special reference to the effects of early eye removal on mitotic activity in the larval tectum. *J. Comp. Neurol.* 156:123–42

Diamond, J., Cooper, E., Turner, C., Macintyre, L. 1976. Trophic regulation of nerve sprouting. *Science* 193:371–77

Edds, M. V., Gaze, R. M., Schneider, G. E., Irwin, N. I. 1979. Specificity and plasticity of retinotectal connections. *Neurosci. Res. Program* 000:000–000

Edwards, P. A. W. 1978. Differential cell adhesion may result from nonspecific interactions between cell surface glycoproteins. *Nature* 271:248–49

Feldman, J. D., Gaze, R. M. 1975. The development of the retinotectal projection in *Xenopus* with one compound eye. *J. Embryol. Exp. Morphol.* 33:775–87

Feldman, J. D., Gaze, R. M., Keating, M. J. 1971. Delayed innervation of the optic tectum during development in *Xenopus laevis. Exp. Brain Res.* 14:16–23

Fraser, S. E. 1979. Late LEO: A new system for the study of neuroplasticity in *Xenopus.* In *Developmental Neurobiology of Vision,* ed. W. Singer, J. Freeman. New York: Plenum

Fraser, S. E., Hunt, R. K. 1978. Neuroplasticity in *Xenopus. Biophys. J.* 21:110a

Freeman, J. A. 1977. Possible regulatory function of acetylcholine receptor in maintenance of retinotectal synapses. *Nature* 269:218–22

Frost, D. O., Schneider, G. E. 1976. Normal and abnormal uncrossed retinal projections in Syrian hamsters as demonstrated by Fink-Heimer and autoradiographic techniques. *Soc. Neurosci. Abst., 6th Ann. Meet.* 2:812

Gaze, R. M. 1960. Regeneration of the optic nerve in Amphibia. *Int. Rev. Neurobiol.* 2:1–40

Gaze, R. M. 1970. *Formation of Nerve Connections.* New York: Academic

Gaze, R. M. 1978. The problem of specificity in the formation of nerve connections. In *Specificity of Embryological Interactions,* ed. D. Garrod, pp. 1–29 London: Chapman & Hall. In press

Gaze, R. M. 1979. Comments on the formation of retinotopic maps. *Curr. Top. Dev. Biol.* Vol. 16. In press

Gaze, R. M., Grant, P. 1978. The diencephalic course of regenerating retinotectal fibers in *Xenopus laevis. J. Embryol. Exp. Morphol.* 44:201–16

Gaze, R. M., Sharma. S. C. 1970. Axial differences in the reinnervation of the goldfish optic tectum by regenerating optic nerve fibers. *Exp. Brain Res.* 10:171–81

Gaze, R. M., Keating, M. J., Chung, S. H. 1974. The evolution of the retinotectal map during development in *Xenopus. Proc. R. Soc. London Ser. B* 185:301–30

George, S. A., Marks, W. B. 1974. Optic nerve terminal arborizations in the frog: Shape and orientation inferred from electrophysiological measurements. *Exp. Neurol.* 42:467–82

Goldberg, S. 1976. Polarization of the avian retina: Ocular transplantation studies. *J. Comp. Neurol.* 168:379–92

Harrison, R. G. 1935. On the origin and development of the nervous system studied by the methods of experimental embryology. *Proc. R. Soc. London Ser. B* 118:155–96

Head, H. 1920. *Studies in Neurology,* Vol. 2. London: Henry Frowde, Hodder & Stoughton, Ltd.

Herring, E. 1913. *Memory: Lectures on the Specific Energies of the Nervous System.* Chicago: Open Court

Hibbard, E. 1967. Visual recovery following regeneration of the optic nerve through oculomotor nerve root in *Xenopus. Exp. Neurol.* 19:350–56

Hollyfield, J. G. 1971. Differential growth of the neural retina in *Xenopus laevis* larvae. *Dev. Biol.* 24:264–86

Holtfreter, J. 1939. Gewebaffinitat, ein Mittel der embryonalen Formbildung. *Arch. Exp. Zellforsch. Besonders Gewebezuecht.* 23:169–209

Hope, R. A., Hammond, B. J., Gaze, R. M. 1976. The arrow model: Retinotectal specificity and map formation in the goldfish visual system. *Proc. R. Soc. London Ser. B* 194:447–66

Horder, T. J. 1974. Changes of fiber pathways in the goldfish optic tract following regeneration. *Brain Res.* 72:41–52

Horder, T. J., Martin, K. A. C. 1978. Morphogenesis as an alternative to chemospecificity in the formation of nerve connections. *Symp. Soc. Exp. Biol.* 32:275–358

Huang, R. T. C. 1978. Cell adhesion mediated by glycolipids. *Nature* 276:624–26

Hunt, R. K. 1975a. Developmental programming for retinotectal patterns. In *Cell Patterning,* ed. R. Porter, J. Rivers, pp. 131–50. Ciba Found. Symp., No. 29 (NS). Amsterdam: Associated Science

Hunt, R. K. 1975b. The cell cycle, cell lineage, and neuronal specificity. In *Cell Cycle and Cell Differentiation,* ed. J. Reinart, H. Holtzer, pp. 43–62. Berlin & New York: Springer

Hunt, R. K. 1976. Position dependent differentiation of neurons. In *Developmental Biology: Pattern Formation: Gene Regulation,* ed. D. MacMahon, C. F. Fox, ICN-UCLA Symp. New York: Benjamin

Hunt, R. K. 1977a. Positional signalling and nerve cell specificity. In *Cell Interactions in Differentiation,* ed. M. Karkinen-Jaaskelainen, L. Saxen, L. Weiss, pp. 97–110. London: Academic

Hunt, R. K. 1977b. Competitive retinotectal mapping in *Xenopus. Biophys. J.* 17:128a

Hunt, R. K. 1978. Genetic control of neural development. In *Cell, Tissue, and Organ Culture in Neurobiology,* ed. S. Fedoroff, L. Hertz, pp. 369–92. New York: Academic

Hunt, R. K. 1979a. Combinatorial specifiers on retinal ganglion cells for retinotectal map assembly in *Xenopus. Nature.* In press

Hunt, R. K. 1979b. Target properties in the optic tectum for retinotectal map assembly in *Xenopus. Nature.* In press

Hunt, R. K., Berman, N. 1975. Patterning of neuronal locus specificities in retinal ganglion cells after partial extirpation of the embryonic eye. *J. Comp. Neurol.* 162:43–70

Hunt, R. K., Frank, E. 1975. Neuronal locus specificity: Transrepolarization of *Xenopus* embryonic retina after the time of axial specification. *Science* 189:563–65

Hunt, R. K., Ide, C. F. 1977. Radial propagation of positional signals for retinotectal patterns in *Xenopus. Biol. Bull.* 153:430–31

Hunt, R. K., Jacobson, M. 1972. Development and stability of positional information in *Xenopus* retinal ganglion cells. *Proc. Natl. Acad. Sci. USA* 69:780–83

Hunt, R. K., Jacobson, M. 1973. Specification of positional information in *Xenopus* retinal ganglion cells. *Proc. Natl. Acad. Sci. USA* 70:507–11

Hunt, R. K., Jacobson, M. 1974. Neuronal specificity revisited. In *Curr. Top. Dev. Biol.* 8:203.

Ide, C. F., Hunt, R. K. 1978. Positional signalling in chimeric *Xenopus* retinae. *Biophys. J.* 21:110a

Ide, C. F., Kosofsky, B. E., Hunt, R. K. 1979. Control of pattern duplication in the retino-tectal system of *Xenopus I. Dev. Biol.* 69:337–60

Jacobson, M. 1968a. Development of neuronal specificity in retinal ganglion cells of *Xenopus. Dev. Biol.* 17:202–18

Jacobson, M. 1968b. Cessation of DNA synthesis in retinal ganglion cells correlated with the time of specification of their central connections. *Dev. Biol.* 17:219–32

Jacobson, M. 1976. Histogenesis of retina in the clawed frog with implications for the pattern of development of retinotectal connections. *Brain Res.* 103:541–45

Jacobson, M. 1978. *Developmental Neurobiology.* New York: Plenum. 2nd ed.

Jacobson, M., Hunt, R. K. 1973. The origins of nerve cell specificity. *Sci. Am.* 228 (2):17–26

Jacobson, M., Levine, R. L. 1975a. Plasticity in the adult frog brain: Filling the visual scotoma after excision or translocation of parts of the optic tectum. *Brain Res.* 88:339–45

Jacobson, M., Levine, R. L. 1975b. Stability of implanted duplicate tectal positional markers serving as targets for optic axons in adult frogs. *Brain Res.* 92:468–71

Jehle, H. 1963. Intermolecular forces and biological specificity. *Proc. Natl. Acad. Sci. USA* 50:516–24

Langley, J. W. 1898. On the union of cranial autonomic (visceral) fibers with the nerve cells of the superior cervical ganglion. *J. Physiol. London* 23:240–70

Lazar, G. 1973. The development of the optic tectum in *Xenopus laevis:* A Golgi study. *J. Anat.* 116:347–55

Levine, R. L., Jacobson, M. 1974. Deployment of optic nerve fibers is determined by positional markers in the frog's tectum. *Exp. Neurol.* 43:527–38

Levine, R. L., Jacobson, M. 1975. Discontinuous mapping of retina onto tectum innervated by both eyes. *Brain Res.* 98:172–76

Ling, R. T., Ide, C. F., Hunt, R. K. 1979. Control of pattern duplication in the retinotectal system of *Xenopus,* II. *Dev. Biol.* 69:361–74

Longley, A. 1978. Anatomical mapping of retino-tectal connections in developing and metamorphosed *Xenopus:* Evidence for changing connections. *J. Embryol. Exp. Morphol.* 45:249–70

MacDonald, N. 1975. Appendix: The development of half eyes in *Xenopus* tadpoles. *J. Comp. Neurol.* 162:13–22

MacDonald, N. 1977. A polar co-ordinate system for positional information in the vertebrate neural retina. *J. Theor. Biol.* 69:153–65

Marchase, R. B. 1977. Biochemical investigations of retinotectal adhesive specificity. *J. Cell. Biol.* 75:237–57

Marchase, R. B., Barbera, A. J., Roth, S. 1975. A molecular approach to retinotectal specificity. In *Cell Patterning,* ed. R. Porter, J. Rivers, pp. 315–27. Ciba Found. Symp. No. 29 (NS). Amsterdam: Associated Science

Martin, K. A. C. 1978. Combination of fibre-fibre competition and regional tectal differences accounting for the results of tectal graft experiments in goldfish. *J. Physiol. London* 276:44–45

Meyer, R. L. 1975. Tests for reguation in the goldfish retinotectal system. *Anat. Rec.* 181:427

Meyer, R. L. 1978. Deflection of selected optic fibers into a denervated tectum in goldfish. *Brain Res.* 155:213–27

Meyer, R. L. 1979a. Retinotectal projection in goldfish to an inappropriate region with a reversal in polarity. *Science.* In press

Meyer, R. L. 1979b. "Extra" optic fibers exclude normal fibers from tectal regions in goldfish. *J. Comp. Neurol.* In press

Meyer, R. L., Sperry, R. W. 1973. Tests for neuroplasticity in the anuran retinotectal system. *Exp. Neurol.* 40:525–39

Potter, H. D. 1972. Terminal arborizations of retinotectal axons. *J. Comp. Neurol.* 144:269–84

Prestige, M. C., Willshaw, D. J. 1975. On a role for competition in the formation or patterned neural connexions. *Proc. R. Soc. London Ser. B* 190:77–98

Ramon y Cajal, S. 1928. *Degeneration and Regeneration of the Nervous System.* Transl. R. M. May, 1959. New York: Hafner

Rho, J. H. 1978. Cell interactions in retinotectal mapping. *Biophys. J.* 21:137a

Rho, J. H., Hunt, R. K. 1979. The retinotectal projection to a surgically prepared double lateral tectum. *Dev. Biol.* In press

Rubin, J., Grant, P. 1979. Fiber patterns in *Xenopus* retina. *J. Comp. Neurol.* In press

Schmidt, J. T. 1978. Retinal fibers alter tectal positional markers during the expansion of the half retinal projection in goldfish. *J. Comp. Neurol.* 177:279–300

Schmidt, J. T., Easter, S. S. 1978. Independent biaxial reorganization of the retinotectal projections: A reassessment. *Exp. Brain Res.* 31:155–62

Schneider, G. E., Jhaveri, S. R. 1974. Neuroanatomical correlates of spared or altered function after brain lesion in the newborn hamster. In *Plasticity and Recovery of Function in the Central Nervous System,* ed. D. G. Stein, J. J. Rosen, N. Butters, pp. 65–109. New York: Academic

Scott, M. Y. 1977. Behavioral tests of compression of retinotectal projection after partial tectal ablation in goldfish. *Exp. Neurol.* 54:579–90

Scott, T. M., Lazar, G. 1976. An investigation into the hypothesis of shifting neuronal relationships during development. *J. Anat.* 121:485–96

Sharma, S. C. 1972. Reformation of the retinotectal projections after various tectal ablations in adult goldfish. *Exp. Neurol.* 34:171–82

Sharma, S. C. 1975. Visual projection in surgically created 'compound' tectum in adult goldfish. *Brain Res.* 93:497–501

Sharma, S. C. 1979. See Edds et al 1979, pp. 000

Sharma, S. C., Gaze, R. M. 1971. The retinotopic organization of visual responses from tectal reimplants in adult goldfish. *Arch. Ital. Biol.* 109:357–66

Sharma, S. C., Romeskie, M. 1977. Immediate "compression" of the goldfish retinal projection to a tectum devoid of degenerating debris. *Brain Res.* 133: 367–70

Sharma, S. C., Tung, Y. L. 1979. Competition between nasal and temporal heteronymous hemiretinal fibers in adult goldfish tectum. *J. Physiol. London.* In press

Skarf, B. 1973. Development of binocular single units in the optic tectum of frogs raised with disparate stimulation to the eyes. *Brain Res.* 51:352–57

Skarf, B., Jacobson, M. 1974. Development of binocularly-driven single units in frogs raised with asymmetrical visual stimulation. *Exp. Neurol.* 42:669–86

Sperry, R. W. 1945. Restoration of vision after uncrossing of optic nerves and after contralateral transposition of the eye. *J. Neurophysiol.* 8:15–28

Sperry, R. W. 1947. Effect of crossing nerves to antagonistic limb muscles in the monkey. *Arch. Neurol. Psychiat. Chicago* 58:452–73

Sperry, R. W. 1950. Neuronal specificity. In *Genetic Neurology,* ed. P. Weiss. Chicago: Univ. Chicago Press

Sperry, R. W. 1963. Chemoaffinity in the orderly growth of nerve fiber patterns and connections. *Proc. Natl. Acad. Sci. USA* 50:703–10

Sperry, R. W. 1965. Embryogenesis of behavioral nerve nets. In *Organogenesis,* ed. R. L. DeHaan, H. Ursprung, pp. 161–86. New York: Holt, Rinehart & Winston

Sperry, R. W., Miner, N. 1949. Formation within sensory nucleus V of synaptic associations mediating cutaneous localization. *J. Comp. Neurol.* 90:403–23

Steinberg, M. S. 1970. Does differential adhesion govern self-assembly processes in

histogenesis? Equilibrium configurations and the emergence of a hierarchy among populations of embryonic cells. *J. Exp. Zool.* 73:395–434

Straznicky, K. 1973. The formation of the optic fiber projection after partial tectal removal in *Xenopus. J. Embryol. Exp. Morphol.* 29:397–409

Straznicky, K. 1979. The acquisition of tectal positional specification in *Xenopus. Proc. Aust. Physiol. Pharmacol. Soc.* In press

Straznicky, K., Gaze, R. M. 1971. The growth of the retina in *Xenopus laevis:* An autoradiographic study. *J. Embryol. Exp. Morphol.* 26:67–79

Straznicky, K., Gaze, R. M. 1972. The development of the tectum in *Xenopus laevis:* An autoradiographic study. *J. Embryol. Exp. Morphol.* 28:87–115

Straznicky, K., Tay, D. 1977. Retinal growth in normal double dorsal and double ventral eyes in *Xenopus. J. Embryol. Exp. Morphol.* 40:175–85

Tello, J. F. 1915. *Probl. Lab. Invest. Biol. Univ. Madrid* 15:101–99

Udin, S. B. 1978. Permanent disorganization of the regenerating optic tract in the frog. *Exp. Neurol.* 58:455–70

von der Malsburg, C., Willshaw, D. J. 1977. How to label nerve cells so that they can interconnect in an orderly fashion. *Proc. Natl. Acad. Sci. USA* 74:5176–78

Weiss, P. 1928. Erregungsspezifitat und Erregungsresonanz. Grundzuge einer theorie der motorischen Nerventatigkeit auf Grund spezifischer Zuordnung ("Abstimmung") zwischen zentraler und peripherer Erregungsform. *Ergeb. Biol.* 3:1

Weiss, P. 1942. Lid-closure reflex from eyes transplanted to atypical locations in *Triturus torosus:* Evidence of a peripheral origin of sensory specificity. *J. Comp. Neurol.* 77:131–69

Yoon, M. G. 1972. Transposition of the visual projection from the nasal hemiretina onto the foreign rostral zone of the optic tectum in goldfish. *Exp. Neurol.* 37:451–62

Yoon, M. G. 1975. Topographic polarity of the optic tectum studied by reimplantation of the tectal tissue in adult goldfish. In *Cold Spring Harbor Symp Quant. Biol.* 40:503–19

Ann. Rev. Neurosci. 1980. 3:353–402
Copyright © 1980 by Annual Reviews Inc. All rights reserved

THE NERVE GROWTH FACTOR: Biochemistry, Synthesis, and Mechanism of Action

❖11546

Lloyd A. Greene

Department of Pharmacology, New York University, School of Medicine,
New York, NY 10016

Eric M. Shooter

Department of Neurobiology, Stanford University, School of Medicine,
Stanford, California 94305

INTRODUCTION

The discovery of nerve growth factor (NGF) in the early 1950s was a landmark in developmental neurobiology. Extending earlier observations (Buecker 1948) that the dorsal root ganglia of a chick embryo grew neurites into a mouse sarcoma 180 implanted on the chick body wall, Levi-Montalcini & Hamburger (1951, 1953) showed that sympathetic ganglia also participated in the nerve ingrowth to the tumor and demonstrated that the effects were due to a specific growth factor produced by the sarcoma. Levi-Montalcini (1975a) has written an interesting account of this early work, and her own highly original, insightful, and continuing work on NGF is discussed at length here and in the reviews listed at the end of this section. Although a great deal has been learned about the biological processes in which NGF is involved and about the growth factor itself, and at least a start has been made toward understanding the mechanisms by which NGF acts, not all of the functions of NGF have yet been clearly defined. It is

353

beyond the scope of this review to go into great detail about these functions; nevertheless a brief review is in order.

It is generally agreed that NGF plays a critical role in the development of sympathetic and some sensory neurons in vertebrates (Levi-Montalcini 1966, Levi-Montalcini & Angeletti 1968). Among the responses of these neurons to NGF are greatly increased neurite outgrowth and increase in cell body size (Levi-Montalcini 1966). In sympathetic ganglia NGF also brings about an increase in the synthesis of tyrosine hydroxylase and dopamine-β-hydroxylase (Thoenen et al 1971). Both in vitro and in vivo experiments have established that NGF plays a necessary role in the survival of sympathetic and some embryonic sensory neurons (Levi-Montalcini & Booker 1960, Levi-Montalcini & Angeletti 1963, 1966, Dolkart & Johnson 1979). NGF may also be critically involved in the maintenance of the differentiated state of these neurons (Hendry 1976, Stöckel & Thoenen 1975, Varon & Bunge 1978). In addition, there is evidence that NGF may have a chemotactic role in guiding the direction of growth of axons (Levi-Montalcini 1976, Campenot 1977, Letourneau 1978, Menesini-Chen et al 1978). Furthermore, the findings that regenerating catecholaminergic neurons in mammalian brain (Björklund & Stenevi 1972, Bjerre et al 1973) and optic nerve fibers in the newt may respond to NGF (Turner & Glaze 1977) and that brain contains specific NGF receptors (Szutowicz et al 1976a,b, Zimmerman et al 1978) suggest possible additional functions for this growth factor in the CNS. Recently, NGF has been shown to cause rat pheochromocytoma cells (Tischler & Greene 1975, Greene & Tischler 1976) and immature adrenal chromaffin cells both in vitro (Unsicker & Chamley 1977, Tischler & Greene 1979) and in vivo (Aloe & Levi-Montalcini 1979) to acquire neuronal characteristics such as neurite outgrowth. This suggests that NGF may play a direct role in directing or altering phenotypic expression of target cells during development and that the factor may affect mitotic cells as well as postmitotic neurons.

Previous reviews have dealt with many aspects of the discovery and characterization of the NGF phenomenon and with the use of NGF as a unique and important tool for probing the development of the peripheral nervous system (Levi-Montalcini 1966, Levi-Montalcini & Angeletti 1968, Hendry 1976, Varon 1975, Levi-Montalcini 1976, Bradshaw & Young 1976, Mobley et al 1977, Varon & Bunge 1978, Barde & Thoenen 1979). This review therefore presents a more detailed discussion of recent developments regarding the biochemistry of NGF and its mechanisms of action, and forms a complement to a number of recent reviews on these same subjects (Server & Shooter 1977, Bradshaw & Frazier 1977, Bradshaw 1978).

MOLECULAR PROPERTIES OF NERVE GROWTH FACTOR

Early History

In the last 30 years much has been learned about the biochemistry of NGF. A few key discoveries serve to document the progress in this area. The finding of a potent source of NGF in the male mouse submaxillary gland and the demonstration that the factor is a protein (Cohen 1960) were important events that finally led to isolation of the purified form of the NGF protein (Varon et al 1967a,b, 1968). At the same time, it was shown that the NGF protein is contained in a larger subunit-containing complex, 7S NGF. A few years later the amino acid sequence of NGF was determined (Angeletti & Bradshaw 1971) and this in turn led to suggestions of homologies between NGF and the insulin family of proteins (Frazier et al 1972, Bradshaw 1978). This knowledge, together with the finding that one of the other subunits in 7S NGF is an arginyl esteropeptidase (Greene et al 1968), led to a model for the biosynthesis of NGF from a pro-NGF (Berger & Shooter 1977, 1978).

The NGF Protein

The biologically active NGF molecule isolated from the adult male mouse submaxillary gland consists of two noncovalently linked, identical peptide chains (Greene et al 1971, Angeletti et al 1971, Pignatti et al 1975). Each chain contains 118 amino acid residues whose sequence has been determined (Angeletti & Bradshaw 1971, Angeletti et al 1973a,b). The chains have NH_2-terminal serine residues, COOH-terminal arginine residues, three internal disulfide bridges, and two and three residues respectively of tyrosine and tryptophan. The chains are subject to two specific and limited proteolytic modifications by enzymes in the submaxillary gland. A carboxypeptidase-B (CPB)-like enzyme releases the COOH-terminal arginine residues (Moore et al 1974) while an endopeptidase cleaves the eighth peptide bond between histidine and methionine residues to release the NH_2-terminal octapeptide sequences (Angeletti et al 1973a, Mobley et al 1976). The extent to which proteolysis occurs at these two sites depends on the method of isolation of the NGF protein. In the original procedure, in which the NGF protein was obtained from purified 7S NGF (Varon et al 1967b, Varon et al 1968), little proteolysis occurred and no more than 10% of the COOH-terminal arginine residues or NH_2-terminal octapeptide sequences were lost (Mobley et al 1976). This preparation is called the β-subunit or βNGF (Varon et al 1968). In procedures where the NGF protein was obtained from partially purified 7S NGF fractions (Bocchini & Angeletti

1969, Mobley et al 1976) substantial proteolysis occurred at both termini. These preparations are described as 2.5S NGF and contain up to ten possible dimeric forms of the NGF dimer (Bradshaw 1978). In this review the terms NGF or NGF protein are used to describe the biologically active subunit of 7S NGF unless it is important to distinguish between the various preparations.

THE STABILITY OF THE NGF DIMER   The NGF dimer is very stable, having a dissociation equilibrium constant of less than 10^{-13} M (Bothwell & Shooter 1978). No dissociation of the NGF dimer into monomers was detected by gel filtration of native NGF at 5 nM or of ^{125}I-NGF at concentrations as low as 5 pM. Only after extensive succinylation did ^{125}I-NGF show appreciable dissociation and even then only after prolonged incubation at concentrations of 10^{-11} M. Removal of the COOH-terminal arginine residues or of a substantial fraction of the NH$_2$-terminal octapeptide sequences did not affect the stability of the dimer. Using a method that measured the rate of exchange of monomers between NGF and bisdes-Arg^{118}-NGF and an estimate of $10^5 M^{-1} s^{-1}$ for the association rate constant, Moore & Shooter (1975) found the dissociation equilibrium constant to be 3×10^{-10} M at pH 4. Because the data indicated that the rate of exchange was much slower at higher pH and because rate constants for protein-protein association as great as 10^7 M^{-1} s^{-1} have been measured, it is reasonable to conclude that the equilibrium constant at neutral pH could be several orders of magnitude smaller than the value measured at pH 4. On the other hand, the equilibrium constant of 10^{-7} M obtained by Young et al (1976) is significantly at variance with the limit of 10^{-13} M. The discrepancy is probably due to the fact that the measurements of Young et al (1976) were obtained by chromatography of NGF on Sephadex G-75 under conditions in which NGF interacts with the gel. As a consequence, not only was an incorrectly low value obtained for the molecular weight, but the molecular weight did not show a dependency on protein concentration as would be expected for an equilibrium reaction (Bothwell & Shooter 1978). It is likely, therefore, that NGF occurs exclusively in the form of the dimer at physiologically relevant concentrations. In this regard it should be noted that covalently cross-linked βNGF (Stach & Shooter 1974) or 2.5S NGF (Pulliam et al 1975) showed the same biological activity in the ganglion bioassay and the same binding characteristics to the NGF receptors respectively as did native NGF. Evidence that the NGF monomer is also biologically active has been presented (Frazier et al 1973a).

THE CHEMISTRY OF THE NGF PROTEIN   The amino acid sequence of the NGF chain is homologous to that of proinsulin in the sections of the chain that correspond to the presumptive B and A chains of insulin (Frazier et al 1972). Moreover one of the interchain disulfide bridges of insulin (B19-A20) is conserved in the NGF structure. This relatedness extends also to the hormone relaxin (Schwabe & McDonald 1977) and to the insulinlike growth factors I and II (Rinderknecht & Humbel 1978a,b), particularly in the regions corresponding to the insulin B chain, and in the conservation of the specific disulfide bridge noted above (Bradshaw 1978, Rinderknecht & Humbel 1978a). The homologies also extend to the three-dimensional structure of the four proteins. The reactivities of the two tyrosine and two of the three tryptophan residues in the NGF chains toward specific chemical reagents are consistent with their being located in regions whose three-dimensional structure is similar to that of the corresponding residues in insulin (Frazier et al 1972). In contrast, Argos (1976), using structure prediction methods, found that the regions in NGF that were predicted to have an α-helical structure were not correlated with the known α-helical regions in insulin. The NGF dimer has been crystalized in hexagonal bipyramids and X-ray diffraction analysis has confirmed the hexagonal symmetry (Wlodawer et al 1975), but the structure analysis is not sufficiently advanced to allow detailed comparisons of the three-dimensional structures of NGF with those of insulin or relaxin.

The receptor-binding site of NGF has not yet been characterized (Bradshaw 1978). Modification of one or both tyrosine residues of NGF, either by nitration (Frazier et al 1973b) or iodination (Herrup & Shooter 1973, Sutter et al 1979), had no effect on biological activity. Similarly, most if not all the lysine residues are nonessential for biological activity, because they can be modified by treatment with dimethylsuberimidate (Stach & Shooter 1974, Pulliam et al 1975), by acetylation or succinylation, or by attachment to CNBr-activated Sepharose beads (Bradshaw et al 1977) without loss of NGF activity. However complete modification of the arginine residues by reaction with 1,2-cyclohexanedione did inactivate NGF (Bradshaw et al 1977). The three tryptophan residues in the NGF monomer are located at residues 21, 76, and 99 respectively (Angeletti et al 1973a) and are converted to oxindole derivatives by N-bromosuccinimide at different rates (Angeletti 1970, Frazier et al 1973b). When sufficient reagent was used to modify only one tryptophan residue per monomer, the reaction occurred exclusively at Trp-21. Increasing amounts of the reagent subsequently modified Trp-99 and Trp-76, in that order (Frazier et al 1973b). Even modification of the most reactive Trp-21 residue altered the biological activity of NGF (Frazier et al 1973b, P. Cohen et al, unpublished). As judged by the response of both embryonic chick sensory neurons and the pheochromocytoma clone PC12, the biological activity of the derivative with Trp-21 modified was 3 to 5% that of native NGF (P. Cohen et al, unpublished). The decrease in biological activity was paralleled by a decrease in the binding to NGF receptors. The affinity of this derivative for binding to the higher affinity NGF receptors on sensory cells was 2.5%, and on PC12 cells, 1.5% that of native NGF. That the residual binding was not due to small amounts of unmodified NGF was evident from a cytotoxic assay for NGF (Zimmerman et al 1978), which placed the maximum level of the latter at 0.2%. The derivative with two tryptophans modified per monomer had only 0.1% of the biological activity of unmodified NGF and a correspondingly low affinity for binding to the NGF receptors. Interestingly, the region in insulin that corresponds to the region in NGF around Trp-21 has been implicated in the receptor binding site (Pullen et al 1976, Bradshaw 1978).

Removal of the COOH-terminal arginine residues had no effect on the biological activity of the NGF protein (Moore et al 1974); the same was true for removal of the NH_2-terminal octapeptide sequence (Mobley et al 1976). These observations have, in turn, led to the development of a number of procedures for the rapid isolation of the NGF protein modified to varying degrees at its NH_2- and COOH-termini (Mobley et al 1976, Jeng & Bradshaw 1978). The endopeptidase responsible for the release of the NH_2-terminal octapeptide sequence has been purified (Wilson & Shooter 1979) and found to be an arginine esteropeptidase identical to submaxillary gland kallikrein (Bothwell et al 1979). The endopeptidase is closely related to the two arginine esteropeptidases (γ-subunit and EGF-binding protein), which are found associated with NGF and EGF proteins respectively in the submaxillary gland (Bothwell et al 1979, loc cit).

The 7S NGF Complex

In the mouse submaxillary gland the NGF dimer is contained in a multisubunit complex referred to as 7S NGF (Varon et al 1967a,b, 1968). The latter contains, besides the NGF dimer, 2 mol of the arginine esteropeptidase γ-subunit, 2 mol of the acidic protein α-subunit (Smith et al 1968), and 1, or more likely 2, zinc ions (Pattison & Dunn 1975). The composition of 7S NGF is therefore $\alpha_2\gamma_2\beta_1$, where β refers to the NGF dimer; because all the subunits have molecular weights close to 26,000, the molecular weight of 7S NGF is approximately 130,000 (Server & Shooter 1977). That 7S NGF contains 2 mol of γ-subunit was demonstrated by the finding that the basic Kunitz pancreatic trypsin inhibitor formed a 2:1 complex with dissociated 7S NGF, whereas it formed a 1:1 complex with isolated γ-subunit (Au & Dunn 1977).

The stability of 7S NGF is markedly increased by the zinc ions. A measure of this stability is provided by the equilibrium constant for the dissociation of γ-subunit from 7S NGF. This constant was determined in a variety of ways, including one that relied on the inhibition of the enzymic activity of the γ-subunit in 7S NGF. This activity is entirely suppressed in 7S NGF even in the absence of zinc ion (Bothwell & Shooter 1978). The value for the equilibrium dissociation constant of 7S NGF in the absence of zinc ion was found to be 10^{-7} M, indicating, in agreement with earlier work (Greene et al 1969, Perez-Polo et al 1972), that 7S NGF is a rather stable complex. The presence of zinc ion greatly stabilized the association of the γ-subunit in 7S NGF, with the equilibrium constant for the dissociation of the γ-subunit from 7S NGF reaching a minimum value of 10^{-12} M at zinc ion concentrations of 1 μM or more (Bothwell & Shooter 1978). Measurement of the enzymatic activity of 7S NGF at well defined free zinc ion concentrations showed that this activity could be accounted for simply in terms of the dissociation of γ-subunit from 7S NGF. This conclusion differs from the alternative explanation proposed by Pattison & Dunn (1976a,b) and Au & Dunn (1977) that removal of zinc ion from 7S NGF alone, without dissociation, was sufficient to allow expression of its esteropeptidase activity; but it is consistent with the observation that the enzymatic activity of the γ-subunit toward the artificial substrate Bz-Arg-nitroanilide was competitively inhibited by βNGF or βNGF and the α-subunit. From this it may be inferred that the binding of NGF and Bz-Arg-nitroanilide to the active site of γ-subunit are mutually exclusive; therefore, in 7S NGF, the γ-subunit cannot bind the artificial arginine amide and cannot be enzymatically active toward this substrate. As anticipated from this model, artificial arginine amide substrates (Bothwell & Shooter 1978) and pancreatic trypsin inhibitor (Au & Dunn 1977) promoted dissociation of 7S NGF through binding to the active site of γ-subunit. The enzymatic activity of the γ-subunit was also inhibited by NGF in the absence of the α-subunit, and the equilibrium constant for the association of the NGF and γ-subunits derived from these measurements was approximately 10^{-7} M. Zinc ion had no effect on this association and so the equilibrium constant for the dissociation of zinc ion from the NGF-γ complex is the same as that for the dissociation of zinc ion from γ-subunit, namely 10^{-6} M (Pattison & Dunn 1975). Because the dissociation constant for zinc ion in 7S NGF is 10^{-11} M (Pattison & Dunn 1975, 1976a,b Bothwell & Shooter 1978), it follows that zinc ion alters the dissociation constant of the α-subunit from 7S NGF by five orders of magnitude, just as it does for γ-subunit. Given that the equilibrium constant for the dissociation of α-subunit from zinc-free 7S NGF is less than 10^{-7} M (Bothwell & Shooter 1978), then the equilibrium constant for dissociation of α-subunit from 7S NGF (in the presence of μM zinc ion) must also be of the order of 10^{-12} M. Zinc ion is clearly important, therefore, in stabilizing the interactions of both α-and γ-subunits with the NGF dimer in 7S NGF. These interactions lead to the inhibition of the enzymatic activity of γ-subunit through the interaction of its active site with βNGF (Greene et al 1969,

Bothwell & Shooter 1978); this has implications in the role of 7S NGF in the biosynthesis of the NGF protein (see below). The interactions also lead to the inhibition of the biological activity of the NGF dimer in 7S NGF. This has been shown in two ways. First, cross-linking of 7S NGF with dimethylsuberimidate produced a nondissociable species that contained the three subunit types in 7S NGF and with the molecular weight of 7S NGF. This species was biologically inactive in the ganglion bioassay (Stach & Shooter 1979). Second, the 7S NGF complex labeled in the NGF subunit with ^{125}I and stabilized in the presence of zinc ions by excess a-and γ-subunits failed to bind to either of the specific NGF receptors on dissociated cells from embryonic chick dorsal root ganglia (Harris-Warrick & Shooter 1978). In contrast, the binding of bisdes arg^{118}-NGF, the derivative that retains the biological activity of NGF but is incapable of forming 7S NGF in the presence of a- and γ-subunits, was unaffected under these conditions. The 7S NGF complex therefore behaves like a typical allosteric protein in which the activities of the subunits are modified by their interactions in the complex. The zinc ion associated with 7S NGF plays a key role in stabilizing these interactions.

THE SIGNIFICANCE OF 7S NGF   The 7S NGF complex is a stable storage form of NGF from which the biologically active factor can be obtained by dissociation at very low protein and/or zinc ion concentrations. The dissociation of 7S NGF can also be promoted by the proteolytic removal of the COOH-terminal arginine residues of the NGF dimer, which renders it incapable of reforming 7S NGF (Moore et al 1974). One example is known in which 7S NGF also acts in the transport of NGF, namely the secretion of 7S NGF in saliva, but the physiological significance of this process is not yet clear. The 7S NGF complex also protects the NGF dimer from the two limited proteolytic cleavages induced by enzymes in the submaxillary gland and possibly elsewhere. Moore et al (1974) have shown that the activity of the CPB-like enzyme against the COOH-terminal arginine residues of NGF was completely inhibited when the latter was in 7S NGF, and Mobley et al (1976) made the same observation for the endopeptidase responsible for releasing the NH$_2$-terminal octapeptide sequences. Because neither cleavage affects the biological activity of the NGF protein, the protection afforded by 7S NGF must be directed either to some other activity of NGF or to the timing of the release of the octapeptide. Finally, there is growing evidence that the 7S NGF complex is formed during the biosynthesis of NGF. The presence in 7S NGF of both an arginine-specific esteropeptidase and an NGF subunit with COOH-terminal arginine residues led to the suggestion that the NGF chains are derived from larger precursor chains by the proteolytic action of the γ-subunit (Angeletti & Bradshaw 1971, see below).

ALTERNATIVE MODELS FOR SUBMAXILLARY GLAND NGF The data presented by Pantazis et al (1977a) on the dissociation of 7S NGF are compatible with the known stability of 7S NGF as described above. Using essentially zinc-free solvents, these authors noted significant dissociation of purified 7S NGF at concentrations of 50 μg/ml (4 X 10^{-7} M) and 10 μg/ml (8 X 10^{-6} M). Given that the equilibrium dissociation constant of zinc-free 7S NGF is 10^{-7} M (Bothwell & Shooter 1978), this would be the anticipated result. Furthermore their observation that NGF, in a dilution of the original submaxillary gland extract and at a concentration of approximately 10^{-9} M (calculated as 7S NGF), was still of high molecular weight is readily explained by the high zinc ion concentrations of gland extracts (Burton et al 1978). Because Bothwell & Shooter (1978) observed that free zinc ion concentrations as low as 10^{-8} M reduced the equilibrium dissociation constant of 7S NGF below 10^{-10} M, it is reasonable to conclude that Pantazis et al (1977a) were analyzing zinc-stabilized 7S NGF in their gland extract and that their results are not in conflict with earlier data. Not only does the mouse submaxillary gland have a high zinc content, but so also does saliva secreted from the gland (Burton et al 1978). It is not surprising, therefore, that Burton et al (1978) found that the exclusive form of NGF secreted in epinephrine-elicited mouse saliva was 7S NGF and that purified 7S NGF could be isolated from this saliva by a single gel filtration step. In keeping with these observations. L. E. Burton et al (unpublished data) showed by specific antibody staining that all three subunits in 7S NGF were present in the same tubule cells in the mouse submaxillary gland.

These results are in contrast to the suggestions of Murphy et al (1977a) that mouse saliva contained new high and low molecular weight forms of NGF. However it is likely from their data that their new high molecular weight form is simply zinc-stabilized 7S NGF and that the low molecular weight form is the NGF dimer lacking one or both COOH-arginine residues. This proteolytic modifaction of the NGF dimer at its COOH-termini significantly decreases its affinity for the other subunits in 7S NGF (Moore et al 1974, Perez-Polo & Shooter 1975). The CPB-like enzyme responsible for this modification has been demonstrated to be present in saliva (Burton et al 1978).

In a recent paper Young et al (1978) suggested that 7S NGF was not the form of NGF in the mouse submaxillary gland but rather that there were six different NGF species. Five of these were unstable and gave rise to the NGF protein on dilution, whereas the sixth, of mol wt 116,000 and present in about 10% of the normal amount of 7S NGF, was stable at approximately 10^{-9} M and contained immunoreactive NGF. With respect to this latter species, when the partial specific volume used by Young et al (1978), 0.69 ml/g, is changed to a value of 0.72 ml/g, which is more consistent with the amino acid composition of 7S NGF, the molecular weight of the stable species is increased to approximately 130,000 or close to that of 7S NGF. Moreover, the lowest concentration used in this work to test the stability of the protein, 10^{-9} M, is still well within the stability limit for zinc-containing 7S NGF as noted earlier. The identity of the five less stable NGF-containing species was not determined. However, it is known from earlier work (cf Server & Shooter 1977) that the dissociation of 7S NGF produces smaller NGF-containing protein complexes. The 7S NGF complex can be dissociated by appropriate dilution in zinc-free buffer or by mild extremes of pH. The 7S NGF complex dissociates at pH's outside the pH range of 5 to 8 and the mechanism of dissociation is different at acid and alkaline pH's (Varon et al 1968). Binary complexes form between the α and NGF and γ and NGF subunits, but not between the α-and γ-subunits (Server & Shooter 1977, Silverman 1978, Bradshaw 1978). The α_2NGF complex has been shown to be more stable at slightly acid pH and the γ_2NGF complex more stable at slightly alkaline pH (Silverman 1978), which suggests that these complexes may be intermediates in the dissociation of 7S NGF at acid and alkaline pH respectively. Also, there is evidence that an intermediate 7S complex lacking one γ-subunit was formed when one of the NGF chains lacked its critical γ-subunit-binding COOH-terminal arginine residue (Silverman 1978). Because the binary complexes are not stabilized by zinc ion (Bothwell & Shooter 1978) and because it is known that zinc-free 7S NGF can dissociate during chromatography on DEAE-cellulose (Varon et al 1968), it is possible that the five unstable species (Young et al 1978) are intermediates in the dissociation of 7S NGF. The issue could be readily resolved by subunit analysis of the new species.

THE MECHANISM OF ZINC ION DISSOCIATION   The dissociation of zinc ion from 7S NGF induced by chelating agents is a two-step process. Pattison & Dunn (1976a,b) found that the zinc ion first bound chelator at a rapid rate that corresponded to the formation of an inner sphere complex between chelator and an aquo-zinc ion. The ternary complex so formed then dissociated at a slower rate with the concomitant production of zinc-free 7S NGF and the chelator-zinc ion complex. Using the activation of the γ-subunit activity as an indicator of the release of zinc ion from 7S NGF (and its subsequent dissociation to produce free γ-subunit) these authors determined that the affinity constant for the 7S NGF-zinc ion complex was 10^{11} M. In contrast, the affinity of zinc ion for the γ-subunit, estimated from inhibition of enzymatic activity, was only 10^6 M (Pattison & Dunn 1975).

THE γ-SUBUNIT   The γ-subunit belongs to the serine protease family and is inhibited by DFP (Greene 1970). Its amino acid sequence shows homology to bovine trypsin and to the EGF-binding protein, the corresponding serine protease associated with EGF (Bradshaw 1978, Thomas et al 1979). The γ-subunit contains attached carbohydrate residues and consists of a number of closely related proteins that are derived by limited proteolysis from a parent enzyme containing two unequal-sized peptide chains (Stach et al 1976, Burton & Shooter 1979, Silverman 1978). One proteolytic cleavage converts the longer peptide chain into two shorter chains, whereas a CPB-like enzyme removes one or more of the original COOH-terminal arginine residues and the new COOH-terminal lysine and penultimate arginine residues generated by the chain cleavage. A maximum of eight species can be produced by these mechanisms and six of these have been separated by a combination of electrophoresis and isoelectric focusing. Although all the various γ-subspecies have the same specific activity toward synthetic substrates (Greene et al 1969), they differ somewhat in their ability to reform 7S NGF with added α-subunit and NGF. Server & Shooter (1976) noted that when an excess of γ-subunit was used in a recombination experiment, the γ^3-subunit (a three chain form) preferentially formed 7S NGF, compared to the γ^1- and γ^2-subunits.

THE α-SUBUNIT   Like the γ-subunit, the α-subunit is heterogeneous, although the basis of this heterogeneity has not yet been completely determined. Each α-subunit contains two unequal-sized peptide chains, one of a pair of 17,300 mol wt chains and one of a pair of 9,300 mol wt chains, and the heterogeneity may result from different combinations of these chains (Server & Shooter 1977). The only known role for the α-subunit in 7S NGF is its property, in conjunction with NGF, of inhibiting the enzymatic activity of γ-subunit and, in conjunction with γ-subunit, of inhibiting the biological activity of NGF. Varon & Raiborn (1972b) have found that the α-subunit increases the yield of neuronal and nonneuronal cells from chick DRG and that this property of the α-subunit is enhanced in 7S NGF.

The Biosynthesis of NGF

One aspect of the study of NGF has progressed rather slowly, namely the question of where and when NGF is synthesized. What cells or tissues supply the NGF that is critical for the survival and development of sympathetic and some sensory neurons? Is NGF carried by the circulation or does it reach the responsive neurons by an alternate route? Do target tissues make the NGF needed for retrograde transport, and if so at what stage of development does synthesis begin? Or do they acquire it from other sources? At least one sympathetic target tissue, the mouse submaxillary gland, contains large amounts of NGF (Cohen 1960), and recent work has confirmed that NGF is synthesized in the gland (Berger & Shooter 1978). However, NGF synthesis in the gland is only significant long after innervation has occurred and it appears to be under the control of testosterone rather than neuronal input (Ishii & Shooter 1975, Berger & Shooter 1978). Moreover other submaxillary glands that are not sexually dimorphic, as in the rat, make no detectable NGF (Bothwell et al 1979). Some target tissues, like rat irides (Johnson et al 1972) and mouse adrenals (Harper et al 1976), synthesize and secrete an NGF-like material in culture. Again, however, this appears to be the property of many cultured cells, including those that are not target organs of sympathetic and sensory neurons. These include primary fibroblasts (Young et al 1975, Saide et al 1975), skeletal muscle cells (Murphy et al 1977b), 3T3 and transformed 3T3 cells (Oger et al 1974), neuroblastoma (Murphy et al 1975), and glioma (Murphy et al 1977c, Longo & Penhoet 1974). On the basis of these observations it has been suggested that circulating NGF arises from multifocal cellular secretion (Murphy et al 1977a), in agreement with earlier views (Levi-Montalcini & Angeletti 1968). In support of this idea, Hendry & Iversen (1973) found that circulating NGF levels in the male mouse fell to low levels after removal of the submaxillary gland, but were eventually restored without concomitant regeneration of the gland. In contrast, Murphy et al (1977d) reported no decline in serum NGF levels after sialectomy. Several studies have emphasized the potential importance of glial cells in NFG production during development (Varon 1975, Hendry 1976, Longo 1978, Murphy et al 1977a, but see also Levi-Montalcini 1975b). At the moment, the de novo synthesis of NGF has only been rigorously proven for the male mouse submaxillary gland (Berger & Shooter 1978) and, as anticipated from previous kinetic studies (Ishii & Shooter 1975), the hormonal modulation of gland NGF levels (Cohen 1960, Caramia et al 1962, Hendry 1972) correlated with direct measurement of NGF synthesis (Berger & Shooter 1978). Under the conditions in which NGF synthesis in the male mouse submaxillary gland could be measured, no synthesis of NGF was detected in other tissues from the normal or castrated and testosteronized mouse, including those that received heavy sympathetic innervation (Berger & Shooter 1978).

THE PRECURSOR OF NGF—INDIRECT EVIDENCE   As noted above, the analysis of the composition and subunit properties of 7S NGF led to the hypothesis that an early product of the NGF gene is a pro-NGF peptide chain which is cleaved by the specific γ-subunit at an internal arginine residue that then becomes the COOH-terminal residue of NGF. A similar suggestion with respect to the biosynthesis of EGF from a precursor pro-EGF was made by Taylor et al (1970, 1974). In this instance the processing would be mediated by EGF-binding protein, the specific esteropeptidase associated with EGF in its high molecular weight protein complex, HMW-EGF, with again release of extrapeptide material from the COOH-terminus of pro-EGF. A considerable amount of indirect evidence supports this model. In both 7S NGF and HMW-EGF, the activity of the esteropeptidase subunits is inhibited (Greene et al 1968, Bothwell & Shooter 1978, Server et al 1976), which suggests that their active sites are occupied by moities of one or more of the other subunits in the complexes. When the COOH-terminal arginine residues of either NGF or EGF were removed, no inhibition of the activity of the corresponding esteropeptidases was observed because the modified growth factors were unable to form their respective high molecular weight complexes (Moore et al 1974, Server et al 1976). This strongly implicates the COOH-terminal arginine residues of NGF and EGF as binding sites for the esteropeptide subunits through their active sites. Finally, although closely related to the γ-subunit in size, peptide chain and amino acid composition, and enzymatic specificity, the EGF-binding protein failed to replace the γ-subunit in 7S NGF (Server et al 1976). This is the expected result if the EGF-binding protein is specific for binding to and cleaving only pro-EGF and not pro-NGF. The fact that the γ-subunit and EGF-binding protein remain attached to their products, NGF and EGF respectively, after the cleavage event, is related to the sequence homologies between the two growth factors and the family of trypsin inhibitors (Hunt et al 1974, Dayhoff 1976).

THE PRECURSOR OF NGF—DIRECT EVIDENCE   When adult male mouse submaxillary glands were incubated in vivo or in vitro with L-(^{35}S) cystine for several hours, radioactivity was incorporated into one major protein that was precipitated with anti-NGF antiserum, migrated with authentic NGF under denaturing conditions in acrylamide gel, and contained the five major cysteine-containing peptides of NGF (Berger & Shooter 1977, 1978). With shorter labeling periods the major radioactive labeled species precipitated with anti-NGF antiserum had a molecular weight of 22,000 (22K species) and contained all the cysteine-containing peptides of NGF. The kinetics of labeling of the 22K and NGF species in continuous or pulse-chase experiments identified the 22K species as the precursor of NGF, pro-NGF (Berger & Shooter 1977). The pro-NGF species in either its complex with anti-NGF antibody or after elution from an SDS-containing gel was converted to NGF by the γ-subunit. However, several other proteolytic enzymes, including trypsin and EGF-binding protein, carried out the same conversion; it is not yet clear whether this is because pro-NGF was not in its appropriate native conformation or the α-subunit is also needed to confer specificity on the cleavage reaction.

The specificity of the cleavage has been demonstrated in the EGF system. In experiments parallel to those noted above, and following earlier work of Turkington et al (1971), who demonstrated EGF synthesis in the submaxillary gland, it has been shown (P. Frey et al, unpublished data) that there is also a larger molecular weight biosynthetic precursor to EGF, pro-EGF. The pro-EGF chain has a molecular weight of approximately 9000 (9K species) and, like EGF, is stable to boiling. It was therefore possible to denature the endogenous proteases in the labeled gland homogenate by boiling and then determine which added protease would convert pro-EGF to EGF. Of the three closely-related esteropeptidases, EGF-binding protein, γ-subunit, and the NGF-endopeptidase (Wilson & Shooter 1979), only EGF-binding protein carried out the conversion. Even trypsin in large excess failed to affect pro-EGF. By specific labeling of pro-EGF on the NH_2-terminal α-amino group, it was also possible to show that cleavage of pro-EGF occurred at its COOH-terminus as predicted.

NGF BIOSYNTHESIS IN CELLS IN CULTURE   The biosynthesis of NGF has also been demonstrated in rat C6 glioma cells in culture (Longo & Penhoet 1974, Longo 1978). Incorporation of radioactive amino acids into a protein that was precipitable with anti-NGF antiserum and had the size of NGF was demonstrated. This material also gave a tryptic peptide pattern, similar but not exactly identical to that of mouse NGF—the differences probably arising from differences in amino acid sequence between mouse and rat NGF, estimated to be about 10% from immunological studies. Among other labeled proteins precipitated by anti-NGF antiserum from the C6 glioma homogenate was one of mol wt 24,000; some preliminary evidence was obtained suggesting that this species corresponded to pro-NGF identified in the mouse submaxillary gland.

The nature of the NGF-like material secreted by mouse L cells (Pantazis et al 1977b) and skeletal muscle cells (Murphy et al 1977b) in culture has been examined. Both materials had molecular weights in the region of 140,000 to 160,000 and there was no difference between intracellular and extracellular NGF-like material of the muscle cells. The relationship of these substances, if any, to 7S NGF is not yet clear, although it is reported that the mouse L cell material, when exposed to denaturing solvents, released a protein indistinguishable from mouse NGF by electrophoretic and chromatographic criteria (Pantazis et al 1977b). This is a well-known property of 7S NGF. The conclusion (Pantazis et al 1977b, Murphy et al 1977b) that these two NGF-like materials are not identical to 7S NGF because they differ from the latter in stability, needs to be reexamined in the light of the known stabilization of 7S NGF by zinc ion (see above).

Summary

The NGF protein can be isolated from the mouse submaxillary gland or saliva in relatively large amounts by rather simple procedures. Because of its ready availability, the protein has been well characterized by chemical and physical methods. Its crystallization opens up the possibility of comparing its three-dimensional structure with those of a family of related proteins, including insulin and relaxin, and of defining its receptor binding site. One of the steps in the biosynthesis of the NGF dimer—the proteolytic cleavage of pro-NGF to NGF peptide chains by an apparently specific trypsin-like enzyme, the γ-subunit—initiates the formation of the multisubunit complex, 7S NGF, in which the NGF protein finally resides. The stability of this complex is greatly enhanced by zinc ions, which are found in high concentration in the gland and saliva. The complex protects the NGF protein from at least two limited proteolytic cleavages. However, the purpose of this protection is not yet understood. Little is known about the specific sites of NGF synthesis, either during development or in the adult animal, although there are indications that they are widespread. The continued characterization of the NGF-like materials secreted by a number of tissues or cells in culture and further studies of biosynthesis of these proteins should, however, provide valuable information on this subject. As research continues along these avenues, increasing attention is also being given to the mechanisms by which NGF influences the survival, development, and maintenance of sympathetic and sensory neurons.

THE MECHANISM OF ACTION OF NGF

In considering NGF's mechanism of action, it is necessary to first define what is meant by actions of NGF. For the purpose of this review *actions* of NGF will be considered to be those effects of the factor that are in themselves the final results of a series of prior causal events and have clear consequences for the biological role and functional behavior of target cells. Among such actions that have been described are maintenance of survival, promotion of neurite outgrowth, cell hypertrophy, induction of neurotransmitter synthesis enzymes, and, possibly, chemotropism, and promotion of differentiation. Such actions of NGF must be distinguished from responses to NGF that are in themselves only *causal* steps in the sequence of events that constitute NGF's mechanism of action. Examples of such causal responses may be changes in cell surface architecture, increased uptake of small molecules, enhanced protein synthesis, and so forth (see below). A second point that is important to consider is that just as NGF may have multiple actions, it is also likely to have multiple mechanisms of action. Thus, a given mechanism can be evaluated only in relationship to the action for which it is proposed.

The material in the following sections is meant to amplify, supplement, and complement material on the same subject matter previously covered in other reviews (Levi-Montalcini & Angeletti 1968, Varon 1975, Mobley et al 1977, Bradshaw 1978). In particular, we review (*a*) the experimental systems used to study the NGF's mechanism of action, (*b*) NGF receptors, (*c*) the roles of RNA and (*d*) protein synthesis, and finally (*e*) a number of steps that could link interaction of NGF with its target cells to its ultimate biological actions.

Experimental Systems

As is the case in many scientific problems, progress in studying the mechanism of action of NGF is greatly dependent on the availability of suitable experimental model systems. These systems are as follows:

1. *Intact animals:* Many of the biological actions of NGF were first demonstrated in chick embryos and neonatal or adult rats (see Levi-Montalcini 1966 for review). Nevertheless, because of their obvious difficulties, in vivo systems have had only limited use for studying the mechanism of action of the factor. However, one particularly important use of intact animals has been to study the retrograde axonal transport of NGF (see below).

2. *In vitro systems:* (*a*) *Primary cultures.* Many of the actions of NGF have also been shown to take place in vitro (see Levi-Montalcini 1966 for review). Most studies on the mechanism of action of the factor have been carried out on cultures of explanted or dissociated sympathetic or dorsal root ganglia from chick embryos or neonatal rodents. Among the well-known advantages of in vitro systems is the possibility of controlling the chemical and cellular environment in a way not possible in vivo. Primary cultures of neurons also have certain experimental disadvantages. First, explanted ganglia are heterogeneous with respect to cell type and this may pose the problem of identifying the cellular origin of a given response to NGF. This disadvantage has been overcome in monolayer cultures by eliminating non-neuronal cells (Mains & Patterson 1973, McCarthy & Partlow 1976). A second problem of primary neurons is that such cells require NGF for survival. This raises difficulties for the use of NGF-untreated cultures as controls (Varon 1975). A final problem is that neurons used for culture in all likelihood have had prior exposure to NGF in vivo. As will be discussed below, exposure to NGF in vivo may have major consequences for the responses to NGF that neurons display when placed in vitro. (*b*) *Continuous cell lines.* Several types of tumor-derived cell lines have been shown to interact with NGF and have potential as model systems for mechanistic studies. Human neuroblastoma cells undergo responses to NGF including neurite outgrowth (Waris et al 1973, Kolber et al 1974,

Perez-Polo et al 1979), but thus far they have had only limited use for NGF studies. There is also evidence that some clones of murine C1300 neuroblastoma cells may specifically bind NGF (Revoltella et al 1974, Diamond et al 1976). However, many clones of C1300 neuroblastoma cells do not show evident morphological responses to NGF (L. A. Greene, unpublished), and reports of NGF responses by such cells are rather limited (Hermetet et al 1972, Brodeur & Goldstein 1976). Certain human melanoma cell lines possess NGF receptors, and there is evidence that NGF can affect the survival of such cells in culture (Fabricant et al 1977). Thus far, perhaps because of the absence of easily measured responses to NGF, melanoma cells have not been used for studies of NGF's mechanism. Another type of tumor that has been shown to respond to NGF is the pheochromocytoma (Tischler & Greene 1975, Tischler et al 1976). Recently, several NGF-responsive lines have been established from a transplantable rat pheochromocytoma. One of these, the PC-G2, shows increased specific activity of tyrosine hydroxylase in response to both NGF and EGF (Goodman & Herschman 1978, Goodman et al 1979) and hence should be useful as a model for this aspect of NGF's action. Another line, the PC12 clone (Greene & Tischler 1976), has become increasingly used to study NGF's mechanism of action (see below); thus, it merits a degree of detailed discussion. When grown in serum-containing medium without exogenous NGF, PC12 cells possess many of the noradrenergic and ultrastructural features associated with pheochromocytomas and with their non-neoplastic counterparts, adrenal medullary chromaffin cells (Greene & Tischler 1976, Greene & Rein 1977a,b, Tischler & Greene 1978). When PC12 cells are exposed to NGF over a time course of several days to a week, they acquire a number of properties of sympathetic neurons including cessation of proliferation, increased somatic size, neurite outgrowth, increased electrical excitability and responsiveness to acetylcholine, and the presence of synaptic-like vesicles (Greene & Tischler 1976, Dichter et al 1977, Tischler & Greene 1978). Other actions of NGF on these cells at the biochemical level include increases in the specific activities of choline acetyltransferase (Greene & Rein 1977c, Schubert et al 1977, Edgar & Thoenen 1978) and acetylcholinesterase (Rieger, Shelanski & Greene, in press) and increased synthesis of a large, external glycoprotein, named the NGF-Inducible Large External glyoprotein, or NILE glycoprotein (McGuire et al 1978). Unlike PC-G2 cells or sympathetic neurons, PC12 cells do not show an increase in tyrosine hydroxylase activity in response to NGF (Greene & Tischler 1976, Edgar & Thoenen 1978). A particularly useful feature of PC12 cells is that the effects of NGF on this system appear to be reversible. Withdrawal of NGF from PC12 cultures results in disintegration of neurites and recommencement of cell proliferation (Greene & Tischler 1976). The PC12 and other NGF-responsive cell lines have a number of potential advantages for studying NGF's mechanism of action. These potential advantages include: (1) the availability of large numbers of homogeneous cells; (2) the ability of such cells to respond to, but not require, NGF (thus providing viable NGF-untreated cells as controls; (3) the opportunity to study the initial effects of NGF on cells previously unexposed to the factor; (4) the possibility of studying the offset, as well as the onset, of the effects of NGF. Tumor-derived cell lines also have the potential disadvantage that some, or all, of their responses to NGF may differ from those of normal tissues. Thus, the final description of the mechanism of action will probably require both in vivo and in vitro studies with normal and neoplastic tissues.

NGF Receptors

As with many other peptide effectors, it is generally agreed that the primary interaction of NGF with its target cells is via specific receptors. Work over the last several years has indicated, however, that this interaction may be complex and that there may be multiple means whereby NGF can interact with its target cells.

SURFACE MEMBRANE RECEPTORS   Binding studies employing (^{125}I)-NGF have revealed specific receptors for NGF on the plasma membranes of cells in sympathetic and dorsal root ganglia (Banerjee et al 1973, Herrup & Shooter 1973, Frazier et al 1973a). Banerjee et al (1973) reported specific binding of (^{125}I)NGF to membrane preparations of rabbit superior cervical ganglia (SCG). The interaction was saturable and had an apparent K_D of about 2×10^{-10} M. Modification of NGF with N-bromosuccinimide (see above) led to losses in biological activity that were accompanied by parallel losses in specific binding (see also Frazier et al 1974a). Specific binding was not displaced by a number of other hormones and was not observed with a variety of non-neuronal tissues (Snyder et al 1974). In addition, specific binding was reduced by treating the membrane preparation with trypsin or with protein-modifying agents and required Ca^{2+} (Banerjee et al 1975). In a subsequent study, Banerjee et al (1976) showed that NGF receptor binding sites from rabbit SCG membranes could be solubilized with Triton X-100 and that the K_D for NGF binding to solubilized receptors was the same as that for binding to receptors in intact membrane preparations. Frazier et al (1974a) studied binding of (^{125}I)NGF to membranes of homogenates of chick embryo sympathetic and dorsal root ganglia (DRG). Specific binding was found to be nonsaturable [up to (^{125}I)NGF concentrations of 10^{-8} M] and to display apparent multiple dissociation constants ranging from $\simeq 10^{-10}$ M to 10^{-6} M. Because excess unlabeled NGF accelerated the dissociation of membrane bound (^{125}I)NGF, it was suggested, in analogy with studies on insulin receptor binding (De Meyts et al 1973), that the apparent multiple dissociation constants for NGF-binding were due to negative cooperative interaction of NGF receptors. Andres et al (1977) found that, like NGF receptors in sympathetic ganglia, NGF receptors in DRG were also solubilized with Triton X-100. In subsequent studies, Frazier et al (1974b,c) reported binding of (^{125}I)NGF to membrane fractions of several peripheral tissues and to brain of chick embryo and rat. The kinetic properties of such binding were similar to the kinetic properties of binding to sympathetic and dorsal root ganglia. Furthermore, specific binding of NGF to peripheral tissues did not appear to be restricted to sympathetic nerve endings, because such binding persisted in preparations of rats whose sympathetic nerve terminals were destroyed by guanethidine treatment. Szutowicz et al (1976a,b) carried out a more detailed study of the subcellular localization of binding of (^{125}I)NGF to chick brain. In the early stages of development, binding was localized to dense subcellular structures, whereas at later stages, binding was mainly to synaptosomal fractions. In contrast to the above studies that employed membrane fragments, Herrup & Shooter (1973) examined binding of (^{125}I)NGF to intact cells mechanically dissociated from chick embryo sensory ganglia. Binding was found to be saturable and to have an apparent K_D of about $1–3 \times 10^{-10}$ M. Insulin and several other peptides did not compete for the NGF-binding site. In a subsequent developmental study, Herrup & Shooter (1975) reported that the loss in biological responsiveness to NGF that chick embryo DRG neurons undergo between days 16 and 18 of incubation was temporally correlated with a large drop in the number of NGF binding sites. Binding of (^{125}I)NGF to intact chick DRG cells has recently been examined in further detail by Sutter et al (1979). These workers employed a membrane filtration technique in their preparation of (^{125}I)NGF that considerably reduced nonspecific binding and thus permitted examination of specific binding at very low concentrations of the factor. With this labeled NGF preparation, both steady state and kinetic data showed two distinct, saturable binding sites rather than the one described earlier (Herrup & Shooter 1973). The apparent K_D's for these two sites were 2×10^{-11} M (site I) and 2×10^{-9} M (site II). These data confirmed the earlier observations of Frazier et al (1974a) that the binding of NGF to embryonic chick DRG cells was heterogeneous. Moreover, in agreement with Frazier et al (1974a), when membrane preparations from chick embryo DRG were used instead of whole cells, the binding was found to be nonsaturable because of the appearance of a much lower affinity binding site with an apparent dissociation constant of approximately 10^{-6} M (R. J. Riopelle et al, unpublished data). Cellular disruption therefore appears to expose internal binding sites for

NGF that have different characteristics from those on the cell surface. Sutter et al (1979) found that one of the main reasons that the two sites (I and II) displayed different affinities was because of their different rate constants of dissociation. Whereas the half-time of release of NGF from site I was approximately 10 min, that from site II was only a few seconds. This behavior allowed the heterogeneity of the binding to be displayed over a wide range of (^{125}I)NGF concentrations. Also the fact that the dissociation of (^{125}I)NGF from the cells was markedly dependent on the degree of receptor occupancy that existed before initiation of dissociation was an argument against the lower affinity site II being generated by a negative cooperative reaction from site I. When the cells were preequilibrated with (^{125}I)NGF at concentrations less than 0.5 X 10^{-11} M, the subsequent dissociation was monophasic with a half-time of approximately 10 min. At higher preequilibration concentrations the dissociation pattern became biphasic and the relative amounts of (^{125}I)NGF, which dissociated slowly or rapidly, correlated with the relative occupancies of sites I and II respectively before dissociation began. Frazier et al's (1974a) observation, that the rate of dissociation of (^{125}I)NGF from its receptors on sympathetic neurons was accelerated by addition of unlabeled NGF, was also found with the DRG cells (Sutter et al 1979). However, acceleration of dissociation was observed not only when excess unlabeled NGF was added to increase receptor occupancy in the dissociation step, but also when the cells were diluted into an NGF-containing solution to produce a decreased receptor occupancy (Sutter et al 1979). The phenomenon of negative cooperativity is insufficient, by itself, to account for this behavior. Evidence that, in the absence of added NGF, dissociation of (^{125}I)NGF is hindered by hydrophobic, polar, or steric interactions with membrane components has recently been obtained by R. J. Riopelle et al (unpublished data). Vigorous stirring of membrane fractions during the dissociation accelerated the release of (^{125}I)NGF by disturbing the inhibiting diffusion barrier.

The presence of cell surface receptors for NGF has recently been studied by means of a serological cytotoxicity assay (Zimmerman et al 1978) in which single cell preparations are incubated with NGF in the presence of anti-NGF and complement. The principle of the assay is that cells having NGF bound to their receptors can be lysed by complement activation following binding of anti-NGF antibodies to the surface bound hormone. Evidence was presented that cytotoxic effects were found only for cells bearing the site I high affinity receptors described by Sutter et al (1979). Binding of NGF to site II, however, did not cause a NGF-mediated cytotoxic reaction, probably because of the short half-time of dissociation of NGF from these sites. By these means, site I NGF receptors were demonstrated on the large NGF-responsive neurons (but not other cell types) present in chick DRG. Such receptors were absent from chick embryo liver and heart, but interestingly, were detected on cells from dissociated chick embryo forebrain and tectum. In the latter case, the cytotoxicity was highest at days 8 and 9 of incubation.

In addition to studies with tissues, specific binding of (^{125}I)NGF has also been examined with several tumor-derived cell lines. Fabricant et al (1977) observed specific binding of (^{125}I)NGF to cultured melanoma cells, but not to a variety of other cultured cell types. The levels of binding varied for different melanoma lines and was highest for those derived from metastatic tumors. The interaction of NGF with one line (A875) was studied in detail and had an apparent K_D of 1×10^{-9} M. Sherwin et al (1979) used an indirect membrane immunofluorescence assay, which also confirmed the presence of NGF receptors on human melanoma cells, but not on a variety of other human cell types. Binding of NGF to PC12 cells has also been recently examined (Herrup & Thoenen 1979, Calissano & Shelanski, in press, G. Landreth and E. M. Shooter, unpublished, D. Ishii, unpublished). At low temperatures (0.5°), low affinity binding to a single site was observed (K_D approx. 10^{-9} M) (Herrup & Thoenen 1979, G. Landreth and E. M. Shooter, unpublished). Specific binding was destroyed by trypsin but not by removal of divalent cations, and prior exposure of the cells to NGF did not change the binding characteristics with respect to either receptor affinity or number (Herrup & Thoenen 1979). At higher temperatures (37°) the binding was more complex and was characterized in naive PC12 cells, as with DRG cells, by the appearance of sites with two different affinities. The lower affinity binding was comparable to that observed at low temperatures and, in addition, a higher affinity site with an apparent K_D of 10^{-10} M was observed (G. Landreth and E. M. Shooter unpublished). When NGF was added to PC12 cells, binding to the low affinity site started immediately and increased rapidly. However there was a delay of about 30 to 60 sec before high affinity binding was observed; after this delay the proportion of high affinity binding increased with time. The appearance of high affinity binding was paralleled, both temporally and quantitatively, by the movement of bound (^{125}I)NGF into a trypsin-resistant state (G. Landreth and E. M. Shooter unpublished). Calissano & Shelanski (in press), in agreement with these findings, found that a portion of the NGF that bound to PC12 cells did not dissociate from the cells in NGF-free medium and could not be removed by exposure to trypsin. This fraction was termed "tightly bound" and within one hour of exposure to NGF, depending on the NGF concentration, it represented a significant fraction of the total NGF bound to the cells. All those data suggest that development of high affinity NGF-binding may represent either changes in the NGF receptor or its localization and/or internalization of the NGF-receptor complex.

An NGF-dependent decrease in cell surface binding, as a consequence of the internalization of NGF by PC12 and its transport intact to the nucleus, has been demonstrated (Yankner & Shooter 1979, see below). Unlike binding to sensory cells (Sutter et al 1979), binding to PC12 cells was accompanied by significant degradation of NGF. Because the degradation was accompanied by down regulation of cell surface receptors and both were inhibited by chloroquine, it is clear that a major part of the degradation occurs in lysosomes after internalization of the NGF-receptor complex (P. Layer and E. M. Shooter, unpublished data). Even these preliminary studies of binding of NGF to PC12 cells suggest that this system will provide much useful information on the interaction of the factor with its target cells.

Other than its apparent susceptibility to trypsin and solubility in Triton X-100, little is known about the properties of the cell surface NGF receptor. Costrini & Bradshaw (1979) examined the apparent mol size of Triton-solubilized receptors from rabbit SCG. The apparent Stokes radius of this protein (71 ± 5 Å) was identical to that of the insulin receptor. This may be of particular interest in light of the structural similarities between NGF and insulin discussed above.

The studies reviewed above raise a number of questions about membrane NGF receptor binding:

1. *Does the presence of membrane receptors correlate with responsiveness to NGF?* The studies do indicate that specific receptors for NGF are present on cell membranes in tissues that respond to NGF and are not present on many cell types that lack NGF responses. The significance of the apparent presence of NGF receptors in brain is not yet clear. There is evidence that NGF can affect CNS regeneration in lower vertebrates (Turner & Glaze 1977) and possibly in mammals (Björklund & Stenevi 1972, but see Menesini-Chen et al 1978), and thus it is not inconceivable that NGF receptors that have been identified in brain (Szutowicz et al 1976a,b, Zimmerman et al 1978) could play a role in development and regeneration of CNS tissue.

2. *If multiple surface membrane receptors are present, what roles do they play in the mechanism of action of NGF?* The evidence that there are kinetically distinct NGF receptors on sensory cells is convincing. Although it appears that the lower affinity receptor is not generated from the higher affinity receptor by negative cooperativity, this does not necessarily rule out a structural relationship between the two forms. P. Cohen et al (unpublished), for example, found that the binding of NGF oxidized at Trp-21 was reduced to the same extent for both sites I and II on sensory cells. Site I has been implicated as the receptor that mediates maintenance of neuronal survival and stimulation of neuronal outgrowth (Sutter et al 1977). Not only was site I detected on neuronal and not on non-neuronal cells from chick embryo DRG, but at the concentrations of NGF required to initiate these

biological activities, only site I would have significant levels of receptor occupancy. Whether the binding sites of lesser affinity mediate other responses to NGF and/or facilitate internalization of the factor remain interesting possibilities (see below).

3. *Is binding of NGF to cell surface receptors sufficient to elicit the biological actions of the factor?* Frazier et al (1973a) reported that NGF covalently coupled to Sepharose beads was capable of eliciting neurite outgrowth from cultured chick embryo DRG neurons. On this basis, it was concluded that NGF could elicit its biological actions exclusively via interaction with surface receptors. However, it has been subsequently pointed out (Bradshaw 1978) that such experiments did not exclude the possibilities that a small fragment of the bead to which NGF was bound was internalized, or that NGF or an active fragment thereof (Mercanti et al 1977) was cleaved from the bead and internalized. Furthermore, even if NGF had not been internalized in such experiments, Frazier's et al interpretation does not take into consideration the possible effects of prior in vivo exposure of the ganglia to NGF (see below). Thus, the question as to whether NGF's biological actions can be triggered exclusively via cell surface receptors alone remains to be resolved.

INTERNALIZATION OF NGF AND RETROGRADE AXONAL TRANSPORT   Work from a number of laboratories indicates that NGF can be internalized by target cells (Hendry et al 1974b, Norr & Varon 1975, Andres et al 1977, Yankner & Shooter 1979), as suggested earlier (Hendry & Iversen 1973, Burnham & Varon 1973). Moreover, it has been well established that NGF can be internalized at nerve endings and retrogradely transported to the cell bodies of DRG and sympathetic neurons (Hendry et al 1974a,b, Stöckel et al 1975a,b, Hendry 1977, Johnson et al 1978, Brunso-Bechtold & Hamburger 1979). The uptake at nerve endings appears to be saturable and not affected by, or to include, a number of other proteins (Hendry et al 1974a, Stöckel & Thoenen 1975a,b). NGF that reaches the cell bodies of sympathetic neurons in this manner is biologically active because it causes selective induction of tyrosine hydroxylase (Paravicini et al 1975, Hendry 1977) and increased perikaryal diameter (Hendry 1977). Also, experiments carried out by Campenot (1977), employing a system in which the media bathing the neurite endings and cell bodies of cultured rat sympathetic neurons were isolated from one another by a diffusion-proof barrier, suggest that retrograde transported NGF can also sustain cell survival. Retrograde transport of NGF appears to have a certain degree of specificity with respect to cell type. (^{125}I)NGF injected into the periphery is selectively taken up and transported by sensory but not motor nerve endings (Stöckel et al 1975a, Stöckel & Thoenen 1975, Brunso-Bechtold &

Hamburger 1979). Also, (^{125}I)NGF taken up by sensory nerve endings reaches the dorsal root ganglia by retrograde transport, but is not transported by orthograde transport to the spinal cord (Stöckel et al 1975b, Brunso-Bechtold & Hamburger 1979). On the other hand, although the known biological effects of NGF on DRG neurons are thought to be restricted to a limited period of development, such uptake and transport by sensory nerves was observed even in adult animals (Stöckel et al 1975b). Moreover, Max et al (1978) recently demonstrated retrograde axonal transport of NGF by neurons of chick and rat ciliary ganglia. Such parasympathetic neurons have no known biological responses to NGF. Also, Ebbott & Hendry (1978) presented evidence for a widespread but selective retrograde axonal transport of (^{125}I)NGF in rat brain. Such findings could either indicate the existence of other potential target cells for NGF or suggest that retrograde transport of NGF is not restricted only to cell types that respond to the factor.

Hendry (1976) and Thoenen et al (1978) have elaborated a model in which NGF is responsible for the trophic influence that end organs have on their innervating sympathetic and sensory neurons and that retrograde axonal transport is the means by which NGF is carried between target organs and neuronal cell bodies. A number of observations are consistent with this model (see Hendry 1976 and Thoenen et al 1978 for reviews). For example, administration of exogenous NGF can prevent the degenerative changes in biochemical (Hendry 1975), morphological (Hendry 1977, Hendry & Campbell 1976), and electrophysiological (Nja & Purves 1978) behavior that sympathetic neurons undergo after being isolated from their target organs by axotomy. On the other hand, Banks & Walter (1977) found that whereas NGF treatment prevented such changes during the first week after axotomy, by three weeks after axotomy exogenous NGF no longer prevented sympathetic neuron degeneration and death. Also, although it has been shown that a variety of cell types can synthesize NGF in vitro (for review see Bradshaw & Young 1976), it has yet to be demonstrated that NGF can be synthesized or stored by target cells and conveyed in biologically active amounts to innervating fibers.

A corollary to the model that NGF acts as a trophic substance by retrograde transport is that target organs that release or synthesize large amounts of NGF can become selectively innervated in comparison with organs that lack NGF (Thoenen et al 1978). Of relevance to this possibility are Campenot's (1977) experiments demonstrating that neurites will selectively grow only in a local environment containing NGF, and Letourneau's (1978) experiments suggesting that sympathetic nerve fibers can show preferential orientation and neurite elongation up NGF gradients. At present, however, there is no systematic evidence for preferential synthesis and/or storage of NGF by sympathetically innervated end organs. Thus, although a specific trophic function for retrogradely transported NGF is an appealing possibility, further work is necessary to test this hypothesis.

An important question yet to be settled is that regarding the fate and possible site of action of NGF that is internalized at nerve endings. The in vitro experiments of Campenot (1977), in which neurites of cultured sympathetic neurons were found to grow only if the local environment contained NGF, suggest the possibility that the effect of NGF on neurite elongation could, in part, be local and not require retrograde transport to the cell body. It is clear, however, that (^{125}I)NGF taken up at nerve endings is transported to the cell body and that the majority of radioactive material that reaches the cell body in this manner co-migrates with native NGF when analyzed by SDS polyacrylamide gel electrophoresis (Stöckel et al 1976, Johnson et al 1978). Autoradiographic studies with the electron microscope (Schwab & Thoenen 1977) indicated that most of the material transported in the axon was associated with vesicles of the smooth endoplasmic reticulum (SER). In the cell body, most of the transported radioactivity was localized to the SER and in vesicles and only a small proportion (about 20%) was associated with secondary lysosomes. There was no evidence for significant accumulation of radioactivity inside the nucleus. These findings were confirmed by the use of horseradish peroxidase-coupled-NGF (Schwab 1977). However, despite the absence of morphologic evidence for association of retrogradely transported NGF with the nucleus, subcellular localization of this material by biochemical means has indicated that 20 to 30% of the transported material in the cell body can be isolated with a nuclear fraction (Stöckel et al 1976, Johnson et al 1978).

The mechanism whereby NGF is internalized is not presently clear. There is evidence that several other peptides undergo a receptor-mediated internalization in which the receptor-peptide complexes formed after binding cluster over coated pits and are then internalized (for review, see Goldstein et al 1979). Recent experiments (J. Schlessinger et al, unpublished) suggest that NGF may be internalized via the same mechanism.

NUCLEAR BINDING OF NGF   The discovery that NGF can be internalized by responsive neurons has led to studies on the possible interaction of the factor with nuclei. Andres et al (1977) found that a portion of (^{125}I)NGF that binds to chick embryo sensory neurons is not solubilized by Triton X-100, a detergent that readily solubilizes plasma membrane NGF receptors (Banerjee et al 1976). Subcellular fractionation indicated that the Triton-resistant binding was localized to the nuclear fraction. Binding to these nuclear receptors was saturable, had an apparent K_D of 2×10^{-10} M, and was not inhibited by excess insulin, cytochrome c, or lysozyme. Only much lower levels of binding were reported for nuclei of chick erythrocytes or optic tectum cells. Further characterization of the binding suggested that the receptor site in the nucleus was the chromatin. Yankner & Shooter (1979), working with PC12 cells in suspension, found that $(^{125}I)NGF$ could be internalized and that within a short time, Triton-insoluble binding to a nuclear fraction could be demonstrated. After longer exposure (17 hr), 60% of the NGF bound to the cell was associated with the nucleus. Two distinct nuclear binding sites were found with apparent K_D's of about 10^{-10} M and 10^{-8} M. Subfractionation of the nucleus indicated that these binding sites were associated with the nuclear membrane, but not with the chromatin. In addition, it has been found that the incubation of PC12 cells with NGF resulted in an initial decrease in the number of cell surface binding sites for NGF (Yankner & Shooter 1979, P. Layer and E. M. Shooter, unpublished, Calissano & Shelanski, in press) and a long-term increase in the receptor capacity of the nucleus (Yankner & Shooter 1979). In the study of Yankner & Shooter (1979), the change in the number of binding sites in the nucleus correlated with the number of NGF molecules that accumulated there. This was interpreted to suggest that NGF may be internalized and transported to the nucleus as a complex with its cell surface receptor. High affinity nuclear receptors were not detected in nuclei of rat liver or chicken heart. These workers also pointed out that although in previous morphological studies (Schwab & Thoenen 1977, Schwab 1977) internalized NGF was not localized within the nucleus, considerable localization occurred in the perinuclear region. Such observations would be consistent with the presence of an NGF receptor on the nuclear membrane rather than on the chromatin.

The above observations clearly indicate that there are multiple sites at which NGF can interact with its target cells: plasma membrane, cytoplasm, and nucleus. Such findings raise the presently unresolved issue as to the possible role of each site of interaction in NGF's mechanism of action. Several possibilities can be envisioned. One is that the actions of NGF require only interaction with a plasma membrane receptor and that subsequent events are carried out by means of a second messenger. A second possibility is that surface membrane receptors function only as carriers to facilitate the internalization of NGF and the primary sites of action of NGF are in the cytoplasm and/or nucleus. A third possibility is that each of the multiple sites of interaction plays a role in NGF's activities. For instance, any given biological action of NGF could require that the factor interact with multiple receptor sites. Alternatively, different responses to NGF could involve different types of receptors (Bradshaw 1978, Yankner & Shooter 1979). Such a concept is consistent with the recent demonstration (Burstein & Greene 1978) that NGF-stimulated neurite outgrowth in PC12 cells requires both transcription-dependent and independent pathways. It would thus be appealing to consider that transcription-independent actions of NGF are triggered via receptors on the surface membrane or in the cytoplasm, and that transcription-dependent actions require interaction with nuclear receptors.

RNA Transcription and the Mechanism of Action of NGF

An important aspect of the mechanism of action of NGF is the role of RNA transcription. In particular, one might pose the following questions. 1. Does NGF stimulate overall RNA synthesis in its target cells? 2. Do any or all of the biological actions of NGF require RNA transcription? 3. Does the action of NGF include and require alterations in the transcription of specific genes?

Early studies with cultured chick embryo DRG were interpreted to show that NGF stimulated incorporation of (^{3}H) uridine into RNA and therefore stimulated transcription of RNA (Angeletti et al 1965). Upon reexamination of this issue, however, Partlow & Larrabee (1971) and Burnham & Varon (1974) found that such effects were likely due to a decline in incorporation by NGF-treated "control" cultures. Moreover, Horii & Varon (1975, 1977) found that the differences in labeling of RNA in NGF-treated and untreated DRG cultures could be accounted for by effects of NGF on uptake, rather than incorporation, of uridine. In the presence of NGF, uptake was maintained at constant levels, whereas in the absence of NGF, uptake steadily declined with time. Such findings thus suggest that NGF maintains but does not stimulate overall RNA synthesis.

To test whether RNA transcription is required for NGF stimulated neurite outgrowth, Partlow & Larrabee (1971) cultured chick embryo sympathetic ganglia in the presence of NGF and high concentrations of actinomycin-D. In spite of the inhibition of nearly all ganglionic RNA synthesis, neurite outgrowth over the next 24 hr was only moderately impaired. Similar results were found with cultured sensory ganglia (Burnham & Varon 1974, Mizel & Bamberg 1976). These results have been widely interpreted to indicate that RNA synthesis is not required for stimulation of neurite outgrowth by NGF. However, one potential problem with such an interpretation is that the ganglia had prior exposure to NGF in vivo before being explanted in vitro. One system in which such preexposure to NGF can be controlled is the PC12 line. Burstein & Greene (1978) used the PC12 line to study neurite outgrowth by cells with or without prior exposure to NGF. For cultures without NGF pretreatment, initiation of neurite outgrowth by NGF was totally suppressed in the presence of low levels of several different inhibitors of RNA synthesis. In other experiments, PC12 cells were pretreated with NGF for 1 to 2 weeks, divested of their neurites by mechanical means, and replated either with no NGF, with NGF alone, or with NGF and RNA synthesis inhibitors. In the absence of NGF, significant levels of neurite regrowth did not occur. If NGF was present, however, most of the passage cells regenerated their neurites within 24 hr (Greene 1977, Burstein & Greene 1978); furthermore, such regeneration was not blocked when RNA synthesis was essentially totally inhibited (Burstein & Greene 1978). Thus, whereas generation of neurites by PC12 cells without previous exposure to NGF required RNA transcription, regeneration of neurites by cells preexposed to NGF did not. In this light, the transcription-independent neurite outgrowth that ganglia undergo in vitro could depend on prior exposure to NGF in vivo.

The above findings have led to the following *priming* model for the role of RNA synthesis in NGF-stimulated neurite outgrowth (Burstein & Greene 1978, Greene et al 1979, 1980). 1. NGF-stimulated neurite outgrowth requires the physical presence of NGF in one or more transcription-independent roles. 2. Neurite outgrowth also requires the presence of one or more materials whose synthesis is transcriptionally regulated by NGF. 3. During initial exposure to NGF and initial generation of neurites, the pool of such material is low and neurite outgrowth is therefore transcription-dependent. However, after cells have been exposed to NGF for a sufficient length of time, they accumulate a pool of such material; such pools can support neurite regeneration, even when transcription is blocked. One critical test of this model is that withdrawal of NGF should result in cessation of its transcriptional effects and a consequent time-dependent diminution through turnover of the pool of material required for neurite outgrowth. In agreement with this prediction, Burstein & Greene (1978) found that when NGF-pretreated PC12 cells were divested of their neurites and then subcultured for various lengths of time before being reexposed to NGF, they underwent a progressive loss (with a half-time of about 16 hr) of the capacity for transcription-independent neurite regeneration.

The process whereby cells acquire the ability for rapid, transcription-independent neurite outgrowth in the presence of NGF has been termed *priming* (Burstein & Greene 1978). Recent further studies with PC12 cells (L. A. Greene, D. E. Burstein and M. Black 1980, in press) have established that priming is not merely a consequence of the effect of NGF on cell replication, does not require either substrate attachment or prior neurite outgrowth, and is reflected not only by the ability of cells to regenerate neurites, but also by the initial rate at which such neurites elongate.

In addition to stimulation of neurite outgrowth, maintenance of survival is another fundamental biological activity of NGF that appears to have a transcription-independent component. Because the work of Partlow & Larrabee (1971) showed that cultured ganglia treated with actinomycin-D were able to extend neurites for 24 hr in the presence of NGF, it would appear to follow that inhibition of transcription also does not block the capacity of NGF to maintain neuronal viability. However, as for neurite regeneration, such experiments do not exclude the possible consequence of prior in vivo exposure to NGF, particularly with respect to RNA synthesis. The role of RNA transcription in maintenance of cell viability has also been studied with the PC12 system (Greene 1978). Although PC12 cells survive without NGF in serum-containing medium, in serum-free medium they die. However, if serum-free medium is supplemented with NGF, the cells survive and extend neurites. Moreover, exposure of serum-free PC12 cultures to levels of an RNA synthesis inhibitor at concentrations that blocked initiation of neurite outgrowth did not interfere with the ability of NGF to maintain cell viability. Whereas such findings do not rule out the possibility that the effect of NGF on survival of PC12 cells is altogether independent of RNA synthesis, they do suggest that the effects of NGF on neurite outgrowth and survival do not share the same transcriptional basis.

A third fundamental action of NGF that has been studied with respect to RNA synthesis is the induction of tyrosine hydroxylase (TOH). Stöckel et al (1974) found that induction of TOH activity in neonatal rats by administration of NGF could be blocked by pretreatment of the animals with actinomycin-D. Also, MacDonnell et al (1977a) compared the rates of synthesis of TOH by superior cervical ganglia explanted in vitro from neonatal rats that had been treated with or without NGF. The explanted ganglia from the NGF-treated animals showed markedly higher rates of synthesis of the enzyme. In addition, enhancement of in vitro synthesis of TOH was abolished if the animals were treated with actinomycin one hr before administration of NGF. In both studies, it was suggested that the induction of TOH by NGF was mediated via a specific stimulation of mRNA synthesis. Other data, however, indicate that the situation may be somewhat more complex. In their study, MacDonnell et al (1977a) noted that the increased rate of synthesis of TOH by ganglia explanted from NGF-treated rats was not abolished if actinomycin-D was present only in the cultures. More recently, using a modified culture system to maintain superior cervical ganglia (SCG) of adult rats, Rohrer et al (1978) demonstrated that NGF could induce TOH activity when administered in vitro. This increase in specific activity was not blocked by agents that inhibited RNA synthesis and on this basis it was concluded that the induction of TOH by NGF does not require transcription. Rohrer et al (1978) suggested that the apparent conflict of their results with earlier studies could be due to differences in culture conditions and to the complexity of the experimental situation in which SCG are treated with NGF in vivo. Another possibility that is consistent with present data is that, like neurite outgrowth, induction of TOH by NGF requires both transcriptional and nontranscriptional actions of NGF. Thus, exposure of SCG to NGF in vivo would bring about transcriptional changes that could prime the neurons so that induction of the enzyme in vitro appears to be insensitive to actinomycin-D. This would also explain why blockade of transcription in vivo with actinomycin-D effectively abolishes subsequent responses of the ganglia to NGF.

In summary, there is evidence that the mechanism of action by which NGF brings about its biological activities may include both transcriptional and nontranscriptional steps. In subsequent sections, we review studies regarding the possible nature of these mechanisms.

Protein Synthesis

The questions posed for RNA synthesis may also be posed for the role of protein synthesis in the mechanism of action of NGF: Does NGF stimulate overall protein synthesis in its target cells? Do the biological actions of NGF require protein synthesis? Do the actions of NGF include and require alterations in the synthesis of specific proteins?

Angeletti et al (1965) presented evidence that NGF stimulated incorporation of (^{3}H) leucine by cultured chick embryo DRG. However, subsequent workers (Partlow & Larrabee 1971, Burnham & Varon 1974) raised the possibility that such effects could be due to deterioration of incorporation by control NGF-untreated ganglia and maintenance of basal protein synthesis in NGF-treated ganglia. Despite this experimental ambiguity, there are some indications that NGF may stimulate overall protein synthesis. For example, the large increase in somatic size and fiber outgrowth that neurons undergo when treated with excess NGF is consistent with such a possibility. However, other factors such as increased uptake of precursors or a decreased rate of protein degradation could also affect the neuronal protein content. Also consistent with stimulation of protein synthesis by NGF are the electron microscopic observations of Schwab & Thoenen (1975). These workers examined rat SCG at various times after in vivo administration of NGF. As early as 6 hr after injection of NGF, significant changes were found in the Nissl substance and Golgi apparatus. These changes were of the type associated with increased cellular protein synthesis.

Evidence regarding the requirement of ongoing protein synthesis for NGF's actions is somewhat clearer. Partlow & Larrabee (1971) found similar dose-response curves for the inhibitory effects of cycloheximide on both leucine incorporation and neurite outgrowth. Luduena (1973), Burnham & Varon (1974), and Mizel & Bamburg (1976) also demonstrated that NGF-stimulated neurite outgrowth by cultured chick DRG cells was blocked in the presence of cycloheximide. In the latter study, the suppression of neurite outgrowth by cycloheximide was shown to be reversible and hence was not due to loss of cell viability. Also, not unexpectedly, the NGF-mediated inductions of TOH and dopamine β hydroxylase (DBH) activities are also blocked by cycloheximide and hence appear to require ongoing protein synthesis (Otten & Thoenen 1977, Otten et al 1977).

The role of protein synthesis in NGF's effect on cell viability has not been systematically examined. Nevertheless, of relevance are Mizel & Bamburg's (1976) observation that inhibition of neurite outgrowth caused by 12 hr of exposure to cycloheximide could be reversed when the drug was removed. Such findings suggest that maintenance of viability by NGF does not require ongoing protein synthesis.

By immunologic techniques, MacDonnell et al (1977a) and Max et al (1978) demonstrated that the increase of TOH activity in NGF-treated SCG was due to enhanced synthesis of enzyme molecules. This is one example of the capacity of NGF to selectively alter the synthesis of a specific protein as well as a case in which an NGF-induced change in protein synthesis has clear biological significance. Another enzyme whose synthesis appears to be regulated by NGF is ornithine decarboxylase (ODC). Mac-Donnell et al (1977b) demonstrated that NGF increases the activity of this enzyme in SCG and that this increase can be prevented by cycloheximide or actinomycin-D. As is discussed below, however, the biological significance of this induction is not clear.

Given the dramatic effects of NGF on neurite outgrowth and the evidence that such effects require transcription, it would appear reasonable to expect that NGF-stimulated fiber outgrowth also includes and requires selective changes in the synthesis of specific proteins. For the reasons discussed above, it is experimentally difficult to study such possible changes with in vivo or in vitro preparations of neurons. The question of specific effects of NGF on protein synthesis have begun to be studied, however, with PC12 cells. McGuire et al (1978) used two-dimensional isoelectric focusing SDS-polyacrylamide gradient gel electrophoresis to compare the patterns of (^{14}C) amino acid-labeled polypeptides synthesized by PC12 cells before and after various times of treatment with NGF. At a resolution of about 1,000 peptides, no qualitative and only a very small number of reproducible quantitative changes in protein synthesis were detected. Such findings are quite remarkable in light of the readily detectable changes in protein synthesis that accompany differentiation in other cell systems (cf Paterson & Bishop 1977). Among the suggestions drawn from such experiments on PC12 cells was that stimulation of neurite outgrowth does not require either major qualitative or quantitative changes in the synthesis of abundant cell proteins and therefore must take place largely via structural rearrangement of the types of proteins that the cells already synthesize prior to NGF treatment. These findings also suggest that if NGF does bring about alterations in the pattern of protein synthesis, such changes affect the less abundant class of cell proteins. Using PC12 cultures, efforts have been made to identify changes in synthesis of such nonabundant molecules. McGuire et al (1978) analyzed the patterns of (^{3}H) fucose and (^{3}H) glucosamine-labeled glycoproteins synthesized by PC12 cells before and after NGF treatment. A large increase was noted in the labeling of a surface-exposed glycoprotein of apparent mol wt 230,000. The effects of NGF on synthesis of this molecule were selectively blocked by low concentrations of camptothecin, an inhibitor of RNA synthesis. At the levels of camptothecin employed, initiation of neurite outgrowth is also blocked in PC12 cultures. It was further demonstrated that this glycoprotein, or NILE glycoprotein, is distinct from the well-known fibronectin glycoprotein. A second glycoprotein that McGuire et al (1978) found to be increased with NGF was of apparent mol wt 25,000 to 30,000. Recent experiments (C. Richter-Landsberg et al, unpublished) indicate that this protein is enriched in cytoskeletal preparations. Using one-dimensional SDS gel electrophoresis, McGuire & Greene (1977, 1979b) also found that NGF selectively increased the capacity of PC12 cells to synthesize a minor cell protein of apparent mol wt 80,000. Synthesis of this molecule was maximally stimulated within 1 to 2 days of NGF treatment, hence preceding neurite outgrowth, and returned to basal levels within 1 to 2 days after withdrawal of NGF. The increased synthesis of this protein in the presence of NGF was also selectively blocked by low levels of camptothecin. Although this and other minor proteins whose syntheses are selectively altered by NGF appear to be of potential interest, it has not yet been possible to assign them roles in NGF's mechanism of action.

In summary, the above findings suggest that (*a*) stimulation of overall protein synthesis by NGF is likely, but not presently unquestionably demonstrated, (*b*) NGF's effects on neurite outgrowth and induction of TOH require ongoing protein synthesis, whereas maintenance of survival may not, and (*c*) NGF can selectively alter the synthesis of certain cell proteins. In the case of increased levels of enzymes such as TOH, the role of protein synthesis in the action of NGF is reasonably clear. However, it has not yet been possible to link NGF's effects on synthesis of selected cell proteins to other biological actions of the factor.

Linking Steps

As noted above, there are a number of demonstrated and suggested effects of NGF that have been proposed to causally link the interaction of NGF with its target cells to NGF's final biological actions. A number of these possible linking mechanisms are reviewed and evaluated below.

CYCLIC AMP   Cyclic AMP (cAMP) has been demonstrated to be a "second messenger" in mediating the responses of cells to a variety of hormones. There is, however, conflicting evidence regarding the possible role of cAMP in the mechanism of action of NGF.

Support for a role of cAMP in mediating some or all of NGF's actions have come from several types of evidence: 1. NGF treatment brings about rapid increases in cAMP levels. Nikodijevic et al (1975) reported a two to threefold increase of cAMP levels of cultured rat SCG in response to NGF. Narumi & Fujita (1978) and Skaper et al (1979) showed a several-fold increase of cAMP in NGF-treated cultured chick dorsal root ganglia. Working with PC12 cultures, Schubert & Whitlock (1977) and Schubert et al (1978) reported a 70% increase in cAMP in response to NGF. In each case, the increases were transitory and peaked within 5 to 10 min after NGF addition. 2. Exposure to cAMP analogues (generally dibutyryl cAMP or dBcAMP), or to agents that raise intracellular levels of cAMP, can mimic the actions of NGF. A number of groups have reported that dBcAMP can elicit neurite outgrowth from cultured chick DRG (Roisen et al 1972, Hier et al 1972, Frazier et al 1973c, Narumi & Fujita 1978). The latter group also reported that exposure of cultured chick embryo DRG to substance P raised intracellular levels of cAMP and elicited fiber outgrowth. Also, Schubert & Whitlock (1977) and Schubert et al (1978) claimed to elicit neurite outgrowth in PC12 cultures with analogues of cAMP as well as with agents that raise intracellular cAMP levels. 3. Yu et al (1978) have found that NGF as well as cAMP analogues, caused an increase in the phosphorylation of a 30,000 mol wt nuclear protein in cultures of rat sympathetic ganglia.

In apparent conflict with the above findings are a number of reports that do not support a role of cAMP in the mechanism of action of NGF. Frazier et al (1973c) found no significant effect of NGF on cAMP levels in cultured chick embryo DRG or on adenylate cyclase activity in broken cell preparations of the ganglia. In addition, Otten et al (1978) reported no significant alteration in the levels of cAMP or cGMP in rat SCG exposed to NGF in vivo or in vitro. Lakshmanan (1978) reported a similar lack of effect of NGF on the cAMP content of organ cultured rat SCG. Finally, Hatanaka et al (1978) were unable to demonstrate an effect of NGF on cAMP or cGMP levels in PC12 cultures. In each of the above studies, the time courses were comparable to those examined by workers who reported positive effects of NGF on cAMP levels. With respect to neurite outgrowth, Frazier et al (1973c) noted that the morphologic response of cultured DRGs to dBcAMP was distinctly different from that promoted by NGF. These workers also found that dBcAMP did not elicit neurite outgrowth from cultured chick embryo sympathetic ganglia. Also, dBcAMP, in a wide range of tested concentrations, fails to support the survival of neurons in cell cultures of dissociated chick embryo sympathetic ganglia (L. A. Greene, unpublished). Furthermore, in contrast to the reports of Schubert & Whitlock (1977) and Schubert et al (1978), others (Greene & Tischler 1976, Greene et al 1979, D. Burstein and L. A. Greene, in preparation) have found that analogues of cAMP or agents that increase intracellular cAMP do not mimic the effects of NGF in promoting neurite outgrowth in PC12 cultures. When tested over a wide range of concentrations, such agents only elicit short cytoplasmic extensions and these are quite distinct from the long, branching neurites elicited by NGF. These agents also failed to prime PC12 cells for NGF-induced rapid, transcription independent neurite growth. In examining the effects of dBcAMP on PC12 cells, D. Burstein and L. A. Greene (in preparation) found that although this agent did not cause initiation of neurite outgrowth, it could elicit regeneration of neurites from cells that had been pretreated or primed with NGF. However, in contrast to the neurites regenerated in presence of NGF, those regenerated in the presence of dBcAMP did not continue to elongate beyond 24 hr of treatment and did not appear in cultures treated with inhibitors of RNA synthesis. Such findings indicate that the effects of cAMP analogues on fiber outgrowth in cultured dorsal root ganglia may be due to prior exposure of the ganglia to NGF in vivo and that NGF and cAMP analogues may stimulate neurite outgrowth by independent mechanisms. Another relevant observation of D. Burstein and L. A. Greene (in preparation) was that, in contrast to NGF, dBcAMP did not cause PC12 cells to alter their synthesis of the 80,000 mol wt protein or of the 230,000 and 25,000 to 30,000 mol wt glycoproteins referred to in a previous section of this review.

In summary, at present there is no conclusive evidence that cAMP functions as a "secondary messenger" for NGF. However, it has not been completely ruled out that cAMP could play a role in some or all of the actions of NGF.

ORNITHINE DECARBOXYLASE   Rapid induction of ornithine decarboxylase has been shown to be associated with the responses of a number of cell types to various growth and differentiation-promoting hormones (see Russell et al 1976 for review). It has been suggested that this enzyme and the polyamines whose synthesis it catalyzes may play a causal role in mediating subsequent cellular responses to such hormones (Russell et al 1976). A number of studies have demonstrated that NGF can cause rapid induction of ODC in various tissues including brain (Roger et al 1974, Lewis et al 1978), rat superior cervical ganglia (MacDonnell et al 1977b), embryonic chick DRG (Lakshmanan 1979), and PC12 cells (Hatanaka et al 1978, Greene & McGuire 1978). These effects appear to require ongoing synthesis of both RNA and protein (MacDonnell et al 1977b, Hatanaka 1978, Green & McGuire 1978). Such findings have led to the hypothesis that induction of ODC may also play a causal role in the mechanism of action of NGF (MacDonnell et al 1977b). Greene & McGuire (1978) have evaluated this possibility with cultures of PC12 cells and chick embryo sympathetic neurons. Treatment of the cultures with drugs that either specifically block ODC activity or that specifically suppress induction of this enzyme did not detectably alter the ability of NGF to promote neurite outgrowth, yield an increase in somatic volume, or maintain survival. On this basis, it was concluded that the above actions of NGF do not require induction of ODC. Thus, induction of ODC could be an effect of NGF that is without further biological consequences. On the other hand, it cannot be presently ruled out that there are other biological actions of NGF in which induction of ODC plays a required role.

ION FLUX   It has been hypothesized that the mechanism of action of NGF might include regulation of ion fluxes (Shelanski 1973, Horii & Varon 1977). Schubert et al (1978) presented evidence that NGF enhanced Ca^{2+} efflux from PC12 cells and that exposure of PC12 cultures to conditions that increased Ca^{2+} mobilization (i.e. depolarization with elevated K^+ and exposure to 100 μM veratridine) elicited neurite outgrowth. On this basis, it was proposed that mobilization of Ca^{2+} mediates the effect of NGF on neurite outgrowth. Other data, however, appear to be in conflict with the results presented by Schubert et al (1978). G. Landreth, P. Cohen, and E. M. Shooter (unpublished), in work with PC12 cells in suspension, found no significant effect of NGF on Ca^{2+} flux. Moreover, the flux experiments reported by Schubert et al (1978) were carried out with high concentrations of 7S NGF. As reviewed above, the 7S preparation of NGF contains a subunit that possesses enzymatic activity and appears to have effects on cell membranes (Greene et al 1971). Also, Greene et al (1979) and L. A. Greene, B. Burstein, and M. Black (in preparation) have found that exposure of PC12 cells to elevated K^+ or to 100 μM veratridine caused neither initiation of neurite outgrowth, neurite regeneration by cells pretreated with NGF, nor priming of the cells for NGF-dependent neurite regeneration.

In examining the regulation of transport of exogenous substrates by NGF (see below), Skaper & Varon (1979a) and Varon & Skaper (1980) noted that the affected transport systems were all Na^+ dependent. It was found that NGF appeared to regulate these systems by regulating the Na^+ gradient required for them. Chick sensory ganglia cells cultured without NGF accrued Na^+; when NGF was reintroduced to the cultured cells, they actively and rapidly (with a response time of minutes) extruded Na^+. On this basis, the above authors proposed that Na^+ extrusion may be an early linking event in the mechanism of action of NGF. It was suggested that such ionic responses to NGF not only could affect substrate transport but could also ultimately lead to effects on survival, cell size, neurite outgrowth, and synthesis of neurotransmitter enzymes. This hypothesis has attractive possibilities and further work is clearly in order to test it as well as to define how Na^+ extrusion could be linked to other events in the sequence of NGF's action.

SUBSTRATE ADHESION   A number of studies have led to emphasis on the role of cell-substrate interaction in regulating the presence and directionality of neurite outgrowth (see Sidman & Wessells 1975, Letourneau 1977 for review). The possibility has thus been raised that a primary action of NGF, particularly with respect to initiating and directing neurite outgrowth, might be to alter the substrate adhesiveness of its target cells (Varon 1975, Letourneau 1977, 1978).

Given the experimental difficulties involved, few direct tests of the adhesion hypothesis have been carried out. Schubert & Whitlock (1977) and Schubert et al (1978) reported that NGF treatment resulted in rapidly detectable (within 10 min) increased adhesion of PC12 cells to one another and to plastic tissue culture dishes. These authors theorized that such alterations in adhesion were in turn responsible for initiation of neurite outgrowth. The interpretation of such experiments, however, has a number of difficulties. First, adhesion to a tissue plastic substrate was studied only over a time span of a few hours. It has been found that, like normal sympathetic neurons (Varon & Raiborn 1972a), PC12 cells treated with NGF for more than about 24 hr show markedly decreased adhesion to tissue culture plastic (Greene & Tischler 1976, Greene et al 1979). Second, plating of PC12 cells on substrates (such as concentrated collagen or polylysine) to which they have high adhesion even in the absence of NGF does not induce neurite outgrowth. Third, PC12 cells undergo responses to NGF even in the absence of substrate adhesion (McGuire et al 1978, Greene et al 1979). Fourth, such changes appear to be nonspecific in that they also occur in response to EGF (H. Herschman et al, unpublished). Thus, further experiments appear necessary to critically evaluate the significance of such findings. Also, it will be especially important to use normal neurons in addition to tumor cells to probe the existence and possible consequences of NGF's effects on substrate adhesion. One particularly suitable experimental system in which to carry out such studies may be the barrier system designed by Campenot (1977), in which various parts or even different neurites of the same cultured neuron can be exposed to different local concentrations of NGF.

EFFECTS ON MICROTUBULE PROTEIN   There is good evidence that the outgrowth of neurites by NGF-responsive neurons requires the formation and maintenance of microtubules (Daniels 1972, Roisen & Murphy 1973). This evidence has raised the possibility that NGF could promote neurite outgrowth by causing an increase in the levels of tubulin protein and/or enhancing assembly of tubulin to form microtubules. In studies with short-term (24 to 72 hr) cultures of chick embryo DRG (Yamada & Wessells 1971, Mizel & Bamburg 1975) and sympathetic ganglia (Mizel & Bamburg 1975) it was concluded that the concentration of microtubule protein did not change in response to NGF treatment. Stöckel et al (1974) also reported that neither the total amount nor the concentration of microtubule protein was altered in the SCG's of neonatal rats in response to NGF treatment in vivo. One interpretation of these studies was that NGF-induced neurite outgrowth involves an increase in polymerization, but not synthesis of tubulin. In this regard, it is of interest that induction of neurite outgrowth in neuroblastoma cultures caused by withdrawal of serum also occurs without a change in tubulin concentration (Mizel & Bamburg 1975, Morgan & Seeds 1975). Before conclusions can be drawn, however, there are several points to consider. The use of cultured ganglia raises the problems of heterogeneity of tissue, deterioration of NGF-untreated controls, and the influence of prior in vivo exposure to NGF. Also, such studies have been carried out for time courses of only up to several days and there has been no experimental verification of whether or not NGF does indeed affect tubulin assembly. The potential importance of the role of synthesis and assembly of tubulin in the mechanism of action of NGF suggests that these questions be reexamined with homogeneous cell systems, during initiation (rather than regeneration), over an extended time course, and using experimental techniques for measuring both total microtubule protein and polymerized tubulin (Pipeleers et al 1977). It will also be important to distinguish whether possible effects of NGF on tubulin/microtubules are primary causes or secondary consequences of neurite outgrowth.

INTERACTION OF NGF WITH FIBROUS PROTEINS   Another closely related model proposed for the mechanism of action of NGF is interaction of the factor with fibrous proteins. Calissano & Cozzari (1974), Calissano et al (1976), Levi et al (1975), and Monaco et al (1977) have presented evidence that, in vitro, NGF (in the 1 to 10 μM range) can form stoichiometric complexes with tubulin, promote the assembly of tubulin, and interact with tubulin and preformed microtubules to cause the formation of larger insoluble "supramolecular" aggregates. Additional in vitro experiments (Monaco et al 1977) demonstrated that preexposure of microtubules to

NGF suppressed the formation of "paracrystalline" arrays in the presence of vinblastine. It was also shown that the destructive in vivo effects of vinblastine on sympathetic ganglia of newborn rats could be prevented by prior administration of exogenous NGF (Chen et al 1977). The latter findings were interpreted as being consistent with the interaction of NGF with microtubules in vivo as well as in vitro. In similar types of in vitro experiments, it was found that NGF (μM range) can form stoichiometric complexes with G actin and favor polymerization of the latter to microfilaments and into ordered paracrystalline arrays that activate myosin ATPase (Calissano et al 1976, Calissano et al 1978). These findings have led to the proposal that NGF interacts in vivo with nonpolymerized tubulin and actin and/or microtubules and microfilaments and that this interaction plays an integral role in the mechanism of action of the factor (Calissano & Cozzari 1974, Calissano et al 1976, Calissano et al 1978). Among the possible consequences of such an interaction that were suggested were specific binding of NGF to target cells, internalization and retrograde transport of NGF, enhanced formation and stabilization of microtubules and microfilaments, and alterations in membrane mobility. Such actions could mediate the nontranscriptional events in neurite outgrowth.

Although the interaction of NGF with fibrous proteins is an intriguing finding, the relationship of this phenomenon to the mechanism of action of NGF has many aspects that are in need of clarification. First, there is the question of concentration. The in vitro experiments reviewed above employed, and appeared to require, NGF at concentrations of 1 to 10 μM; the biological actions of NGF generally require much lower concentrations. Second, there is the question of specificity. For instance, in addition to NGF, a number of other basic proteins have been shown to promote in vitro assembly of tubulin (Lee et al 1978). Finally, it remains to be demonstrated that the in vitro effects of NGF on tubulin and actin take place in vivo. Experiments on binding of (^{125}I)NGF to and within its target cells have not indicated localization of binding to filamentous structures.

UPTAKE OF SMALL MOLECULES   There is evidence that NGF can alter the uptake and accumulation of small molecules by its target cells. Levi & Lattes (1968) provided evidence that NGF enhanced the ability of cultured chick embryo DRG's to accumulate acidic, but not other types of amino acids. Horii & Varon (1975, 1977) demonstrated that NGF maintained the ability of cells from dissociated chick embryo dorsal root ganglia to take up and accumulate uridine, cytidine and guanosine as well as 2-deoxyglucose and α-aminoisobutyric acid. Uptake of leucine was not affected. In experiments with (^{3}H) uridine (Horii & Varon 1975, 1977) and (^{3}H)2-deoxyglucose (Skaper & Varon 1979b) it was found that if NGF was withdrawn or was never present in the incubation medium, uptake of these small molecules by cultured DRG cells progressively declined with time. Moreover, when NGF was reintroduced to the DRG dissociates within 6 hr, transport was rapidly restored (in less than 10 min) to the levels of uptake displayed by cells that had not been deprived of the factor. As discussed above, these effects on transport appear to be regulated via effects of NGF on intracellular Na^+ levels (Skaper & Varon 1979a). The abilities of NGF to maintain or restore transport were not prevented by actinomycin-D or cycloheximide and thus required neither ongoing RNA nor protein synthesis. These findings—that NGF can regulate the uptake of certain small molecules by its target cells—led Horii & Varon (1977) to speculate that such effects on uptake could in turn "regulate the extent of anabolic activities of the cell and, thus, control in the longer run survival, proliferation, and/or enlargement, neurite elongation, production of specific enzymes, and so on." The ability of PC12 cells to respond to, but not require, NGF, makes them a potentially useful system in which to study further the effects of NGF on uptake of small molecules. McGuire & Greene (1979a) found that NGF increased (rather than maintained) the rate at which PC12 cells transported α-aminoisobutyric acid. The onset of this stimulation was detectable within 15 min of NGF administration. Similar increases in uptake were also induced by exposure to elevated levels of serum. However, because serum treatment did not induce neurite outgrowth in this system, it was concluded that increased uptake in itself was not sufficient to trigger all of the effects of NGF. Nevertheless, such considerations do not rule out the possibility that modulation of uptake may play a necessary role in one or several of the actions of NGF. Particularly inviting possibilities (Horii & Varon 1977, Skaper & Varon 1979b) are that modulation of small molecule transport is the primary mechanism by which NGF maintains cell survival and causes cellular hypertrophy.

CELL SURFACE ARCHITECTURE   Because one site of interaction of NGF with its target cells is the plasma membrane, it is possible that one link in the mechanism of action of the factor includes alterations in cell surface architecture. It has been reported (Schubert & Whitlock 1977, Schubert et al 1978) that within 10 to 20 min after PC12 cells are exposed to NGF, changes occur in their attachment rate to tissue culture plastic, agglutinability by certain lectins, and cell-cell adhesion. Connolly et al (1979a, b), employing scanning electron microscopy, found that NGF treatment caused a rapidly onsetting, highly synchronized, reproducible sequence of changes in the surface architecture of PC12 cells. Within 1 min after exposure to NGF, large ruffles appeared on the dorsal surfaces of the cells. These dorsal ruffles were prominent by 3 min and almost completely disappeared by 7 min of treatment. Ruffles were also observed at the cell periphery by 3 min; these were prominent by 7 min and were absent by 15 min. During this time period, microvilli, which were present in large numbers before NGF treatment, disappeared. Between about 15 10 45 min of treatment the cell surface remained smooth. After this time, large blebs began to appear. These were present on most cells at 2 hrs and were gone by 4 hrs. By 6 to 7 hrs of NGF treatment, the microvilli began to reappear and by 10 hrs these once again covered the cell surface. At present it is difficult to evaluate whether the above types of rapid changes in the cell surface play a causal, required role in the mechanism of action of NGF. For example, such changes have only been described thus far in a tumor cell line and occur well before appreciable neurite outgrowth commences. Nevertheless, one can speculate that such change in the cell surface may be related to rearrangement of NGF receptors, internalization of NGF, increased uptake of small molecules, or alterations in the interaction of NGF-treated cells with their physical environment.

CONCLUSIONS AND PERSPECTIVES

It is no doubt evident to the reader that the mechanism of action of NGF is far from being definitively described. Nevertheless, a good deal of real and exciting progress has been made over the last five years. Recent studies have reinforced the likelihood that the multiple actions of NGF involve multiple mechanisms of action. In particular, the demonstration of multiple means by which NGF can interact with its target cells raises the possibility that different actions of the factor may be mediated via several distinct receptor sites (Bradshaw 1978, Yankner & Shooter 1979). This notion is also consistent with the demonstration that at least some of the actions of NGF require both transcriptional and nontranscriptional events. There remains, however, much experimental work to conclusively establish (or disprove) this hypothesis as well as to clearly define the steps that link association of NGF with its target cell and NGF's final biological actions. These goals will be materially aided by the recent establishment of several NGF-responsive cell lines. One particularly attractive prospect that is possible with such cell lines is the use of genetic techniques to uncover and dissect the various steps in NGF's mechanism of action.

ACKNOWLEDGMENTS

Support during preparation of this review was provided for L. A. Greene by grants from the National Institutes of Health (NS 11557 and NS 12200), Sloan Foundation, and National Foundation—March of Dimes and for E.

M. Shooter by grants from the National Institutes of Health (NS 04270), the National Science Foundation (BNS 77–00088), and the Sloan Foundation. The authors would like to thank Mrs. Vicky Shroff for her patience and skill in preparing the finished manuscript and their colleagues for many helpful discussions.

Literature Cited

Aloe, L., Levi-Montalcini, R. 1979. Nerve growth factor-induced transformation of immature chromaffin cells in vivo into sympathetic neurons: Effect of antiserum to nerve growth factor. *Proc. Natl. Acad. Sci. USA* 76:1246–50

Andres, R. Y., Jeng, I., Bradshaw, R. A. 1977. Nerve growth factor receptors: Identification of distinct classes in plasma membranes and nuclei of embryonic dorsal root neurons. *Proc. Natl. Acad. Sci. USA* 74:2785–89

Angeletti, P. U., Gandini-Attardi, D., Toschi, G., Salvi, M. I., Levi-Montalcini, R. 1965. Metabolic aspects of the effect of nerve growth factor on sympathetic and sensory ganglia: Protein and ribonucleic acid synthesis. *Biochim. Biophys. Acta* 95:111–20

Angeletti, R. H. 1970. The role of the tryptophan residues in the activity of the nerve growth factor. *Biochim. Biophys. Acta* 214:478–81

Angeletti, R. H., Bradshaw, R. A. 1971. Nerve growth factor from mouse submaxillary gland: Amino acid sequence. *Proc. Natl. Acad. Sci. USA* 68:2417–20

Angeletti, R. H., Bradshaw, R. A., Wade, R. D. 1971. Subunit structure and amino acid composition of mouse submaxillary gland nerve growth factor. *Biochemistry* 10:463–69

Angeletti, R. H., Hermodson, M. A., Bradshaw, R. A. 1973b. Amino acid sequences of mouse 2.5S nerve growth factor. II. Isolation and characterization of the thermolytic and peptic peptides and the complete covalent structure. *Biochemistry* 12:100–15

Angeletti, R. H., Mercanti, D. Bradshaw, R. A. 1973a. Amino acid sequences of mouse 2.5S nerve growth factor. I. Isolation and characterization of the soluble tryptic and chromotryptic peptides. *Biochemistry* 12:90–99

Argos, P. 1976. Prediction of the secondary structure of mouse nerve growth factor and its comparison with insulin. *Biochem. Biophys. Res. Commun.* 70:805–9

Au, M. J., Dunn, M. F. 1977. Reaction of the basic trypsin inhibitor from bovine pancreas with the chelator-activated 7S nerve growth factor esteropeptidase. *Biochemistry* 16:3958–66

Banerjee, S. P., Cuatrecasas, P., Snyder, S. 1975. Nerve growth factor receptor binding: Influence of ions and proteins reagents. *J. Biol. Chem.* 250:1427–33

Banerjee, S. P., Cuatrecasas, P., Snyder, S. 1976. Solubilization of nerve growth factor receptor of rabbit superior cervical ganglia. *J. Biol. Chem.* 251:5680–85

Banerjee, S. P., Snyder, S. H., Cuatrecasas, P., Greene, L. A. 1973. Binding of nerve growth factor in sympathetic ganglia. *Proc. Natl. Acad. Sci. USA* 70:2519–23

Banks, B. E. C., Walter, S. J. 1977. The effects of postganglionic axotomy and nerve growth factor on the superior cervical ganglia of developing mice. *J. Neurocytol.* 6:287–97

Barde, Y. A., Thoenen, H. 1979. Physiology of nerve growth factor. *Physiol. Rev.* In press

Berger, E. A., Shooter, E. M. 1977. The biosynthesis and processing of pro βNGF, a biosynthetic precursor to β-nerve growth factor. *Proc. Natl. Acad. Sci. USA* 74:3647–51

Berger, E. A., Shooter, E. M. 1978. The biosynthesis of β nerve growth factor in mouse submaxillary gland. *J. Biol. Chem.* 243:804–10

Bjerre, B., Björklund, A., Stenevi, U. 1973. Stimulation of growth of new axonal sprouts from lesioned monoamine neurones in adult rat brain by nerve growth factor. *Brain Res.* 60:161–76

Björklund, A., Stenevi, U. 1972. Nerve growth factor: Stimulation of regenerative growth of central noradrenergic neurons. *Science* 175:1251–53

Bocchini, V., Angeletti, P. U. 1969. The nerve growth factor: Purification as a 30,000-Molecular-weight protein. *Proc. Natl. Acad. Sci. USA* 64:787–94

Bothwell, M. A., Shooter, E. M. 1978. Thermodynamics of the interaction of the subunits of 7S nerve growth factor. *J. Biol. Chem.* 253:8458–64

Bothwell, M. A., Wilson, W. H., Shooter, E. M. 1979. The relationship between glandular kallikrein and growth factor-processing proteases of mouse submaxillary gland. *J. Biol. Chem.* 254: 7287–94

Bradshaw, R. A. 1978. Nerve growth factor. *Ann. Rev. Biochem.* 47:191–216

Bradshaw, R. A., Frazier, W. A. 1977. Hormone receptors as regulators of hormone action. In *Current Topics in Cellular Regulation,* ed. B. L. Horecker, E. R. Stadtman, 12:1–37. New York: Academic. 280 pp.

Bradshaw, R. A., Jeng, I., Andres, R., Pulliam, M. W., Silverman, R. E., Rubin, J., Jacobs, J. W. 1977. The structure and function of nerve growth factor. In *Endocrinology: Proc. Vth Int. Congr. Endocrinol.,* ed. H. V. T. James, 2:206–12. Amsterdam: Excerpta Medica. 616 pp.

Bradshaw, R. A., Young, M. 1976. Nerve growth factor—recent developments and perspectives. *Biochem. Pharmocol.* 25:1445–49

Brodeur, G. M., Goldstein, M. N. 1976. Histochemical demonstration of an increase in acetylcholinesterase in established lines of human and mouse neuroblastomas by nerve growth factor. *Cytobios.* 16:133–38

Brunso-Bechtold, J. K., Hamburger, V. 1979. Retrograde transport of nerve growth factor in chicken embryo. *Proc. Natl. Acad. Sci. USA* 76:1494–96

Bueker, E. D. 1948. Implantation of tumors in the hind limb field of the embryonic chick and the developmental response of the lumbosacral nervous system. *Anat. Rec.* 102:369–89

Burnham, P., Varon, S. 1973. In vitro uptake of active nerve growth factor by dorsal root ganglia of embryonic chick. *Neurobiology* 3:232–45

Burnham, P. A., Varon, S. 1974. Biosynthetic activities of dorsal root ganglia in vitro and the influence of nerve growth factor. *Neurobiology* 4:57–70

Burstein, D. E., Greene, L. A. 1978. Evidence for both RNA-synthesis-dependent and independent pathways in stimulation of neurite outgrowth by nerve growth factor. *Proc. Natl. Acad. Sci. USA* 75:6059–63

Burton, L. E., Shooter, E. M. 1979. Heterogeneity of the trypsin-like subunit of 7S nerve growth factor. *Fed. Proc.* 36:718

Burton, L. E., Wilson, W. H., Shooter, E. M. 1978. Nerve growth factor in mouse saliva. Rapid isolation procedures for and characterization of 7S nerve growth factor. *J. Biol. Chem.* 253:7807–12

Calissano, P., Cozzari, C. 1974. Interaction of nerve growth factor with the mouse brain neurotubule protein(s). *Proc. Natl. Acad. Sci. USA* 71:2131–35

Calissano, P., Levi, A., Alema, S., Chen, J. S., Levi-Montalcini, R. 1976. Studies on the interaction of the nerve growth factor with tubulin and actin. In *26th Colloquium-Mosbach 1975. Molecular Basis of Motility,* ed. L. Heilmeyer, pp. 186–202. Berlin: Springer-Verlag

Calissano, P., Monaco, G., Castellani, L., Mercanti, D., Levi, A. 1978. Nerve growth factor potentiates actomyosin adenosinetriphosphatase. *Proc. Natl. Acad. Sci. USA* 75:2210–14

Calissano, P., Shelanski, M. L. 1980. Interaction of nerve growth factor with pheochromocytoma cells-evidence for tight binding and sequestration. *Neuroscience.* In press

Campenot, R. B. 1977. Local control of neurite development by nerve growth factor. *Proc. Natl. Acad. Sci. USA* 74:4516–19

Caramia, F., Angeletti, P. U., Levi-Montalcini, R. 1962. Experimental analysis of the mouse submaxillary salivary gland in relationship to its nerve growth factor content. *Endocrinology* 70:915–22

Chen, M. G., Chen, J. S., Calissano, P., Levi-Montalcini, R. 1977. Nerve growth factor prevents vinblastine destructive effects on sympathetic ganglia in newborn mice. *Proc. Natl. Acad. Sci. USA* 74:5539–63

Cohen, S. 1960. Purification of a nerve growth promoting protein from the mouse salivary gland and its neurocytotoxic antiserum. *Proc. Natl. Acad. Sci. USA* 46:302–11

Connolly, J. L., Greene, L. A., Viscarello, R. R. 1979a. Rapid sequential surface changes of PC12 pheochromocytoma cells in response to nerve growth factor. *Fed. Proc.* 38:1430

Connolly, J. L., Greene, L. A., Viscarello, R. R., Riley, W. D. 1979b. Rapid sequential changes in surface morphology of PC12 pheochromocytoma cells in response to nerve growth factor. *J. Cell Biol.* 82:820–27

Costrini, N. V., Bradshaw, R. A. 1979. Binding characteristics and apparent molecular size of detergent-solubilized nerve growth factor receptor of sympathetic ganglia. *Proc. Natl. Acad. Sci. USA* 76:3242–45

Daniels, M. P. 1972. Colchicine inhibition of nerve fiber formation in vitro. *J. Cell Biol.* 53:164–76

Dayhoff, M. O. 1976. Survey of new data and computer methods of analyses. In *Atlas of Protein Sequence and Structure*, ed. M. O. Dayhoff, pp. 1–10. Washington: Nat. Biomed. Res. Found.

De Meyts, P., Roth, J., Neville, D. M. Jr., Gavin, J. R. III., Lesniak, M. A. 1973. Insulin interactions with its receptors: Experimental evidence for negative cooperativity. *Biochem. Biophys. Res. Commun.* 55:154–61

Diamond, L., Revoltella, R., Bertolini, L. 1976. Differences between murine C1300 neuroblastoma clones detected by rosette formation with nerve growth factor coated sheep red blood cells. *Brain Res.* 118:453–59

Dichter, M. A., Tischler, A. S., Greene, L. A. 1977. Nerve growth factor-induced increase in electrical excitability and acetylcholine sensitivity of a rat pheochromocytoma cell line. *Nature* 268: 501–504

Dolkart, P., Johnson, E. M. 1979. Experimental auto-immune model of nerve growth factor deprivation: Effects on developing peripheral sympathetic and sensory neurons. *Proc. Natl. Acad. Sci. USA.* In press

Ebbott, S., Hendry, I. 1978. Retrograde transport of nerve growth factor in the rat central nervous system. *Brain Res.* 139:160–63

Edgar, D. H., Thoenen, H. 1978. Selective enzyme induction in a nerve growth factor—responsive pheochromocytoma cell line (PC12). *Brain Res.* 154:186–90

Fabricant, R. N., De Larco, J. E., Todaro, G. J. 1977. Nerve growth factor receptors on human melanoma cells in culture. *Proc. Natl. Acad. Sci. USA* 74:565–69

Frazier, W. A., Angeletti, R. H., Bradshaw, R. A. 1972. Nerve growth factor and insulin. *Science* 176:482–88

Frazier, W. A., Boyd, L. F., Bradshaw, R. A. 1973a. Interaction of the nerve growth factor with surface membranes: Biological competence of insolubilized nerve growth factor. *Proc. Natl. Acad. Sci. USA* 70:2931–35

Frazier, W. A., Boyd, L. F., Bradshaw, R. A. 1974a. Properties of the specific binding of ^{125}I nerve growth factor to responsive peripheral neurons. *J. Biol. Chem.* 249:5513–19

Frazier, W. A., Boyd, L. F., Pulliam, M. W., Szutowicz, A., Bradshaw, R. A. 1974b. Properties and specificity of binding sites for ^{125}I nerve growth factor in embryonic heart and brain. *J. Biol. Chem.* 249:5918–23

Frazier, W. A., Boyd, L. F., Szutowicz, A., Pulliam, M. W., Bradshaw, R. A. 1974c. Specific binding sites for ^{125}I nerve growth factor in peripheral tissues and brain. *Biochem. Biophys. Res. Commun.* 57:1096–1103

Frazier, W. A., Hogue-Angeletti, R. A., Sherman, R., Bradshaw, R. A. 1973b. Topography of mouse 2.5S nerve growth factor. Reactivity of tyrosine and tryptophan. *Biochemistry* 12: 3281–93

Frazier, W. A., Ohlendorf, C. E., Boyd, L. F., Aloe, L., Johnson, E. M., Ferrendelli, J. A., Bradshaw, R. A. 1973c. Mechanism of action of nerve growth factor and cyclic AMP on neurite outgrowth in embryonic chick sensory ganglia: Demonstration of independent pathways of stimulation. *Proc. Natl. Acad. Sci. USA* 70:2448–52

Goldstein, J. L., Andersen, R. G. W., Brown, M. S. 1979. Coated pits, coated vesicles, and receptor-mediated endocytosis. *Nature* 279:679–85

Goodman, R., Chandler, C., Herschman, H. R. 1979. Pheochromocytoma cell lines as models of neuronal differentiation. In *Hormones and Cells in Culture, Cold Spring Harbor Symp. Cell Prolifer.*, ed. R. Roth, G. Sato, Vol. 6. Cold Spring Harbor: Cold Spring Harbor Lab. In press

Goodman, R., Herschman, H. R. 1978. Nerve growth factor—mediated induction of tyrosine hydroxylase in a clonal pheochromocytoma cell line. *Proc. Natl. Acad. Sci. USA* 75:4587–90

Greene, L. A. 1970. *Studies of the γ subunit of mouse 7S nerve growth factor.* PhD thesis. Univ. of Calif., San Diego. 120 pp.

Greene, L. A. 1977. A quantitative bioassay for nerve growth factor (NGF) activity employing a clonal pheochromocytoma cell line. *Brain Res.* 133:350–53

Greene, L. A. 1978. Nerve growth factor prevents the death and stimulates neuronal differentiation of clonal PC12 pheochromocytoma cells in serum-free medium. *J. Cell Biol.* 78:747–55

Greene, L. A., Burstein, D. E., Black, M. M. 1980. The priming model for the mechanism of action of nerve growth factor: Evidence derived from clonal PC12 pheochromocytoma cells. In *Tissue Culture in Neurobiology*, ed. E. Giacobini, A. Vernadakis. New York: Raven. In press

Greene, L. A., Burstein, D. E., McGuire, J. C., Black, M. 1979. Cell culture studies on mechanism of action of nerve growth factor. *Soc. Neurosci. Symp.* 4:153–71

Greene, L. A., McGuire, J. C. 1978. Induction of ornithine decarboxylase by nerve growth factor dissociated from effects on survival and neurite outgrowth. *Nature* 276:191–94

Greene, L. A., Rein, G. 1977a. Release, storage and uptake of catecholamines by a clonal cell line of nerve growth factor (NGF) responsive pheochromocytoma cells. *Brain Res.* 129:247–63

Greene, L. A., Rein, G. 1977b. Release of (^{3}H)-norepinephrine from a clonal line of pheochromocytoma (PC12) by nicotinic cholinergic stimulation. *Brain Res.* 138:521–28

Greene, L. A., Rein, G. 1977c. Synthesis, storage and release of acetylcholine by a noradrenergic pheochromocytoma cell line. *Nature* 268:349–51

Greene, L. A., Shooter, E. M., Varon, S. 1968. Enzymatic activities of mouse nerve growth factor and its subunits. *Proc. Natl. Acad. Sci. USA* 60:1383–88

Greene, L. A., Shooter, E. M., Varon, S. 1969. Subunit interaction and enzymatic activity of mouse 7S nerve growth factor. *Biochemistry* 8:3735–41

Greene, L. A., Tischler, A. S. 1976. Establishment of a noradrenergic clonal line of rat adrenal pheochromocytoma cells which respond to nerve growth factor. *Proc. Natl. Acad. Sci. USA* 73:2424–28

Greene, L. A., Varon, S., Piltch, A., Shooter, E. M. 1971. Substructure of the β subunit of mouse 7S nerve growth factor. *Neurobiology* 1:37–48

Harper, G. P., Pearce, F. L., Vernon, C. A. 1976. Production of nerve growth factor by the mouse adrenal medulla. *Nature* 261:251–53

Harris-Warrick, R. M., Shooter, E. M. 1978. 7S nerve growth factor does not bind to NGF receptors on embryonic chick sensory neurons. *Neuroscience* 4:514 (Abstr.)

Hatanaka, H., Otten, U., Thoenen, H. 1978. Nerve growth factor mediated selective induction of ornithine decarboxylase in rat pheochromocytoma; a cyclic AMP-independent process. *FEBS Lett.* 92:313–16

Hendry, I. A. 1972. Developmental changes in tissue and plasma concentrations of the biologically active species of nerve growth factor in the mouse, by using a two site radioimmunoassay. *Biochem. J.* 128:1265–72

Hendry, I. A. 1975. The response of adrenergic neurons to axotomy and nerve growth factor. *Brain Res.* 94:87–97

Hendry, I. A. 1976. Control in the development of the vertebrate sympathetic nervous system. In *Review of Neuroscience,* ed. S. Ephrenpreis, I. J. Kopin, 2:149–94. New York: Raven

Hendry, I. A. 1977. The effect of the retrograde axonal transport of nerve growth factor on the morphology of adrenergic neurons. *Brain Res.* 134:213–23

Hendry, I. A., Campbell, J. 1976. Morphometric analysis of rat superior cervical ganglion after axotomy and nerve growth factor treatment. *J. Neurocytol.* 5:351–60

Hendry, I. A., Iversen, L. L. 1973. Reduction in the concentration of nerve growth factor in mice after sialectomy and castration. *Nature* 243:500–4

Hendry, I. A., Stach, R., Herrup, K. 1974a. Characteristics of the retrograde axonal transport system for nerve growth factor in sympathetic nervous system. *Brain Res.* 82:117–28

Hendry, I. A., Stöckel, K., Thoenen, H., Iversen, L. L. 1974b. The retrograde axonal transport of nerve growth factor. *Brain Res.* 68:103–21

Hermetet, J. C., Ciesielski-Treska, J., Mandel, P. 1972. Effect of nerve growth factor on neuroblast cultures from C1300 neuroblastoma. *C. R. Soc. Biol. Paris* 166:1120–33

Herrup, K., Shooter, E. M. 1973. Properties of the NGF receptor of avian dorsal root ganglia. *Proc. Natl. Acad. Sci. USA* 70:3384–88

Herrup, K., Shooter, E. M. 1975. Properties of the β-nerve growth factor receptor in development. *J. Cell Biol.* 67:118–25

Herrup, K., Thoenen, H. 1979. Properties of the nerve growth factor receptor of a clonal line of rat pheochromocytoma (PC12) cells. *Exp. Cell Res.* 121:71–78

Hier, D. B., Arnason, B. G. W., Young, M. 1972. Studies on the mechanism of action of nerve growth factor. *Proc. Natl. Acad. Sci. USA* 69:2268–72

Horii, Z.-I., Varon, S. 1975. Nerve growth factor-induced rapid activation of RNA labeling in dorsal root ganglionic dissociates from chick embryos. *J. Neurosci. Res.* 1:361–75

Horii, Z.-I., Varon, S. 1977. Nerve growth factor action on membrane permeation to exogenous substrates in dorsal root ganglion dissociates from the chick embryo. *Brain Res.* 124:121–33

Hunt, L. T., Barker, W. C., Dayhoff, M. O. 1974. Epidermal growth factor: Internal duplication and probable relationship to pancreatic secretory trypsin inhibitor. *Biochem. Biophys. Res. Commun.* 60:1020–28

Ishii, D. N., Shooter, E. M. 1975. Regulation of nerve growth factor synthesis in mouse submaxillary glands by testosterone. *J. Neurochem.* 25:843–51

Jeng, I., Bradshaw, R. A. 1978. Preparation of nerve growth factor. In *Research Methods in Neurochemistry,* ed. N. Marks, R. Rodnight, 4:265–88. New York: Plenum. 442 pp.

Johnson, D. G., Silberstein, S. D., Hanbauer, I., Kopin, I. J. 1972. The role of nerve growth factor in the ramification of sympathetic nerve fibers into the rat iris in organ culture. *J. Neurochem.* 19:2025–29

Johnson, E. M. Jr., Andres, R. Y., Bradshaw, R. A. 1978. Characterization of the retrograde transport of nerve growth factor (NGF) using high specific activity (^{125}I)NGF. *Brain Res.* 150:319–31

Kolber, A. R., Goldstein, M. N., Moore, B. W. 1974. Effect of nerve growth factor on the expression of colchicine-binding activity and 14–3–2 protein in an established line of human neuroblastoma. *Proc. Natl. Acad. Sci. USA* 71:4203–7

Lakshmanan, J. 1978. Is there a second messenger for nerve growth factor-induced phosphatidylinositol turnover in rat superior cervical ganglia? *Brain Res.* 157:173–77

Lakshmanan, J. 1979. Involvement of cytoskeletal structures in nerve growth factor-mediated induction of ornithine decarboxylase. *Biochem. J.* 178:245–48

Lee, J. C., Tweedy, N., Timasheff, S. 1978. In vitro reconstitution of calf brain microtubules: Effects of macromolecules. *Biochemistry* 17:2783–90

Letourneau, P. C. 1977. Regulation of neuronal morphogenesis by cell-substratum adhesion. In *Society for Neuroscience Symposium,* ed. W. M. Cowan, J. A. Ferendelli, pp. 67–81. Bethesda: Soc. Neurosci. 461 pp.

Letourneau, P. C. 1978. Chemotactic response of nerve-fiber elongation to nerve growth factor. *Dev. Biol.* 66:183–96

Levi, A., Cimino, M., Mercanti, D., Chen, J. S., Calissano, P. 1975. Interaction of nerve growth factor with tubulin. Studies on binding and induced polymerization. *Biochim. Biophys. Acta* 399:50–60

Levi, G., Lattes, M. G. 1968. Effect of a nerve growth factor on the transport of amino acids in spinal ganglia from chick embryos. *Life Sci.* 7:827–34

Levi-Montalcini, R. 1966. The nerve growth factor, its mode of action on sensory and sympathetic nerve cells. *Harvey Lect.* 60:217–59

Levi-Montalcini, R. 1975a. NGF: An unchartered route. In *The Neurosciences: Paths of Discovery,* ed. F. G. Worden, J. P. Swazey, G. Adelman, pp. 243–65. Cambridge, MIT. 622 pp.

Levi-Montalcini, R. 1975b. A new role for the glial cell? *Nature* 253:687

Levi-Montalcini, R. 1976. The nerve growth factor: Its role in growth, differentiation and function of the sympathetic adrenergic neurons. In *Progress in Brain Research,* ed. M. A. Corner, D. F. Swaab, 45:235–56 Amsterdam: Elsevier. 489 pp.

Levi-Montalcini, R., Angeletti, P. U. 1963. Essential role of the nerve growth factor in the survival and maintenance of dissociated sensory and sympathetic embryonic nerve cells in vitro. *Devel. Biol.* 7:653–657

Levi-Montalcini, R., Angeletti, P. U. 1966. Immunosympathectomy. *Pharmacol. Rev.* 18:619–28

Levi-Montalcini, R., Angeletti, P. U. 1968. Nerve growth factor. *Physiol. Rev.* 48:534–69

Levi-Montalcini, R., Booker, B. 1960. Destruction of the sympathetic ganglia in mammals by an antiserum to the nerve growth protein. *Proc. Natl. Acad. Sci. USA* 46:384–91

Levi-Montalcini, R., Hamburger, V. 1951. Selective growth-stimulating effects of mouse sarcoma on the sensory and sympathetic nervous system of the chick embryo. *J. Exptl. Zool.* 116:321–62

Levi-Montalcini, R., Hamburger, V. 1953. A diffusable agent of mouse sarcoma producing hyperplasia of sympathetic ganglia and hyperneurotization of viscera in the chick embryo. *J. Exptl. Zool.* 123:233–78

Lewis, M. E., Lakshmanan, J., Nagaiah, K., MacDonnell, P. C., Guroff, G. 1978. Nerve growth factor increases activity of ornithine decarboxylase in rat brain. *Proc. Natl. Acad. Sci. USA* 75:1021–23

Longo, A. M. 1978. Synthesis of nerve growth factor in rat glioma cells. *Dev. Biol.* 65:260–70

Longo, A. M., Penhoet, E. E. 1974. Nerve growth factor in rat glioma cells. *Proc. Natl. Acad. Sci. USA* 71:2347–49

Luduena, M. A. 1973. Nerve cell differentiation in vitro. *Dev. Biol.* 33:268–84

MacDonnell, P. C., Nagaiah, K., Lakshmanan, J., Guroff, G. 1977b. Nerve growth factor increases activity of ornithine decarboxylase in superior cervical ganglia of young rats. *Proc. Natl. Acad. Sci. USA* 74:4681–84

MacDonnell, P. C., Tolson, N., Guroff, G. 1977a. Selective de novo synthesis of tyrosine hydroxylase in organ cultures of rat superior cervical ganglia after in vivo administration of nerve growth factor. *J. Biol. Chem.* 252:5859–63

Mains, R. E., Patterson, P. H. 1973. Primary cultures of dissociated sympathetic neurons. I. Establishment of long-term growth in culture and studies of differentiated properties. *J. Cell Biol.* 59:329–45

Max, S. R., Schwab, M., Dumas, M., Thoenen, H. 1978. Retrograde axonal transport of nerve growth factor in the ciliary ganglion of the chick and rat. *Brain Res.* 159:411–15

McCarthy, K. D., Partlow, L. M. 1976. Preparation of pure neuronal and non-neuronal cultures from embryonic chick sympathetic ganglia: A new method based on both differential cell adhesiveness and the formation of homotypic neuronal aggregates. *Brain Res.* 114:391–414

McGuire, J. C., Greene, L. A. 1977. NGF alters specific protein synthesis in rat PC12 cells. *Neuroscience* 7:526 (Abstr.)

McGuire, J. C., Greene, L. A. 1979a. Rapid stimulation by nerve growth factor of amino acid uptake by clonal PC12 pheochromocytoma cells. *J. Biol. Chem.* 254:3362–67

McGuire, J. C., Greene, L. A. 1979b. Nerve growth factor stimulation of specific protein synthesis by rat PC12 pheochromocytoma cells. *Neuroscience.* In press

McGuire, J. C., Greene, L. A., Furano, A. V. 1978. NGF stimulates incorporation of glucose or glucosamine into an external glycoprotein in cultured rat PC12 pheochromocytoma cells. *Cell* 15:357–65

Menesini-Chen, M. G., Chen, J. S., Levi-Montalcini, R. 1978. Sympathetic nerve fibers ingrowth in the central nervous system of neonatal rodents upon intracerebral NGF injections. *Arch. Ital. Biol.* 116:53–84

Mercanti, D., Butler, R., Revoltella, R. 1977. A tryptic digestion fragment of nerve growth factor with nerve growth promoting activity. *Biochim. Biophys. Acta* 496:412–19

Mizel, S. B., Bamburg, J. R. 1975. Studies on the action of nerve growth factor. II. Neurotubule protein levels during neurite outgrowth. *Neurobiology* 5:283–90

Mizel, S. B., Bamburg, J. R. 1976. Studies on the action of nerve growth factor. III.

Role of RNA and protein synthesis in the process of neurite outgrowth. *Dev. Biol.* 49:20–28

Mobley, W. C., Schenker, A., Shooter, E. M. 1976. Characterization and isolation of proteolytically modified nerve growth factor. *Biochemistry* 15:5543–51

Mobley, W. C., Server, A. C., Ishii, D. N., Riopelle, R. J., Shooter, E. M. 1977. Nerve growth factor. *N. Engl. J. Med.* 297:1096–1104, 1149–58, 1211–18

Monaco, G., Calissano, P. Mercanti, D. 1977. Effect of NGF on in vitro preformed microtubules: Evidence for a protective action against vinblastine. *Brain Res.* 129:265–74

Moore, J. B., Jr., Mobley, W. C., Shooter, E. M. 1974. Proteolytic modification of the β nerve growth factor protein. *Biochemistry* 13:833–40

Moore, J. B. Jr., Shooter, E. M. 1975. The use of hybrid molecules in a study of the equilibrium between nerve growth factor monomers and dimers. *Neurobiology* 5:369–81

Morgan, J. L., Seeds, N. W. 1975. Tubulin constancy during morphological differentiation of mouse neuroblastoma cells. *J. Cell Biol.* 67:136–45

Murphy, R. A., Oger, J., Saide, J. D., Blanchard, M. H., Aranson, B. G. W., Hogan, C., Pantazis, N. J., Young, M. 1977c. Secretion of nerve growth factor by central nervous system glioma cells in culture. *J. Cell Biol.* 72:769–73

Murphy, R. A., Pantazis, N. J., Arnason, B. G. W., Young, M. 1975. Secretion of a nerve growth factor by mouse neuroblastoma cells in culture. *Proc. Natl. Acad. Sci. USA* 72:1895–98

Murphy, R. A., Saide, J. D., Blanchard, M. H., Young, M. 1977a. Molecular properties of the nerve growth factor secreted in mouse saliva. *Proc. Natl. Acad. Sci. USA* 74:2672–76

Murphy, R. A., Saide, J. D., Blanchard, M. H., Young, M. 1977d. Nerve growth factor in mouse serum and saliva: Role of the submandibular gland. *Proc. Natl. Acad. Sci. USA* 74:2330–33

Murphy, R. A., Singer, R. H., Saide, J. D., Pantazis, N. J., Blanchard, M. H. Byron, K. S., Arnason, B. G. W. 1977b. Synthesis and secretion of a high molecular weight form of nerve growth factor by skeletal muscle cells in culture. *Proc. Natl. Acad. Sci. USA* 74:4496–4500

Narumi, S., Fujita, T. 1978. Stimulatory effects of substance P and nerve growth factor (NGF) on neurite outgrowth in embryonic chick dorsal root ganglia. *Neuropharmacology* 17:73–76

Nikodijevic, B., Nikodijevic, O., Yu, M. W., Pollard, H., Guroff, G. 1975. The effect of nerve growth factor on cyclic AMP levels in superior cervical ganglia of the rat. *Proc. Natl. Acad. Sci. USA* 72: 4769–71

Nja, A., Purves, D. 1978. The effects of nerve growth factor and its antiserum on synapses in the superior cervical ganglion of the guinea pig. *J. Physiol.* 277:53–75

Norr, S. C., Varon, S. 1975. Dynamic, temperature-sensitive association of ^{125}I-nerve growth factor in vitro with ganglionic and nonganglionic cells from embryonic chick. *Neurobiology* 5:101–18

Otten, U., Hatanaka, H., Thoenen, H. 1978. Role of cyclic nucleotides in NGF-mediated induction of tyrosine hydroxylase in rat sympathetic ganglia and adrenal medulla. *Brain Res.* 140:385–89

Otten, U., Schwab, M., Gagnon, C., Thoenen, H. 1977. Selective induction of tyrosine hydroxylase and dopamine β-hydroxylase by nerve growth factor: Comparison between adrenal medulla and sympathetic ganglia of adult and newborn rats. *Brain Res.* 133:291–303

Otten, U., Thoenen, H. 1977. Effect of glucocorticoids on nerve growth factor-mediated enzyme induction in organ cultures of rat sympathetic ganglia: Enhanced response and reduced time requirement to initiate enzyme induction. *J. Neurochem.* 29:69–75

Oger, J., Arnason, B. G. W., Pantazis, N., Lehrich, J., Young, M. 1974. Synthesis of nerve growth factor by L and 3T3 cells in culture. *Proc. Natl. Acad. Sci. USA* 71:1554–58

Pantazis, N. J., Blanchard, M. H., Arnason, B. G. W., Young, M. 1977b. Molecular properties of the nerve growth factor secreted by L. cells. *Proc. Natl. Acad. Sci. USA* 74:1492–96

Pantazis, N. J., Murphy, R. A., Saide, J. D., Blanchard, M. H., Young, M. 1977a. Dissociation of the 7S nerve growth factor complex in solution. *Biochemistry* 16:1525–30

Paravicini, U., Stöckel, K., Thoenen, H. 1975. Biological importance of retrograde axonal transport of nerve growth factor in adrenergic neurons. *Brain Res.* 84:279–91

Partlow, L. M., Larrabee, M. G. 1971. Effects of a nerve growth factor, embryonic age and metabolic inhibitors on synthesis of ribonucleic acid and protein in embry-

onic sensory ganglia. *J. Neurochem.* 18:2101–18

Paterson, B. M., Bishop, J. O. 1977. Changes in the mRNA population of chick myoblasts during myogenesis in vitro. *Cell* 12:751–65

Pattison, S. E., Dunn, M. F. 1975. On the relationship of zinc ion to the structure and function of the 7S nerve growth factor protein. *Biochemistry* 14:2733–39

Pattison, S. E., Dunn, M. F. 1976a. On the mechanism of divalent metal ion chelation induced activation of the 7S nerve growth factor esteropeptidase. Activation by 2,2,',2"-terpyridine and by 8-hydroxyquinoline-5-sulfonic acid. *Biochemistry* 15:3691–96

Pattison, S. E., Dunn, M. F. 1976b. On the mechanism of divalent metal ion chelation induced activation of 7S nerve growth factor esteropeptidase. Thermodynamics and kinetics of activation. *Biochemistry* 15:3696–3700

Perez-Polo, J. R., De Jong, W. W. W., Straus, D., Shooter, E. M. 1972. The physical and biological properties of 7S and βNGF from the mouse submaxillary gland. In *Functional and Structural Proteins of the Nervous System,* ed. A. N. Davison, P. Mandel, I. G. Morgan, pp. 91–97. New York: Plenum. 286 pp.

Perez-Polo, J. R., Shooter, E. M. 1975. The preparation and properties of nerve growth factor protein at alkaline pH. *Neurobiology* 5:329–38

Perez-Polo, J. R., Werrbach, K. W., Castiglioni, E. T. 1979. A human clonal cell line model of differentiating neurons. *Dev. Biol.* 71:341–55

Pignatti, P. F., Baker, M. E., Shooter, E. M. 1975. Solution properties of beta nerve growth factor protein and some of its derivatives. *J. Neurochem.* 25:155–59

Pipeleers, D. G., Pipeleers-Marichal, M. A., Sherline, P., Kipnis, D. M. 1977. A sensitive method for measuring polymerized and depolymerized forms of tubulin in tissues. *J. Cell Biol.* 74:341–50

Pullen, R. A., Lindsay, D. G., Wood, S. P., Tickle, I. J., Blundell, T. L. Wollmer, A., Krail, G., Brandenburg, D., Zahn, H., Gliemann, J., Gammeltoft, S. 1976. Receptor-binding region of insulin. *Nature* 259:369–73

Pulliam, M. W., Boyd, L. F., Baglan, N. C., Bradshaw, R. A. 1975. Specific binding of covalently cross-linked mouse nerve growth factor to responsive peripheral neurons. *Biochem. Biophys. Res. Commun.* 67:1181–89

Revoltella, R., Bertolini, L., Pediconi, M., Vigneti, E. 1974. Specific binding of nerve growth factor (NGF) by murine C1300 neuroblastoma cells. *J. Exp. Med.* 140:437–51

Rieger, F., Shelanski, M. L., Greene, L. A. 1980. The effects of nerve growth factor on acetylcholinesterase and its multiple forms in cultures of rat PC12 pheochromocytoma cells: Increased total specific activity and appearance of the 16S molecular form. *Dev. Biol.* In press

Rinderknecht, E., Humbel, R. E. 1978a. The amino acid sequence of human insulin-like growth factor I and its structural homology with pro-insulin. *J. Biol. Chem.* 253:2769–76

Rinderknecht, E., Humbel, R. E. 1978b. Primary structure of human insulin-like growth factor II. *FEBS Lett.* 89:283–86

Roger, L. J., Schauberg, S. M. Fellows, R. F. 1974. Growth and lactogenic hormone stimulation of ornithine decarboxylase in fetal rat brain. *Endocrinology* 95:904–11

Rohrer, H., Otten, U., Thoenen, H. 1978. On the role of RNA synthesis in the selective induction of tyrosine hydroxylase by nerve growth factor. *Brain Res.* 159:436–39

Roisen, F. J., Murphy, R. A. 1973. Neurite development in vitro. II. The role of microfilaments and microtubules in dibutyryl adenosine 3',5'-cyclic monophosphate and nerve growth factor stimulated maturation. *J. Neurobiol.* 4:397–412

Roisen, F. J., Murphy, R. A., Braden, W. G. 1972. Neurite development in vitro. I. The effects of adenosine 3'5'-cyclic monophosphate (cyclic AMP). *J. Neurobiol.* 3:347–68

Russell, D. H., Byus, C. V., Manen, C. A. 1976. Proposed model of major sequential biochemical events of a trophic response. *Life Sci.* 19:1297–1306

Saide, J. D., Murphy, R. A., Canfield, R. E., Skinner, J., Robinson, D. B., Arnason, B. G. W., Young, M. 1975. Nerve growth factor in human serum and its secretion by human cells in culture. *J. Cell Biol.* 67:376a

Schubert, D. S., Heinemann, S., Kiddokoro, Y. 1977. Cholinergic metabolism and synapse formation by a rat nerve cell line. *Proc. Natl. Acad. Sci. USA* 74:2579–83

Schubert, D. S., LaCorbiere, M., Whitlock, C., Stallcup, W. 1978. Alterations in the surface properties of cells responsive to nerve growth factor. *Nature* 273:718–23

Schubert, D. S., Whitlock, C. 1977. Alteration of cellular adhesion by nerve growth factor. *Proc. Natl. Acad. Sci. USA* 74:4055–58

Schwab, M. E. 1977. Ultrastructural localization of a nerve growth factor-horseradish peroxidase (NGF-HRP) coupling product after retrograde axonal transport in adrenergic neurons. *Brain Res.* 130:190–96

Schwab, M. E., Thoenen, H. 1975. Early effects of nerve growth factor on adrenergic neurons: An electron microscopic morphometric study of the rat superior cervical ganglion. *Cell Tissue Res.* 158:543–53

Schwab, M., Thoenen, H. 1977. Selective trans-synaptic migration of tetanus toxin after retrograde axonal transport in peripheral sympathetic nerves: A comparison with nerve growth factor. *Brain Res.* 122:459–74

Schwabe, C., McDonald, J. K. 1977. Relaxin: A disulfide homolog of insulin. *Science* 197:914–15

Server, A. C., Shooter, E. M. 1976. A comparison of the arginine esteropeptidases associated with the nerve and epidermal growth factor. *J. Biol. Chem.* 25:165–73

Server, A. C., Shooter, E. M. 1977. Nerve growth factor. *Adv. Protein Chem.* 31:339–409

Server, A. C., Sutter, A., Shooter, E. M. 1976. A modification of the epidermal growth factor affecting the stability of its high molecular weight complex. *J. Biol. Chem.* 251:1188–96

Shelanski, M. L. 1973. Chemistry of the filaments and tubules of brain. *J. Histochem. Cytochem.* 21:529–39

Sherwin, S. A., Sliski, A. H., Todaro, G. J. 1979. Human melanoma cells have both nerve growth factor and nerve growth factor specific receptors on their cell surface. *Proc. Natl. Acad. Sci. USA* 76:1288–92

Sidman, R. L., Wessells, N. K. 1975. Control of direction of growth during the elongation of neurites. *Exp. Neurol.* 48(2):237–51

Silverman, R. E. 1978. *Interactions within the mouse nerve growth factor complex.* PhD thesis. Washington Univ., St. Louis, Missouri, 128 pp.

Skaper, S. D., Bottenstein, J. E., Varon, S. 1979. Effects of nerve growth factor on cyclic AMP levels in embryonic chick dorsal root ganglia following factor deprivation. *J. Neurochem.* 32:1845–51

Skaper, S. D., Varon, S. 1979a. Sodium dependence of nerve growth factor-regulated hexose transport in chick embryo sensory neurons. *Biochem. Biophys. Res. Commun.* 88:563–68

Skaper, S. D., Varon, S. 1979b. Nerve growth factor action on 2-deoxy-D-glucose transport in dorsal root ganglionic dissociates from chick embryo. *Brain Res.* 163:89–100

Smith, A. P., Varon, S., Shooter, E. M. 1968. Multiple forms of the nerve growth factor protein and subunits. *Biochemistry* 7:3259–68

Snyder, S. H., Banerjee, S. P. Cuatrecasas, P., Greene, L. A. 1974. The nerve growth factor receptor: Demonstration of specific binding in sympathetic ganglia. In *Dynamics of Degeneration and Growth in Neurons.* ed. K. Fuxe, pp. 347–57. Oxford: Pergamon. 608 pp.

Stach, R. W., Server, A. C., Pignatti, P. F., Piltch, A., Shooter, E. M. 1976. Characterization of the γ-subunits of the 7S nerve growth factor complex. *Biochemistry* 15:1455–61

Stach, R. W., Shooter, E. M. 1974. The biological activity of cross-linked β nerve growth factor protein. *J. Biol. Chem.* 249:6668–74

Stach, R. W., Shooter, E. M. 1979. The biological activity of cross-linked 7S nerve growth factor. *J. Neurochem.* In press

Stöckel, K., Guroff, G., Schwab, M., Thoenen, H. 1976. The significance of retrograde axonal transport for the accumulation of systemically administered nerve growth factor (NGF) in the rat superior cervical ganglion. *Brain Res.* 109:261–84

Stöckel, K., Schwab, M., Thoenen, H. 1975a. Comparison between the retrograde axonal transport of nerve growth factor and tetanus toxin in motor, sensory and adrenergic neurons. *Brain. Res.* 99:1–16

Stöckel, K., Schwab, M., Thoenen, H. 1975b. Specificity of retrograde transport of nerve growth factor (NGF) in sensory neurons: A biochemical and morphological study. *Brain Res.* 89:1–14

Stöckel, K., Solomon, F., Paravicini, U., Thoenen, H. 1974. Dissociation between effects of nerve growth factor on tyrosine hydrolase and tubulin synthesis in sympathetic ganglia. *Nature* 250:150–51

Stöckel, K., Thoenen, H. 1975a. Retrograde axonal transport of nerve growth factor (NGF): Specificity and biological importance. *Brain Res.* 85:337–41

Stöckel, K., Thoenen, H. 1975b. Specificity and biological importance of retrograde axonal transport of nerve growth factor. *Proc. VIth Int. Congr. Pharmacol.* 2:285–96

Sutter, A., Riopelle, R. J., Harris-Warrick, R. M., Shooter, E. M. 1977. Characterization of two distinct classes of high affinity binding sites for nerve growth factor on sensory ganglia cells from chick embryos. *Neuroscience* 3:461 (Abstr.)

Sutter, A., Riopelle, R. J., Harris-Warrick, R. M., Shooter, E. M. 1979. Nerve growth factor receptors: Characterization of two distinct classes of binding sites on chick embryo sensory ganglia cells. *J. Biol. Chem.* 254:5972–82

Szutowicz, A., Frazier, W. A., Bradshaw, R. A. 1976a. Subcellular localization of nerve growth factor receptors: Thirteen-day chick embryo brain. *J. Biol. Chem.* 251:1516–23

Szutowicz, A., Frazier, W. A., Bradshaw, R. A. 1976b. Subcellular localization of nerve growth factor receptors: Developmental correlations in chick embryo brain. *J. Biol. Chem.* 251:1524–28

Taylor, J. M., Cohen, S., Mitchell, W. M. 1970. Epidermal growth factor—high and low molecular weight forms. *Proc. Natl. Acad. Sci. USA* 67:164–71

Taylor, J. M., Mitchell, W. M., Cohen, S. 1974. Characterization of the high molecular weight form of epidermal growth factor. *J. Biol. Chem.* 249:3198–3203

Thoenen, H., Angeletti, P. U., Levi-Montalcini, R., Kettler, R. 1971. Selective induction by nerve growth factor of tyrosine hydroxylase and dopamine-β-hydroxylase in rat superior cervical ganglion. *Proc. Natl. Acad. Sci. USA* 68:1598–1602

Thoenen, H., Schwab, M., Otten, U. 1978. Nerve growth factor as a mediator of information between effector organs and innervating neurons. *Symp. Soc. Dev. Biol.* 35:101–18

Thomas, K. E., Silverman, R. E., Jeng, I., Baglan, N. C., Bradshaw, R. A. 1979. Purification, characterization and partial amino acid sequence of the γ-protease of mouse 7S nerve growth factor. *Fed. Proc.* 38:324

Tischler, A. S., Dichter, M. A., Biales, B., DeLellis, R. A., Wolfe, H. W. 1976. Neural properties of cultured human endocrine tumor cells of proposed neural crest origin. *Science* 192:902–4

Tischler, A. S., Greene, L. A. 1975. Nerve growth factor-induced process formation by cultured rat pheochromocytoma cell. *Nature* 258:341–42

Tischler, A. S., Greene, L. A. 1978. Morphologic and cytochemical properties of a clonal line of rat adrenal pheochromocytoma cells which respond to nerve growth factor. *Lab. Invest.* 39:77–89

Tischler, A. S., Greene, L. A. 1979. Phenotypic plasticity of pheochromocytoma and normal adrenal medullary cells. In *Histochemistry and Cell Biology of Autonomic Neurons, SIF Cells and Paraneurons*, ed. O. Eränkö. New York: Raven. In press

Turkington, R. W., Males, J. L., Cohen, S. 1971. Synthesis and storage of epithelial-epidermal growth factor in submaxillary gland. *Cancer Res.* 31:252–56

Turner, J. E., Glaze, K. A. 1977. Regenerative repair in the severed optic nerve of the newt (*Triturus viridenscens*): Effect of nerve growth factor. *Exp. Neurol.* 57:687–97

Unsicker, K., Chamley, J. H. 1977. Growth characteristics of postnatal rat adrenal medulla in culture. *Cell Tissue Res.* 177:247–68

Varon, S. S. 1975. Nerve growth factor and its mode of action. *Exp. Neurol.* 43(2): 75–92

Varon, S. S., Bunge, R. P. 1978. Trophic mechanisms in the peripheral nervous system. *Ann. Rev. Neurosci.* 1:327–61

Varon, S., Nomura, J., Shooter, E. M. 1967a. Subunit structure of a high molecular weight form of the nerve growth factor from mouse submaxillary gland. *Proc. Natl. Acad. Sci. USA* 57:1782–89

Varon, S., Nomura, J., Shooter, E. M. 1967b. The isolation of the mouse nerve growth factor protein in a high molecular weight form. *Biochemistry* 6:2202–9

Varon, S., Nomura, J., Shooter, E. M. 1968. Reversible dissociation of the mouse nerve growth factor protein into different subunits. *Biochemistry* 7:1296–303

Varon, S., Raiborn, C. 1972a. Dissociation, fractionation and culture of chick embryo sympathetic ganglionic cells. *J. Neurocytol.* 1:211–21

Varon, S., Raiborn, C. 1972b. Protective effect of mouse 7S nerve growth factor protein and its alpha subunit on embryonic sensory ganglionic cells during dissociation. *Neurobiology* 2:183–96

Varon, S., Skaper, S. D. 1980. Short-latency effects of nerve growth factor: An ionic view. In *Tissue Culture in Neurobiology*, ed. E. Giacobini, A. Vernadakis. New York: Raven. In press

Waris, T., Rechardt, L., Waris, P. 1973. Differentiation of neuroblastoma cells induced by nerve growth factor in vitro *Experientia* 29:1128–29

Wilson, W., Shooter, E. M. 1979. Structural modification of the NH_2-terminus of nerve growth factor: Purification and characterization of βNGF endopeptidase. *J. Biol. Chem.* 254:6002–9

Wlodawer, A., Hodgson, K. O., Shooter, E. M. 1975. Crystallization of nerve growth factor from mouse submaxillary glands. *Proc. Natl. Acad. Sci. USA* 72:777–79

Yamada, K. M., Wessells, N. K. 1971. Axonal elongation. Effect of nerve growth factor on microtubule protein. *Exp. Cell Res.* 66:346–52

Yankner, B. A., Shooter, E. M. 1979. Nerve growth factor in the nucleus: Interaction with receptors on the nuclear membrane. *Proc. Natl. Acad. Sci. USA* 76:1269–73

Young, M., Oger, J., Blanchard, M. H., Asdourian, H., Amos, H., Arnason, B. G. W. 1975. Secretion of a nerve growth factor by primary chick fibroblast cultures. *Science* 187:361–62

Young, M., Saide, J. D., Murphy, R. A., Arnason, B. G. W. 1976. Molecular size of nerve growth factor in dilute solution. *J. Biol. Chem.* 251:459–64

Young, M., Saide, J. D., Murphy, R. A., Blanchard, M. H. 1978. Nerve growth factor: Multiple dissociation products in homogenates of the mouse submandibular gland. Purification and molecular properties of the intact undissociated form of the protein. *Biochemistry* 17:1490–98

Yu, M. W., Hori, S., Tolson, N., Huff, K., Guroff, G. 1978. Increased phosphorylation of a specific nuclear protein in rat superior cervical ganglia in response to nerve growth factor. *Biochem. Biophys. Res. Commun.* 81:941–45

Zimmerman, A., Sutter, A., Samuelson, J., Shooter, E. M. 1978. A serological assay for the detection of cell surface receptors of nerve growth factor. *J. Supramol. Struct.* 9:351–61

AUTHOR INDEX

403

SUBJECT INDEX

A

N-Acetyl-d-glucosamine
retino-tectal cell adhesion
and, 309
Adenosine
synaptic function and, 14–18
Alzheimer's disease
see Dementia of Alzheimer
type
4-Aminopyridine
potassium channels blocked
by, 155
Amphetamine
neurotoxic action on DA
neurons of, 178
Amphetamines, halogenated
toxic effect on 5-HT neurons,
180
Antibodies
cell recognition by, 312, 313
Antipsychotic drugs
see Dyskinesia from
antipsychotic drugs
ATP
bacterial motility and, 47

B

Bacterial chemotaxis in relation
to neurobiology, 43–75
additivity desensitization and
potentiation, 58–60
bacterial memory and
response regulator
model, 51–54
bacterial system description,
45, 46
chemotaxis studies, 45, 46
processing systems and, 45
proteins as "brain" of
bacteria, 45
receptors and, 45
behavioral response, 46
swimming and tumbling,
46, 48
conclusions and
extrapolations, 68–72
adaptation and, 68, 69
bacterium-neuron
homologies, 71, 72
genes relative
contributions, 69, 70
heredity, environment and
chance, 70, 71
homologues with neuron
activity, 69
inheritance patterns
homologs, 70

integration in processing
system, 68
long-term and short-term
memory, 71–72
methylation role, 69
schizophrenia homologs,
70
functions of components,
55–58
flagellar protein, 58
genes and receptor
activity, 56–58
induced association model,
57
membrane proteins, 58
receptors, 55–57
signaling system, 57
individuality and chance,
65–68
chemotactic response,
66–68
Poisson variation and, 66,
67
receptor induction, 66
introduction, 43–45
comparison of neurons
with bacteria, 44
signal processing in a
single cell, 44
membrane potential role, 63,
64
chemotactic system and,
64
external pH and, 64
memory processes
diagrammed, 51–53
nonchemotactic mutants,
53, 54
phosphorylation
uncoupled, 64
response regulator, 51–54
swimming patterns and, 63
tumbling role in, 51
tunneling and, 64
memory time optimization,
54, 55
enzymes and bacterial
memory, 54, 55
locomotion and, 55
methylation and
demethylation in the
adaptive response, 60
s-adenosylmethionine and,
60
motor apparatus, 46–49
ATP role in, 47
flagellar bundle function,
47, 48
flagellar rotation, 47, 48
flagellar structure, 46, 47

flagellin protein of
flagellae, 47
mutant locomotion, 48, 49
polymorphic flagellae, 48
mutant alteration of
behavior, 61–63
chU function model, 61,
62
inversion of behavioral
response, 61, 62
override mechanisms, 64, 65
decreased membrane
potentials, 65
methionine deprivation
and, 64
plasticity of bacterial cell, 65
galactose receptor
production, 65
sensing system complexity,
49–51
compounds sensed, 49
proteins of processing
system, 50, 51
receptor types and
numbers, 49, 50
signal processing genes of,
50, 51
Bicuculline
blocking of substantia nigra
by, 249
Brain
membrane receptors for
NGF in, 374
Brain slices in vitro for study
of synaptic function, 1–19
concluding remarks, 18, 19
hippocampus advantages, 2,
3
synaptic circuitries in, 3
introduction, 1, 2
physiology of slice
preparation, 3–8
electrode placement in
slices, 4
intracellular recordings, 5
long-term potentiation, 7
pair-pulse facilitation, 5, 7
"population spike" in, 5
potentials from
stimulation, 5
pyramidal projection to
slice, 4, 5
repetitive stimulation, 7
typical records reproduced,
6
synaptic plasticity in
hippocampal slices, 8–18
adenosine possible role,
14–18

416

CUMULATIVE INDEXES

CONTRIBUTING AUTHORS, VOLUMES 1–3

CHAPTER TITLES, VOLUMES 1–3

ORDER FORM   ANNUAL REVIEWS INC.

Please list on the order blank on the reverse side the volumes you wish to order and whether you wish a standing order (the latest volume sent to you automatically each year). Volumes not yet published will be shipped in month and year indicated. Prices subject to change without notice. Out of print volumes subject to special order.

NEW TITLES FOR 1981

ANNUAL REVIEW OF NUTRITION   ISSN 0199-9885
    Vol. 1 (avail. July 1981): $20.00 (USA), $21.00 (elsewhere) per copy

INTELLIGENCE AND AFFECTIVITY: Their Relationship During Child Development
    A monograph, translated from a course of lectures by Jean Piaget   ISBN 0-8243-2901-5
    Avail. Feb. 1981   Hard cover: $8.00 (USA), $9.00 (elsewhere) per copy

SPECIAL PUBLICATIONS

ANNUAL REVIEWS REPRINTS: CELL MEMBRANES, 1975–1977   ISBN 0-8243-2501-X
    A collection of articles reprinted from recent *Annual Review* series
    Published 1978   Soft cover: $12.00 (USA), $12.50 (elsewhere) per copy

ANNUAL REVIEWS REPRINTS: IMMUNOLOGY, 1977–1979   ISBN 0-8243-2502-8
    A collection of articles reprinted from recent *Annual Review* series
    Published 1980   Soft cover: $12.00 (USA), $12.50 (elsewhere) per copy

THE EXCITEMENT AND FASCINATION OF SCIENCE, VOLUME 1   ISBN 0-8243-1602-9
    A collection of autobiographical and philosophical articles by leading scientists
    Published 1965   Clothbound: $6.50 (USA), $7.00 (elsewhere) per copy

THE EXCITEMENT AND FASCINATION OF SCIENCE, VOLUME 2: Reflections by Eminent Scientists
    Published 1978   Hard cover: $12.00 (USA), $12.50 (elsewhere) per copy   ISBN 0-8243-2601-6
                     Soft cover: $10.00 (USA), $10.50 (elsewhere) per copy   ISBN 0-8243-2602-4

THE HISTORY OF ENTOMOLOGY   ISBN 0-8243-2101-7
    A special supplement to the *Annual Review of Entomology* series
    Published 1973   Clothbound: $10.00 (USA), $10.50 (elsewhere) per copy

ANNUAL REVIEW SERIES

Annual Review of ANTHROPOLOGY   ISSN 0084-6570
    Vols. 1–8 (1972–79): $17.00 (USA), $17.50 (elsewhere) per copy
    Vol. 9 (1980): $20.00 (USA), $21.00 (elsewhere) per copy
    Vol. 10 (avail. Oct. 1981): $20.00 (USA), $21.00 (elsewhere) per copy

Annual Review of ASTRONOMY AND ASTROPHYSICS   ISSN 0066-4146
    Vols. 1–17 (1963–79): $17.00 (USA), $17.50 (elsewhere) per copy
    Vol. 18 (1980): $20.00 (USA), $21.00 (elsewhere) per copy
    Vol. 19 (avail. Sept. 1981): $20.00 (USA), $21.00 (elsewhere) per copy

Annual Review of BIOCHEMISTRY   ISSN 0066-4154
    Vols. 28–48 (1959–79): $18.00 (USA), $18.50 (elsewhere) per copy
    Vol. 49 (1980): $21.00 (USA), $22.00 (elsewhere) per copy
    Vol. 50 (avail. July 1981): $21.00 (USA), $22.00 (elsewhere) per copy

Annual Review of BIOPHYSICS AND BIOENGINEERING   ISSN 0084-6589
    Vols. 1–9 (1972–80): $17.00 (USA), $17.50 (elsewhere) per copy
    Vol. 10 (avail. June 1981): $20.00 (USA), $21.00 (elsewhere) per copy

Annual Review of EARTH AND PLANETARY SCIENCES   ISSN 0084-6597
    Vols. 1–8 (1973–80): $17.00 (USA), $17.50 (elsewhere) per copy
    Vol. 9 (avail. May 1981): $20.00 (USA), $21.00 (elsewhere) per copy

Annual Review of ECOLOGY AND SYSTEMATICS   ISSN 0066-4162
    Vols. 1–10 (1970–79): $17.00 (USA), $17.50 (elsewhere) per copy
    Vol. 11 (1980): $20.00 (USA), $21.00 (elsewhere) per copy
    Vol. 12 (avail. Nov. 1981): $20.00 (USA), $21.00 (elsewhere) per copy

Annual Review of ENERGY   ISSN 0362-1626
    Vols. 1–4 (1976–79): $17.00 (USA), $17.50 (elsewhere) per copy
    Vol. 5 (1980): $20.00 (USA), $21.00 (elsewhere) per copy
    Vol. 6 (avail. Oct. 1981): $20.00 (USA), $21.00 (elsewhere) per copy

Annual Review of ENTOMOLOGY   ISSN 0066-4170
    Vols. 7–25 (1962–80): $17.00 (USA), $17.50 (elsewhere) per copy
    Vol. 26 (avail. Jan. 1981): $20.00 (USA), $21.00 (elsewhere) per copy

Annual Review of FLUID MECHANICS   ISSN 0066-4189
    Vols. 1–12 (1969–80): $17.00 (USA), $17.50 (elsewhere) per copy
    Vol. 13 (avail. Jan. 1981): $20.00 (USA), $21.00 (elsewhere) per copy

Annual Review of GENETICS   ISSN 0066-4197
    Vols. 1–13 (1967–79): $17.00 (USA), $17.50 (elsewhere) per copy
    Vol. 14 (1980): $20.00 (USA), $21.00 (elsewhere) per copy
    Vol. 15 (avail. Dec. 1981): $20.00 (USA), $21.00 (elsewhere) per copy

(continued on reverse)

Annual Review of MATERIALS SCIENCE   ISSN 0084-6600
   Vols. 1–9 (1971–79): $17.00 (USA), $17.50 (elsewhere) per copy
   Vol. 10 (1980): $20.00 (USA), $21.00 (elsewhere) per copy
   Vol. 11 (avail. Aug. 1981): $20.00 (USA), $21.00 (elsewhere) per copy

Annual Review of MEDICINE: Selected Topics in the Clinical Sciences   ISSN 0066-4219
   Vols. 1–3, 5–15, 17–31 (1950–52, 1954–64, 1966–80): $17.00 (USA), $17.50 (elsewhere) per copy
   Vol. 32 (avail. Apr. 1981): $20.00 (USA), $21.00 (elsewhere) per copy

Annual Review of MICROBIOLOGY   ISSN 0066-4227
   Vols. 15–33 (1961–79): $17.00 (USA), $17.50 (elsewhere) per copy
   Vol. 34 (1980): $20.00 (USA), $21.00 (elsewhere) per copy
   Vol. 35 (avail. Oct. 1981): $20.00 (USA), $21.00 (elsewhere) per copy

Annual Review of NEUROSCIENCE   ISSN 0147-006X
   Vols. 1–3 (1978–80): $17.00 (USA), $17.50 (elsewhere) per copy
   Vol. 4 (avail. Mar. 1981): $20.00 (USA), $21.00 (elsewhere) per copy

Annual Review of NUCLEAR AND PARTICLE SCIENCE   ISSN 0066-4243
   Vols. 10–29 (1960–79): $19.50 (USA), $20.00 (elsewhere) per copy
   Vol. 30 (1980): $22.50 (USA), $23.50 (elsewhere) per copy
   Vol. 31 (avail. Dec. 1981): $22.50 (USA), $23.50 (elsewhere) per copy

Annual Review of PHARMACOLOGY AND TOXICOLOGY   ISSN 0362-1642
   Vols. 1–3, 5–20 (1961–63, 1965–80): $17.00 (USA), $17.50 (elsewhere) per copy
   Vol. 21 (avail. Apr. 1981): $20.00 (USA), $21.00 (elsewhere) per copy

Annual Review of PHYSICAL CHEMISTRY   ISSN 0066-426X
   Vols. 10–21, 23–30 (1959–70, 1972–79): $17.00 (USA), $17.50 (elsewhere) per copy
   Vol. 31 (1980): $20.00 (USA), $21.00 (elsewhere) per copy
   Vol. 32 (avail. Nov. 1981): $20.00 (USA), $21.00 (elsewhere) per copy

Annual Review of PHYSIOLOGY   ISSN 0066-4278
   Vols. 18–42 (1956–80): $17.00 (USA), $17.50 (elsewhere) per copy
   Vol. 43 (avail. Mar. 1981): $20.00 (USA), $21.00 (elsewhere) per copy

Annual Review of PHYTOPATHOLOGY   ISSN 0066-4286
   Vols. 1–17 (1963–79): $17.00 (USA), $17.50 (elsewhere) per copy
   Vol. 18 (1980): $20.00 (USA), $21.00 (elsewhere) per copy
   Vol. 19 (avail. Sept. 1981): $20.00 (USA), $21.00 (elsewhere) per copy

Annual Review of PLANT PHYSIOLOGY   ISSN 0066-4294
   Vols. 10–31 (1959–80): $17.00 (USA), $17.50 (elsewhere) per copy
   Vol. 32 (avail. June 1981): $20.00 (USA), $21.00 (elsewhere) per copy

Annual Review of PSYCHOLOGY   ISSN 0066-4308
   Vols. 4, 5, 8, 10–31 (1953, 1954, 1957, 1959–80): $17.00 (USA), $17.50 (elsewhere) per copy
   Vol. 32 (avail. Feb. 1981): $20.00 (USA), $21.00 (elsewhere) per copy

Annual Review of PUBLIC HEALTH   ISSN 0163-7525
   Vol. 1 (1980): $17.00 (USA), $17.50 (elsewhere) per copy
   Vol. 2 (avail. May 1981): $20.00 (USA), $21.00 (elsewhere) per copy

Annual Review of SOCIOLOGY   ISSN 0360-0572
   Vols. 1–5 (1975–79): $17.00 (USA), $17.50 (elsewhere) per copy
   Vol. 6 (1980): $20.00 (USA), $21.00 (elsewhere) per copy
   Vol. 7 (avail. Aug. 1981): $20.00 (USA), $21.00 (elsewhere) per copy

To ANNUAL REVIEWS INC., 4139 El Camino Way, Palo Alto, CA 94306 USA   (Tel. 415-493-4400)

Please enter my order for the following publications:
(Standing orders: indicate which volume you wish order to begin with)

_____, Vol(s). _____ Standing order ☐

_____, Vol(s). _____ Standing order ☐

_____, Vol(s). _____ Standing order ☐

_____, Vol(s). _____ Standing order ☐

Amount of remittance enclosed $_____   California residents please add applicable sales tax.
Please bill me ☐   Prices subject to change without notice.

SHIP TO (include institutional purchase order if billing address is different)

Name _____

Address _____

_____

_____ Zip Code _____

Signed _____ Date _____

☐ Please add my name to your mailing list to receive a free copy of the current Prospectus each year.
☐ Send free brochure listing contents of recent back volumes for *Annual Review(s)* of _____

An Annual Reviews Monograph

INTELLIGENCE AND AFFECTIVITY: Their Relationship During Child Development

by Jean Piaget

Based on a course of lectures delivered in 1953, this monograph
offers Piaget's only extensive statement on the subject of affectivity.

Translated and edited by T. A. Brown and C. E. Kaegi.
Originally published in French in *Bulletin de Psychologie*, Vol. 7, 1954.

ANNUAL REVIEW OF NUTRITION (Volume 1 to be published July 1981)

*Planned Topics (Note: List is tentative; some articles may fail to appear.
Final contents information available Feb. 1981.)*

Regulation of Energy Balance • Nutritional Significance of Fructose and Sugar Alcohol
Recently Recognized Amino Acids • Folic Acid, Vitamin B_{12}, and Labile Methyl Metabolism
Iatrogenic Nutrient Deficiencies • Metabolism and Nutritional Significance of Carotenoids
Formation and Mode of Action of Flavoproteins • Taurine Metabolism
Dietary Choline: Biochemistry, Physiology, and Pharmacology • Trace Elements in Milk
Dietary Bioavailability of Iron • Iron Absorption and Transport in Microorganisms
Epidemiological Studies of Health Effects of Water from Different Sources
The Germ-Free Animal As a Model in Nutritional Studies • Nutrition and Immunity
Nutritional Conditioning of Athletes • Nutrients That Are Not: "Vitamins" B_{15} and B_{17}
Newly Observed Deficiency Syndromes • Physiology of Appetite Regulation
Metabolism of Pigments and Food Colors • Chemical Senses and Digestive Metabolism
The Role of Nutrition in Toxicology • The Moral Dimension of the World's Food Supply
Sodium and Potassium • Cultural Nutrition: Anthropological and Geographical Themes